continued on back

Statistics and
Experimental Design

in Engineering and the Physical Sciences

A WILEY PUBLICATION IN APPLIED STATISTICS

Statistics and Experimental Design

in Engineering and the Physical Sciences

volume **II**,
second edition

NORMAN L. JOHNSON

FRED C. LEONE

John Wiley & Sons, New York • London • Sydney • Toronto

Library of Congress Cataloging in Publication Data: (Revised)

Johnson, Norman Lloyd.
    Statistics and experimental design in engineering
and the physical sciences.

    (Wiley series in probability and mathematical
statistics)
    "A Wiley-Interscience publication."..
    Includes bibliographical references and index.
    1. Mathematical statistics. 2. Experimental
design. I. Leone, Fred C., joint author. II. Ti-
tle.
QA276.J59 1976   519.5   76-28337
ISBN 0–471–01757–4 (v. 2)

Printed in the United States of America

10 9 8 7 6 5 4 3 2

Preface to the First Edition

This book is intended to give students and research workers in science and, particularly, engineering a sound understanding of, and facility with, basic statistical techniques. To this end some fundamental statistical theory has been included, and a competent knowledge of engineering mathematics is needed for a proper understanding of the text. However, the primary object has been to give an intelligent understanding of the ways in which statistical methods can be useful, and the arguments underlying their use. The book is not intended to be encyclopedic or academically "advanced." There is no interest in elaboration for its own sake, or in mathematical rigor or elegance as ends in themselves. These features appear occasionally where they are needed to provide adequate flexibility, to guard against erroneous conclusions, or to clarify the argument. But they are included only to serve the main aim of providing a sound and thorough training in useful application of statistical techniques.

A reader with less mathematical skill than the usual acquaintance with elementary calculus can still derive a great deal of useful information from this textbook. The several necessary proofs can be studied primarily from the viewpoint of the statement of the problem and the conclusions of the theory. A great many examples are included in order that the student or research worker may see more readily the specific statistical techniques under consideration. The examples included in this book are in no way an attempt to make this a "cook book." On the contrary, their purpose is to emphasize and elucidate.

Some sections are marked with a star and printed in smaller type. These may be omitted at a first reading, but should not therefore be regarded as unimportant. Usually they contain either special topics or alternative modes of treatment.

At the end of each chapter a large variety of exercises is presented. These are purposely of varying degrees of difficulty in order that readers of heterogeneous background and interest may have a choice from which to work out the principles learned in the chapter. Examples and exercises are drawn from many fields of science and engineering. Answers are provided at the end of the book for approximately ten exercises in every chapter (except Chapter 1). Some references are given to guide students to suitable further reading. These references are not intended to be at all exhaustive, and they would be insufficient for those readers primarily interested in statistical theory.

The material in this book should cover two or possibly three semesters of applied statistics at the upper-division undergraduate level. It is also suitable for graduates whose major subject is not statistics.

The first volume can be made the basis for a semester's course. However, it may be found desirable to postpone Chapter 11, and possibly Chapter 10, for later consideration.

An Appendix to Volume I contains a number of Tables which are useful in statistical work. A few of these Tables are not needed until Volume II is studied, but the Tables have been collected in one place for convenience.

We should like to express our deep appreciation to the many people whose assistance and encouragement helped to make this volume possible. To families and friends whose patience was a vital factor, in completing this book, we extend sincere thanks. We especially want to acknowledge the assistance of the secretaries at Case Institute of Technology and University College, London, Mrs. Margaret Laczko and Mrs. Marie Fuller (Case) and Mrs. Joan Seaborne (University College), as well as others who materially assisted in the preparation of the manuscript.

For permission to reproduce data, thanks are extended to the editors of *Annals of Mathematical Statistics, Applied Statistics, Biometrics, Biometrika, Bulletin of the International Statistical Institute, Industrial Quality Control, Journal of the American Statistical Association, Journal of the Royal Statistical Society, Technometrics,* and other journals, not primarily concerned with statistical matters.

We would like especially to thank Professor E. S. Pearson and the Biometrika Trustees for permission to make extensive use of some of the tables in *Biometrika Tables for Statisticians,* Volume I, by E. S. Pearson and H. O. Hartley. We are indebted to the late Sir Ronald A. Fisher, F.R.S., Cambridge, to Dr. Frank Yates, F.R.S., Rothamstead, and to Messrs. Oliver and Boyd Ltd., Edinburgh, for permission to reprint Tables Nos. III, VI, and XXIII from their *Statistical Tables for Biological, Agricultural and Medical Research.* We are also indebted to O. L. Davies and to Messrs. Oliver and Boyd Ltd., Edinburgh, for permission to reprint Tables

E, E.1, G, and H from their book *Design and Analysis in Industrial Experiments*.

July 1964   N. L. JOHNSON
                                                                          F. C. LEONE

Preface to the Second Edition

This edition incorporates a substantial number of changes from the first (1964) edition. The basic philosophy of our presentation remains as stated in the preface to the first volume of that edition. The book is not intended to be encyclopedic or academically "advanced." There is no interest in elaboration for its own sake, or in mathematical rigor or elegance as ends in themselves. These features do appear occasionally where they are needed to provide adequate flexibility, to guard against erroneous conclusions, or to clarify the argument; but they are included only to serve the main aim of providing a sound and thorough training in useful application of statistical techniques.

The changes we have made fall under the following main headings:

1. Removal of material that is now of less relevance (for example, the Doolittle method of solving normal equations in Chapter 12).

2. Inclusion of new material, either because it broadens the scope of the treatment of a topic (for example, the material on "discounting" in Chapter 11) or because it reflects new developments since the first edition was written (for example, some of the results on sequential analysis in Chapter 16). We have also taken the opportunity to add a number of new illustrative examples.

3. Rearrangement of material both within and between chapters. In particular, we have split our original Chapter 17 into two chapters, one on response surfaces and the other on multivariate analysis. In each of the new chapters we have introduced new material. There are many other less obvious instances of rearrangement. We would especially mention that the order of the exercises at the end of each chapter has been revised to make exercises on similar topics close neighbors. In addition a considerable number of new exercises has been added and a few of the original exercises have been removed.

4. A major feature of this edition is the greatly extended coverage of solutions to exercises. Not only has the number of exercises with solutions

been much increased, but many of the solutions themselves discuss the problem in greater detail. This should prove of value especially to students using the book on their own.

5.   We have included tables at the end of Volume 2 as well as at the end of Volume 1. Each volume should now contain tables appropriate to the material in it.

In addition to all the individuals mentioned in the preface to the first edition, we wish to express thanks to June Maxwell (University of North Carolina) and Mrs. Dorothy Zimmermann for their assistance in preparing the manuscript of the second edition.

Chapel Hill, North Carolina   N. L. JOHNSON
Washington, D.C.   F. C. LEONE
May 1976

Contents

CHAPTER 13

Analysis of Variance (I): Construction and Use of Model, Design, and Analysis

13.1 INTRODUCTION

The term "analysis of variance" is given to a wide range of standard statistical techniques. Nearly all practicing statisticians use, in some form or other, an "analysis of variance" technique, though there may be considerable difference in the amount of complexity of the methods required in different kinds of work. The breadth of application of the analysis of variance arises from the fact that it is not only a group of analytical techniques, but a flexible method of constructing stastistical models for experimental material, which has been found to be of value in many different circumstances. The form of model that is used can be expressed as

$$(\text{observed value}) = \sum [\text{parameters representing assignable effects}]$$

$$+ \sum [\text{random variables representing assign-}$$

$$\text{able effects}]$$

$$+ [\text{random variables representing unassign-}$$

$$\text{able (residual) effects}] \tag{13.1}$$

"Assignable effects" means effects resulting from the operation of changes in recognizable conditions. In any investigation there are a number of factors—for example, pressure, temperature, time—which *might* have some

593

effect on our observations. Such of these factors as we decide to recognize formally, and allow for in our calculations, correspond to "assignable" effects. It is a matter of experience that the more factors we introduce, the less the unassignable (residual) variation that is left unaccounted for, *but* there is nearly always *some* such residual variation remaining. In any actual investigation the residual variation contains elements that could be accounted for almost completely by introducing additional "assignable effects," but that (it is hoped) are relatively of such small importance that it is not worthwhile trying to eliminate them. Whether all residual variation could be eliminated with sufficiently exhaustive enumeration of possible factors, or whether there would always be a residue of "irreducible chaos" are questions which we, fortunately, are not called upon to answer. All we need to know is that we can make useful investigation even when there is residual variation, whatever the sources of the latter. The "analysis of variance" is a systematic method which helps us to do this.

In applying the analysis of variance (as in most applications of statistical techniques) we must first construct an appropriate statistical model, and frame the questions we wish to investigate in terms of this model. This means that we must first understand the structure, or pattern, of the experimental data as clearly as possible. The appropriate forms of analysis then follow—though it should be realized that, even for the same set of data, different techniques are needed to answer different questions.

In this chapter and succeeding chapters we develop analysis of variance techniques appropriate to a considerable variety of experimental patterns. To become proficient in applying the analysis of variance, however, we require more than a mere knowledge of available techniques. It is also essential to be able to recognize the pattern of an experiment and to select the technique(s) most suitable to the problem(s) posed, from the available repertoire. We also consider a number of illustrations of these different techniques.

13.2 RESIDUAL VARIATION

We have just discussed the nature of residual variation; we are now going to consider *how* this variation is to be represented in the model. Evidently it will be suitably represented by some combination of random variables and parameters, different for different observed values. The analysis of variance is based on certain plausible assumptions about these random variables and parameters, which follow.

(1) *The expected value of each residual random variable is zero.* This simply means that all variation in expected values is taken care of by the parameters (and possibly random variables) representing assignable

effects. Even though we (usually) do not *know* the appropriate values of these quantities, we need only suppose that they are defined so as to represent variation in expected values. This assumption is not likely to be incorrect in any applied problem.

(2) *The residual random variables are mutually independent.* Often this is also a very reasonable assumption. It is not so clearly correct as (1), however. The meaning of the assumption is that there is no link between different observations that is not accounted for by the terms representing assignable effects. If individuals are chosen at random and measured separately, the assumption is quite reasonable, but situations can arise in which one observation can affect a later observation. For example, in a routine series of estimations of a physical quantity there is sometimes a tendency to balance a "high" observed value by "low" values to produce an average close to that expected, or desired.

(3) *The residual random variables all have the same standard deviation.* This "homoscedasticity" is a really critical assumption. It is much less likely to be valid than either (1) or (2), yet formal analysis of variance is based on this assumption. In fact, an essential feature of many techniques is the estimation of this (presumed common) standard deviation. Many circumstances can vitiate this assumption. It is quite likely that different methods of measurement can produce variations of different magnitude; higher expected values are often associated with higher standard deviation. Therefore it is important to consider possible variations in standard deviation before applying analysis of variance to any given set of data.

(4) *The residual random variables are each normally distributed.* On the whole, of the 4 assumptions this is the least likely to be valid. If the observation is restricted in range of possible values (such as weight which must be positive, or a count which must be an integer), the assumption is certainly incorrect. Usually it can only be a more or less close approach to the actual situation. But, fortunately, a substantial part of analysis of variance can be developed without using this assumption, which is needed only to justify the use of certain formally precise tests of significance and estimation formulas. Furthermore, it can be shown that a fair amount of departure from exact normality can be tolerated with little practical effect on the properties of standard analysis of variance procedures.

To summarize, it is with respect to assumption (3) that the most care is needed. Theoretical assessments of the effect of variation in standard deviation (heteroscedasticity) can help sometimes. In other cases, it is possible to transform the original observations (by taking logarithms, square roots, etc.) so that the assumptions are more nearly satisfied for the transformed variables. Methods of achieving this end are discussed later in this chapter.

13.3   FACTORS

At this point it is convenient to introduce some conventional terms which help describe experimental patterns in a concise manner. The basic concept is that of a *factor*: a quality, or property, according to which the data are classified. If observations are made on the products from each of a number of machines on successive days when operated by different shifts of workers, we are concerned with 3 (at least) factors—machine, shift, and day. Each factor appears at a number of different *levels*. "Levels" is a general term used to describe the particular property defining each category in the classification. Thus the levels of the factor *machine* might be No. 1 (machine), No. 2, No. 3, and so on; the levels of the factor *shift* might be early, middle, and late (or perhaps day and night); and the levels of the factor *day* might be September 3, 1977, September 4, 1977, and so on.

The *structure* or *pattern* of an experiment (usually called the experimental design) is described by the factors appearing in it and the way in which the various levels of the different factors are combined. Also of importance are the characters measured on each individual and the way in which they are supposed to be related to each other and to the levels of the factors present in the experiment.

This terminology is deliberately framed in a rather abstract form. Although this may make it more difficult to understand at first, it enables many different kinds of data to be summarized under relatively few different headings. In any particular case, of course, factors, levels, and characters correspond to real features of the situation under consideration.

13.4   ONE-WAY CLASSIFICATION

The simplest form of experimental pattern we discuss arises when we are comparing a number of levels of a single factor, and when the same character is measured on each individual. It is very easy to think of examples of this kind of pattern of data, but for the sake of definiteness we refer, in our discussion, to an investigation in which the effects of k different treatments for increasing periods of natural sleep are to be compared. We suppose that out of a number (N) of persons, n_1 are given the first treatment, n_2 the second treatment, and so on. (Of course, $\sum_{t=1}^{k} n_t = N$ if we use all the individuals.) We further suppose that we observe some character, x, representing the effect of the treatment. For our present purposes the precise nature of this character—whether it is period of sleep averaged over a week, 2 weeks, or 1 month, or increase in average period of sleep over some such period, or some more complicated index

—is not important (though it may well be important in regard to the particular investigation). We are, for the moment, concentrating on the pure *structure* of the data. In the present case this may be represented diagrammatically as in Figure 13.1. Here the large crosses represent single observations, and the levels (particular treatments) of the factor, treatment, are called by the generic term, group. For the data represented by Figure 13.1, we would have $n_1 = 3, n_2 = 4, n_3 = 2, \ldots, n_t = 3, \ldots, n_k = 1$.

Since we have only one factor to allow for, the model is of the form

observed value = [term representing effect of appropriate

level of the factor] + [random residual]   (13.2)

GROUP	1	2	3	...	t	...	k
	×	×	×		×		×
	×	×	×		×		
	×	×			×		
		×					

Figure 13.1   One-way classification.

A convenient notation for the observed values is X_{ti}, where $t (= 1, 2, \ldots, k)$ denotes the group (that is, factor level) and $i (= 1, 2, \ldots, n_t)$ identifies the observation within the group. Because there is a separate random residual for each observation, it is convenient to use the similar notation, Z_{ti}, to represent these random variables. The assumptions described in Section 13.2 can then be summarized as

(i) $E(Z_{ti}) = 0$ for all t and i.
(ii)   The Z_{ti}'s are mutually independent.
(iii) $\text{Var}(Z_{ti}) = \sigma^2$ for all t and i.
(iv) Z_{ti} is normally distributed.

For the first term on the right side of (13.2) we have the choice of using a parameter or a random variable. This is an important choice. We are not free to make our choice simply on grounds of conveneince; rather we must try to choose a model which will (a) *represent the real situation* as faithfully as possible, and (b) *help us to answer the question* to which we are asked to suggest replies. In a real problem this choice of model is effectively made when the method of choosing the sample and the aims of the investigation are decided upon.

In the special case we are considering, the use of parameters would mean that we are regarding the k treatments just as k specific treatments; if random variables were used we would be regarding them rather as samples chosen from a larger number of possible treatments. The latter approach would be reasonable, for example, if the k "treatments" were in fact k batches selected from routine production by a standard method. It is important to recognize that the *same* set of data may be regarded from 2 *different* aspects, according to the questions it is hoped to answer. For example, although we may primarily be interested in our k treatments as a sample representing the variability to be expected from one preparation to another, we may also be interested, for immediate application, in comparisons between the effectiveness of the particular treatments actually used in the investigation. The first of these approaches calls for random variables to represent variation between different levels of the factor; the second requires that parameters be used.

We now proceed to the symbolic representation of the 2 types of model we have discussed so far.

If parameters are used to denote factor level effects the model is called *parametric, fixed*, or *systematic* (also *Model I*).

In the case of a one-way classification this means that the average value in the tth group is denoted by a constant value ξ_t. The model is then

$$X_{ti} = \xi_t + Z_{ti} \tag{13.3}$$

For convenience in the algebra, we usually represent ξ_t in the form $\xi + \gamma_t$, where the γ_t's now represent deviations of the ξ_t's from the "average" value ξ. Since there are $k+1$ parameters $\xi, \gamma_1, \gamma_2, \ldots, \gamma_k$ to represent the k parameters $\xi_1, \xi_2, \ldots, \xi_k$, one linear condition can be imposed on the γ_t's. This we take as $\sum_{t=1}^{k} n_t \gamma_t = 0$, which means that ξ is a weighted mean of the ξ_t's. In fact

$$\xi = N^{-1} \sum_{t=1}^{k} n_t \xi_t \quad \text{where} \quad N = \sum_{t=1}^{k} n_t$$

The model is then

$$X_{ti} = \xi + \gamma_t + Z_{ti}$$

(The introduction of the n_t's, which depend on the experimental arrangement, into the linear condition imposed on the γ_t's, may be a little confusing, but it is merely a matter of conveniently simplifying the algebra. The essential structure of the model is unaffected.)

If random variables are used to denote factor level effects, the model is called *random* or *component-of-variance* (also *Model II*). In this case the model for the one-way classification is

$$X_{ti} = \xi + U_t + Z_{ti} \tag{13.4}$$

where the U_t's are mutually independent random variables each with expected value 0, and variance σ_U^2, and are usually supposed to be normally distributed. Further, the U_t's are supposed to be independent of the Z_{ti}'s.

13.5   THE ANALYSIS—PARAMETRIC MODEL

Although a general method of approach can be given for parametric models (this method is described in Section 13.16) we initially develop the analysis for each pattern of data on its own merits. The techniques presented are put forward as reasonable procedures to use, without claiming any general optimum properties for them.

Consider, now, the problem of investigating possible differences between groups. This means that within the limitations of the restrictions placed on the Z_{ti}'s (in particular, the supposed equality of variances), we are concerned with comparisons among the values γ_t if the parametric model (13.3) is appropriate, or with the value of the "between groups" standard deviation σ_U, if the component-of-variance model (13.4) is valid.

Firstly, we suppose that the parametric model

$$X_{ti} = \xi + \gamma_t + Z_{ti} \tag{13.5}$$

is appropriate. It is natural to calculate the arithmetic means for each group

$$\overline{X}_{1.} = n_1^{-1} \sum_{i=1}^{n_1} X_{1i}, \dots, \overline{X}_{t.} = n_t^{-1} \sum_{i=1}^{n_t} X_{ti}, \dots, \overline{X}_{k.} = n_k^{-1} \sum_{i=1}^{n_k} X_{ki}$$

From the model it follows that

$$\overline{X}_{t.} = \xi + \gamma_t + \overline{Z}_{t.} \tag{13.6}$$

where $\overline{Z}_{t.} = n_t^{-1} \sum_{i=1}^{n_t} Z_{ti}$.

Similarly if $\overline{X}_{..} = N^{-1} \sum_{t=1}^{k} n_t \overline{X}_{t.}$ is the mean of all the observations (the "overall" mean), then

$$\overline{X}_{..} = \xi + \overline{Z}_{..} \tag{13.7}$$

where $\overline{Z}_{..} = N^{-1} \sum_{t=1}^{k} n_t \overline{Z}_{t.}$ (since $\sum_{t=1}^{k} n_t \gamma_t = 0$).

If we are interested in testing the hypothesis $\gamma_1 = \gamma_2 = \cdots = \gamma_k = 0$ (that is, $\gamma_t = 0$ for all t) it is natural to calculate $\overline{X}_{t.} - \overline{X}_{..}$. From (13.6) and (13.7),

$$\overline{X}_{t.} - \overline{X}_{..} = \gamma_t + \overline{Z}_{t.} - \overline{Z}_{..} \qquad (13.8)$$

It follows from assumptions of Section 13.2 that the expected value of $\overline{X}_{t.} - \overline{X}_{..}$ is equal to γ_t, and the standard deviation is proportional to σ. To estimate σ it is necessary to get rid of all other parameters by constructing a function of the observations which can be expressed in terms of the Z_{ti}'s alone. In the present case this can be done quite simply by calculating the differences

$$X_{ti} - \overline{X}_{t.} = Z_{ti} - \overline{Z}_{t.} \qquad (13.9)$$

[from (13.5) and (13.6)].

We now note that since $X_{ti} - \overline{X}_{..} = (\overline{X}_{t.} - \overline{X}_{..}) + (X_{ti} - \overline{X}_{t.})$ the sum of squares of deviations of X_{ti} from the overall mean $\overline{X}_{..}$ can be split up into 2 parts. One part is the sum of squares of the quantities appearing in (13.8); the other is the sum of squares of the quantities appearing in (13.9). In each case the sum is over "all observations," but, since $(\overline{X}_{t.} - \overline{X}_{..})^2$ does not depend on i, it is repeated n_t times and appears as $n_t(\overline{X}_{t.} - \overline{X}_{..})^2$. The splitting up corresponds to the algebraic identity

$$\sum_{t=1}^{k} \sum_{i=1}^{n_t} \left(X_{ti} - \overline{X}_{..} \right)^2 = \sum_{t=1}^{k} n_t \left(\overline{X}_{t.} - \overline{X}_{..} \right)^2 + \sum_{t=1}^{k} \sum_{i=1}^{n_t} \left(X_{ti} - \overline{X}_{t.} \right)^2 \qquad (13.10)$$

$$\left(\text{Note that } \sum_{t=1}^{k} \sum_{i=1}^{n_t} \left(\overline{X}_{t.} - \overline{X}_{..} \right)\left(X_{ti} - \overline{X}_{t.} \right) = \sum_{t=1}^{k} \left(\overline{X}_{t.} - \overline{X}_{..} \right) \sum_{i=1}^{n_t} \left(X_{ti} - \overline{X}_{t.} \right) = 0 \right)$$

It is customary to give the quantities in this equation names, and the equation may be expressed

Total sum of squares = [Between Groups sum of squares]

+ [Within Groups sum of squares]

The expected value of the Between Groups sum of squares is

$$E\left[\sum_{t=1}^{k} n_t \left(\overline{Z}_{t.} - \overline{Z}_{..} + \gamma_t \right)^2 \right] = E\left[\sum_{t=1}^{k} n_t \left(\overline{Z}_{t.} - \overline{Z}_{..} \right)^2 \right] + \sum_{t=1}^{k} n_t \gamma_t^2$$

[using assumption (i) of Section 13.4].

Using assumptions (*i*), (*ii*), and (*iii*) of Section 13.4 we find

$$E\left[\sum_{t=1}^{k} n_t\left(\bar{X}_{t.} - \bar{X}_{..}\right)^2\right] = (k-1)\sigma^2 + \sum_{t=1}^{k} n_t\gamma_t^2 \qquad (13.11)$$

The Within Groups sum of squares is equal to $\sum_{t=1}^{k}\sum_{i=1}^{n_t}(Z_{ti} - \bar{Z}_{t.})^2$ and its expected value is equal to

$$\sigma^2 \sum_{t=1}^{k} (n_t - 1) = (N-k)\sigma^2$$

The multipliers of σ^2 (in the expected values) are called the "degrees of freedom" of the corresponding sums of squares. So the Between Groups sum of squares has $(k-1)$ degrees of freedom, and the Within Groups sum of squares has $(N-k)$ degrees of freedom.

In order to make the expected values comparable we divide each sum of squares by its number of degrees of freedom, forming a "mean square." Thus the Between Groups mean square is $(k-1)^{-1}\sum_{t=1}^{k} n_t(\bar{X}_{t.} - \bar{X}_{..})^2$, the expected value of Between Groups mean square is $\sigma^2 + (k-1)^{-1}\sum_{t=1}^{k} n_t\gamma_t^2$, the Within Groups mean square is $(N-k)^{-1}\sum_{t=1}^{k}\sum_{i=1}^{n_t}(X_{ti} - \bar{X}_{t.})^2$, and the expected value of Within Groups mean square is σ^2.

The "expected value of mean square" is an important concept in the analysis of variance. It is often abbreviated to EMS. From time to time we will follow this practice. Considerable excess of the Between Groups over the Within Groups mean square can reasonably be ascribed to the existence of some nonzero γ_t's—that is, to some real differences between groups, in regard to expected values.

So far we have not used assumption (*iv*) of Section 13.4. The results so far obtained do not depend for their validity on the normality of the distribution of the Z_{ti}'s. However, if we do introduce this further assumption it can be used in the construction of a formal test of the hypothesis of no difference between groups ($\gamma_1 = \gamma_2 = \cdots = \gamma_k = 0$) with a precisely calculable significance level. For in this case it can be shown that the ratio

$$R = \frac{\text{Between Groups mean square}}{\text{Within Groups mean square}}$$

has a noncentral F distribution (see Section 8.11) with $k-1$, $N-k$ degrees of freedom and noncentrality parameter $\sigma^{-2}\sum_{t=1}^{k} n_t\gamma_t^2$. If the hypothesis $\gamma_1 = \gamma_2 = \cdots = \gamma_k = 0$ is true, the distribution is a central F distribution with $k-1$, $N-k$ degrees of freedom. Hence if the hypothesis is rejected whenever R is greater than the $100(1-\alpha)$ percent point of this distribution, that is,

$$R > F_{k-1,N-k,1-\alpha}$$

then the significance level of the test will be α. Using the noncentral F distribution (with noncentrality parameter $\sigma^{-2}\Sigma_{t=1}^{k}n_t\gamma_t^2$) the power of the test to detect any given set of γ_t's (not all 0) can be found. Table H of the Appendix contains charts from which the power can be read off for certain cases.

Example 13.1   Four rubber compounds were tested for tensile strength. Rectangular samples were prepared and pulled in a longitudinal direction. The purpose of the test was (1) to determine the strength of each of these samples in the hopes of choosing the superior one, and (2) to obtain an estimate of the testing error. A preliminary sample of 16 specimens, 4 for each compound, was prepared. One of the specimens of compound A was judged defective and removed before the tensile test. The data of Table 13.1 are the results of the test. X_{ti} are measured in pounds per square inch.

TABLE 13.1

TENSILE STRENGTH OF 4 COMPOUNDS

A	B	C	D
3210	3225	3220	3545
3000	3320	3410	3600
3315	3165	3320	3580
	3145	3370	3485

The data are exhibited graphically in Figure 13.2. It is clear that D differs markedly from A, B, and C but not so definite that C differs from A and B.

Before analyzing the data it is best to simplify the values (if a computer is not available) by subtracting 3000 and dividing by 5. Table 13.2 presents these coded values. Note that the mean square ratio will be unchanged, but the residual variance for the coded values is of course, $1/25$ of the original residual variance.

The model is a parametric model defined by the equations

$$X_{ti} = \xi + \gamma_t + Z_{ti}$$

where $i=1,\ldots,n_t$, $t=1,2,3,4$; and $n_1=3$, $n_2=4$, $n_3=4$, and $n_4=4$. The calculation is as follows:

$$\sum_{i=1}^{3} X_{1i}=105 \qquad \sum_{i=1}^{4} X_{3i}=264 \qquad \sum_{t=1}^{4}\sum_{i=1}^{n_t} X_{ti}=982$$

$$\sum_{i=1}^{4} X_{2i}=171 \qquad \sum_{i=1}^{4} X_{4i}=442 \qquad \sum_{t=1}^{4}\sum_{i=1}^{n_t} X_{ti}^2=81162$$

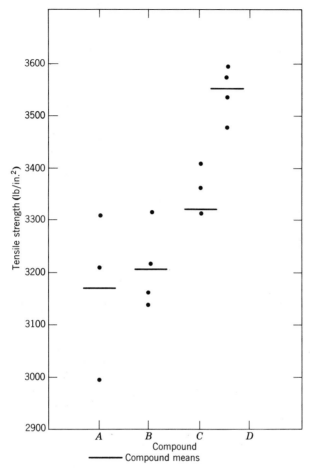

Figure 13.2 Tensile strength of 4 compounds.

TABLE 13.2

CODED TENSILE STRENGTH

A	B	C	D
42	45	44	109
0	64	82	120
63	33	64	116
	29	74	97

603

The Between Compounds mean square is

$$(k-1)^{-1} \sum_{t=1}^{k} n_t (\bar{X}_{t.} - \bar{X}_{..})^2 = \frac{1}{3} \left[\sum_{t=1}^{k} n_t^{-1} \left(\sum_{i=1}^{n_t} X_{ti} \right)^2 - \frac{1}{15} \left(\sum_{t=1}^{k} \sum_{i=1}^{n_t} X_{ti} \right)^2 \right]$$

$$= \frac{1}{3} \left[\frac{1}{3}(105)^2 + \frac{1}{4}(171)^2 + \frac{1}{4}(264)^2 \right.$$

$$\left. + \frac{1}{4}(442)^2 - \frac{1}{15}(982)^2 \right]$$

$$= \frac{1}{3}[77250.25 - 64288.27] = \frac{1}{3}[12961.98]$$

$$= 4320.66$$

The Within Compounds mean square is

$$(N-k)^{-1} \sum_{t=1}^{k} \sum_{i=1}^{n_t} (X_{ti} - \bar{X}_{t.})^2 = \frac{1}{11} \left[\sum_{t=1}^{k} \sum_{i=1}^{n_t} X_{ti}^2 - \sum_{t=1}^{k} n_t^{-1} \left(\sum_{i=1}^{n_t} X_{ti} \right)^2 \right]$$

$$= \frac{1}{11}[81162 - 77250.25] = \frac{3911.75}{11} = 355.61$$

Finally, the value of the mean square ratio is $4320.66/355.61 = 12.15$. Since $12.15 > F_{3,11,0.999}$ we say that there is strong evidence of differences in mean tensile strength among the 4 compounds.

Separate 95 percent confidence intervals for mean tensile strength for each of the compounds are given by

$$\bar{X}_{t.} \pm \frac{t_{11,0.975}S}{\sqrt{n_t}}$$

Note that the number of degrees of freedom is that of the Within Compounds sum of squares since our estimate of σ^2 is based on these 11 degrees of freedom out of the total of 14 available.

13.6 THE ANALYSIS—COMPONENT-OF-VARIANCE MODEL

The model is now

$$X_{ti} = \xi + U_t + Z_{ti} \tag{13.12}$$

In this case we use the same formal analysis as for the parametric model.

That is, we calculate the same sums of squares:

$$\text{Between Groups sum of squares} = \sum_{t=1}^{k} n_t \left(\overline{X}_{t.} - \overline{X}_{..} \right)^2$$

$$\text{Within Groups sum of squares} = \sum_{t=1}^{k} \sum_{i=1}^{n_t} \left(X_{ti} - \overline{X}_{t.} \right)^2$$

It is easy to see that $X_{ti} - \overline{X}_{t.} = Z_{ti} - \overline{Z}_{t.}$ just as in the parametric model [see (13.9)]. The expected value of the Within Groups sum of squares is therefore $(N-k)\sigma^2$, as before. However, there is a different situation for the Between Groups sum of squares. Expressing this in terms of the right side of the model (13.12), we find

$$\sum_{t=1}^{k} n_t \left(\overline{X}_{t.} - \overline{X}_{..} \right)^2 = \sum_{t=1}^{k} n_t \left(U_t - \overline{U} + \overline{Z}_{t.} - \overline{Z}_{..} \right)^2$$

where

$$\overline{U} = N^{-1} \sum_{t=1}^{k} n_t U_t$$

Since the U's and the Z's are mutually independent, and each has expected value 0,

$$E \left[\sum_{t=1}^{k} n_t \left(U_t - \overline{U} \right) \left(\overline{Z}_{t.} - \overline{Z}_{..} \right) \right] = 0$$

Hence

$$E \left[\sum_{t=1}^{k} n_t \left(\overline{X}_{t.} - \overline{X}_{..} \right)^2 \right] = E \left[\sum_{t=1}^{k} n_t \left(U_t - \overline{U} \right)^2 \right] + E \left[\sum_{t=1}^{k} n_t \left(\overline{Z}_{t.} - \overline{Z}_{..} \right)^2 \right]$$

From (13.11),

$$E \left[\sum_{t=1}^{k} n_t \left(\overline{Z}_{t.} - \overline{Z}_{..} \right)^2 \right] = (k-1)\sigma^2$$

Also,

$$\sum_{t=1}^{k} n_t \left(U_t - \overline{U} \right)^2 = \sum_{t=1}^{k} n_t U_t^2 - N\overline{U}^2$$

Since $E(U_t)=0$ and $\text{var}(U_t)=\sigma_U^2$,

$$E(U_t^2)=\sigma_U^2 \quad \text{and} \quad E(\overline{U}^2)=\sigma_U^2 \sum_{t=1}^{k} \left(\frac{n_t}{N} \right)^2$$

Hence the expected value of the Between Groups sum of squares is

$$(k-1)\sigma^2 + \left[\sum_{t=1}^{k} n_t - N \sum_{t=1}^{k} \left(\frac{n_t}{N} \right)^2 \right] \sigma_U^2$$

$$= (k-1)\sigma^2 + \left[N - \sum_{t=1}^{k} \frac{n_t^2}{N} \right] \sigma_U^2 \qquad (13.13)$$

Dividing by the appropriate numbers of degrees of freedom, we obtain the expected value of Between Groups mean square,

$$= \sigma^2 + (k-1)^{-1} \left[N - \frac{\sum\limits_{t=1}^{k} n_t^2}{N} \right] \sigma_U^2$$

and the expected value of Within Groups mean square, σ^2.

The coefficient of σ_U^2 is necessarily positive, and so considerable excess of the Between Groups over the Within Groups mean square can reasonably be ascribed to a nonzero value of σ_U, that is, to some real variation from group to group.

It will be noticed that the position is the same as that in the analysis of the parametric model, except that the interpretation of "variation between groups" is different. The formal calculations are exactly the same, and if $\sigma_U = 0$, the ratio

$$R = \frac{\text{Between Groups mean square}}{\text{Within Groups mean square}}$$

should have the (central) F distribution with $k-1$, $N-k$ degrees of freedom, just as in the parametric case. The formal test of significance (this time of the hypothesis $\sigma_U = 0$) therefore uses the same significance limit $F_{k-1, N-k, 1-\alpha}$ as in the parametric case.

There is, however, a difference in the *noncentral* distribution of R. In the parametric case this was a noncentral F distribution. In the component-of-variance case the distribution is, in general, rather complicated. However, in the special case of equal numbers of observations in each group $(n_1 = n_2 = \cdots = n_k = n$, say) the noncentral distribution is very simply obtained.

In this case the Between Groups sum of squares is

$$n \sum_{t=1}^{k} \left[\left(U_t + \bar{Z}_{t.} \right) - \left(\bar{U} + \bar{Z}_{..} \right) \right]^2 = n \sum_{t=1}^{k} \left(Y_t - \bar{Y} \right)^2 \qquad (13.14)$$

where

$$Y_t = U_t + \bar{Z}_{t.}; \quad \bar{Y} = k^{-1} \sum_{t=1}^{k} Y_t$$

The Y_t's are mutually independent normal variables each with zero expected value and variance $\sigma_U^2 + \sigma^2 n^{-1}$. Hence the Between Groups sum of squares is distributed as

$$n \chi^2 \left(\sigma^2 n^{-1} + \sigma_U^2 \right) = \chi^2 \left(\sigma^2 + n \sigma_U^2 \right)$$

the χ^2 having $(k-1)$ degrees of freedom. Furthermore, the Y_t's are independent of the within group deviations $Z_{ti} - \bar{Z}_{t.}$, since they depend only on the U_t's and the $\bar{Z}_{t.}$'s. Hence R is distributed as the ratio

$$\frac{\left[\chi_{k-1}^2 \left(\sigma^2 + n \sigma_U^2 \right) \right] / (k-1)}{\chi_{N-k}^2 \sigma^2 / (N-k)}$$

where the two χ^2's are mutually independent. The noncentral distribution of R is thus a *central* F with $(k-1)$, $(N-k)$ degrees of freedom, multiplied by $1 + n(\sigma_U / \sigma)^2$. This simple result makes it possible to evaluate the operating characteristic (or the power function), of the test in a straight-forward manner.

It should be noted that the comparison of expected values of mean square does not make use of assumption (*iv*) (the normality of the random variables). This assumption is needed only when exact distributions are required as in calculating the power.

Although the choice of model appears to make little difference in the present case (one-way classification), we shall soon encounter (in Sections 13.10 and 13.12) cases where the choice of model actually affects the analysis of the data.

Example 13.2 Suppose we are satisfied that the component-of-variance model (13.12)

$$X_{ti} = \xi + U_t + Z_{ti}$$

applies to our problem. We wish to take random samples of size n from each of k groups, calculate the mean square ratio,

$$R = \frac{\text{Between Groups mean square}}{\text{Within Groups mean square}}$$

and reject the product examined if $R > R_0$ where R_0 is a suitably chosen number, so that

(i) the probability of rejection is at least $1 - \beta$ when $\dfrac{\sigma_U^2}{\sigma^2} > \delta_1^2$

(ii) the probability of rejection is at most α when $\dfrac{\sigma_U^2}{\sigma^2} < \delta_0^2 (<\delta_1^2)$

This means that we want R_0 to satisfy the inequalities

$$\Pr\left[R > R_0 \middle| \frac{\sigma_U^2}{\sigma^2} = \delta_1^2 \right] \geqslant 1 - \beta$$

$$\Pr\left[R > R_0 \middle| \frac{\sigma_U^2}{\sigma^2} = \delta_0^2 \right] \leqslant \alpha$$

Since R is distributed as $F_{k-1, k(n-1)} \cdot (1 + n\sigma_U^2/\sigma^2)$, these inequalities imply that

$$\Pr\left[F_{k-1, k(n-1)} > R_0 \left(1 + n\delta_1^2\right)^{-1} \right] \geqslant 1 - \beta$$

$$\Pr\left[F_{k-1, k(n-1)} > R_0 \left(1 + n\delta_0^2\right)^{-1} \right] \leqslant \alpha$$

In turn, these inequalities imply

$$F_{k-1, k(n-1), \beta} \geqslant R_0 \left(1 + n\delta_1^2\right)^{-1}$$

and

$$F_{k-1, k(n-1), 1-\alpha} \leqslant R_0 \left(1 + n\delta_0^2\right)^{-1}$$

Combining these inequalities we have

$$\left(1 + n\delta_0^2\right) F_{k-1, k(n-1), 1-\alpha} \leqslant R_0 \leqslant \left(1 + n\delta_1^2\right) F_{k-1, k(n-1), \beta}$$

In order to be able to find a value R_0 satisfying both these inequalities we must have

$$\frac{F_{k-1,k(n-1),1-\alpha}}{F_{k-1,k(n-1),\beta}} \leqslant \frac{1+n\delta_1^2}{1+n\delta_0^2}$$

As n increases, the right side increases from 1 to $(\delta_1/\delta_0)^2$; the left side decreases to $\chi_{k-1,1-\alpha}^2/\chi_{k-1,\beta}^2$ [because as n increases $F_{k-1,k(n-1)}$ tends to be distributed as $\chi_{k-1}^2/(k-1)$]. If $\delta_0^2 > 0$, therefore, and

$$\chi_{k-1,1-\alpha}^2/\chi_{k-1,\beta}^2 > \left(\frac{\delta_1}{\delta_0}\right)^2$$

it is impossible to satisfy the required conditions. We therefore have to choose k big enough to have

$$\frac{\chi_{k-1,1-\alpha}^2}{\chi_{k-1,\beta}^2} < \left(\frac{\delta_1}{\delta_0}\right)^2$$

There is thus a *minimum possible* number of groups to satisfy the required conditions. A few such minimum values of k are shown in Table 13.3.

TABLE 13.3

MINIMUM NUMBER OF GROUPS—COMPONENT-OF-VARIANCE MODEL

$\alpha = \beta = 0.05$		$\alpha = \beta = 0.01$	
δ_1/δ_0	MINIMUM NUMBER OF GROUPS	δ_1/δ_0	MINIMUM NUMBER OF GROUPS
1.5	35	1.5	68
2	14	2	25
2.5	9	2.5	16
3	7	3	12

There must be at least the minimum number of groups shown, however large the number of observations per group.

It is interesting to note that for the similar problem of discrimination when a *parametric* model applies, it is *always* possible to attain a required degree of sensitivity by taking n sufficiently large, whatever be the number of groups.

13.6.1 ANOVA Table

The analysis of variance (ANOVA) table containing the sums of squares, degrees of freedom, mean square, expectations, etc., presents the initial analysis in a compact form. This is summarized for both the parametric (I) and component-of-variance (II) models in Table 13.4.

TABLE 13.4
Analysis of Variance (ANOVA) Table for a One-Way Classification—Models I and II

SOURCE	SUM OF SQUARES* (S.S.)	DEGREES OF FREEDOM (D.F.)	MEAN SQUARE (M.S.)	MEAN SQUARE RATIO	EXPECTED VALUE OF MEAN SQUARE Model I	Model II
Between Groups	$S_1 = \sum_{t=1}^{k} n_t(\bar{X}_{t\cdot} - \bar{X}_{..})^2 = \sum_t \left(\frac{X_{t\cdot}^2}{n_t}\right) - \frac{X_{..}^2}{N}$	$k-1$	$\dfrac{S_1}{(k-1)}$	$\dfrac{S_1(N-k)}{S_e(k-1)}$	$\sigma^2 + \dfrac{\sum_{t=1}^{k} n_t\gamma_t^2}{k-1}$	$\sigma^2 + \dfrac{\left(N - \sum_t n_t^2/N\right)\sigma_U^{2\dagger}}{k-1}$
Within Groups	$S_e = \sum_{t=1}^{k} \sum_{i=1}^{n_t}(X_{ti} - \bar{X}_{t\cdot})^2$ $= \sum_t \sum_i X_{ti}^2 - \sum_t \left(\frac{X_{t\cdot}^2}{n_t}\right)$	$\sum_{t=1}^{k}(n_t - 1) = N - k$	$\dfrac{S_e}{(N-k)}$		σ^2	σ^2
Total	$S = \sum_{t=1}^{k} \sum_{i=1}^{n_t}(X_{ti} - \bar{X}_{..})^2 = \sum_t \sum_i X_{ti}^2 - \frac{X_{..}^2}{N}$					

*We shall use the symbols: $X_{t\cdot} = \sum_{i=1}^{n_t} X_{ti}, X_{..} = \sum_{t=1}^{k} \sum_{i=1}^{n_t} X_{ti}$

$\bar{X}_{t\cdot} = n_t^{-1}\sum_{i=1}^{n_t} X_{ti} = n_t^{-1}X_{t\cdot}, \bar{X}_{..} = N^{-1}\sum_{t=1}^{k} \sum_{i=1}^{n_t} X_{ti} = N^{-1}X_{..}$

†If the n_t are equal this reduces to $\sigma^2 + n\sigma_U^2$

This kind of tabular representation is customarily used to set out the results of analysis of variance calculations. We describe appropriate ANOVA tables for each of the experimental patterns as we consider them. ANOVA tables present, among other things, calculated values of criteria for a number of formal significance tests. To assist appraisal of the results of these tests the following conventional notation is used:

*	denotes significance level between 0.01 and 0.05
**	denotes significance level between 0.001 and 0.01
***	denotes significance level less than 0.001

(Occasionally * is called "significant," ** is called "highly significant," and *** is called "very highly significant.")

These symbols are usually placed immediately to the right of the mean square ratios or mean squares to which they refer. They are only intended to help in appreciating the general nature of the results of the analysis, and should not be used in an automatic manner, as a substitute for careful thought about interpretation.

The first use of the notation in this book is found in Table 13.7.

13.6.2 Estimation of Between Groups Variance Component

If $n_1 = n_2 = \cdots = n_k = n$, then a $100(1 - \alpha)$ percent confidence interval for the ratio σ_U^2/σ^2 is easily constructed, using the relationship

$$\Pr\left[\left(1 + \frac{n\sigma_U^2}{\sigma^2}\right)F_{k-1,N-k,\alpha/2} < R < \left(1 + \frac{n\sigma_U^2}{\sigma^2}\right)F_{k-1,N-k,1-\alpha/2} \,\middle|\, \frac{\sigma_U^2}{\sigma^2}\right] = 1 - \alpha$$

The inequalities can be rearranged to give the confidence limits

$$n^{-1}\left[\frac{R}{F_{k-1,N-k,1-\alpha/2}} - 1\right], \quad n^{-1}\left[\frac{R}{F_{k-1,N-k,\alpha/2}} - 1\right]$$

for the ratio σ_U^2/σ^2.

However, it is often useful to have confidence limits for σ_U^2 alone, and this is a much less straightforward problem, even when the n_i's are all the same. A number of methods of constructing *approximate* confidence intervals for σ_U^2 have been proposed. The method now to be described was suggested by J. S. Williams [13.13]. This gives an interval with a confidence coefficient which, although not known exactly, is guaranteed to be at least $100(1 - 2\alpha)$ percent.

The interval is obtained by combining two confidence intervals. One is that given above, for σ_U^2/σ^2. The other is the $100(1-\alpha)$ percent interval for $\sigma^2 + n\sigma_U^2$:

$$\left(\frac{S_1}{\chi^2_{k-1,1-\alpha/2}} , \frac{S_1}{\chi^2_{k-1,\alpha/2}} \right)$$

Whatever the value of σ^2, if both $A < \sigma_U^2/\sigma^2 < B$ and $A_1 < \sigma^2 + n\sigma_U^2 < B_1$, then σ_U^2 must lie between the limits

$$AA_1(nA+1)^{-1} \quad \text{and} \quad BB_1(nB+1)^{-1} \tag{13.15}$$

Taking

$$A = n^{-1}\left[R\left(F_{k-1,N-k,1-\alpha/2} \right)^{-1} - 1 \right]$$

$$B = n^{-1}\left[R\left(F_{k-1,N-k,\alpha/2} \right)^{-1} - 1 \right]$$

$$A_1 = \frac{S_1}{\chi^2_{k-1,1-\alpha/2}}$$

$$B_1 = \frac{S_1}{\chi^2_{k-1,\alpha/2}}$$

the formulas (13.15) give the required confidence limits for σ_U^2.

The theoretical argument is

$$\Pr\left[AA_1(nA+1)^{-1} < \sigma_U^2 < BB_1(nB+1)^{-1} \right]$$

$$\geqslant \Pr\left[\left(A < \frac{\sigma_U^2}{\sigma^2} < B \right) \cap \left(A_1 < \sigma^2 + n\sigma_U^2 < B_1 \right) \right]$$

$$\geqslant \Pr\left[A < \frac{\sigma_U^2}{\sigma^2} < B \right] + \Pr\left[A_1 < \sigma^2 + n\sigma_U^2 < B_1 \right] - 1$$

$$= 1 - \alpha + 1 - \alpha - 1 = 1 - 2\alpha$$

Of course, if a limit is negative it can be replaced by 0 without reducing the confidence coefficient, because σ_U^2 cannot be negative.

Example 13.3 Suppose we have obtained the analysis of variance table shown below, calculated from measurements on each individual in random samples of equal size from batches in routine production. We wish to estimate the batch-to-batch ("between batches") component of variance.

SOURCE	DEGREES OF FREEDOM	SUM OF SQUARES	MEAN SQUARE
Batches	14	28.8	2.057
Within Batches	60	41.7	0.695
Total	74	70.5	

We construct a confidence interval with minimum confidence coefficient 98 percent. Using the notation of this section, we have

$$k = 15; \quad n = 5; \quad N = 75$$

$$S_1 = 28.8; \quad S_e = 41.7; \quad R = \frac{2.057}{0.695} = 2.960$$

$$F_{14,60,0.01} = F_{60,14,0.99}^{-1} = 0.417; \quad F_{14,60,0.99} = 3.18$$

$$\chi_{14,0.01}^2 = 4.660; \quad \chi_{14,0.99}^2 = 29.14$$

$$A = \tfrac{1}{5}\left[2.960 \times (3.18)^{-1} - 1\right] < 0$$

$$B = \tfrac{1}{5}\left[2.960 \times (0.417)^{-1} - 1\right] = 1.221$$

$$B_1 = \frac{28.8}{4.660} = 6.18$$

There is no need to calculate A_1, because the lower limit will be 0, as $A < 0$. The upper limit is

$$\frac{1.221 \times 6.18}{5 \times 1.221 + 1} = 1.06$$

An unbiased estimate of the between batches component of variance is $\tfrac{1}{5}(2.057 - 0.695) = 0.27$.

If we had from previous production a very reliable (i.e., accurate) estimate of 0.75 for the within batch variance (σ^2) we could regard σ^2 as "known" and use the 98 percent limits for $5\sigma_U^2 + \sigma^2$:

$$\frac{S_1}{\chi_{14,0.99}^2} = 0.988 \quad \text{and} \quad \frac{S_1}{\chi_{14,0.01}^2} = 6.18$$

to obtain (exact) 98 percent confidence limits for σ_U^2:

$$\tfrac{1}{5}(0.988 - 0.75) = 0.048$$

and

$$\tfrac{1}{5}(6.18 - 0.75) = 1.10$$

(Note that "exact" refers to the 98 percent confidence coefficient.)

*13.7   THE ANALYSIS—RANDOMIZATION MODEL

We now come to consider a third kind of model, rather different in nature from the 2 previous models, parametric and component-of-variance. This new model is based, essentially, on 2 assumptions—one about the way in which the investigation is carried out, and the other about the way in which observed values for a given individual change when it is moved from one group to another. The first assumption is

(I) Each possible arrangement of individuals in the experimental pattern is as likely as any other.

This means, for example, that in the problem described in Section 13.4, each possible assignment of n_1 individuals (out of N) to Drug 1, n_2 to Drug 2,...,n_k to Drug k is as likely as any other. This can be effected by using tables of random numbers, or some equivalent procedure, and is called *randomization*. (It should be noted that randomization may not always be physically possible. Unless it is possible, *and* effected, randomization theory, as developed in this section, is not necessarily applicable.) The second assumption is

(II) If the value observed for a particular individual in group i is X_i, the value that would be observed if the individual were in group j would be $X_i + (\gamma_j - \gamma_i)$, where γ_j, γ_i are fixed constants.

Assumption II implies that if the B's are all 0 (the "null" hypothesis of no difference between groups is true), then each individual will give the *same* observed value, wherever it is placed in the experiment.

It will be noticed that this assumption is much more stringent than the "null" hypothesis assumed in either Model I or Model II. In those models the null hypothesis simply states that there is no difference *on the average* between values observed in the different groups. The randomization "null" hypothesis states that each separate individual gives *exactly the same* observed value, whatever be the group to which it is assigned.

What, then, do we get in return for the extra trouble taken to satisfy (I) and the rather severe requirements of (II)? The most important advantage is that we are more confident of the nature of the random variables representing residual variation, and (in principle, at any rate) we can carry out a test of significance without introducing the assumption (*iv*) of normality of distribution.

Suppose, in fact, that the null hypothesis is true, so that there is no difference between groups. Then the set $\{x_{ti}\}$ of observed values is one out of $N! \prod_{t=1}^{k} (n_t!)^{-1}$ equally likely arrangements of the N available values. [Note that the *same* set of N values will be observed [on assumption (II)] if the null hypothesis is valid, whatever the assignment of individuals to groups.] We could, given sufficient patience and time, enumerate each of these possible arrangements and calculate the mean square ratio R for each of them. Out of the $N! \prod_{t=1}^{k} (n_t!)^{-1}$ values we have, in fact, observed 1 particular value. A test of significance, at significance level 100α percent, can be carried out by regarding as significant (of departure from the null hypothesis) any of the largest 100α percent of values of R. It is clear that the probability of obtaining such a value of R is 100α percent, and we do not need to introduce any special form of distribution to demonstrate this. In practice, however, the task of evaluating all possible values of R and so constructing the *randomization distribution* appropriate to each particular set of data would be excessive. Fortunately, it is found that a simple modification of normal theory usually gives a good approximation. Even a direct use of the F distribution with $(k-1)$, $(N-k)$ degrees of freedom gives fairly satisfactory results.

Let us now consider the form of model implied by assumptions (I) and (II). Since, under the null hypothesis the observed values $\{X_{ti}\}$ will be just one arrangement of N quantities $u_1, u_2, \ldots, u_N$ it follows that X_{ti} takes the value u_j with probability N^{-1} $(j = 1, \ldots, N)$. Hence we can write

$$X_{ti} = \xi + Z'_{ti} \qquad \text{with } \xi = N^{-1} \sum_{j=1}^{N} u_j \tag{13.16}$$

where $\Pr[Z'_{ti} = u_j - \xi] = N^{-1}$. If follows that $E(Z'_{ti}) = 0$ and

$$\text{var}(Z'_{ti}) = N^{-1} \sum_{j=1}^{N} (u_j - \xi)^2$$

The variables Z_{ti} therefore satisfy conditions (*i*) and (*iii*) of Section 13.4. However, they do not satisfy conditions (*ii*) or (*iv*). The latter condition cannot be satisfied because the Z's are *discrete* variables; condition (*ii*) is not satisfied because, for example,

$$\Pr\{Z'_{t'i'} = u_j - \xi \,|\, Z'_{ti} = u_j - \xi\} = 0 \qquad \text{if } t \neq t' \text{ or } i \neq i'$$

If (13.16) is completed by inserting the fixed group "effects" γ_t we have

$$X_{ti} = \xi + \gamma_t + Z'_{ti} \tag{13.17}$$

which is formally very similar to (13.5) except that the Z''_{ti}'s do not possess exactly the same properties as the Z_{ti}'s, and they arise as a consequence of assumptions (I) and (II) rather than from general speculation about the nature of residual variation. From the model (13.16) it is possible to derive the moments of the Within Groups and Between Groups sum of squares. The process is rather tedious and will not be given in full detail here. We note that

Total sum of squares = (Within Groups sum of squares)

+ (Between Groups sum of squares)

$$= \sum_{j=1}^{N} (u_j - \xi)^2 \tag{13.18}$$

and this sum is the same for *all* possible arrangements of the N available values. Hence we need consider only 1 of the 2 sums of squares.

We choose to consider the Between Groups sum of squares. If the null hypothesis is valid, then from (13.16) this sum of squares is equal to

$$\sum_{t=1}^{k} n_t \left(\bar{Z}'_{t.} - \bar{Z}'_{..} \right)^2 = \sum_{t=1}^{k} n_t \bar{Z}'^2_{t.} - N\bar{Z}'^2_{..}$$

where $\bar{Z}'_{t.} = n_t^{-1} \sum_{i=1}^{n_t} Z'_{ti}$; $\bar{Z}'_{..} = N^{-1} \sum_{t=1}^{k} n_t \bar{Z}'_{t.}$. Since $\bar{Z}'_{..}$ is simply the arithmetic mean of all the Z''_{ti}'s, it is always equal to 0 [because $\Sigma(u_j - \xi) = 0$], whatever the arrangement. It can be shown that the expected value of the Between Groups sum of squares is

$$\sum_{t=1}^{k} n_t E\left(\bar{X}'^2_{t.} \right) = (k-1)\left[(N-1)^{-1} \sum_{j=1}^{N} (u_j - \xi)^2 \right] \tag{13.19}$$

From (13.18) and (13.19) the expected value of the Within Groups sum of squares is

$$(N-k)\left[(N-1)^{-1} \sum_{j=1}^{N} (u_j - \xi)^2 \right] \tag{13.20}$$

Comparison of these results with the null hypothesis expected values for Models I and II shows that they are of exactly the same form, with $(N-1)^{-1}\sum_{j=1}^{N}(u_j - \xi)^2$ replacing σ^2. [Note that $N^{-1}\sum_{j=1}^{N}(u_j - \xi)^2$ is in fact the variance of the observed values.]

If we consider the statistic

$$Y = \frac{\text{Between Groups sum of squares}}{\text{Total sum of squares}}$$

then from (13.18) and (13.20)

$$E(Y) = \frac{(k-1)}{(N-1)}$$

Straightforward though lengthy analysis shows that

$$\text{var}(Y) = \frac{2(k-1)(N-k)}{(N+1)(N-1)^2} \left\{ 1 - \theta \frac{K_4}{K_2^2} \right\} \tag{13.21}$$

where

$$K_2 = (N-1)^{-1} \sum_{j=1}^{N} (u_j - A)^2$$

$$K_4 = (N-1)^{-1}(N-2)^{-1}(N-3)^{-1}$$

$$\times \left[N(N+1) \sum_{j=1}^{N} (u_j - A)^4 - 3(N-1) \left(\sum_{j=1}^{N} (u_j - A)^2 \right)^2 \right]$$

and

$$\theta = N^{-1} + \frac{N+1}{2(k-1)(N-k)} \left(k^2 N^{-1} - \sum_{t=1}^{k} n_t^{-1} \right)$$

$E(Y)$ and var(Y) agree with the corresponding values for Models I and II, if the degrees of freedom of the latter are each multiplied by

$$d = (N-1)^{-1} \left[(N+1)(1 - \theta K_4 K_2^{-2})^{-1} - 2 \right]$$

Hence as a practical method of applying the randomization distribution, without the necessity to evaluate the distribution exactly, we compare the observed ratio R with an appropriate percentage point of the F distribution with

$$d(k-1), \quad d(N-k)$$

degrees of freedom.

Very often (especially if N is large and no n_t's are small) d is very close to 1, and we are effectively using the same limits as for Models I and II ("normal theory").

Example 13.4 Suppose we have the following data from a one-way experimental design.

Group	I	II	III
	107	96	98
	105	101	104

The analysis of variance for this experiment is shown below:

SOURCE	DEGREES OF FREEDOM	SUM OF SQUARES	MEAN SQUARE
Between groups	2	58.3	29.2
Residual (within groups)	3	32.5	10.8
Total	5	90.8	

The upper 25, 10, and 5 percent points of the F distribution with 2,3 degrees of freedom are 2.28, 5.46, and 9.55, respectively. Therefore, on normal theory, the observed mean square ratio, 2.69 would be very unlikely to be judged significant.

To construct the randomization distribution of the mean square ratio we have to consider all possible arrangements of the 6 observed values 96, 98, 101, 104, 105, and 107 among the 6 positions in the experiment.

There are $6!/(2!)^3 = 90$ ways of dividing the 6 observed values into 3 groups of 2 each. These fall into sets of 6 ways of arrangement by permuting group order, each of which gives the same analysis of variance. There are, then, only 15 essentially different arrangements. The other 14 are itemized in Table 13.5.

The fourth column, minus $6\bar{X}^2$, gives the Between Groups sum of squares. Since the Total sum of squares is unchanged by mere rearrangement of the observations, the Within Groups sum of squares is obtained immediately as the difference

Within Groups sum of squares = (Total sum of squares)

− (Between Groups sum of squares)

The last column gives the other 14 possible, and, on the randomization distribution, equally likely, values of the mean square ratio. The observed value (2.69) is exceeded by 3 of the 15 possible values. So the randomization test gives a significance level of 20 to 27 percent.

It is interesting to note that the normal theory 25 percent point (2.28) is exceeded by 4 out of 15 values (27%), the 10 percent point (5.46) is exceeded by 2 out of 15 values (13 percent), and the 5 percent point (9.55) by 1 out of 15 values (7 percent). In view of the small size of this experimental design, there is quite a good correspondence between this randomization F distribution and the normal theory F distribution.

13.8   TWO FACTORS—HIERARCHAL (NESTED) CLASSIFICATION AND CROSS-CLASSIFICATION

In the one-way classification there is only 1 factor. The data are categorized according to the different levels of this single factor, and no other type of classification is considered.

If 2 (or more) factors are present, however, we must give some attention to the way in which these factors are related. Fortunately, for 2 factors there are only 2 different types of relationship that we need consider for most practical purposes. It is important to recognize the type of relationship that exists in any given case, so that the appropriate analysis can be applied.

The 2 types of relationship are *hierarchal* (or *nested*) *classification* and *cross classification*. The essential difference between them is that in the first each level of one factor (the *main group* factor) is associated with a different set of levels of the second factor (the *subgroup* factor), whereas in

TABLE 13.5
ANALYSIS FOR A RANDOMIZATION MODEL

I	II	III	CODED WITH ORIGIN AT 100 $2\sum_{t=1}^{3} \bar{X}_{t\cdot}^{2}$	SUMS OF SQUARES		MEAN SQUARE RATIO
				BETWEEN GROUPS	WITHIN GROUPS	
107 104	105 101	98 96	96.5	76.33	14.5	7.90
107 104	105 98	101 96	69.5	49.33	41.5	1.78
107 104	105 96	101 98	61.5	41.33	49.5	1.25
107 101	105 104	98 96	90.5	70.53	20.5	5.15
107 101	105 98	104 96	36.5	16.33	74.5	0.33
107 101	105 96	104 98	34.5	14.33	76.5	0.28
107 98	105 104	101 96	57.5	37.33	53.5	1.05
107 98	105 101	104 96	30.5	10.33	80.5	0.19
107 98	105 96	104 101	25.5	5.33	85.5	0.09
107 96	105 104	101 98	45.5	25.33	65.5	0.58
107 96	105 101	104 98	24.5	4.33	86.5	0.08
107 96	105 98	104 101	21.5	1.33	89.5	0.02
107 105	104 101	98 96	102.5	82.33	8.5	14.53
107 105	104 96	101 98	72.5	52.33	38.5	2.04

the second *each* level of one factor can be combined with *all* levels of the other factor. The factors in the nested classification are ordered "hierarchally" (as *main* and *subgroup* factors), but the factors in a cross-classification cannot be so ordered.

This general description can be made more explicit by considering a special case. If samples of the work of 3 operators on each of 4 machines (12 operators in all) are taken, in the absence of further information there is no reason to link any operator on one machine with any operator on another machine. The (subgroup) factor "Operators" thus has 12 levels, and 3 levels of this factor are associated with each of the 4 levels of the (main group) factor "Machines." The pattern can be represented diagrammatically as in Figure 13.3. Each cross represents 1 observation; for example, there are 2 observations for the first operator on the first machine and 3 for the second operator on this machine.

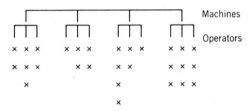

Figure 13.3   Two-factor hierarchal (nested) classification.

If, however, we knew that of the 3 operators assigned to any one machine one had 2 years' experience, one 4 years' experience, and one 6 years' experience, the pattern would then be regarded as a *cross-classification* of the 3-level factor "Experience" and the 4-level factor "Machines." The pattern could be represented in a "*two-way table*" as in Figure 13.4. The crosses representing observations now appear in the "cells" of this table. If there are the same number of observations (*n*, say) in each cell of the table, the analysis is considerably simplified. Such a pattern is termed

| | MACHINE | | | |
EXPERIENCE	I	II	III	IV
2 Years	× ×	×	× × × ×	× × ×
4 Years	× × ×	× ×	× ×	× × ×
6 Years	× ×	× ×	×	× × ×

Figure 13.4   Two-factor cross-classification.

an *n-replicate of a 3 × 4 cross-classification*. If $n = 1$ the pattern is termed a (*complete*) *3 × 4 cross-classification*. It can happen that no observations are available for some cells of the table. In this chapter, however, we assume that there is at least 1 observation in each cell.

We now turn to the construction of models appropriate to these two forms of classification. First we consider only parametric models (Model I).

In the case of a *nested* classification the model provides a separate expected value for each subgroup level in the pattern. This is broken up into

(*a*) an overall average, ξ;
(*b*) an average deviation from ξ, for the *t*th main group (γ_t, say);
(*c*) an average deviation from ($\xi + \gamma_t$), for the *i*th subgroup *in the* t*th main group* (δ_{ti}, say).

To this value, $\xi + \gamma_t + \delta_{ti}$, is added the random residual term Z_{tij}. The Z_{tij}'s are assumed to satisfy conditions like (*i*) to (*iv*) of Section 13.4. The model is thus

$$X_{tij} = \xi + \gamma_t + \delta_{ti} + Z_{tij} \tag{13.22}$$

where $j = 1, 2, \ldots, n_{ti}$ (the number of observations in the *i*th subgroup of the *t*th main group)

$i = 1, 2, \ldots, m_t$ (the number of subgroups in the *t*th main group)

$t = 1, 2, \ldots, k$ (the number of levels of the main group factor)

We impose conditions on the γ_t's and δ_{ti}'s analogous to those imposed on the γ_t's in the parametric model for one-way classification. These conditions are

$$\sum_{t=1}^{k} N_{t.}\gamma_t = 0; \quad \sum_{i=1}^{m_t} n_{ti}\delta_{ti} = 0 \qquad \text{for } t = 1, 2, \ldots, k$$

$$\left(\text{Note that } N_{t.} = \sum_{i=1}^{m_t} n_{ti}. \right)$$

If the γ_t's and δ_{ti}'s are replaced by random variables U_t, V_{ti}, we have the component-of-variance model,

$$X_{tij} = \xi + U_t + V_{ti} + Z_{tij} \tag{13.23}$$

where it is assumed that the U's, V's, and Z's are all mutually independent, that $E(U_t) = 0, E(V_{ti}) = 0$ for all t and i that $\text{var}(U_t) = \sigma_U^2$ and $\text{var}(V_{ti}) = \sigma_V^2$ for all t and all i, and (in order to apply formal tests of significance) that each U, V, and Z is normally distributed.

If the δ_{ti}'s are replaced by random variables V_{ti}, but the parameters γ_t remain in the model, it is no longer purely parametric or purely component-of-variance. Models of this kind

$$X_{tij} = \xi + \gamma_t + V_{ti} + Z_{tij} \qquad (13.24)$$

are called *mixed*. The model

$$X_{tij} = \xi + U_t + \delta_{ti} + Z_{tij}$$

is also a "mixed" model, but it is only rarely that a model of this kind is likely to be appropriate.

Cross-classification data with 2 factors are also represented by a symbol X_{tij} with 3 suffixes. The suffix t represents level of one factor, i represents level of the other factor, and j represents the order among the n_{ti} observations for which the tth level of the first factor is combined with the ith level of the second factor.

In the case of a cross-classification the suffix i represents the *same* level of the second factor, whatever the level of the first factor (and, of course, conversely for j). If the effects of changes in factor levels were simply additive we could represent the expected value of X_{tij} by

$$E(X_{tij}) = \xi + \rho_t + \kappa_i \qquad (t = 1, \ldots, r; \ i = 1, \ldots, c; \ j = 1, \ldots, n_{ti})$$

Conditions such as $\sum_{t=1}^{r} \rho_t = 0 = \sum_{i=1}^{c} \kappa_i$ can be imposed on the ρ's and the κ's. (The symbols ρ and κ (and r and c) have been used to help envision the data arranged in a table as in Figure 13.4, with 1 factor represented by rows, the other by columns.)

However, the formula $\xi + \rho_t + \kappa_i$ is, in general, not adequate to represent any possible set of rc expected values. It will only be adequate if the difference in expected values between any 2 levels, i and i', of one factor is the *same* whatever be the level of the other factor. To allow for deviation from this (so-called "additive") model we must introduce a further set of parameters $(\rho\kappa)_{ti}$. Since we already have the parameters $\xi, \rho_1, \ldots, \rho_r$, and $\kappa_1, \ldots, \kappa_c$, there are $1 + r + c + rc$ parameters in all to represent rc expected values. Hence in addition to the 2 relationships

$$\sum_{t=1}^{r} \rho_t = 0 = \sum_{i=1}^{c} \kappa_i$$

$(r + c - 1)$ further linear conditions can be imposed. The natural conditions to impose are

$$\sum_{t=1}^{r} (\rho\kappa)_{ti} = 0 \qquad \text{for all } i$$

$$\sum_{i=1}^{c} (\rho\kappa)_{ti} = 0 \qquad \text{for all } t$$

(The condition $\sum_{i=1}^{c}(\rho\kappa)_{ri} = 0$ is implied by the other $r + c - 1$ conditions.) The model is then

$$X_{tij} = \xi + \rho_t + \kappa_i + (\rho\kappa)_{ti} + Z_{tij} \tag{13.25}$$

The terms $(\rho\kappa)_{ti}$ are said to represent the *interaction* between the 2 factors.

The component-of-variance model for an $r \times c$ cross-classification is obtained by replacing $\rho_t, \kappa_i, (\rho\kappa)_{ti}$ by U_t, V_i, W_{ti}, respectively, where the U's, V's, and W's are mutually independent normal random variables, each with 0 expected value, and $\text{var}(U_t) = \sigma_U^2$ for all t, $\text{var}(V_i) = \sigma_V^2$ for all i, $\text{var}(W_{ti}) = \sigma_W^2$ for all t and all i. The component-of-variance model is then

$$X_{tij} = \xi + U_t + V_i + W_{ti} + Z_{tij} \tag{13.26}$$

Mixed models of the form

$$X_{tij} = \xi + \rho_t + \kappa_i + W_{ti} + Z_{tij} \tag{13.27}$$

(that is, with random interaction, but parametric average effects in each factor) are quite often appropriate, but models with parametric interaction and random factor effects are rarely used, though formally possible.

Randomization theory can also be applied, in suitable cases, to hierarchal classifications and to cross-classifications.

We now proceed to develop the formal analyses appropriate to these forms of classification. Each type of analysis is discussed in a separate section.

Example 13.5 As another illustration of the comparison of a nested and a crossed experiment, consider the sampling of 4 carloads of crude rubber. If the purpose of the experiment is to distinguish between particular positions of loading in the cars, then we have a crossed classification. The first factor is carloads and the second is positions in the car.

We construct the model

$$X_{tij} = \xi + \kappa_t + \pi_i + (\kappa\pi)_{ti} + Z_{tij}$$

where $t(=1,\ldots,4)$ denotes carloads and $i(=1,2,3)$ denotes fixed positions in the car.

On the other hand if we are primarily interested in the variability among and within carloads, there being no systematic arrangement in positions, we have a nested or hierarchal model. Three random samples within each car might be selected. The first sample in carload 1 is in no way related to the first in any other car (except that it happened to be the first in each carload). The model in this case would be

$$X_{tij} = \xi + \kappa_t + V_{ti} + Z_{tij}$$

In the second case we can estimate variance within each carload. In the first case, since there is a definite position effect to be allowed for, we use the variance within position–car combinations to estimate residual variance.

Note that in this case there are 2 possible models. Which one is appropriate depends on the way the data were obtained and the questions to be answered by the analysis.

13.9 ANALYSIS OF NESTED CLASSIFICATIONS— PARAMETRIC MODEL

Starting from the model in (13.22), namely,

$$X_{tij} = \xi + \gamma_t + \delta_{ti} + Z_{tij}$$

we calculate the *subgroup means*

$$\overline{X}_{ti.} = \xi + \gamma_t + \delta_{ti} + \overline{Z}_{ti.} \tag{13.28}$$

the *main-group means*

$$\overline{X}_{t..} = \xi + \gamma_t + \overline{Z}_{t..} \tag{13.29}$$

and the *overall mean*

$$\overline{X}_{...} = \xi + \overline{Z}_{...} \tag{13.30}$$

The $\overline{Z}$'s are the same functions of the Z's as the corresponding $\overline{X}$'s are of the X's; the conditions $\sum_{t=1}^{k} N_t \gamma_t = 0$; $\sum_{i=1}^{m_t} n_{ti} \delta_{ti} = 0$ have been used in deriving (13.29) and (13.30).

From these equations we find

$$\overline{X}_{t..} - \overline{X}_{...} = \gamma_t + \overline{Z}_{t..} - \overline{Z}_{...} \tag{13.31}$$

$$\overline{X}_{ti.} - \overline{X}_{t..} = \delta_{ti} + \overline{Z}_{ti.} - \overline{Z}_{t..} \tag{13.32}$$

$$X_{tij} - \overline{X}_{ti.} = Z_{tij} - \overline{Z}_{ti.} \tag{13.33}$$

From (13.31) we see that the deviations of group means from the overall mean give information about the γ_t's; from (13.32) the deviations of subgroup means from the mean of the main group to which they belong give information about the δ_{ti}'s. In each case there is also a random effect represented by the terms in Z's; to make some allowance for these we need an estimate of σ^2. This can be provided from the deviations of individual observations from the mean of subgroup to which they belong, as can be seen from (13.33).

The statistics by which the information available is used are derived from the following partition of the total sum of squares:

$$
\sum_{t=1}^{k} \sum_{i=1}^{m_t} \sum_{j=1}^{n_{ti}} \left(X_{tij} - \overline{X}_{...} \right)^2 = \sum_{t=1}^{k} N_{t.} \left(\overline{X}_{t..} - \overline{X}_{...} \right)^2
$$

$$
+ \sum_{t=i}^{k} \sum_{i=1}^{m_t} n_{ti} \left(\overline{X}_{ti.} - \overline{X}_{t..} \right)^2
$$

$$
+ \sum_{t=1}^{k} \sum_{i=1}^{m_t} \sum_{j=1}^{n_{ti}} \left(X_{tij} - \overline{X}_{ti.} \right)^2 \qquad (13.34)
$$

The sums of squares on the right-hand side are called the Between Main Groups, Between Subgroups Within Main Groups, and Residual (Within Subgroups) sums of squares, respectively. Their expected values are:

Between Main Groups:
$$(k-1)\sigma^2 + \sum_{t=1}^{k} N_{t.}\gamma_t^2$$

Between Subgroups Within Main Groups:
$$\left(\sum_{t=1}^{k} m_t - k \right)\sigma^2 + \sum_{t=1}^{k} \sum_{i=1}^{m_t} n_{ti}\delta_{ti}^2$$

Residual:
$$\left(N_{..} - \sum_{t=1}^{k} m_t \right)\sigma^2$$

where
$$N_{..} = \sum_{t=1}^{k} N_{t.} = \sum_{t=1}^{k} \sum_{i=1}^{m_t} n_{ti}$$

The numbers of degrees of freedom are thus $(k-1)$, $(\sum_{t=1}^{k}m_t - k)$, and $(N_{..} - \sum_{t=1}^{k}m_t)$, respectively, and the expected values of the mean squares are

$$E\left[\frac{1}{k-1}\sum_{t=1}^{k}N_{t.}\left(\bar{X}_{t..}-\bar{X}_{...}\right)^2\right]=\sigma^2+\frac{1}{k-1}\sum_{t=1}^{k}N_{t.}\gamma_t^2 \tag{13.35}$$

$$E\left[\frac{1}{\sum_{t=1}^{k}m_t-k}\sum_{t=1}^{k}\sum_{i=1}^{m_t}n_{ti}\left(\bar{X}_{ti.}-\bar{X}_{t..}\right)^2\right]=\sigma^2+\frac{1}{\sum_{t=1}^{k}m_t-k}\sum_{t=1}^{k}\sum_{i=1}^{m_t}n_{ti}\delta_{ti}^2$$

$$\tag{13.36}$$

and

$$E\left[\frac{1}{N_{..}-\sum_{t=1}^{k}m_t}\sum_{t=1}^{k}\sum_{i=1}^{m_t}\sum_{j=1}^{n_{ti}}\left(X_{tij}-\bar{X}_{ti.}\right)^2\right]=\sigma^2 \tag{13.37}$$

Comparison of the Between Main Groups mean square with the Residual mean square can be used to test the hypothesis $\gamma_1=\gamma_2=\cdots=\gamma_k=0$ (that is, constant average value in each main group). Comparison of the Between Subgroups Within Main Groups mean square with the Residual mean square can be used to test the hypotheses "$\delta_{ti}=0$ for all t and all i" (no difference between averages for different subgroups in each mean group). If it is desired to test the hypotheses $\delta_{ti}=0$ for $i=1,2,\ldots,m_t$, separately for each t, the mean squares

$$\frac{1}{m_t-1}\sum_{i=1}^{m_t}n_{ti}\left(\bar{X}_{ti.}-\bar{X}_{t..}\right)^2$$

can be calculated *separately* for each value of t. Each of these mean squares [expected value equal to $\sigma^2+(m_t-1)^{-1}\sum_{i=1}^{m_t}n_{ti}\delta_{ti}^2$] can then be compared with the Residual mean square.

If the Z's are normally distributed then all the sums of squares mentioned above are mutually independent, and the mean square ratios can be compared with significance limits obtained from the appropriate F distributions.

Example 13.6 As an illustration of a nested model in which the terms are parametric, consider the following problems. A company manufactures bricks which are used for furnace linings. Two plants produce these bricks. Each plant has 3 storage locations in which the water content of the brick is kept at a minimum.

For a particular brick it is important that the best combination of brick production and storage be obtained. In this situation the mathematical model is

$$X_{tij} = \xi + \pi_t + \lambda(\pi)_{ti} + Z_{tij}$$

where $t(=1,2)$ denotes plants (π) and $i(=1,2,3)$ denotes location (λ) in this plant and X_{tij} is measured in percent moisture of the brick. Furthermore,

$$\sum_t \pi_t = 0, \ \sum_i \lambda(\pi)_{ti} = 0, \text{ but in general } \sum_t \lambda(\pi)_{ti} \neq 0$$

$t(=1,2)$ denotes plants and $i(=1,2,3)$ denotes locations *within each plant*. Note that it is meaningless to refer to "location 2," since location 2 in the first plant is not related to "location 2" in the other. Further, we assume that in this problem we are not interested in variance among locations. Rather we want to determine whether particular locations are significantly different from the others, and if there is a difference, which locations are the better ones.

13.10 ANALYSIS OF NESTED CLASSIFICATIONS— COMPONENT-OF-VARIANCE MODEL

The same sums of squares are calculated as for the parametric model. However the model is as given in (13.23)

$$X_{tij} = \xi + U_t + V_{ti} + Z_{tij}$$

and in place of (13.31) and (13.32) we have

$$\overline{X}_{t..} - \overline{X}_{...} = U_t - \overline{U} + \overline{V}_{t.} - \overline{V}_{..} + \overline{Z}_{t..} - \overline{Z}_{...} \tag{13.38}$$

and

$$\overline{X}_{ti.} - \overline{X}_{t..} = V_{ti} - \overline{V}_{t.} + \overline{Z}_{ti.} - \overline{Z}_{t..} \tag{13.39}$$

where

$$\overline{U} = N_{..}^{-1} \sum_{t=1}^{k} N_{t.} U_t; \quad \overline{V}_{t.} = N_{t.}^{-1} \sum_{i=1}^{m_t} n_{ti} V_{ti}; \quad \overline{V}_{..} = N_{..}^{-1} \sum_{t=1}^{k} N_{t.} \overline{V}_{t.}$$

Equation (13.33) remains unchanged, so the mean square

$$\left(\sum_{t=1}^{k} \sum_{i=1}^{m_t} \sum_{j=1}^{n_{ti}} \left(X_{tij} - \overline{X}_{ti.} \right)^2 \right) \div \left(N_{..} - \sum_{t=1}^{k} m_t \right)$$

can again be used as a Residual mean square to estimate σ^2. The expected values of the other mean squares are

$$E\left[\frac{1}{k-1}\sum_{t=1}^{k}N_{t.}\left(\bar{X}_{t..}-\bar{X}_{...}\right)^2\right]$$

$$=\sigma^2+\frac{\sum_{t=1}^{k}\left(N_{t.}^{-1}-N_{..}^{-1}\right)\sum_{i=1}^{m_t}n_{ti}^2}{k-1}\sigma_V^2+\frac{N_{..}-N_{..}^{-1}\sum_{t=1}^{k}N_{t.}^2}{k-1}\sigma_U^2 \quad (13.40)$$

and

$$E\left[\frac{1}{\sum\limits_{t=1}^{k}m_t-k}\sum_{t=1}^{k}\sum_{i=1}^{m_t}n_{ti}\left(\bar{X}_{ti.}-\bar{X}_{t..}\right)^2\right]=\sigma^2+\frac{N_{..}-\sum\limits_{t=1}^{k}N_{t.}^{-1}\sum\limits_{i=1}^{m_t}n_{ti}^2}{\sum\limits_{t=1}^{k}m_t-k}\sigma_V^2$$

$$(13.41)$$

We see immediately that whereas the hypothesis $\sigma_V=0$ can be tested by comparing the Between Subgroups Within Main Groups mean square with the Residual mean square, the hypothesis $\sigma_U=0$ cannot be tested by simple comparison of the Between Main Groups mean square with the Residual mean square, because the expected value of the Between Main Groups mean square depends on σ_V as well as σ_U.

If $n_{ti}=n$ for all t and all i (so that each subgroup contains the same number of observations) then $N_{t.}=m_t n$ and the right sides of (13.40) and (13.41) become

$$\sigma^2+n\sigma_V^2+\left[\frac{\left(\sum m_t\right)^2-\sum m_t^2}{(k-1)\sum m_t}\right]\cdot n\sigma_U^2 \quad \text{and} \quad \sigma^2+n\sigma_V^2$$

respectively. Hence in this case the hypothesis $\sigma_U=0$ (no variation between the expected values over different main groups) can be tested by comparison of the Between Main Groups mean square with the Between Subgroups Within Main Groups mean square—*not* with the Residual mean square. [Note that $(\sum m_t)^2-\sum m_t^2=\sum\sum m_t m_{t'}>0$.]

Here, then, is the first case where the change of model from parametric to component-of-variance actually affects the test procedures which we employ.

If the n_{ti}'s are not all equal then no simple test of the hypothesis $\sigma_U=0$ is available.

It is possible, however, to combine the Between Subgroups Within Main Groups and the Residual sum of squares to produce a statistic which has expected value

$$\sigma^2 + \left[(k-1)^{-1} \sum_{t=1}^{k} \left(N_{t.}^{-1} - N_{..}^{-1} \right) \sum_{i=1}^{m_t} n_{ti}^2 \right] \sigma_V^2$$

This statistic is, in fact

$$\theta \left[\frac{\sum_{t=1}^{k} \sum_{i=1}^{m_t} n_{ti} \left(\bar{X}_{ti.} - \bar{X}_{t.} \right)^2}{\sum_{t=1}^{k} m_t - k} \right] + (1-\theta) \left[\frac{\sum_{t=1}^{k} \sum_{i=1}^{m_t} \sum_{j=1}^{n_{ti}} \left(X_{tij} - \bar{X}_{ti.} \right)^2}{N_{..} - \sum_{t=1}^{k} m_t} \right] \quad (13.42)$$

where

$$\theta = \left[\frac{\sum_{t=1}^{k} m_t - k}{k-1} \right] \left[\frac{\sum_{t=1}^{k} \left(N_{t.}^{-1} - N_{..}^{-1} \right) \sum_{i=1}^{m_t} n_{ti}^2}{N_{..} - \sum_{t=1}^{k} N_{t.}^{-1} \sum_{i=1}^{m_t} n_{ti}^2} \right]$$

If the Between Main Groups mean square is substantially in excess of this value it can reasonably be ascribed to the existence of a nonzero value of σ_U.

Even if it be assumed that the U's, V's, and Z's are all normally distributed, it is not possible to derive a simple exact significance test using the ratio of the Between Main Groups mean square to (13.42). An approximate procedure is described in Example 13.7.

It should be noted that it is the randomness or nonrandomness of the term representing Between Subgroups Within Main Groups variation which determines the form of the test procedures. This is not affected by the nature of the term representing Main Groups.

Example 13.7 As part of a classification investigation on grades of 60 mm mortar, 5 lots of ammunition were tested. Each lot was tested on 2 different days, owing to the size of the experiment. (Taken from "An Application of the Design of Experiments in the Surveillance of Ammunition," by Jerome R. Johnson, *Proceedings of the First Conference on the Design of Experiments in Army Research, Development and Testing,* OOR Report 57-1, June, 1957.) The character is the range measured in yards. The data are given in Table 13.6. The model equation [see

TABLE 13.6
Observed Ranges (in Yards) for Shell, HE, M49A2 for 60 mm Mortar Fired with Standard Charge at an Elevation of 45°

Lots	MA-1-53		MA-1-82		MA-1-363		MA-1-604		MA-1-613	
Date	24 Oct 52	31 Oct 52	17 Dec 52	12 Jan 53	19, 20 Aug 53	8, 9 Sep 53	30 Dec 54	10 Jan 55	20 Jan 55	31 Jan 55
	1890	2057	1925	2112	1988	1932	1967	1980	2000	2110
	1863	2028	1903	2083	1876	1862	2021	2025	1769	1983
	1927	1964	2043	2096	1874	1863	2014	1983	1885	2098
	1830	1955	1957	2078	1914	1927	2019	1862	2004	2084
	1803	1976	1946	2031	1882	1907	2002	2041	1904	2015
	1951	1996	1940	2084	1774	1763	2128	2001	1865	1978
	1995	2057	1916	2017	1872	1841	1949	1970	1927	2098
	1967	1979	1967	2035	1822	1914	1904	2053	1972	2124
	1934	2010	1958	1978	1891	1837	2029	1978	1886	2077
	1897	2037	1879	2045	1855	1911	1989	1940	2019	2036
	1869	2013	1995	2002	1809	1866	2052	1980	1884	2060
	1847	1990	1980	2078	1894	1797	2042	1921	1990	2141
	1882	2015	1934	2118	1870	1983	1835	2002	1884	2075
	1965	1975	1990	2017	1910	1873	1970	1969	1938	2074
	1954	1934	1906	2107	1970	1907	2028	1849	1950	2003
	1973	2074	1985	2005	1980	1923	2000	2030	1999	2077
	1870	2071	2000	2094	1885	1962	2041	2006	1987	2110
	1894	1993	1951	1993	1775	1859	2033	1999	1823	2061
	1927	1943	1979	2020	1871	1993	1996	2006	1951	2099

TABLE 13.7

Analysis of Variance Table for Range of 60 mm Mortar Ammunition

SOURCE OF VARIATION	SUM OF SQUARES	DEGREES OF FREEDOM	MEAN SQUARE	MEAN SQUARE RATIO	EXPECTED VALUE OF MEAN SQUARE
Between Lots	369,254	4	92,314	1.25	$\sigma^2 + 19\sigma_D^2 + [38\sigma_\lambda^2]$
Between Days within Lots	369,941	5	73,988	25.75**	$\sigma^2 + 19\sigma_D^2$
Within Days	517,069	180	2873		σ^2
Total	1,256,264	189			

(13.24)] is

$$X_{tij} = \xi + \lambda_t + D_{ti} + Z_{tij}$$

where the λ_t's represent parametric terms referring to the 5 lots and the D_{ti}'s are random independent normal variables, each with the distribution $N(0, \sigma_D)$. The Z_{tij}'s are independent random normal variables, each with the distribution $N(0, \sigma)$.

The data of Table 13.6 are analyzed in Table 13.7. Note (from the EMS column) that we test the hypothesis of no variation among days ($\sigma_D = 0$) by calculating the ratio of Between Days Within Lots mean square to Within Days mean square, but the hypothesis of no average difference between lots ($\lambda_t = 0$ for all t) is tested by calculating the ratio of Between Lots mean square to Between Days Within Lots mean square. The first ratio is $73,988/2873$ and is significant at the 0.01 level (designated by **). There is no significant difference among lots, but there is evidence that σ_D differs from 0.

We may estimate the Among Days variance. Since $\hat{\sigma}^2 = 2873$ is an unbiased estimate of the Residual variance, then

$$\hat{\sigma}_D^2 = \frac{73,988 - 2873}{19} = 3743$$

is an unbiased estimate of the Among Days variance. Note also that the term $38\sigma_\lambda^2$ at the end of the first row is in brackets. This is to indicate that it is not really a variance, since the factor "Lots" is represented by a parametric term.

13.11   ANALYSIS OF CROSS-CLASSIFICATIONS— PARAMETRIC MODEL

We now use the model given in (13.25), namely,

$$X_{tij} = \xi + \rho_t + \kappa_i + (\rho\kappa)_{ti} + Z_{tij}$$

where

$$\sum_{t=1}^{r} \rho_t = 0 = \sum_{i=1}^{c} \kappa_i; \quad \sum_{i=1}^{c} (\rho\kappa)_{ti} = 0 = \sum_{t=1}^{r} (\rho\kappa)_{ti}$$

for all t and all i.

We first express in terms of the model the

Row means: $\qquad \bar{X}_{t..} = \xi + \rho_t + \bar{Z}_{t..}$

Column means: $\qquad \bar{X}_{.i.} = \xi + \kappa_i + \bar{Z}_{.i.}$

Cell means: $\qquad \bar{X}_{ti.} = \xi + \rho_t + \kappa_i + (\rho\kappa)_{ti} + \bar{Z}_{ti.}$

Overall mean: $\qquad \bar{X}_{...} = \xi + \bar{Z}_{...}$

It is possible to separate the various groups of parameters from each other by using the statistics

$$\overline{X}_{t..} - \overline{X}_{...} = \rho_t + \overline{Z}_{t..} - \overline{Z}_{...} \tag{13.43}$$

$$\overline{X}_{.i.} - \overline{X}_{...} = \kappa_i + \overline{Z}_{.i.} - \overline{Z}_{...} \tag{13.44}$$

$$\overline{X}_{ti.} - \overline{X}_{t..} - \overline{X}_{.i.} + \overline{X}_{...} = (\rho\kappa)_{ti} + \overline{Z}_{ti.} - \overline{Z}_{t..} - \overline{Z}_{.i.} + \overline{Z}_{...}. \tag{13.45}$$

In order to obtain an estimator of the common variance σ^2 of each residual random variable Z_{tij}, we use the quantities

$$X_{tij} - \overline{X}_{ti.} = Z_{tij} - \overline{Z}_{ti.} \tag{13.46}$$

The total sum of squares $\sum_{t=1}^{r}\sum_{i=1}^{c}\sum_{j=1}^{n_{ti}}(X_{tij} - \overline{X}_{...})^2$ can first be split up into Between Cells and Within Cells sums of squares:

$$\sum_{t=1}^{r}\sum_{i=1}^{c}\sum_{j=1}^{n_{ti}}\left(X_{tij} - \overline{X}_{...}\right)^2$$

$$= \sum_{t=1}^{r}\sum_{i=1}^{c}n_{ti}\left(\overline{X}_{ti.} - \overline{X}_{...}\right)^2 + \sum_{t=1}^{r}\sum_{i=1}^{c}\sum_{j=1}^{n_{ti}}\left(X_{tij} - \overline{X}_{ti.}\right)^2$$

From (13.46) and (13.47) it can be seen that

$$E[\text{Within Cells sum of squares}] = (N_{..} - rc)\sigma^2$$

(since we are supposing $n_{ti} \geq 1$ for all t and all i).

Hence the number of degrees of freedom of the Within Cells sums of squares is $(N_{..} - rc)$ and the mean square

$$(N_{..} - rc)^{-1}\sum_{t=1}^{r}\sum_{i=1}^{c}\sum_{j=1}^{n_{ti}}\left(X_{tij} - \overline{X}_{ti.}\right)^2 \tag{13.48}$$

can be used as a Residual mean square, giving an unbiased estimator of σ^2.

The remaining sums of squares can be obtained by splitting up the Between Cells sum of squares into 3 parts only if the n_{ti}'s satisfy certain conditions. Generally these conditions are that the n_{ti}'s are *proportional*, that is, $n_{ti}/n_{t'i}$ is independent of i, or, equivalently, $n_{ti}/n_{ti'}$ is independent of t. Here we discuss only the simpler case in which all n_{ti}'s are equal, with a common value n, say. (For further details, see *Topics in Intermediate Statistical Methods*, by T. A. Bancroft, Iowa State University Press, 1968.)

Then the Between Cells sum of squares breaks up in the following way:

$$n \sum_{t=1}^{r} \sum_{i=1}^{c} (\overline{X}_{ti.} - \overline{X}_{...})^2 = nc \sum_{t=1}^{r} (\overline{X}_{t..} - \overline{X}_{...})^2 + nr \sum_{i=1}^{c} (\overline{X}_{.i.} - \overline{X}_{...})^2$$

Between Rows   Between Columns

$$+ n \sum_{t=1}^{r} \sum_{i=1}^{c} (\overline{X}_{ti.} - \overline{X}_{t..} - \overline{X}_{.i.} + \overline{X}_{...})^2 \quad (13.49)$$

Interaction

The appropriateness of the names given to the sums of squares on the right side of (13.49) is confirmed by (13.43) to (13.45). These latter also aid us in finding the following expected values:

For Between Rows sum of squares: $(r-1)\sigma^2 + nc \sum_{t=1}^{r} \rho_t^2$

For Between Columns sum of squares: $(c-1)\sigma^2 + nr \sum_{i=1}^{c} \kappa_i^2$

For Interaction sum of squares: $(r-1)(c-1)\sigma^2 + n \sum_{t=1}^{r} \sum_{i=1}^{c} (\rho\kappa)_{ti}^2$

Hence the numbers of degrees of freedom of the 3 sums of squares are $(r-1)$, $(c-1)$ and $(r-1)(c-1)$, respectively, and the expected values of the mean squares are:

Between Rows mean square: $\sigma^2 + (r-1)^{-1}nc \sum_{t=1}^{r} \rho_t^2$

Between Columns mean square: $\sigma^2 + (c-1)^{-1}nr \sum_{i=1}^{c} \kappa_i^2$

Interaction mean square: $\sigma^2 + (r-1)^{-1}(c-1)^{-1}n \sum_{t=1}^{r} \sum_{i=1}^{c} (\rho\kappa)_{ti}^2$

The hypothesis $\rho_1 = \rho_2 = \cdots = \rho_r = 0$ can be tested by calculating the ratio of the Between Rows mean square to the Residual (13.48) mean square; the hypothesis of no average column differences can be tested using the ratio of the Between Columns mean square to the Residual mean square; and the hypothesis of additivity, or no interaction, can be tested using the ratio of the Interaction mean square to the Residual mean square. If the residual Z_{tij}'s are supposed to be normal, then the F distributions with the appropriate numbers of degrees of freedom: $(r-1), rc(n-1); (c-1), rc(n-1);$ and $(r-1)(c-1), rc(n-1)$, respectively, can be used to give significance limits for the ratios.

The case $n = 1$ calls for special comment. In this case there is no need for the third suffix, j, in the model (13.25). This can now be written

$$X_{ti} = \xi + \rho_t + \kappa_i + (\rho\kappa)_{ti} + Z_{ti} \quad (t = 1, \ldots, r; \ i = 1, \ldots, c) \qquad (13.50)$$

There will now be no Residual sum of squares and it will not be possible to estimate σ^2 separately from the parameter values. It is thus not possible to assess whether an observed value of a mean square is remarkably large or not, because the standard of comparison provided by the Residual mean square is not available.

In these circumstances it is usual to proceed as if $(\rho\kappa)_{ti}$ were 0 (for all t and all i) and use the model

$$X_{ti} = \xi + \rho_t + \kappa_i + Z_{ti} \qquad (13.51)$$

That is, we assume that row and column effects combine *additively*. Although this may well not be precisely correct, it does enable us to apply a valid form of analysis in which the Interaction mean square

$$\frac{\sum\limits_{t=1}^{r} \sum\limits_{i=1}^{c} \left(X_{ti} - \overline{X}_{t.} - \overline{X}_{.i} + \overline{X}_{..} \right)^2}{(r-1)(c-1)}$$

is used as the Residual mean square. If there are, in reality, nonzero interaction terms, these are [in our model (13.51)] included in the residual Z's, thus swelling the value of σ^2 and making the analysis correspondingly less accurate. Nevertheless, useful conclusions can still be drawn from analyses of this kind, where no analysis at all would have been possible if we had not been willing "to use the Interaction as Residual." The procedure described by this phrase is of quite frequent occurrence in practice, and is often described in words similar to those used here.

A very common form of experimental design in which $n = 1$ is known as *Randomized Blocks*. Here we are primarily interested in the effect of changes in level of a single factor ϕ. The Blocks represent a factor in which we are not primarily interested, but which is known (or expected) to have a substantial effect on the variable (or variables) to be measured. In order to be in a position to allow for this, each level of the factor ϕ is combined with each level of the Blocks factor, giving a cross-classification, with $n = 1$, in the 2 factors, ϕ and Blocks. Originally this kind of design was devised for use in agricultural experiments. The Blocks were then, in fact, blocks of land in an experimental area (for example, a field) and each block was divided into as many plots as there were levels of the factor ϕ to be considered. (The term "plot" is used generally to denote an experimental unit, even in nonagricultural investigations.) There have been many elaborations and modifications of this simple design. Some of them are described in Chapter 15.

It can be seen that the analysis developed in this section allows the effects of changes in level of ϕ to be studied, unaffected by changes in levels of Blocks (for example, fertility in an agricultural experiment) using the model

$$X_{ti} = \xi + \gamma_t + \phi_i + Z_{ti} \qquad (13.52)$$

where t represents "block" level and i represents level of ϕ.

Even with $n = 1$ it is possible to test for some specific kinds of interactions, as in Section 13.14.7.

Example 13.8 Two randomized block experiments, each comparing the yields of the same 5 varieties of wheat, are carried out simultaneously on 2 neighboring fields. The observed yields (in pounds per plot) are shown in Table 13.8.

We wish to carry out an analysis of the data for each experiment separately and to comment on (*a*) the comparability and consistency of the results in the 2 experiments; and (*b*) the relative suitability of the 2 fields for randomized block experiments. (Taken from B.Sc. General Examination, University of London, 1952.)

TABLE 13.8

YIELD OF WHEAT

				VARIETY		
	BLOCK	A	B	C	D	E
	1	61	60	64	69	83
Experiment I	2	65	66	66	80	83
(Plot size: 1/40 acre)	3	60	55	68	72	70
	4	75	70	80	80	89
	1	64	74	69	73	75
Experiment II	2	60	60	63	63	70
(Plot size: 1/50 acre)	3	54	51	59	50	64
	4	48	55	62	67	65

The separate analyses are shown in Table 13.9.

Since the 2 experiments use different plot sizes, the observations are not directly comparable. In each experiment the Variety mean square is significant, and so it is useful to present tables of varietal means. To make these comparable, both are presented in *pounds per acre*. (Note that the estimates of the standard deviations have to appropriately adjusted.) The results are shown in Table 13.10.

TABLE 13.9

ANOVA OF YIELD OF WHEAT

SOURCE	DEGREES OF FREEDOM	SUM OF SQUARES		MEAN SQUARE	
		Exp. I	Exp. II	Exp. I	Exp. II
Blocks	3	553.2	649.0	184.4	216.3
Varieties	4	905.2	316.7	226.3	79.2
Residual	12	160.8	214.5	13.4	17.9
Total	19	1619.2	1180.2		

TABLE 13.10

ESTIMATED YIELDS (LBS/ACRE)

VARIETY	A	B	C	D	E	ESTIMATED STANDARD DEVIATION OF ESTIMATES
Experiment I	2610	2510	2780	3010	3250	$73.2 \left(= 40 \sqrt{\dfrac{160.8}{(12)(4)}} \right)$
Experiment II	2837	3000	3162	3162	3425	$105.7 \left(= 50 \sqrt{\dfrac{214.5}{(12)(4)}} \right)$

To inquire whether the experiments are *comparable*, the Residual mean squares should be compared, allowance being made for the different plot sizes. If each set of yields were expressed in pounds per acre, the residual mean squares would be

$$\text{Experiment I} \quad 40^2 \times \frac{160.8}{12}$$

$$\text{Experiment II} \quad 50^2 \times \frac{214.5}{12}$$

The ratio (II to I) is equal to 2.08, and is compared with the F distribution with 12, 12 degrees of freedom. It is appropriate to use a 2-sided test in the present case.

Consistency is checked by inquiring whether the 2 experiments lead to similar conclusions. The generally higher yields estimated from the second experiment do not affect this; it is differences between mean yields that are of importance. The highest yields are obtained with variety E in both experiments. A and B are the 2 lowest yielding varieties in both experiments. In other respects there does not appear to be any clear conclusion to be drawn from the data provided.

Figure 13.5 shows the observed average yields *per acre* for each variety in each experiment. (Note that the mean yields have to be multiplied by 40 in Experiment I

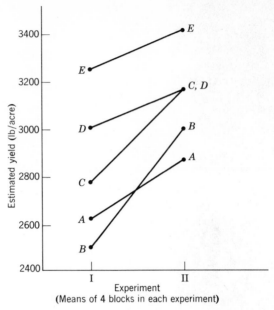

Figure 13.5 Estimated yields for five varieties. (Means of 4 blocks in each experiment.)

and 50 in Experiment II.) The diagram indicates that (*i*) Experiment II gave higher yields than I and (*ii*) there may be some interaction, because the *C* and *B* lines are not parallel to the *A*, *D*, and *E* lines.

13.12 ANALYSIS OF CROSS-CLASSIFICATIONS— COMPONENT-OF-VARIANCE MODEL

We consider only the case $n_{ti} = n$ for all t and all i. The model is that given in (13.26), namely,

$$X_{tij} = \xi + U_t + V_i + W_{ti} + Z_{tij}$$

In place of (13.43) to (13.45) we find

$$\bar{X}_{t..} - \bar{X}_{...} = \left(U_t - \bar{U} \right) + \left(\bar{W}_{t.} - \bar{W}_{..} \right) + \left(\bar{Z}_{t..} - \bar{Z}_{...} \right) \qquad (13.53)$$

$$\bar{X}_{.i.} - \bar{X}_{...} = \left(V_i - \bar{V} \right) + \left(\bar{W}_{.i} - \bar{W}_{..} \right) + \left(\bar{Z}_{.i.} - \bar{Z}_{...} \right) \qquad (13.54)$$

$$\bar{X}_{ti.} - \bar{X}_{t..} - \bar{X}_{.i.} + \bar{X}_{...} = \left(W_{ti} - \bar{W}_{t.} - \bar{W}_{.i} + \bar{W}_{..} \right) + \left(\bar{Z}_{ti.} - \bar{Z}_{t..} - \bar{Z}_{.i.} + \bar{Z}_{...} \right)$$

$$(13.55)$$

where

$$\overline{U} = r^{-1} \sum_{t=1}^{r} U_t; \quad \overline{V} = c^{-1} \sum_{i=1}^{c} V_i; \quad \overline{W}_{t.} = c^{-1} \sum_{i=1}^{c} W_{ti}$$

$$\overline{W}_{.i} = r^{-1} \sum_{t=1}^{r} W_{ti}; \quad \overline{W}_{..} = (rc)^{-1} \sum_{t=1}^{r} \sum_{i=1}^{c} W_{ti}$$

Equation (13.46), $X_{tij} - \overline{X}_{ti.} = Z_{tij} - \overline{Z}_{ti.}$, is still correct, and so the Residual mean square

$$(N_{..} - rc)^{-1} \sum_{t=1}^{r} \sum_{i=1}^{c} \sum_{j=1}^{n} \left(X_{tij} - \overline{X}_{ti.} \right)^2$$

can again be used to estimate σ^2, the variance of each z_{tij}. From (13.53) to (13.55) we find the following expected values of *mean squares*:

Between Rows: $\sigma^2 + n\sigma_W^2 + nc\sigma_U^2$
Between Columns: $\sigma^2 + n\sigma_W^2 + nr\sigma_V^2$
Interaction: $\sigma^2 + n\sigma_W^2$

The situation is similar to that in a hierarchal classification when a component-of-variance model is appropriate. The hypothesis $\sigma_W = 0$ can be tested by comparison of the Interaction mean square with the Residual mean square; but the hypotheses $\sigma_U = 0$, $\sigma_V = 0$ should be tested by comparison of the Between Rows or Between Columns (respectively) mean squares with the Interaction mean square.

Just as in the case of the hierarchal classification, also, it is the randomness of the term representing interaction in the model which makes the essential difference to the analysis. The formal test procedure is unaffected if the main row or column effects are changed from parametric to randomterms, or *vice versa* (giving a mixed model).

It is useful to remember that if the Between Rows (or Columns) mean square is compared with the Residual mean square when a component-of-variance model is appropriate, the possible effect of a nonzero interaction variance σ_W^2 can only be to *increase* the average size of the ratio to the Residual mean square. So a verdict of *nonsignificance* can be relied upon in this case. Similarly, if comparison of the Between Rows (or Columns) mean square with the Interaction mean square gives a *significant* verdict, this can be relied upon, since even if a parametric model with nonzero $(\rho\kappa)_{ti}$ terms were appropriate, this could only reduce (on the average) the apparent relative magnitude of the Between Rows (or Columns) mean square. Similar remarks apply to hierarchal classifications.

The question often arises, "If the Interaction mean square is not significant is it proper to *pool* the Interaction with the Residual mean square?" If the Interaction mean square is not significant this does not imply that σ_W^2 is actually equal to 0. However, if we "accept" the hypothesis that $\sigma_W^2 = 0$, then we have 2 estimates of σ^2. Pooling of the 2 into a single estimate would then appear to provide an estimator with a greater number of degrees of freedom, which could be used as the new Residual mean square. Pooling of Interaction and Residual mean squares should be done with caution, and certainly not if the original mean square ratio exceeds 2. Pooling is accomplished by adding the sums of squares involved and dividing by the sum of the degrees of freedom, to obtain a new estimate of the Residual mean square. An illustration of the bias introduced by pooling is shown in the following example.

Example 13.9   In a 3×3 cross-classification with 2 replications the Interaction and Residual sums of squares are pooled if the Interaction mean square is not significant at the 5 percent level. Assuming that there is really no interaction present find the expected value of the modified Residual mean square.

Let us denote the Interaction and Residual sums of square by S_I, S_R, respectively. Then, if there is really no interaction,

$$S_I \text{ is distributed as } \chi_4^2 \sigma^2$$

and

$$S_R \text{ is distributed as } \chi_9^2 \sigma^2$$

where σ^2 is the variance of the residual term in the model. S_I and S_R are mutually independent. The mean square ratio is

$$\frac{\tfrac{1}{4} S_I}{\tfrac{1}{9} S_R}$$

It is significant at the 5 percent level if

$$\frac{\tfrac{9}{4} S_I}{S_R} > F_{4,9,0.95} = 3.63$$

Therefore, the modified Residual mean square is

$$\tfrac{1}{9} S_R \qquad \text{if } S_I > k S_R$$

and

$$\tfrac{1}{13}(S_I + S_R) \qquad \text{if } S_I < k S_R$$

where $k = \tfrac{4}{9} \times 3.63 = 1.613$.

The expected value of the modified Residual mean square is

$$\iint_{S_I > k S_R} \tfrac{1}{9} S_R p(S_R, S_I)\, dS_R \, dS_I + \iint_{S_I < k S_R} \tfrac{1}{13}(S_R + S_I) p(S_R, S_I)\, dS_R \, dS_I \quad (13.56)$$

We put $x_1 = S_R \sigma^{-2}$; $x_2 = S_I \sigma^{-2}$. Then

$$\int \int_{S_I > kS_R} S_R p(S_R, S_I) \, dS_R \, dS_I$$

$$= \sigma^2 \int_0^\infty \int_{kx_1}^\infty x_1 \left[\frac{x_1^{7/2} e^{-x_1/2}}{2^{9/2} \Gamma(9/2)} \right] \cdot \left[\frac{x_2 e^{-x_2/2}}{4} \right] dx_2 \, dx_1$$

$$= \sigma^2 \int_0^\infty x_1 \left[\frac{x_1^{7/2} e^{-x_1/2}}{2^{9/2} \Gamma(9/2)} \right] \cdot (1 + \tfrac{1}{2} k x_1) e^{-kx_1/2} \, dx_1$$

$$= \frac{\sigma^2}{2^{9/2} \Gamma(9/2)} \left[2^{11/2} (1+k)^{-11/2} \Gamma(11/2) + k 2^{13/2} (1+k)^{-13/2} \Gamma(13/2) \right]$$

$$= \tfrac{1}{2} \sigma^2 \left[18(1+k)^{-11/2} + 99 k (1+k)^{-13/2} \right]$$

Therefore the first integral in (13.56) is equal to

$$\sigma^2 (1+k)^{-11/2} \left[1 + \frac{11}{2} k (1+k)^{-1} \right]$$

Evaluating the second integral in a similar way, and adding, we find the expected value of the modified Residual mean square is

$$\sigma^2 \left\{ 1 - \frac{11}{12} k^2 (1+k)^{-13/2} \right\} = 0.972 \sigma^2$$

There is thus a negative bias of about 3 percent.

13.13   CALCULATION AND PRESENTATION OF ANALYSES OF VARIANCE

In the calculation of sums of squares the basic formula is

$$\sum_{l=1}^n \left(X_l - \overline{X}_. \right)^2 = \sum_{l=1}^n X_l^2 - n\overline{X}_.^2 = \sum_{i=1}^n X_i^2 - \frac{X_.^2}{n}$$

Applying this formula repeatedly we obtain the following computing formulas:

One-Way Classification:

Between Groups:
$$\sum_{t=1}^k n_t \overline{X}_{t.}^2 - N\overline{X}_{..}^2 = \sum_{t=1}^k \left(\frac{X_{t.}^2}{n_t} \right) - \frac{X_{..}^2}{N}$$

Residual
(Within Groups):

$$\sum_{t=1}^{k} \sum_{i=1}^{n_t} X_{ti}^2 - \sum_{t=1}^{k} n_t \overline{X}_{t.}^2$$

$$= \sum_{t=1}^{k} \sum_{i=1}^{n_t} X_{ti}^2 - \sum_{t=1}^{k} \left(\frac{X_{t.}^2}{n_t} \right)$$

Hierarchal Classification:

Between Main
Groups:

$$\sum_{t=1}^{k} N_{t.} \overline{X}_{t..}^2 - N_{..} \overline{X}_{...}^2 = \sum_{t=1}^{k} \left(\frac{X_{t..}^2}{N_{t.}} \right) - \frac{X_{...}^2}{N_{..}}$$

Between Subgroups
Within Main Groups:

$$\sum_{t=1}^{k} \sum_{i=1}^{m_t} n_{ti} \overline{X}_{ti.}^2 - \sum_{t=1}^{k} N_{t.} \overline{X}_{t..}^2$$

$$= \sum_{t=1}^{k} \sum_{i=1}^{m_t} \left(\frac{X_{ti.}^2}{n_{ti}} \right) - \sum_{t=1}^{k} \left(\frac{X_{t..}^2}{N_{t.}} \right)$$

Residual
(Within Subgroups):

$$\sum_{t=1}^{k} \sum_{i=1}^{m_t} \sum_{j=1}^{n_{ti}} X_{tij}^2 - \sum_{t=1}^{k} \sum_{i=1}^{m_t} n_{ti} \overline{X}_{ti}^2$$

$$= \sum_{t=1}^{k} \sum_{i=1}^{m_t} \sum_{j=1}^{n_{ti}} X_{tij}^2 - \sum_{t=1}^{k} \sum_{i=1}^{m_t} \left(\frac{X_{ti.}^2}{n_{ti}} \right)$$

Cross Classification:

Between Rows:

$$nc \sum_{t=1}^{r} \overline{X}_{t..}^2 - nrc \overline{X}_{...}^2 = \sum_{t=1}^{r} \frac{X_{t..}^2}{nc} - \frac{X_{...}^2}{nrc}$$

Between Columns:

$$nr \sum_{i=1}^{c} \overline{X}_{.i.}^2 - nrc \overline{X}_{...}^2 = \sum_{i=1}^{c} \frac{X_{.i.}^2}{nr} - \frac{X_{...}^2}{nrc}$$

Residual
(Within Cells):

$$\sum_{t=1}^{r} \sum_{i=1}^{c} \sum_{i=1}^{n} X_{tij}^2 - n \sum_{t=1}^{r} \sum_{i=1}^{c} \overline{X}_{ti.}^2$$

$$= \sum_{t=1}^{r} \sum_{i=1}^{c} \sum_{j=1}^{n} X_{tij}^2 - \sum_{t=1}^{r} \sum_{i=1}^{c} \frac{X_{ti.}^2}{n}$$

The Interaction (Rows $\times$ Columns) sum of squares is obtained by subtracting the above 3 quantities from the total sum of squares

$$\sum_{t=1}^{r} \sum_{i=1}^{c} \sum_{j=1}^{n} \left(X_{tij} - \overline{X}_{...}\right)^2 = \sum_{t=1}^{r} \sum_{i=1}^{c} \sum_{j=1}^{n} X_{tij}^2 - nrc\overline{X}_{...}^2$$

$$= \sum_{t=1}^{r} \sum_{i=1}^{c} \sum_{j=1}^{n} X_{tij}^2 - \frac{X_{...}^2}{nrc}$$

If $n = 1$ then the Residual (Within Cells) sum of squares is not calculated and the Residual (Interaction) sum of squares is calculated as

Total − Between Rows − Between Columns

It is useful to remember that the sums of squares are unaltered if every observed value is increased or decreased by a constant amount. By subtracting (or, occasionally, adding) a suitable constant the observed figures can often be reduced in magnitude, making the necessary calculations much less cumbersome.

Ratios of mean squares are unaffected if every observed value is *multiplied* by a constant. This can sometimes be helpful, but it is necessary to remember that if the Residual mean square is to be used to estimate the variance (σ^2) of the Z's, the figure obtained *should be in the correct units*.

A very convenient way of setting out the calculations is by means of an *Analysis of variance* table. Analysis of variance (ANOVA) tables have already been given in Section 13.6.1 and Example 13.3. These apply to one-way classifications. Here we give standard forms of ANOVA tables appropriate to each of the models so far discussed.

ANOVA tables usually have 4 columns. [The expected value of mean square (EMS) column, as given in Table 13.4, is not usually written explicitly; the mean square ratios are not always inserted as an extra column.]

Column 1 (headed "Source") indicates where the variation, estimated in each row of the table, comes from. (In the typical tables shown below, general and rather abstract terms are used in this column; in any particular application more concrete, picturesque descriptions can and should be given.)

Column 2 gives the number of degrees of freedom.

Column 3 contains the numerical values of the sum of squares. (Here the original formulas obtained in the test, as opposed to computational formulas just presented, are given.)

Column 4 contains the numerical value of the mean square (equal to column 3 divided by column 2).

TABLE 13.11
ANALYSIS OF VARIANCE TABLE OF A ONE-WAY CLASSIFICATION

SOURCE	DEGREES OF FREEDOM	SUM OF SQUARES
Between Groups	$k-1$	$\sum\limits_{t=1}^{k} n_t(\overline{X}_{t.} - \overline{X}_{..})^2$
Residual		
(Within Groups)	$N-k$	$\sum\limits_{t=1}^{k}\sum\limits_{i=1}^{n_t} (X_{ti} - \overline{X}_{t.})^2$
Total	$N-1$	$\sum\limits_{t=1}^{k}\sum\limits_{i=1}^{n_t} (X_{ti} - \overline{X}_{..})^2$

TABLE 13.12
ANALYSIS OF VARIANCE TABLE OF A HIERARCHAL (NESTED) CLASSIFICATION

SOURCE	DEGREES OF FREEDOM	SUM OF SQUARES
Between Main Groups	$k-1$	$\sum\limits_{t=1}^{k} N_{t.}(\overline{X}_{t..} - \overline{X}_{...})^2$
Between Subgroups Within Main Groups	$\sum\limits_{t=1}^{k} m_t - k$	$\sum\limits_{t=1}^{k}\sum\limits_{i=1}^{m_t} n_{ti}(\overline{X}_{ti.} - \overline{X}_{t..})^2$
Residual (Within) Subgroups)	$N_{..} - \sum\limits_{t=1}^{k} m_t$	$\sum\limits_{t=1}^{k}\sum\limits_{i=1}^{m_t}\sum\limits_{j=1}^{n_{ti}} (X_{tij} - \overline{X}_{ti.})^2$
Total	$N_{..} - 1$	$\sum\limits_{t=1}^{k}\sum\limits_{i=1}^{m_t}\sum\limits_{j=1}^{n_{ti}} (X_{tij} - \overline{X}_{...})^2$

The order of columns 2 and 3 is sometimes reversed. In Tables 13.11 to 13.13 the first three columns only are shown. Symbols are used in columns 2 and 3; in practice the appropriate numerical values obtained by calculation would be shown in these positions.

The descriptions under "Source" are often abbreviated by omission of the words "Between" and (or) "Interaction"; also the description of source of the Residual sum of squares is sometimes omitted. As an example, if all these forms of abbreviation were used in the cross-classification table, the row titles would read (in order): Rows, Columns, Rows × Columns, Residual. In this case, if $n = 1$, there will, of course, be no Within Cells sum of squares, and the previous row (Rows × Columns) will be called "Residual."

TABLE 13.13

ANALYSIS OF VARIANCE TABLE OF A CROSS CLASSIFICATION

SOURCE	DEGREES OF FREEDOM	SUM OF SQUARES
Between Rows	$r - 1$	$nc \sum_{t=1}^{r} \left(\overline{X}_{t..} - \overline{X}_{...} \right)^2$
Between Columns	$c - 1$	$nr \sum_{i=1}^{c} \left(\overline{X}_{.i.} - \overline{X}_{...} \right)^2$
Interaction Rows × Columns	$(r - 1)(c - 1)$	$n \sum_{t=1}^{r} \sum_{i=1}^{c} \left(\overline{X}_{ti.} - \overline{X}_{t..} - \overline{X}_{.i.} + \overline{X}_{...} \right)^2$
Residual (Within Cells)	$rc(n - 1)$	$\sum_{t=1}^{r} \sum_{i=1}^{c} \sum_{j=1}^{n} \left(X_{tij} - \overline{X}_{ti.} \right)^2$
Total	$rcn - 1$	$\sum_{t=1}^{r} \sum_{i=1}^{c} \sum_{j=1}^{n} \left(X_{tij} - \overline{X}_{...} \right)^2$

It is often very helpful to supplement the standard analysis of variance table by estimates of the values of the parameters in the model. In the case of parametric models this often takes the form of estimates of expected values and estimates of standard errors of these estimates. In a one-way classification, for example, the group expected values $\xi + \gamma_t$, can be estimated by the statistics $\overline{X}_{t.}$. The standard deviation of $\overline{X}_{t.}$ is $\sigma / \sqrt{n_t}$ and this can be estimated by

$$\sqrt{(\text{Residual mean square})/ n_t}$$

(The estimator is not quite unbiased, but the bias is not of serious import.)

In the case of a component-of-variance model there is no logical reason for giving tables of statistics like observed group means since these do not estimate parameters in the model. On the other hand, we can try to obtain estimates of the variances of the different components. For example, in a one-way classification [see (13.13)],

$E(\text{Between Groups mean squares}) - E(\text{Within Groups mean square})$

$$= \left(N - N^{-1} \sum_{t=1}^{k} n_t^2 \right)(k - 1)^{-1} \sigma_U^2$$

and so the statistic

$$\frac{[(\text{Between Groups mean squares}) - (\text{Withing Groups mean square})]}{\left(N - N^{-1}\sum_{t=1}^{k} n_t^2\right)(k-1)^{-1}}$$

is an unbiased estimator of σ_U^2. (It should be noted that this estimator can give negative values. If these, when they occur, are replaced by 0, we are certain to get a value nearer the true value of σ_U^2, though the modified estimator will no longer be unbiased—that is, its average value in a long series of experiments, will not tend to σ_U^2.)

Example 13.10 In a study of the determination of thermal expansion coefficients of sintered titanium alloy compacts, 16 compacts were prepared (Table 13.14).

TABLE 13.14

THERMAL EXPANSION COEFFICIENTS

TREATMENT	ALLOY			
	A	B	C	D
1	4.78	3.84	5.82	4.57
	4.28	5.28	5.77	5.44
2	4.46	4.73	4.76	4.30
	4.79	3.36	3.31	3.86
Totals	18.31	17.21	19.66	18.17

These represented 4 alloys, or compositions, each prepared in 2 different ways (treatment 1 and 2). The mathematical model of this crossed experiment with all terms parametric is

$$X_{ijk} = \xi + \alpha_i + \tau_j + (\alpha\tau)_{ij} + Z_{ijk}$$

with

$$i = 1, \ldots, 4; j = 1, 2; k = 1, 2$$

where

$$\sum_i \alpha_i = \sum_j \tau_j = \sum_i (\alpha\tau)_{ij} = \sum_j (\alpha\tau)_{ij} = 0$$

The computational formulas together with the numerical values are presented in Table 13.15. To test for significance in a completely crossed and parametric model,

TABLE 13.15
ANOVA TABLE ON THERMAL EXPANSION

SOURCE	SUM OF SQUARES		D.F.	MEAN SQUARE
Alloys	$S_A = \frac{1}{4} \sum\limits_{i=1}^{4} X_{i\cdot\cdot}^2 - \frac{1}{16} X_{\cdots}^2$	$= 0.7623$	3	0.2541
Treatments	$S_T = \frac{1}{8} \sum\limits_{j=1}^{2} X_{\cdot j\cdot}^2 - \frac{1}{16} X_{\cdots}^2$	$= 2.4103$	1	2.4103
Alloys ×Treatments	$S_{AT} = \frac{1}{2} \sum\limits_{i} \sum\limits_{j} X_{ij\cdot}^2 - \frac{1}{4} \sum\limits_{i} X_{i\cdot\cdot}^2$			
	$\qquad - \frac{1}{8} \sum\limits_{j} X_{\cdot j\cdot}^2 + \frac{1}{16} X_{\cdots}^2$	$= 1.8171$	3	0.6057
Residual	$S_e = $ difference	$= 3.6825$	8	0.4603
Total	$S = \sum\limits_{i} \sum\limits_{j} (X_{ijk} - \bar{X}_{\cdots})^2$			
	$\quad = \sum \sum X_{ijk}^2 - \frac{1}{16} X_{\cdots}^2$	$= 8.6722$	15	

each of the mean squares is tested against the residual. These ratios are

$$\text{Alloy:} \qquad \frac{0.2541}{0.4603} < 1 \text{ not significant}$$

$$\text{Treatment:} \quad 5.24 < F_{1,8,0.95} \, (= 5.32)$$

Hence treatment effect is just below the 5 percent level of significance. The interaction of alloy and treatment effects does not prove significant (ratio is 1.32).

Since there appears to be a real difference between treatments we gave a table of estimates of treatment means together with an estimate of their standard error

Treatment:	1	2
Estimated Mean:	4.97	4.20

Estimated standard error of each treatment mean is $\sqrt{0.4603/8} = 0.24$.

Figure 13.6, on which the means of the 2 observations for each combination of alloy and treatment are plotted, brings out a feature of the data not clearly indicated by the analysis of variance. The response for Alloy A changes little from treatment 1 to treatment 2, and there is a marked decrease for each of the other 3 alloys (reflected in the significant treatment effect). This represents an interaction type of effect, which though insufficient to achieve significance, might be of importance for later investigation.

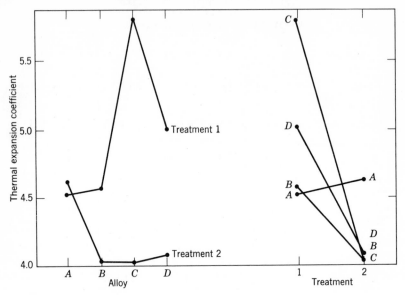

Figure 13.6   Means of two observations for each Alloy–Treatment combination.

13.14   ANCILLARY TECHNIQUES

13.14.1   Tables of Means

Although the analysis of variance table presents the results of calculations in a useful form, and the formal significance tests give us information on the structure of the populations being investigated, these standard processes are often insufficient to complete our analysis of the data. In particular, we are often interested in pursuing the actual sources of differences indicated by significant results of comparison of mean square ratios with the F distribution.

As an example, consider the case of a one-way classification with k group and n observations per group, where the parametric model

$$X_{ti} = \xi_t + Z_{ti}$$

is appropriate. It is, of course, of interest to know that the data provide evidence against the validity of the hypothesis $\xi_1 = \xi_2 = \ldots = \xi_k$. But we often would like to investigate, further, just where the differences are really marked, or conversely, whether some of the ξ's are, so far as we can tell, equal (or nearly equal) to each other.

The statistics available to help us in this investigation are the observed group means $\overline{X}_1, \overline{X}_2, \ldots, \overline{X}_k$ and an estimate of the standard deviation

$(\sigma/\sqrt{n})$ of each of these observed means. It is convenient to present these statistics in the form of a "table of means" as shown below.

Group:	1	2	3	...	t	...	k
Mean:	$\bar{X}_1.$	$\bar{X}_2.$	$\bar{X}_3.$	...	$\bar{X}_t.$	...	$\bar{X}_k.$

$$\text{Estimated standard deviation}\left(=\sqrt{\frac{\text{Residual mean square}}{n}}\right)$$

There are tables of this kind in Examples 13.8 and 13.10.

It is a good idea to present such a table in *all* cases where a parametric model is appropriate. (If the numbers in each group are unequal, then a separate estimated standard deviation $= \sqrt{\text{Residual mean square}/n_t}$ should be given for each group.) Often it will be helpful to the reader to change from the units in which observations are recorded to some universally understood units. For example, a yield recorded in grams per cubic centimeter may be changed to pounds per cubic foot. In any case (whether units have been changed or not) the units used in the table of means should be clearly stated, and care should be taken that the *estimated standard deviations are in the same units.*

13.14.2 Simultaneous Confidence Intervals

It is possible to give a confidence interval for the difference between 2 expected values ξ_t and $\xi_{t'}$—*chosen before analyzing the data*—by a direct application of the method developed in Section 7.4.3. The $1-\alpha$ confidence limits for $\xi_t - \xi_{t'}$ are

$$\bar{X}_t. - \bar{X}_{t'}. \pm t_{\nu,1-\alpha/2}\sqrt{(\text{Residual mean square})\left(\frac{1}{n_t}+\frac{1}{n_{t'}}\right)} \qquad (13.57)$$

where ν is the number of degrees of freedom ($N-k$ in the case of one-way classification) of the Residual sum of squares. Suppose, however, that we wish to make confidence interval statements about *each* of the $\binom{k}{2}$ different differences $\{\xi_t - \xi_{t'}\}$. If we use limits of type (13.57) we can expect that about $100(1-\alpha)$ percent of the intervals will contain the corresponding value. However, we may wish to consider the $\binom{k}{2}$ intervals *as a set* and to control the probability that *every one* of the confidence interval statements is correct. If we require the probability of this to be $100(1-\alpha)$ percent we are evidently imposing a more stringent condition than simply requiring a confidence coefficient of $100(1-\alpha)$ percent for each separate confidence interval. This greater stringency is "paid for" by making the length of each interval longer, according to the method now to be described.

Consider the ratio

$$Q_{k,\nu} = \frac{\text{Range}(\overline{X}_{1.}, \overline{X}_{2.}, \ldots, \overline{X}_{k.})}{\text{Estimated standard deviation of each } \overline{X}}$$

$$= \frac{W}{S/\sqrt{n}}$$

where $S = (\text{Residual mean square})^{1/2}$. The variable S is distributed as

$$\sigma \nu^{-1/2} \cdot (\chi \text{ with } \nu \text{ degrees of freedom})$$

where ν is the Residual degrees of freedom.

If $\xi_1 = \xi_2 = \cdots = \xi_k$ then

$$\sqrt{n}\, W = \text{Range}\left(\sqrt{n}\, \overline{X}_{1.}, \sqrt{n}\, \overline{X}_{2.}, \ldots, \sqrt{n}\, \overline{X}_{k.}\right)$$

is distributed, independently of S, as $\sigma \cdot (\text{range of } k \text{ independent normal variables each with 0 mean and unit standard deviation})$. The distribution of $Q_{k,\nu}$ then depends only on k and ν (and not on σ) and is termed the *Studentized range* distribution.

Upper percentage points of this distribution have been tabulated giving values $q_{k,\nu,1-\alpha}$, such that

$$\Pr\left[Q_{k,\nu} < q_{k,\nu,1-\alpha}\right] = 1 - \alpha \qquad \text{for } \alpha = 0.01 \quad \text{and} \quad \alpha = 0.05 \quad (13.58)$$

Values are given in Table *I* of the Appendix.

In the general case the ratio of the range of the quantities $\overline{X}_{1.} - \xi_1$, $\overline{X}_{2.} - \xi_2, \ldots, \overline{X}_{k.} - \xi_k$ to $(S\sqrt{n})$ is distributed as $Q_{k,\nu}$ and (13.58) can be written in the form

$$\Pr\left[\text{Range}(\overline{X}_{1.} - \xi_1, \ldots, \overline{X}_{k.} - \xi_k) < q_{k,\nu,1-\alpha} S/\sqrt{n}\right] = 1 - \alpha \quad (13.59)$$

This, in turn, can be expressed as

$$\Pr\left[\overline{X}_{t.} - \overline{X}_{t'.} - \frac{q_{k,\nu,1-\alpha} S}{\sqrt{n}} < \xi_t - \xi_{t'} < \overline{X}_{t.} - \overline{X}_{t'.} + \frac{q_{k,\nu,1-\alpha} S}{\sqrt{n}},\right.$$

$$\left. \text{for all } t \neq t'\right] = 1 - \alpha$$

In other words, the *set* of confidence intervals

$$\overline{X}_{t.} - \overline{X}_{t'.} - \frac{q_{k,\nu,1-\alpha}S}{\sqrt{n}} < \xi_t - \xi_{t'} < \overline{X}_{t.} - \overline{X}_{t'.} + \frac{q_{k,\nu,1-\alpha}S}{\sqrt{n}} \qquad (13.60)$$

for all $t \neq t'$ will be correct with probabilty $1 - \alpha$

Although the standard confidence interval (13.57) is of width $2t_{\nu,1-\alpha/2}S\sqrt{2/n}$ the intervals (13.60) are each of width $2q_{k,\nu,1-\alpha}S/\sqrt{n}$. The ratio of the former to the latter is

$$\frac{\sqrt{2}\, t_{\nu,1-\alpha/2}}{q_{k,\nu,1-\alpha}}$$

Some numerical values of this ratio (for $\alpha = 0.05$) are shown in Table 13.16. It can be seen that the greater stringency of (13.60) is "paid for" by a width about 50 percent greater than that of the corresponding interval (13.57).

TABLE 13.16

RATIO OF WIDTH OF STANDARD CONFIDENCE INTERVAL TO WIDTH OF SIMULTANEOUS CONFIDENCE INTERVAL

			k			
ν	3	4	5	6	7	8
8	0.807	0.720	0.667	0.631	0.604	0.582
10	0.812	0.728	0.676	0.642	0.615+	0.594
12	0.817	0.734	0.683	0.649	0.623	0.602
15	0.821	0.739	0.690	0.655+	0.630	0.610
20	0.824	0.745	0.696	0.663	0.639	0.618
24	0.827	0.748	0.700	0.668	0.643	0.624
30	0.829	0.751	0.702	0.671	0.647	0.627
60	0.832	0.756	0.711	0.680	0.656	0.637

(For $k = 2$, the ratio is always equal to 1)

It should be noted that the method of constructing simultaneous confidence intervals described above applies only when the parameters ($\xi_t - \xi_{t'}$ in this case) are estimated by statistics ($\overline{X}_{t.} - \overline{X}_{t'.}$ in this case), which are mutually dependent in a particular manner. A different method of construction is necessary when a set of parameters $\{\theta_t\}$ is estimated by a set of mutually independent normal random variables $\{Y_t\}$, for example, $\{\xi_t\}$ and $\{\overline{X}_{t.}\}$ in the situation discussed in this section. We then need to use the distribution of

$$Q'_{k,\nu} = \max_t \frac{|Y_t - \theta_t|}{S_\nu}$$

where $Y_1, Y_2, \ldots, Y_k$ are mutually independent normal variables with means $\theta_1, \theta_2, \ldots, \theta_k$, respectively, and common standard deviation σ, and S_ν^2 is distributed, independently of the Y_t's, as $\chi_\nu^2 \sigma^2 / \nu$. Since

$$\left(\frac{|Y_t - \theta_t|}{S_\nu} \right)^2 \text{ is distributed as } \frac{\chi_1^2}{\chi_\nu^2 / \nu}$$

where the 2 χ^2's are mutually independent, the upper significance limits of $Q'_{k,\nu}$ can be calculated as the square roots of values given in Table 19 of Pearson and Hartley [13.10]. Some of these values are reproduced in Section 15.2.5. By taking square roots of these values, sets of simultaneous confidence limits are calculated as $Y_t - q'_{k,\nu,1-\alpha} S_\nu < \theta_t < Y_t + q'_{k,\nu,1-\alpha} S_\nu$ $(t = 1, 2, \ldots, k)$.

13.14.3 Tests of Homoscedasticity

Condition (*iii*) of Section 13.4 should be satisfied (at least approximately) for standard analysis of variance methods to be applicable. If one is not convinced that this condition is satisfied a test of homoscedasticity of the residual variability should be applied. For example, in a 1-way classification with model

$$X_{ti} = \xi + \gamma_t + Z_{ti} \qquad \left[\text{see } (13.5) \right]$$

we may wish to test whether $\text{var}(Z_{ti}) = \sigma_t^2$ varies from group (that is to test the hypothesis $\sigma_1^2 = \sigma_2^2 = \cdots = \sigma_k^2$). We can use the components

$$S_t = \sum_{i=1}^{n_t} \left(X_{ti} - \bar{X}_{t.} \right)^2$$

of the Within Group sum of squares to test this hypothesis. S_t is distributed as $[\chi^2 \text{ with } (n_t - 1) \text{ degrees of freedom}] \times \sigma_t^2$ if the Z's are normally distributed. The test described in Section 8.4 may be used (or if a more robust test is required the test described in Section 9.6.5 can be used).

In order to cast serious doubt on the applicability of standard method there should be substantial evidence of a considerable degree of heteroscedasticity. For this reason a very low (numerically) value of the significance level (perhaps 0.1 percent) is often used in this test. If the sample sizes $n_1, \ldots, n_k$ are large, even more lenient levels may be used.

The application of the likelihood ratio test of equality of variances calls for rather heavy calculation, expecially if there are more than a few groups. Alternative methods, using criteria which are easier to compute, have been developed. The 2 best-known are

(a) *Cochran's test* which uses as criterion (maximum $S_t)/\Sigma_{t=1}^{k} S_t$;

(b) a test using the criterion (maximum S_t)/(minimum S_t).

These tests can be applied only when each estimate of variance is based on the same number of degrees of freedom (ν) which will be the case if $n_1 = n_2 = \cdots = n_k$. Tables of significance limits for each of these tests are given in Table 13.17.

TABLE 13.17

UPPER 1% POINTS OF SHORT CUT TESTS OF HOMOSCEDASTICITY*

ν = DEGREES OF FREEDOM OF EACH ESTIMATE		NUMBER OF VARIANCES						
		4	5	6	7	8	9	10
	2	0.864	0.788	0.722	0.664	0.615	0.573	0.536
(a)	3	0.781	0.696	0.626	0.568	0.521	0.481	0.447
	4	0.721	0.633	0.564	0.508	0.463	0.425	υ.393
Cochran's	6	0.641	0.553	0.487	0.435	0.393	0.359	0.331
Test	8	0.590	0.504	0.440	0.391	0.352	0.321	0.294
	10	0.554	0.470	0.408	0.362	0.325	0.295	0.270
	2	729	1036	1362	1705	2063	2432	2813
(b)	3	120	151	184	216	249	281	310
Max S_t	4	49	59	69	79	89	97	106
Min S_t	6	19.1	22	25	27	30	32	34
	8	11.7	13.2	14.5	15.8	16.9	17.9	18.9
	10	8.6	9.6	10.4	11.1	11.8	12.4	12.9

* Reproduced with permission from Eisenhart, Hastay and Wallis, *Techniques of Statistical Analysis*, McGraw-Hill (New York and London), 1947; and from Pearson and Hartley, *Biometrika Tables*, Cambridge University Press, 1958.

These tests, as well as the likelihood test, have a similar lack of robustness to departures from normality. If substantial departures are likely, special care should be taken in using the tests. It is useful to remember that high values of the shape factor α_4 (or β_2) tend to increase apparent significance, low values (<3) to decrease it.

Example 13.11 Before carrying out an analysis of variance of a single factor experiment it was decided to investigate the validity of the assumption of homoscedasticity. There were 7 levels of this factor, each with 9 replicates. The variances were as follows: 6.24, 5.16, 6.34, 8.26, 5.93, 5.74, 5.86. By means of Cochran's test we test the hypothesis of whether the largest variance is significantly different from the others. We calculate the ratio

$$\frac{S_t(\max)}{\sum_t S_t} = \frac{8.26}{45.53} = 0.190$$

Comparing this to the upper 1 percent critical value for 7 variances with $\nu = 8(0.391)$ we conclude that we do not have enough evidence to reject the hypothesis of equal variances.

*13.14.4 Transformation of Variables

If it *is* decided that standard methods are inapplicable, consideration may be given to the possiblity of transforming the original data so that the transformed variables may be more reasonably subjected to standard forms of analysis.

Formally we seek to replace observed variables X_{ti} by values of a mathematical function of X_{ti}, say, $g(X_{ti})$. We would like to select our function $g(X_{ti})$ so that the new variables $Y_{ti} = g(X_{ti})$ are more nearly homoscedastic.

We use the method of "statistical differentials" to suggest appropriate forms for the function $g(X)$. Expanding $g(X)$ by Taylor's series about $X = E(X)$ we have

$$g(X) = g\big[E(X) + \{X - E(X)\} \big]$$

$$= g[E(X)] + \{X - E(X)\} g'[E(X)] + \frac{1}{2!} \{X - E(X)\}^2 g''[E(X)] + \dots$$

Taking expected values on both sides of this equation we find

$$E[g(X)] \doteq g(E(X)) + \tfrac{1}{2}\mathrm{var}(X) \cdot g''[E(X)]$$

This is only an approximate equation for 2 reasons. First, because later terms in the expansion are omitted, and second, because in the process of taking expected values X is allowed to vary over its whole range of variation, and for at least part of the range the series may not converge. However, using this result leads to

$$g(X) - E[g(X)] \doteq \{X - E(X)\} g'[E(X)] + \text{terms of second and higher}$$
$$\text{order in } \{X - E(X)\}$$

Squaring and taking expected values we get

$$\mathrm{var}[g(X)] \doteq \mathrm{var}(X) \cdot \{ g'[E(X)] \}^2 \tag{13.61}$$

Although (13.61) is only approximate it serves as a useful basis for suggesting suitable transformations in cases where the variance of the original variables X is an (approximately) known function of $E(X)$.

If

$$\text{var}(X) \doteq h[E(X)]$$

then (13.61) shows that $g(X)$ has approximately constant variance if

$$h[E(X)]\{g'[E(X)]\}^2 = \text{constant}$$

This means that the function $g(X)$ must be such that

$$g'(X) \propto [h(X)]^{-1/2}$$

or

$$g(X) \propto \int^X [h(x)]^{-1/2} dx \tag{13.62}$$

For example, if the standard deviation of X is proportional to its expected value, then $h(x)$ is proportional to x^2 and (13.62) gives

$$g(X) \propto \int^X x^{-1} dx = \ln X$$

Therefore, we consider using the transformation $\ln X$ (or $\log_{10} X$, since $\log_{10} X$ is proportional to $\ln X$).

As a further example consider a Poisson-type original variable X. In this case the *variance* is equal (or proportional) to the expected value, and so $h(x)$ is equal (or proportional) to x, and (13.62) gives

$$g(X) \propto \int^X x^{-1/2} dx = \sqrt{X}$$

Therefore we consider using the transformation $\sqrt{X}$.

A number of common cases are summarized in Table 13.18.

It must be realized that these are only approximately suitable transformations. More detailed study has produced relatively slight modifications aimed at improving the effectiveness of the transformations. Some of these modified transformations are also shown in Table 13.18, based on a table given by M. S. Bartlett, "The Use of Transformations," *Biometrics*, **3** (1957).

*13.14.5 Short-Cut Methods and Separation of Means

A number of methods have been evolved for reducing the amount of calculation needed to apply analysis of variance tests. In particular, some attention has been given to avoidance of calculation of sums of squares. Here we describe one such method, based on the use of sample ranges as applied to simple one-way classifications. The method has been extended to more complicated designs (H. O. Hartley

TABLE 13.18

TRANSFORMATIONS TO EQUALIZE VARIANCES
(λ is a constant in this table)

VARIANCE IN TERMS OF MEAN, θ	TRANSFORMATION	APPROXIMATE VARIANCE ON NEW SCALE	RELEVANT DISTRIBUTION
θ $\lambda^2\theta$	$\begin{cases} \sqrt{x} \text{ (or } \sqrt{x + 3/8} \\ \text{for small integers)} \end{cases}$	0.25 $0.25\lambda^2$	Poisson Empirical
$2\theta^2/(n-1)$	$\ln x$	$2/(n-1)$	Sample variances
$\lambda^2\theta^2$	$\begin{cases} \ln x, \ln(x+1) \\ \log_{10} x, \log_{10}(x+1) \end{cases}$	$\left.\begin{matrix} \lambda^2 \\ 0.189\lambda^2 \end{matrix}\right\}$	Empirical
$\theta(1-\theta)/n$	$\begin{cases} \sin^{-1}\sqrt{x} \text{ (radians)} \\ \sin^{-1}\sqrt{x} \text{ (degrees)} \end{cases}$ $\text{(or } \sin^{-1}\sqrt{\dfrac{x+3/8}{n+3/4}}$ for $x/a/$ small integer)	$\left.\begin{matrix} 0.25/n \\ 821/n \end{matrix}\right\}$	Binomial
$\lambda^2\theta^2(1-\theta)^2$	$\ln[x/(1-x)]$	λ^2	Empirical
$(1-\theta^2)^2/(n-1)$	$\frac{1}{2}\ln[(1+x)/(1-x)]$	$1/(n-3)$	Sample correlations
$\theta + \lambda^2\theta^2$	$\lambda^{-1}\sinh^{-1}[\lambda\sqrt{x}]$	0.25	Negative binomial

[13.7] gives a good account of these), but it becomes progressively more difficult to apply as the design becomes more elaborate. We restrict ourselves to the case of equal numbers of observations in each group.

If the second approximation of Section 6.9 is used then the Between Groups mean square could be replaced by the range of group means and the Within Groups mean square by mean range within groups and the ratio

$$\frac{\sqrt{n} \ \text{range}(\overline{X}_{1.},\ldots,\overline{X}_{k.})}{k^{-1}\sum_{t=1}^{k} [\text{range in } t\text{th group}]} \tag{13.63}$$

would be approximately distributed as

$$\left[\frac{c_2(k)}{c_2(n)} \right] F_{v_2(k),kv_2(n)}$$

if there are no differences among groups. [Here $c_2(n), v_2(n)$ are used to denote values of c_2, v_2 in Section 6.9 as functions of the sample size, n.] If the denominator in (13.63) is replaced by the square root of the Within Groups mean square then the ratio would be distributed as the studentized range $Q_{k,k(n-1)}$ if there were no differences among groups.

This form of ratio does not offer much reduction in amount of calculation, but it is the basis for a heuristic method of sorting out the groups into sets which appear to be homogeneous within themselves but different from one set to another. A significant F ratio merely indicates that there are some differences between groups. We very often want to know just *where* these differences are. No unique "best" method exists for doing this, but one useful method is based on the statistic

$$\frac{\sqrt{n} \text{ (range of group means)}}{\sqrt{\text{Residual mean square}}}$$

First, the range of the complete set of k means is used. If this is significant, then each of the 2 sets of $(k-1)$ means obtained by omitting the 2 extreme values is tested. If one is significant and the other is not, the extreme value in the significant set is "split off" as being distinct, and the process terminates. If, however, both sets are significant, then both extreme means are omitted and the process continued with the remaining set of $(k-2)$ means.

This method has been modified by Duncan [13.5]. He uses a significance level depending on the number of means being tested. In fact, if the "nominal level" is α then for a set of p means the significance limit q_{p,v,α_p} with $\alpha_p = 1 - (1-\alpha)^{p-1}$ is used. Further, assignment to groups is made according to the following rule:

"Any two groups are regarded as different if the Q value for the set of which their means are the extreme values is significant, *except* that no two means are regarded as different if any set (of consecutive means) to which they belong gives a nonsignificant Q."

This method can produce overlapping different sets (as in Example 13.12). The interpretation of such results is that although the existence of different sets is indicated, data are insufficient to justify definite assignment of some groups to single sets.

A number of alternative methods of attacking the same problem have been suggested. Since possible conditions vary very widely, no one method can be said to be "optimum." Hence we describe just one kind of method which seems to give reasonable results in practice. This is based on a method suggested by J. W. Tukey [13.11].

We start by arranging the group means in order of magnitude. Then we test neighboring means for significant difference. The distribution of such differences is very complicated, even when there is no real difference between group means, but as a rough approximation differences greater than

$$\frac{\sqrt{2}\, t_{\nu,1-\alpha/2} S_\nu}{\sqrt{n}} \qquad (13.64)$$

(where S_ν^2 is a Residual mean square based on ν degrees of freedom) are regarded as significant. Then each set of neighboring means is tested for outliers using the criterion

$$\frac{\sqrt{n}\ \text{maximum}\ |\bar{X}_{t.} - \text{mean of}\ \bar{X}_{t.}\text{'s in set}|}{S_\nu} \qquad (13.65)$$

Significance limits for this criterion are given in Pearson and Hartley [13.10], p.173. The process is continued until no more outliers are split off. (If more than 2 outliers are split off from the same side of a set they constitute a new set which should be tested separately for outliers.)

In the methods just described the nominal significance level (α) does not have a direct interpretation in terms of probability of error. By increasing α a greater number of apparently "different" sets is obtained, and conversely.

Example 13.12 In the development of adhesives, 10 different types were being tested simultaneously. These adhesives are used in a fiber glass product in laminations. A tensile test was performed to determine their bond strength. Five specimens were provided for each adhesive. Hence the total sample size was 50. In the analysis of variance a significant difference among adhesive means was found. The residual variance of 1.59 had 40 degrees of freedom. The individual means (arranged in descending order) were 21.6, 21.3, 19.6, 17.2, 16.3, 15.4, 15.3, 15.1, 14.4, 14.2. Can we partition these means into separate classes?

Using Tukey's [13.11] method, the critical difference [using (13.64) with $\alpha = 0.05$] is

$$\frac{\sqrt{2}\, t_{40,0.975} S_\nu}{\sqrt{n}} = (\sqrt{2}\,)(2.021)\sqrt{\frac{1.59}{5}}$$

$$= 1.61$$

Hence we separate means 21.3 and 19.6, and 19.6 and 17.2. We have next to consider the set of means 17.2 down to 14.2. The mean of these 7 means is 15.41. The maximum deviation from this value is $17.2 - 15.41 = 1.79$. The estimated standard deviation of each mean is $\sqrt{1.59/5} = 0.564$.

The ratio $1.79/0.564 = 3.17$ is then compared to the distribution of extreme studentized deviate (Pearson and Hartley [13.10], p. 173, with $n = 7$, $\nu = 40$). The appropriate upper 5 percent point is 2.37, and so the observed value is significant and 17.2 is split off from the set of 7 means. The mean of the remaining 6 means is 15.12. The extreme deviation from this value is $16.3 - 15.12 = 1.18$. The studentized extreme deviate value is

$$\frac{1.18}{0.564} = 2.09$$

The appropriate upper 5 percent point in this case ($n = 6$, $\nu = 40$) is 2.28, and so the observed value is not significant.

The results of applying the procedure are then:

$$
\begin{aligned}
&21.6 \text{ and } 21.3 && \text{form one set,} \\
&19.6 && \text{forms a second "set,"} \\
&17.2 && \text{forms a third "set," and}
\end{aligned}
$$

16.3, 15.4, 15.3, 15.1, 14.4, 14.2 form a fourth "set"

Using Duncan's method with nominal significance level $\alpha = 0.05$ we need the following values:

p	2	3	4	5	6	7	8	9	10
$q_{p,40,\alpha_p}$	2.86	3.01	3.10	3.17	3.22	3.27	3.30	3.33	3.35
$q_{p,40,\alpha_p}\sqrt{1.59/5}$	1.61	1.70	1.75	1.79	1.82	1.84	1.86	1.88	1.89

The last line gives the significance limit for ranges of sets of p consecutive means.

We note that (*i*) all ranges with 21.6 as the upper limit are significant except 21.6 to 21.3; (*ii*) all ranges with 21.3 as upper limit are significant; and (*iii*) all ranges with 19.6 as upper limit are significant. Hence 21.6, 21.3 form one "set" and 19.6 forms a second "set."

Proceeding, we find (*iv*) all ranges with 17.2 as upper limit are significant except 17.2 to 16.3; (*v*) ranges 16.3 to 17.4, 16.3 to 14.3 are significant; and (*vi*) no other ranges are significant. So the final assignment to "sets" is

1st: 21.6, 21.3
2nd: 19.6
3rd: 17.2, 16.3
4th: 16.3, 15.4, 15.3, 15.1
5th: 15.4, 15.3, 15.1, 14.4, 14.3

Note that the sets in this case are not disjoint. There is some overlap, reflecting the fact that we have insufficient information to make completely disjoint assignments.

*13.14.6 Examination of Residuals

In any model of the form (13.1) it is useful to study the values of the estimated residuals obtained by subtracting estimated values of the first two terms on the right hand side from the corresponding observed values. If there is a clearly discernible pattern among these residuals it may reflect some inadequacy in the model, e.g., omission of a more or less important assignable cause. Apart from patterns, the form of distribution may indicate departures from standard assumptions. For example, relatively large (positive or negative) residuals in one (or more) groups indicates departure from homoscedasticity; close examination may reveal clues to departures from normality, and so on.

Calculation of individual residuals is somewhat tedious, but most ANOVA computer programs include (usually as an optional feature) calculation and presentation of residuals.

Example 13.13 For the data of Experiment I in Example 13.8, the residuals

$$\left(X_{ti} - (\hat{\xi} + \hat{\rho}_t + \hat{\kappa}_i)\right) = X_{ti} - \bar{X}_{t.} - \bar{X}_{.i} + \bar{X}_{..})$$

are given in Table 13.19. We note that the signs of the residuals (in the same block) for varieties A and C and for varieties B and E agree. If deemed of sufficient importance, inquiry might be made to ascertain if there are similarities between A and C, or between B and E, that might be useful. Before these effects, which are suggested by the data, could be established statistically, results of further experiments designed to test this particular point would be needed.

TABLE 13.19
RESIDUALS

BLOCK	VARIETY				
	A	B	C	D	E
1	-0.85	0.65	-2.10	-2.85	5.15
2	-1.45	2.05	-4.70	3.55	0.55
3	0.55	-1.95	4.30	2.55	-5.45
4	1.75	-0.75	2.50	-3.25	-0.25

*13.14.7 Testing for a Product Term Interaction

J. W. Tukey [13.12] has constructed a test of the hypothesis $\theta = 0$ in the model

$$X_{ti} = \xi + \rho_t + \kappa_i + \theta \rho_t \kappa_i + Z_{ti}$$

$[t = 1,\ldots,r;\ \ i = 1,\ldots,c;\ \ \Sigma \kappa_i = 0;\ Z_{ti}$'s independent normal variables; $E(Z_{ti}) = 0$; $\mathrm{var}(Z_{ti}) = \sigma^2]$. This corresponds to a cross-classification with $n = 1$ (cf. p. 635), where it is not possible to test for a general interaction term of form $(\rho \kappa)_{ti}$.

Tukey's method, in essence, consists of calculating the sample regression of the "residuals"

$$X_{ti} - \hat{\xi} - \hat{\rho}_t - \hat{\kappa}_i = X_{ti} - \bar{X}_{t.} - \bar{X}_{.i} + \bar{X}_{..}$$

on

$$\hat{\rho}_t \hat{\kappa}_i = \left(\bar{X}_{t.} - \bar{X}_{..}\right)\left(\bar{X}_{.i} - \bar{X}_{..}\right)$$

We get, as an estimator of θ,

$$\hat{\theta} = \frac{\displaystyle\sum_t \sum_i \left(\bar{X}_{t.} - \bar{X}_{..}\right)\left(\bar{X}_{.i} - \bar{X}_{..}\right)\left(X_{ti} - \bar{X}_{t.} - \bar{X}_{.i} + \bar{X}_{..}\right)}{\displaystyle\sum_t \sum_i \left(\bar{X}_{t.} - \bar{X}_{..}\right)^2\left(\bar{X}_{.i} - \bar{X}_{..}\right)^2}$$

[The denominator is $\Sigma_t(\bar{X}_{t.} - \bar{X}_{..})^2 \Sigma_i(\bar{X}_{.i} - \bar{X}_{..})^2$.]

The expected value of $\hat{\theta}$ is θ and its variance, conditional on $\{\bar{X}_{t.} - \bar{X}_{..}, \bar{X}_{.i} - \bar{X}_{..}\}$ being fixed, is

$$\sigma^2 \left[\sum_t \left(\bar{X}_{t.} - \bar{X}_{..} \right)^2 \sum_i \left(\bar{X}_{.i} - \bar{X}_{..} \right)^2 \right]^{-1}$$

The test is constructed by splitting up the standard residual sum of squares

$$S_R = \sum_t \sum_i \left(X_{ti} - \bar{X}_{t.} - \bar{X}_{.i} + \bar{X}_{..} \right)^2$$

into

$$S_\theta = \hat{\theta}^2 \sum_t \left(\bar{X}_{t.} - \bar{X}_{..} \right)^2 \sum_i \left(\bar{X}_{.i} - \bar{X}_{..} \right)^2$$

with 1 degree of freedom (sum of squares "Due to θ") and $S_R - S_\theta$ with $(r-1)(c-1)-1$ degrees of freedom ((new) Residual sum of squares).

Mandel [13.8] has shown how to test for significance of interaction terms of type $\theta' \rho'_t \kappa'_i$, or even $\theta' \rho'_t \kappa'_i + \theta'' \rho''_t \kappa''_i + \cdots$, where ρ'_t, κ'_i are not necessarily equal to ρ_t, κ_i, respectively.

13.15   LINEAR COMPARISONS

In Chapter 12 we discussed orthogonality and orthogonal components. We now consider a method of partitioning sums of squares with, say, $(p-1)$ degrees of freedom into $(p-1)$ separate sums of squares, each with 1 degree of freedom. We concern ourselves with purely parametric models and, further, distinguish between quantitative and qualitative factors. For the quantitative factors we have, of course, some more or less precise measurements. For example, temperature at 30, 50, 70°C; time at 10 min, 1 hr, 3 hr; pH at 3.90, 4.35, 5.25, 6.73; density, frequency, size and so forth. For qualitative factors, on the other hand, we have levels (1), (2), (3), etc., but no method of *ordering* these levels on some scale. For example, laboratories A, B, and C; company $X, Y, Z, \ldots$; grade of steel $XB, UR, VV, \ldots$; thermometer $1, 2, 3, \ldots$—measuring instrument, place, machine—all these are usually qualitative factors.

13.15.1   Quantitative Factors

One way of separating the $p-1$ degrees of freedom of comparison between p levels of a factor, is to employ coefficients of orthogonal polynomials as introduced in Section 12.5.3. The resulting sums of squares are orthogonal, but it is not always possible to assign a natural meaning to them. If the levels are quantitatively defined and are in fact equally spaced, the different components can be interpreted as sums of squares due to linear, quadratic, etc., components of regression of response on the factor. For example, in an experiment where temperature is at the levels 30, 50, and 70°C, we may be interested in the linear and quadratic effect of temperature. Suppose that the model equation is

$$X_{ti} = \xi + \tau_t + Z_{ti}$$

and the totals of the observations for each level of temperature are X_1, X_2, and X_3, respectively. The orthogonal coefficients for 3 levels are

$$\begin{array}{llll}
\text{Linear:} & -1 & 0 & 1 \\
\text{Quadratic:} & 1 & -2 & 1
\end{array}$$

The linear component of the sum of squares due to temperature is

$$\frac{\left(\sum\limits_{i=1}^{3} a_i X_{i.}\right)^2}{n\sum\limits_{i=1}^{3} a_i^2} = \frac{(-1\cdot X_{1.} + 0\cdot X_{2.} + 1\cdot X_{3.})^2}{n\left[(-1)^2 + 0^2 + (1)^2\right]} = \frac{(X_{3.} - X_{1.})^2}{2n}$$

where n is the number of replicates for each level, if the n_i are equal. The quadratic component is

$$\frac{\left(\sum\limits_{i=1}^{3} b_i X_{i.}\right)^2}{n\sum\limits_{i=1}^{3} b_i^2} = \frac{(1\cdot X_{1.} - 2\cdot X_{2.} + 1\cdot X_{3.})^2}{n\left[(1)^2 + (-2)^2 + (1)^2\right]} = \frac{(X_{1.} - 2X_{2.} + X_{3.})^2}{6n}$$

(It should be noted that the standard tables for orthogonal polynomials apply only when the levels are *equally spaced*. In other cases special calculation of coefficients is necessary.)

In general for an *equal number of replicates*, the procedure is as follows:

(1) Plan the experiment so that there is a constant difference between adjacent levels (on some scale) of the levels of the quantitative factor in question.
(2) Determine the number of levels for the factor and obtain the orthogonal coefficients for the linear, quadratic,... effects.
(3) Calculate the individual sum of squares for each of the components

$$\frac{\left(\sum\limits_{i} a_i X_{i.}\right)^2}{n\sum\limits_{i} a_i^2} \qquad i = 1,\dots,p \quad \text{(the number of levels of the factor)}$$

where $X_{i.}$ is the total of the ith level of the factor, a_i are the orthogonal coefficients, and n is the number of replicates. Values of the orthogonal polynomials for 3 to 7 levels are given in Table 13.20.

TABLE 13.20
VALUES OF ORTHOGONAL POLYNOMIALS
FOR 3, 4, 5, 6, 7 EQUALLY SPACED POINTS

(3)

ξ_1	ξ_2
−1	+1
0	−2
+1	+1
D^* = 2	6

(4)

ξ_1	ξ_2	ξ_3
−3	+1	−1
−1	−1	+3
+1	−1	−3
+3	+1	+1
20	4	20

(5)

ξ_1	ξ_2	ξ_3	ξ_4
−2	+2	−1	+1
−1	−1	+2	−4
0	−2	0	+6
+1	−1	−2	−4
+2	+2	+1	+1
10	14	10	70

(6)

ξ_1	ξ_2	ξ_3	ξ_4	ξ_5
−5	+5	−5	+1	−1
−3	−1	+7	−3	+5
−1	−4	+4	+2	−10
+1	−4	−4	+2	+10
+3	−1	−7	−3	−5
+5	+5	+5	+1	+1
70	84	180	28	252

(7)

ξ_1	ξ_2	ξ_3	ξ_4	ξ_5
−3	+5	−1	+3	−1
−2	0	+1	−7	+4
−1	−3	+1	+1	−5
0	−4	0	+6	0
+1	−3	−1	+1	+5
+2	0	−1	−7	−4
+3	+5	+1	+3	+1
28	84	6	154	84

*D equals sum of squares in the column above it.

663

Example 13.14   In the study of polyesterification of fatty acids with glycols, the effect of catalyst concentration (κ) and temperature (τ) on the percent conversion (X_{ijk}) of esterification process was investigated. A total sample of size 18 was prepared with 3 levels of concentration (4×10^{-4}, 8×10^{-4}, and 16×10^{-4}) and 3 levels of temperature (145, 200, and 225°C), with 2 replicates for each combination. The mathematical model is

$$X_{ijk} = \xi + \kappa_i + \tau_j + (\kappa\tau)_{ij} + Z_{ijk}$$

[where $\Sigma_i \kappa_i = \Sigma_j \tau_j = \Sigma_i (\kappa\tau)_{ij} = \Sigma_j (\kappa\tau)_{ij} = 0$] with $i = 1,2,3$; $j = 1,2,3$; and $k = 1,2$. The data are given in Table 13.21. We can separate individual degrees of freedom as linear and quadratic for both temperature and concentration. (We say "quadratic," although this may include higher powers which cannot be separately estimated from these data.) To have the levels of concentration on a scale with a constant distance between them, we can take the $\log_{10}$(or ln) of each level. This would give us -3.39794, -3.09691, and -2.79588. The ANOVA table is presented as Table 13.22. The sums of squares for each single degree of freedom of the main effects are obtained by the following (for concentration—coded)

$$\text{Linear:} \quad \frac{\left[(-1)X_{1..} + (1)X_{3..}\right]^2}{6\left[(1)^2 + (1)^2\right]}$$

$$\text{Quadratic:} \quad \frac{\left[(1)X_{1..} + (-2)X_{2..} + (1)X_{3..}\right]^2}{6\left[(1)^2 + (-2)^2 + (1)^2\right]}$$

Similar formulas are used for the temperature sums of squares.

Since the factors of the crossed experiment [κ_i, τ_j, and $(\kappa\tau)_{ij}$] are all parameters, each mean square is divided by the residual mean square as shown in the last column of Table 13.22. As a result of the ANOVA we can safely say that the effect of concentration is strongly linear in the transformed logarithmic scale. That is, log

TABLE 13.21
PER CENT CONVERSION

CONCENTRATION

TEMPERATURE	4×10^{-4}	8×10^{-4}	16×10^{-4}	TOTALS
175°C	67.4 66.2	73.4 75.5	79.6 81.1	443.2
200°C	82.8 85.3	86.2 89.0	93.3 90.1	526.7
225°C	90.5 93.1	92.8 93.8	98.7 99.8	568.7
Totals	485.3	510.7	542.6	1538.6

TABLE 13.22
ANOVA OF PER CENT CONVERSION

SOURCE		SUM OF SQUARES	DF	MEAN SQUARE	MEAN SQUARE RATIO
Concentration	Linear	273.61	1	273.61	119.0***
	Quadratic	1.17	1	1.17	0.51
Temperature	Linear	1312.52	1	1312.52	570.7***
	Quadratic	47.84	1	47.84	20.8**
Conc. × Temp.		30.55	4	7.64	3.32
Residual		20.73	9	2.30	
Total		1686.42	17		

concentration and percent conversion display a linear relationship. The effect of temperature is both linear and quadratic, although the linear effect is much stronger. Finally, the interaction is small compared with these effects. It is just inside the 5 percent significance level, when compared with the residual.

The interaction can be split up into 4 components with 1 degree of freedom each representing, respectively,

(Linear Concentration)	×	(Linear Temperature)
(Quadratic Concentration)	×	(Linear Temperature)
(Linear Concentration)	×	(Quadratic Temperature)
(Quadratic Concentration)	×	(Quadratic Temperature)

As an example, the (Linear Concentration)×(Quadratic Temperature) effect is the linear component with respect to concentration of the variation of the quadratic component of variation of response with temperature.

At concentration level i, the quadratic temperature effect is measured by

$$Y_i = \frac{X_{i1.} - 2X_{i2.} + X_{i3.}}{\left\{2\left[1^2 + (-2)^2 + 1^2\right]\right\}^{1/2}}$$

and so the (Linear Concentration)×(Quadratic Temperature) component sum of squares is

$$\frac{1}{2}(Y_1 - Y_3)^2 = \frac{1}{24}(X_{11.} - 2X_{12.} - X_{31.} + 2X_{32.} - X_{33.})^2$$

Similarly the (Quadratic Concentration)×(Quadratic Temperature) component is

$$\frac{1}{6}(Y_1 - 2Y_2 + Y_3)^2 = \frac{1}{72}(X_{11.} - 2X_{12.} + X_{13.} - 2X_{21.} + 4X_{22.}$$
$$- 2X_{23.} + X_{31.} - 2X_{32.} + X_{33.})^2$$

Carrying out the necessary calculations, we obtain the following splitting up of the interaction sum of squares, using obvious abbreviations

SOURCE	DEGREES OF FREEDOM	SUM OF SQUARES
$L_{conc.} \times L_{temp.}$	1	18.60
$L_{conc.} \times Q_{temp.}$	1	6.41
$Q_{conc.} \times L_{temp.}$	1	5.41
$Q_{conc.} \times Q_{temp.}$	1	0.14
Conc. $\times$ Temp.	4	30.55

On comparison with the residual mean square (2.70 with 9 degrees of freedom) we find that the (Linear Concentration) $\times$ (Linear Temperature) component of interaction is significant at $2\frac{1}{2}$ percent level ($18.60/2.30 = 8.09 > F_{1,9,0.975} = 7.21$), although the interaction as a whole is not significant.

13.15.2 Qualitative Factors

If the factor being studied is qualitative we may be interested in particular linear comparisons of levels of this factor. For example, suppose we are testing the weight loss, due to abrasion, of 5 rubber compounds. Compound 1 is the standard, compounds 2 and 3 have one ingredient in common, and compounds 4 and 5 have another ingrediant in common. We may want to compare

Compound 1	versus 2, 3, 4, and 5
Compounds 2 and 3	versus 4 and 5
Compound 2	versus 3
Compound 4	versus 5

For this factor at 5 levels, there are 4 degrees of freedom. We know, then, that we can have exactly 4 independent (1 degree of freedom) linear comparisons. Coefficients of linear functions appropriate to measuring the effects described above are shown in Table 13.23. Not all choices (which we should make when the experiment is planned and before it is executed) of 4 comparisons would give us *orthogonal* comparisons that are also orthogonal to the sum of all observations. If they are, then their sums of squares add up to the Total sum of squares of this factor.

Recall that for orthogonality (as shown in Chapter 12) we must choose the coefficients, such that the sum of the coefficients for each effect is 0 and the sum of cross products, pairwise over each 2 effects, is also 0. In terms of the compounds above we have the coefficients in Table 13.23.

Extending the notation used in Section 13.15.1, $a_1 = 4, a_2 = -1, a_3 = -1, a_4 = -1, a_5 = -1$; $b_1 = 0, b_2 = 1, \ldots$; $c_1 = 0, c_2 = 1, \ldots$; $d_1 = 0, \ldots, d_5 = -1$.

TABLE 13.23
COEFFICIENTS FOR A SET OF FOUR LINEAR
COMPARISONS*

			LEVEL		
COMPARISON	1	2	3	4	5
[a] 1 vs 2, 3, 4, 5	4	−1	−1	−1	−1
[b] 2, 3 vs 4, 5	0	1	1	−1	−1
[c] 2 vs 3	0	1	−1	0	0
[d] 4 vs 5	0	0	0	1	−1

* These coefficients are the *smallest* integers for the comparisons. One could have chosen, for example, 20, −5, −5, −5, −5, for the first set.

For orthogonality, we require

$$\sum_i a_i = \sum_i b_i = \sum_i c_i = \sum_i d_i = 0$$

$$\sum_i a_i b_i = \sum_i a_i c_i = \cdots = \sum_i c_i d_i = 0$$

These conditions are, in fact, satisfied by the numbers in Table 13.23. The sum of squares for standard (1) versus the other four is

$$SS[a] = \frac{\left(\sum_i a_i X_{i.} \right)^2}{n \sum_i a_i^2} \tag{13.66}$$

If the number of elements in level i is equal to n_i and all of the n_i are not equal, then in place of (13.66) we use

$$SS[a] = \frac{\left(\sum_i n_i^{-1} a_i X_{i.} \right)^2}{\sum_i n_i^{-1} a_i^2} \tag{13.67}$$

Suppose that we are interested in only $r < p - 1$ of the available orthogonal linear comparisons. Then we simply enumerate these appropriate r sums of squares and lump the rest together. The sums of squares would then be

$$SS[a] \text{ with 1 d.f.}$$
$$SS[b] \text{ with 1 d.f.}$$
$$- - - - - - - - -$$
$$SS[r] \text{ with 1 d.f.}$$

$SS_{\text{remaining}} = SS_{\text{total of factor}} - \Sigma SS[i]$ with $(p - 1 - r)$ degrees of freedom.
Note that if the n_{ij}'s are unequal then in general $SS[a], SS[b], \ldots, SS[r]$ are not independent, nor is the last equation above correct. (If the n_{ij}'s are *proportional*, that is, the ratios $n_{1j} : n_{2j} \cdot - \cdot$ do not depend on j, independence, and the formula for SS remaining, are still valid.)

Example 13.15 In Example 13.10, let us suppose that Alloy B is known to be markedly different from the other alloys in its composition, whereas alloys A and C are somewhat similar except for the concentration of one ingredient. In the planning stage of the design we ask whether the linear comparisons of B versus the others and of A versus C are significant. Since these 2 linear comparisons are orthogonal we can obtain sums of squares for these single degrees of freedom together with those for the third comparison (D versus A and C in which we have no interest). These 3 sums of squares add to the total sum of squares attributed to alloys. The (orthogonal) coefficients are

	$X_{1.}$	$X_{2.}$	$X_{3.}$	$X_{4.}$
[a] B vs. A, C, D	-1	3	-1	-1
[b] A vs. C	1	0	-1	0
[c] D vs. A, C	-1	0	-1	2

The alloy (A) sum of squares in the ANOVA table (Table 13.15) can now be split up as shown in Table 13.24. We calculate, for example, the sum of squares for B versus A, C, D. This is

$$\frac{[-1(18.31) + 3(17.21) - 1(19.66) - 1(18.17)]^2}{4[(-1)^2 + (3)^2 + (-1)^2 + (-1)^2]} = \frac{(4.51)^2}{48} = 0.42375$$

TABLE 13.24
ANOVA (ISOLATING LINEAR COMPARISONS)

SOURCE	SUM OF SQUARES	DF	MEAN SQUARE	MEAN SQUARE RATIO
[a]	0.424	1	Same	< 1.00
Alloys [b]	0.228	1	"	"
[c]	0.111	1	"	"
Total alloys	0.762	3		

*13.16 THE GENERAL LINEAR HYPOTHESIS

The purely parametric models which have been introduced in this chapter are all included in the following *general linear model*.

$X_1, X_2, \ldots, X_N$ are random variables representing the observed values. The expected value of X_i is a linear function

$$c_{i1}\theta_1 + c_{i2}\theta_2 + \cdots + c_{is}\theta_s$$

of s distinct parameters $\theta_1, \theta_2, \ldots, \theta_s$ in the sense that no fewer than s parameters can be used, the values of which are unknown. The coefficients c_{ij}, on the other hand, are known numbers. Further the X_i's are mutually independent and normally distributed with common (but unknown) variance σ^2. [The theory can be extended without difficulty to the case when the X's are correlated and have unequal variances, provided the variance–covariance matrix (see Section 5.11.2) is known, apart from a factor of proportionality.]

Each of the special parametric models discussed in this chapter is included in this general model which can be written explicitly

$$X_i = c_{i1}\theta_1 + c_{i2}\theta_2 + \cdots + c_{is}\theta_s + Z_i \quad (i = 1, 2, \ldots, N) \tag{13.68}$$

where Z_i is distributed $N(0, \sigma)$ and the Z's are mutually independent.

Further, the hypotheses tested (apart from the hypothesis of equality of variance) are particular cases of the *general linear hypothesis*. The *general linear hypothesis* of order p states that the s parameter values $\theta_1, \theta_2, \ldots, \theta_s$ satisfy p "distinct" linear equations (i.e., not expressible in fewer than p equations)

$$b_{l1}\theta_1 + b_{l2}\theta_2 + \cdots + b_{ls}\theta_s = b_l \quad (l = 1, 2, \ldots, p) \tag{13.69}$$

where the b's are known numbers. Both sets of (13.68) and (13.69) are supposed expressed in the most economical form possible. That is to say, (13.68) cannot be expressed in terms of fewer than s different parameters, and (13.69) cannot be reduced to fewer than p linear equations.

Thus in a one-way classification the parametric model [see (13.5)]

$$X_{ti} = \xi + \gamma_t + Z_{ti} \quad (t = 1, \ldots, k; \ i = 1, \ldots, n_t)$$

is not in the standard form (13.68) because there are $k + 1$ parameters $\xi, \gamma_1, \ldots, \gamma_k$, but only k are needed. The form

$$X_{ti} = \xi_t + Z_{ti}$$

is, however, in standard form. The coefficients c_{ij} are all either 0 or 1, and there are k parameters $\xi_1, \xi_2, \ldots, \xi_k$; so $s = k$.

The hypothesis

$$\xi_1 = \xi_2 = \cdots = \xi_k$$

can be expressed as $\xi_1 - \xi_2 = 0, \xi_2 - \xi_3 = 0, \ldots, \xi_{k-1} - \xi_k = 0$. There are $(k-1)$ distinct linear conditions here; each b_{lj} is zero, 1 or -1, and each b_l is 0. The hypothesis is of order $p = k - 1$ since no fewer than $(k - 1)$ equations suffice to give all the conditions.

The concept of the general linear hypothesis enables us to unify a good deal of the theory of analysis of variance for purely parametric models. Its greatest usefulness, however, is that application of the likelihood ratio method for composite hypotheses (see Section 7.8) to tests of the general linear hypotheses gives us a straightforward way of constructing test procedures in cases where the heuristic type of approach we have employed is not a very natural line of attack.

The likelihood ratio test of a general linear hypothesis, H_p of order p, is obtained by the following steps

1. Form the function

$$S = \sum_{i=1}^{N} \left(X_i - c_{i1}\theta_1 - c_{i2}\theta_2 - \cdots - c_{is}\theta_s \right)^2$$

2. Minimize S with respect to $\theta_1, \theta_2, \ldots, \theta_s$. The result is called the *absolute* minimum of S and is denoted by S_a.
3. Minimize S with respect to $\theta_1, \theta_2, \ldots, \theta_s$ subject to the condition that H_p is true. The result is called the *relative* minimum of S and is denoted by S_r.
4. Since S_r is a minimum of S over a more restricted field of variation than is S_a, S_r cannot be less than S_a. Calculate

$$S_b = S_r - S_a \geqslant 0$$

5. The likelihood ratio test criterion is then equivalent to the ratio

$$\frac{S_b/p}{S_a/(N-s)}$$

If H_p is true, this is distributed as $F_{p, N-s}$. Generally the ratio is distributed as a noncentral $F'_{p, N-s}$ with noncentrality parameter

$$\sigma^{-2} \sum_{i=1}^{N} \left(E^*(X_i) - c_{i1}\theta_1 - \cdots - c_{is}\theta_s \right)^2$$

where $E^*(X_i)$ is the expected value of X_i if H_p is true.

The important thing to notice is that the analysis of variance tests are (for purely parametric models) likelihood ratio tests. In the ANOVA table

S_b is the sum of squares for "departure from H_p," and p is the number of degrees of freedom of S_b;
S_a is the Residual sum of squares, and $N - s$ is the number of degrees of freedom of S_a.

Note that whatever the hypothesis H_p, the same mean square $S_a/(N-s)$ appears in the denominator of the test criterion. This draws attention to a characteristic feature of the analysis of variance for purely parametric models—namely, that the *same* Residual mean square is used in *all* tests. For mixed and component-of-variance models, however, we have seen that this is not the case.

*13.16.1 Cross-Classification with Unequal Replication—Parametric Model

In Section 13.11 we described the analysis for a two-way $r \times c$ cross-classification with n replications for each of the rc possible combinations of factor levels. We now apply the general linear hypothesis to construct an analysis appropriate to the more general case when there are n_{ti} replications of the tth row factor level combined with the ith column factor level. The numbers n_{ti} can now take any values; some of them may be 0, but there must be at least one nonzero m_{ti} in each row and each column, and 2 nonzero values in some.

The model is still given by (13.25); that is,

$$X_{tij} = \xi + \rho_t + \kappa_i + (\rho\kappa)_{ti} + Z_{tij}$$

with

$$\sum_{t=1}^{r} \rho_t = 0 = \sum_{i=j}^{c} \kappa_i; \quad \sum_{i=1}^{c} (\rho\kappa)_{ti} = 0 = \sum_{t=1}^{r} (\rho\kappa)_{ti} \quad \text{for all } t \text{ and } i$$

Note that in terms of the general theory, we have

$$N_{..} = \sum_{t=1}^{r} \sum_{i=1}^{c} n_{ti}$$

and

$$S = \sum_{t=1}^{r} \sum_{i=1}^{c} \sum_{j=1}^{n_{ti}} \left[X_{tij} - \xi - \rho_t - \kappa_i - (\rho\kappa)_{ti} \right]^2$$

(If $n_{ti} = 0$, there is, of course, no corresponding term in S.)

It should be noted that the expression

$$E(X_{tij}) = \xi + \rho_t + \kappa_i + (\rho\kappa)_{ti}$$

is not in the standard form (13.68). The number of *distinct* parameter(s) is equal to the number of occupied cells ($n_{ti} > 0$) in the table. If no n_{ti}'s are zero, this is equal to rc.

It is easy to show that the absolute minimum of S is

$$S_a = \sum_{t=1}^{r} \sum_{i=1}^{c} \sum_{j=1}^{n_{ti}} \left(X_{tij} - \bar{X}_{ti.} \right)^2$$

that is, the "within cells" sum of squares. The number of degrees of freedom is $(N_{..} - K)$, where K = number of occupied cells, and the Residual mean square is

$$\frac{S_a}{N_{..} - K}$$

just as in (13.48) with rc replaced by K.

Now suppose we wish to test the hypothesis H that there is no interaction, that is, $(\rho\kappa)_{ti} = 0$ for all t and all i. There are K linear relations here, but they are not all distinct. There are $r + c - 1$ distinct linear relations

$$\left(\sum_{t=1}^{r} (\rho\kappa)_{ti} = \sum_{i=1}^{c} (\rho\kappa)_{ti} = 0 \right)$$

among the $(\rho\kappa)_{ti}$'s, and so there are really only $K - (r + c - 1)$ distinct relations specified by H. This hypothesis is therefore a linear hypothesis of order $p = K - r - c + 1$. If $K = rc$ (that is, $n_{ti} > 0$ for all t and i), then

$$p = (r - 1)(c - 1)$$

We find the relative minimum, S_r, of S by minimizing

$$S' = \sum_{t=1}^{r} \sum_{i=1}^{c} \sum_{j=1}^{n_{ti}} (X_{tij} - \xi - \rho_t - \kappa_i)^2$$

subject to the conditions,

$$\sum_{t=1}^{r} \rho_t = 0 = \sum_{i=1}^{c} \kappa_i$$

Applying the method of Lagrange multipliers we form the quantity

$$F = S' + 2\alpha \sum_{t=1}^{r} \rho_t + 2\beta \sum_{i=1}^{c} \kappa_i$$

We then have to solve for ξ, ρ_t, κ_i, α, and β, the equations

$$\frac{1}{2} \frac{\partial F}{\partial \xi} = 0 = X_{...} - N_{..}\hat{\xi} - \sum_{t=1}^{r} N_{t.}\hat{\rho}_t - \sum_{i=1}^{c} N_{.i}\hat{\kappa}_i \tag{13.70}$$

$$\frac{1}{2} \frac{\partial F}{\partial \hat{\rho}_t} = 0 = X_{t..} - N_{t.}\hat{\xi} - N_{t.}\hat{\rho}_t - \sum_{i=1}^{c} n_{ti}\hat{\kappa}_i - \alpha \quad (t = 1, \ldots, r) \tag{13.71}$$

$$\frac{1}{2} \frac{\partial F}{\partial \hat{\kappa}_i} = 0 = X_{.i.} - N_{.i}\hat{\xi} - \sum_{t=1}^{r} n_{ti}\hat{\rho}_t - N_{.i}\hat{\kappa}_i - \beta \quad (i = 1, \ldots, c) \tag{13.72}$$

where

$$N_{t.} = \sum_{i=1}^{c} n_{ti}; \ N_{.i} = \sum_{t=1}^{r} n_{ti}$$

Adding the r equations (13.71) and comparing with (13.70), we see that $\alpha = 0$; and similarly adding (13.72) we see that $\beta = 0$. Then, from (13.71), we have

$$\hat{\xi} + \hat{\rho}_t = N_{t.}^{-1} X_{t..} - N_{t.}^{-1} \sum_{i=1}^{c} n_{ti}\hat{\kappa}_i \tag{13.73}$$

[One could also proceed from (13.72) and insert formulas for $\hat{\xi}+\hat{\kappa}_i$ in (13.71). This is done in Example 13.16.]

Inserting this formula for $\hat{\xi}+\hat{\rho}_t$ in (13.72), we have

$$\sum_{t=1}^{r} n_{ti} N_t^{-1}\left[X_{t..} - \sum_{i=1}^{c} n_{ti} \hat{\kappa}_i\right] + N_{.i}\hat{\kappa}_i = X_{.i.}$$

or

$$N_{.i}\hat{\kappa}_i - \sum_{i=1}^{c}\left[\sum_{t=1}^{r}\left(\frac{n_{ti}n_{ti'}}{N_t}\right)\hat{\kappa}_{i'}\right] = X_{.i.} - \sum_{t=1}^{r} n_{ti}\bar{X}_{t..} \qquad (13.74)$$

These c equations for $\hat{\kappa}_1, \hat{\kappa}_2, \ldots, \hat{\kappa}_c$ can be reduced to $(c-1)$ equations for $\hat{\kappa}_1, \hat{\kappa}_2, \ldots, \hat{\kappa}_{c-1}$ by using the condition $\Sigma_{i=1}^{c}\hat{\kappa}_i = 0$.

These $(c-1)$ equations are solved for $\hat{\kappa}_1, \ldots, \hat{\kappa}_c$, and then $\hat{\xi}+\hat{\rho}_t$ is calculated for (13.74). Finally, S_r is calculated from the formula

$$S_r = \sum_{t=1}^{r}\sum_{i=1}^{c}\sum_{j=1}^{n_{ti}} X_{tij}\left(X_{tij} - \hat{\xi} - \hat{\rho}_t - \hat{\kappa}_i\right)$$

$$= \sum_{t=1}^{r}\sum_{i=1}^{c}\sum_{j=1}^{n_{ti}} X_{tij}^2 - \sum_{t=1}^{r}(\hat{\xi}+\hat{\rho}_t)X_{t..} - \sum_{i=1}^{c}\hat{\kappa}_i X_{.i.}$$

If a separate estimate of ξ is required, the formula

$$\hat{\xi} = r^{-1}\sum_{t=1}^{r}(\hat{\xi}+\hat{\rho}_t)$$

can be used.

The heaviest part of the work is the solution of the $(c-1)$ simultaneous equations (13.74). Since the row and column factors enter symmetrically into the analysis, we could interchange ρ and κ in the analysis and obtain $(r-1)$ simultaneous equations in ρ to solve. Evidently it is convenient to use whichever factor has the smaller number of levels. For example, in a 4 (row)$\times$9 (column) cross-classification, we need solve only 3 simultaneous equations for the row effects.

Once S_r has been obtained, we calculate the "Row$\times$Column" sum of squares $S_r - S_a = S_b$ and the mean square ratio $[S_b/(K-r-c+1)]/[S_a/(N-K)]$ which can be compared with percentage points of the $F_{K-r-c+1, N-K}$ distribution.

It may so happen that the n_{ij}'s are such that we cannot solve (13.40) to (13.72) completely. In these cases it is not possible to estimate all (perhaps any) of the parameters separately.

Example 13.16 An investigation was planned to study the effect of both curing time and composition on the tensile strength of rubber compounds. Two curing times were proposed (T_1 and T_2) and 3 mixes (A, B, and C) were prepared. The data of Table 13.25 presents the result of the tensile test. Owing to failure in some

TABLE 13.25
TENSILE STRENGTH OF RUBBER COMPOUNDS

| | COMPOUND | | | |
TIME	A	B	C	Totals
T_1	3350	3730		10250
	3170			
T_2		3550	3200	10240
		3490		
Total	6520	10770	3200	20490

of the original preparations, an equal number of replicates does not appear for each combination of curing time and composition.

The mathematical model is

$$X_{tij} = \xi + \tau_t + \kappa_i + (\tau\kappa)_{ti} + Z_{tij}$$

The absolute minimum sum of squares (S_a) is

$$\tfrac{1}{2}(3350 - 3170)^2 + \tfrac{1}{2}(3550 - 3490)^2 = 18000$$

This sum of squares has 2 degrees of freedom.

Now, assuming that $(\tau\kappa)_{ti} = 0$ (for all t and all i) we obtain (in the way described above) the following equations for $\hat{\tau}_1$ and $\hat{\tau}_2$

$$N_t \hat{\tau}_t - \sum_{t'=1}^{2} \left[\sum_{i=1}^{3} \left(\frac{n_{ti} n_{t'i}}{N_{.i}} \right) \hat{\tau}_{t'} \right] = X_{t..} - \sum_{i=1}^{3} n_{ti} \bar{X}_{.i} \quad (t = 1,2)$$

For these data $t = 1,2$; $i = 1,2,3$; $n_{11} = 2, n_{12} = 1, n_{13} = 0$; $n_{21} = 0, n_{22} = 2, n_{23} = 1$; $N_{1.} = 3, N_{2.} = 3, N_{.1} = 2, N_{.2} = 3, N_{.3} = 1$, and $N_{..} = 6$. The equations to be solved are

$$3\hat{\tau}_1 - \hat{\tau}_1 \left(\frac{2 \cdot 2}{2} + \frac{1 \cdot 1}{3} + \frac{0 \cdot 0}{1} \right) - \hat{\tau}_2 \left(\frac{2 \cdot 0}{2} + \frac{1 \cdot 2}{3} + \frac{0 \cdot 1}{1} \right)$$

$$= 10250 - 2(3260) - 1(3590) - 0(3200) \quad \text{for } t = 1$$

$$3\hat{\tau}_2 - \hat{\tau}_1 \left(\frac{0 \cdot 2}{2} + \frac{2 \cdot 1}{3} + \frac{1 \cdot 0}{1} \right) - \hat{\tau}_2 \left(\frac{0 \cdot 0}{2} + \frac{2 \cdot 2}{3} + \frac{1 \cdot 1}{1} \right)$$

$$= 10240 - 0(3260) - 2(3590) - 1(3200) \quad \text{for } t = 2.$$

These equations reduce to

$$\tfrac{2}{3}\hat{\tau}_1 - \tfrac{2}{3}\hat{\tau}_2 = 140$$

$$-\tfrac{2}{3}\hat{\tau}_1 + \tfrac{2}{3}\hat{\tau}_2 = -140$$

These equations are identical, as we should expect, since $\hat{\tau}_1$ and $\hat{\tau}_2$ satisfy the condition $\hat{\tau}_1 = -\hat{\tau}_2$.

Hence $\frac{4}{3}\hat{\tau}_1 = 140$ or $\hat{\tau}_1 = 105$
and $\hat{\tau}_2 = -105$

The equations for $\hat{\xi} + \hat{\kappa}_i$ are

$$2(\hat{\xi} + \hat{\kappa}_1) + 2\hat{\tau}_1 = 6520, \quad \text{whence} \quad \hat{\xi} + \hat{\kappa}_1 = 3155$$

$$3(\hat{\xi} + \hat{\kappa}_2) + \hat{\tau}_1 + 2\hat{\tau}_2 = 10770, \quad \text{whence} \quad \hat{\xi} + \hat{\kappa}_2 = 3625$$

$$\hat{\xi} + \hat{\kappa}_3 + \hat{\tau}_2 = 3200, \quad \text{whence} \quad \hat{\xi} + \hat{\kappa}_3 = 3305$$

If it is desired to estimate ξ separately, we have

$$\hat{\xi} = \tfrac{1}{3}(3155 + 3625 + 3305) = 3361.7.$$

We cannot test the hypothesis of "no interaction" with the present data, because we have $K=4$, $r=2$, $c=3$, and so the number of degrees of freedom of the interaction sum of squares is $4 - 2 - 3 + 1 = 0$.

Note that if the observation (3730) for time T_1 and compound B were absent it would not be possible to solve the equations separately for $\hat{\xi}$, $\hat{\kappa}_1$, $\hat{\kappa}_2$, $\hat{\kappa}_3$, $\hat{\tau}_1$, and $\hat{\tau}_2$.

In the case when 1 of the 2 factors has only 2 levels [e.g., take $r-2$ in (13.25)], there are some useful explicit formulas. The average difference between the 2 levels of the row factor is $\rho_1 - \rho_2 = 2\rho_1 = \Delta$, say. The best linear unbiased estimator of this difference is

$$\hat{\Delta} = \frac{\displaystyle\sum_{i=1}^{c} w_i\left(\bar{X}_{1i.} - \bar{X}_{2i.}\right)}{\displaystyle\sum_{i=1}^{c} w_i}$$

where $w_i = n_{1i} n_{2i}/(n_{1i} + n_{2i})$.

The sum of squares for Rows (adjusted for Columns) is

$$\hat{\Delta}^2 \sum_{i=1}^{c} w_i$$

and the Interaction sum of squares is

$$\sum_{i=1}^{c} w_i\left(\hat{\Delta}_i - \hat{\Delta}\right)^2$$

where $\hat{\Delta}_i = \bar{X}_{1i.} - \bar{X}_{2i.}$.

The methods described in this section can be extended in a straightforward manner to 3 (or more)-way cross-classifications. The calculations are more complicated but computer programs can be constructed to do them.

[It is necessary to bear in mind that when some of the n_i's are 0 it may not be possible to estimate each of the parameters separately because the sets of equations (13.70) to (13.72) may not have unique solutions.]

*13.16.2 Missing Values

However well an experiment is planned there is always the possibility that something will go awry in its execution. In particular cases one or more observations may be lacking because of accident or misfortune.

As a result an originally balanced design, for which a standard form of analysis would have been appropriate, becomes unbalanced, and special forms of analysis must be developed. The general linear hypothesis approach is very useful in such cases, provided a parametric model is appropriate.

Consider a randomized block design with b blocks and f treatment levels. This is just a $b \times f$ cross-classification with one replication. If some observations are missing we have a cross-classification, as in Section 13.16.1, with some n_{ti}'s equal to 1, and some 0. The analysis could be carried out by the method developed in Section 13.16.1, but the special circumstances make it possible for rather simpler techniques to be used.

Suppose there is just one observation missing. Without loss of generality we can suppose that it is the observation for the first treatment level in the first block. The fact that an observation is missing does not, of course, affect the structure of the mathematical model, so we have, as in (13.52)

$$X_{ti} = \xi + \gamma_t + \phi_i + Z_{ti} \quad (t = 1, \ldots, b; \; i = 1, \ldots, f)$$

except that X_{11} is missing.

In this case the number of observations is $N = bf - 1$ and the number of distinct parameters is $1 + (b - 1) + (t - 1) = b + t - 1$.

We have $S = \Sigma\Sigma(X_{ti} - \xi - \gamma_t - \phi_i)^2$, where $\Sigma\Sigma = $ "summation over all observations." We have to minimize S subject to the conditions

$$\sum_{t=1}^{b} \gamma_t = 0 = \sum_{i=1}^{f} \phi_i.$$

We form the quantity

$$\Phi = S + 2\alpha \sum_{t=1}^{b} \gamma_t + 2\beta \sum_{i=1}^{f} \phi_i$$

and solve the equations

$$-\frac{1}{2} \frac{\partial \Phi}{\partial \xi} = 0 = X_{..} - (bf - 1)\hat{\xi} + \hat{\gamma}_1 + \hat{\phi}_1 \tag{13.75}$$

$$-\frac{1}{2} \frac{\partial \Phi}{\partial \gamma_1} = 0 = X_{1.} - (f - 1)(\hat{\xi} + \hat{\gamma}_1) + \hat{\phi}_1 - \alpha \tag{13.76}$$

$$-\frac{1}{2} \frac{\partial \Phi}{\partial \gamma_t} = 0 = X_{t.} - f(\hat{\xi} + \hat{\gamma}_t) - \alpha \quad (t = 2, \ldots, b) \tag{13.77}$$

$$-\frac{1}{2} \frac{\partial \Phi}{\partial \phi_1} = 0 = X_{.1} - (b - 1)(\hat{\xi} + \hat{\phi}_1) + \hat{\gamma}_1 - \beta \tag{13.78}$$

$$-\frac{1}{2} \frac{\partial \Phi}{\partial \phi_i} = 0 = X_{.i} - b(\hat{\xi} + \hat{\phi}_i) - \beta \quad (i = 2, \ldots, f) \tag{13.79}$$

Adding (13.76) and the $(b-1)$ equation, (13.77), and comparing with (13.75), we see that $\alpha = 0$; similarly, it can be shown that $\beta = 0$. Then combining (13.75), (13.76), and (13.78), we have

$$\hat{\xi} + \hat{\gamma}_1 + \hat{\phi}_1 = \frac{fX_{1.} + bX_{.1} - X_{..}}{(b-1)(f-1)} \qquad (13.80)$$

This can be regarded as an estimate of the missing value. (It is in fact an unbiased estimator of its expected value.)

S_a can be calculated by inserting $\hat{\xi} + \hat{\gamma}_1 + \hat{\phi}_1$ in the missing plot and analyzing the completed set of data as if it were an ordinary complete randomized block experiment. S_a is then the Residual (Block × "Treatment") sum of squares in this analysis. It should be noted, however, that S_a has $(b-1)(f-1)-1$, and not $(b-1)(f-1)$ degrees of freedom.

Suppose we wish to test the hypothesis that there are no differences among treatments, that is, $\phi_1 = \phi_2 = \cdots = \phi_f = 0$.

This is a linear hypothesis of order $f-1$. The relative minimum sum of squares, S_r, is the minimum of

$$\sum\sum (X_{ti} - \xi - \gamma_t)^2$$

with respect to ξ and γ.

It is easy to see that

$$S_r = \sum\sum \left(X_{ti} - \bar{X}_{t.}\right)^2$$

Then

$$S_b = S_r - S_a = \sum\sum \left(X_{ti} - \bar{X}_{..}\right)^2 - \sum_t n_t\left(\bar{X}_{t.} - \bar{X}_{..}\right)^2 - S_a$$

where

$$n_t = f \quad (t > 1) \quad \text{but} \quad n_1 = f - 1$$

S_b is thus calculated as

(Total sum of squares)—(Crude Between Blocks sum of squares)—(Residual) = "Adjusted Between Treatments sum of squares."

The ratio $[S_b/(f-1)]/[S_a/(bf-b-f)]$ is compared with the F distribution with $f-1, bf-b-f$ degrees of freedom.

If the hypothesis $\gamma_1 = \gamma_2 = \cdots = \gamma_b = 0$ is to be tested, we calculate

$$S_b' = (\text{Total sum of squares})—(\text{Crude Between Treatments}$$

$$\text{sum of squares})—(\text{Residual sum of squares})$$

$$= \text{"Adjusted Between Blocks sum of squares"}$$

$[S_b'/(b-1)]/[S_a/(bf-b-f)]$ should be distributed as $F_{b-1,bf-b-f}$ if there is no difference among blocks.

Notice that we cannot exhibit S_b, S_b', and S_a in a single table adding up to the Total sum of squares. Two separate tables are needed as shown in Table 13.26.

TABLE 13.26
ANOVA FOR RANDOMIZED BLOCK EXPERIMENT WITH MISSING VALUES

SOURCE	SUM OF SQUARES	SOURCE	SUM OF SQUARES
Crude Blocks	$\sum_t n_t(\bar{X}_{t.} - \bar{X}_{..})^2$	Adjusted Blocks	By difference
Adjusted Treatments	By difference	Crude Treatments	$\sum_i n_i'(\bar{X}_{.i} - \bar{X}_{..})^2$
Residual	S_a	Residual	S_a
Total	$\sum\sum (X_{ti} - \bar{X}_{..})^2$	Total	$\sum\sum (X_{ti} - \bar{X}_{..})^2$

If more than one value is missing, the missing values can be estimated using (13.80) iteratively. That is, all values but one are guessed; that one is then estimated using (13.80), and the estimate used to adjust the value of one of the guessed values, and so on cyclically until stability is reached.

Formulas useful in connection with the analysis of some other designs are shown in the next section. (Among these designs are some which are dealt with in Chapters 14 and 15 of this book.)

*13.16.3 Formulas for Missing Plot Analyses

(1) RANDOMIZED BLOCK: X_{11} missing, b "blocks," f "treatments."

Estimate of missing value is $\hat{X}_{11} = (bX_{1.} + fX_{.1} - X_{..})/(b-1)(f-1)$. S.E. of difference between mean of a treatment with a missing value and mean of any other treatment is

$$\sqrt{\sigma^2\left[\frac{2}{b} + \frac{f}{b(b-1)(f-1)}\right]}$$

(2) $n \times n$ LATIN SQUARE $X_{(111)}$ missing. Estimate of missing value is

$$\hat{X}_{(111)} = \frac{n(X_{(1..)} + X_{(.1.)} + X_{(..1)}) - 2X_{(...)}}{(n-1)(n-2)}$$

S. E. of difference between mean of treatment with missing value and mean of any other treatment is $\{\sigma^2[(2/n) + 1/(n-1)(n-2)]\}^{1/2}$.

N.B. Corrected Treatment sum of squares

= Crude Treatment sum of squares

$$-\frac{1}{(n-1)^2(n-2)^2}\left[(n-1)X_{(..1)} + X_{(.i.)} + X_{(.1.)} - X_{(...)}\right]^2.$$

First Row Missing. (Assuming treatment i missing from column i.) Estimates of column and treatment effects are obtained from the equations

$$n(n-1)\hat{\xi} + n(n-2)\hat{\kappa}_i = (n-1)X_{(.i.)} + X_{(..i)}$$

$$n(n-1)\hat{\xi} + n(n-2)\hat{\tau}_i = (n-1)X_{(..i)} + X_{(.i.)}$$

$$\left(\text{whence} \quad \hat{\xi} + \hat{\tau}_i = \frac{X_{(...)}}{n(n-1)(n-2)} + \frac{(n-1)X_{(..i)} + X_{(.i.)}}{n(n-2)} \right)$$

Comparisons of treatment effects can be made by calculating differences between quantities

$$\frac{(n-1)X_{(..i)} + X_{(.i.)}}{n(n-2)}$$

The standard error of each difference is

$$\sqrt{\sigma^2 \cdot \frac{2(n-1)}{n(n-2)}}$$

The sum of squares for Columns *confounded with Treatments* is given by the usual "Between Groups" formula

$$(n-1)^{-1} \left[\sum_{i=1}^{n} X_{(.i.)}^2 - n^{-1} X_{(...)}^2 \right]$$

The sum of squares for Treatments, *adjusted to eliminate Column effects*, is

$$(n-1)^{-1}(n-2)^{-1} \left[n^{-1} \sum_{i=1}^{n} \{ (n-1)X_{(..i)} + X_{(.i.)} \}^2 - X_{(...)}^2 \right]$$

(This is a Youden Square—see Section 14.5.)

Analyses have been developed for Latin squares with *one row and one column missing* (see [13.14]) and with *two rows and two columns missing* (see [13.15]). These and similar analyses can all be obtained by applying the general method described in Section 13.16. Analyses are usually worked out for standard cases (for example, with $x_{(111)}$ missing, as described above); these analyses can be made applicable to particular cases by appropriate renumbering of rows, columns, and treatments.

(3) CONFOUNDED FACTORIAL EXPERIMENTS (*same effect confounded in all replications*). Use randomized block procedure, applied only to blocks containing treatment combination(s) with missing value(s).

(4) PARTIAL CONFOUNDING.   No treatment comparison to be confounded in more than one replication. Let

k = no. of units per block

r = no. of replicates

t = no. of treatments

B = total yield of all plots in same *block* as missing plot

T = total yield of all plots with same *treatment* as missing plot

R = total yield of all plots in *same replicate* as missing plot

S_B = total yield, in *other* replicates of all *blocks* containing treatment which was on missing plot

G = total of all yields in the experiment

U = total yield of all *other* treatments appearing in block with missing unit

V = total yield, in *other* replications, of all *blocks* containing the treatment on the missing unit, but excluding that treatment

Then

Estimate of missing yield

$$= \frac{kt(r-1)T + tr(r-1)B + kG + tV - krR - t(r-1)U - trS_B}{(r-1)[t(r-1)(k-1) - (t-k)]}$$

This applies to 2^3, 2^4, 2^5, 2^6, 3^3, 3^4, 4^2, and 4×2^2 factorials.

(5) BALANCED INCOMPLETE BLOCKS. Let

$$k = \text{no. of plots per block}$$

$$r = \text{no. of replicates}$$

$$t = \text{no. of treatments}$$

Then

$$\text{Estimate of missing yield} = \frac{tr(k-1)B + k(t-1)Q - (t-1)Q'}{(k-1)[tr(k-1) - k(t-1)]}$$

where

B = total of block containing missing observation

$Q = k \times$ (total yield of missing treatment)

  − (total yield on all blocks where missing treatment occurs)

Q' = sum of "Q" values for all treatments in the block with the missing value

13.17   REGRESSION ANALYSIS—ONE INDEPENDENT VARIABLE

The calculations needed in fitting linear regression equations have already been discussed, in some detail, in Chapter 12. Here we consider how to treat the problems arising in this kind of work by the methods of the analysis of variance.

First we recall a few formulas which were obtained in Chapter 12. If we fit a regression of form $E[Y|x] = \alpha + \beta x$, using a set of n pairs of values (x_j, Y_j) $(j = 1, 2, \ldots n)$, then, using the standard analysis, the estimator of β is

$$B = \frac{\sum (x_j - \bar{x})(Y_j - \bar{Y})}{\sum (x_j - \bar{x})^2}$$

and by (12.6) the estimator of α is

$$D = \bar{Y} - B\bar{x}$$

The model used in Chapter 12 can be written in the form

$$Y_j = \alpha + \beta x_j + Z_j \tag{13.81}$$

where the Z_j's satisfy conditions $(i) - (iv)$ of Section 13.4. Then

$$E(A) = \alpha; \quad E(B) = \beta;$$

$$\text{var}(\bar{Y}) = \frac{\sigma^2}{n}; \quad \text{var}(B) = \sigma^2 \left[\sum (x_j - \bar{x})^2 \right]^{-1}; \quad \text{cov}(\bar{Y}, B) = 0$$

In view of the normality [condition (iv)] of the Z_j's, the last result implies that $\bar{Y}$ and B are mutually independent.

In Figure 13.7 the fitted regression line $E[Y|x] = A + Bx$ is represented by the straight line AB. This line passes through the point G, with coordinates $(\bar{x}, \bar{Y})$—the "center of gravity" of the set of observed points, of which P_j, with coordinates (x_j, Y_j), is a typical member.

The perpendicular to the x axis from P_j meets the x axis at the point X with coordinates $(x_j, 0)$; it meets the fitted regression line AB at K, and the line through G parallel to the x axis at H. The coordinates of K are $(x_j, A + Bx_j)$, and of H, $(x_j, \bar{Y})$.

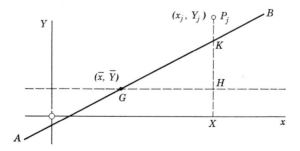

Figure 13.7 A fitted regression line.

The length XP_j can be split up into XH, HK, and KP_j, or, in algebraic terms,

$$Y_j = \bar{Y} + \left(A + Bx_j - \bar{Y}\right) + \left(Y_j - A - Bx_j\right)$$

$$= \bar{Y} + B\left(x_j - \bar{x}\right) + \left(Y_j - A - Bx_j\right)$$

Hence the total sum of squares, $\Sigma(Y_j - \bar{Y})^2$ can be written

$$\Sigma\left(Y_j - \bar{Y}\right)^2 = \Sigma\left[B\left(x_j - \bar{x}\right) + \left(Y_j - A - Bx_j\right)\right]^2$$

$$= B^2 \Sigma\left(x_j - \bar{x}\right)^2 + \Sigma\left(Y_j - A - Bx_j\right)^2 \qquad (13.82)$$

since

$$2B\Sigma\left(x_j - \bar{x}\right)\left(Y_j - A - Bx_j\right) = 2B\left[\Sigma\left(x_j - \bar{x}\right)\left(Y_j - \bar{Y}\right) - B\Sigma\left(x_j - \bar{x}\right)^2\right] = 0$$

The first of these sums of squares, $B^2\Sigma(x_j - \bar{x})^2$, is the sum of quantities like $(HK)^2$, and measures the amount of variation of the Y's "explained" by the fitted regression line. It is, therefore, called the sum of squares, "Due to (Linear) Regression of Y on x" or, more briefly, "(Linear) Regression," or still more briefly, "Slope," referring to the fact that B is the slope of the fitted line.

Since $E(B) = \beta$ it follows that the expected value of the Slope sum of squares is

$$E\left[B^2 \sum_{j=1}^{n} \left(x_j - \bar{x}\right)^2\right] = \sigma^2 + \beta^2 \sum_{j=1}^{n} \left(x_j - \bar{x}\right)^2 \qquad (13.83)$$

and the number of degrees of freedom of this sum of squares is 1 (the coefficient of σ^2).

The other sum of squares, $\Sigma(Y_j - A - Bx_j)^2$, is the sum of quantities like $(KP_j)^2$—squares of deviations of observed values of Y_j from the values predicted by the fitted regression. So this sum of squares may be called "About (Linear) Regression," though, as we shall now see, it is, in fact, used as a Residual sum of squares.

Using (13.81), we see that

$$Y_j - A - Bx_j = Z_j - (A - \alpha) - (B - \beta)x_j$$

Since $E(A) = \alpha$, $E(B) = \beta$, but the distributions of A and B do not otherwise depend on the values of α and β, it follows that $\Sigma(Y_j - A - Bx_j)^2$ is a quadratic function of the Z's, and so its expected value is a multiple of σ^2. Putting $E[\sum_{j=1}^{n}(Y_j - A - Bx_j)^2] = \lambda\sigma^2$, we now need to find λ. This can be effected by using the equation

$$E\left[\sum_{j=1}^{n}\left(Y_j - \bar{Y}\right)^2\right] = E\left[B^2\sum_{j=1}^{n}\left(x_j - \bar{x}\right)^2\right] + E\left[\sum_{j=1}^{n}\left(Y_j - A - Bx_j\right)^2\right]$$

(13.84)

Since λ does not depend on β we can take $\beta = 0$. From (13.82) and (13.83)

$$(n-1)\sigma^2 = \sigma^2 + \lambda\sigma^2$$

so $\lambda = n - 2$. The About (Linear) Regression sum of squares thus has $(n-2)$ degrees of freedom. The results so far obtained are utilized in forming the analysis of variance table shown in Table 13.27.

TABLE 13.27

ANALYSIS OF VARIANCE OF LINEAR REGRESSION

SOURCE	DEGREES OF FREEDOM	SUM OF SQUARES	MEAN SQUARE
Linear Regression	1	$B^2\sum_{j=1}^{n}(x_j - \bar{x})^2$	$B^2\sum_{j=1}^{n}(x_j - \bar{x})^2$
Residual (About Linear Regression)	$n-2$	$\sum_{j=1}^{n}(Y_j - A - Bx_j)^2$	$\left[\sum_{j=1}^{n}(Y_j - A - Bx_j)^2\right]/(n-2)$
Total	$n-1$	$\sum_{j=1}^{n}(Y_j - \bar{Y})^2$	

The mean square ratio

$$\frac{(n-2)B^2\sum_{j=1}^{n}\left(x_j - \bar{x}\right)^2}{\sum_{j=1}^{n}\left(Y_j - A - Bx_j\right)^2} = R$$

(13.85)

is compared with the F distribution with $1, n-2$ degrees of freedom to test the hypothesis $\beta = 0$.

A test of the hypothesis $\beta = \beta_0$ can be constructed, replacing B by $(B - \beta_0)$ in (13.85).

$$\sqrt{F} = \sqrt{(n-2)} \cdot (B - \beta_0) \sqrt{\frac{\sum (x_j - \bar{x})^2}{\sum (Y_j - A - Bx_j)^2}} \qquad (13.86)$$

is distributed as t with $n - 2$ degrees of freedom if $\beta = \beta_0$. Hence

$$\Pr\left[-t_{n-2,1-\gamma/2} \leqslant \sqrt{F} \leqslant t_{n-2,1-\gamma/2} \,\middle|\, \beta_0 \right] = 1 - \gamma$$

Rearranging the inequalities we obtain the $100(1 - \gamma)$ percent confidence interval for β, which is given in Section 12.2.5.

If $\beta_0 = 0$, (13.86) can be written in the form

$$\sqrt{F} = \frac{R\sqrt{n-2}}{\sqrt{(1 - R^2)}} \qquad (13.87)$$

where $R = \sum (x_j - \bar{x})(Y_j - \bar{Y}) / \sqrt{\sum (x_j - \bar{x})^2 \sum (Y_j - \bar{Y})^2}$ is the sample correlation coefficient between x and Y. Since $\sqrt{F}$ is distributed as t with $(n - 2)$ degrees of freedom, the distribution of R can be deduced from (13.86) using the methods developed in Section 5.12. The result obtained is

$$p_R(r) \propto (1 - r^2)^{(n-4)/2} \qquad (-1 \leqslant r \leqslant 1) \qquad (13.88)$$

Note that this is the distribution of R when $\beta = 0$, that is, when Y does not depend on x, so that the population correlation coefficient, ρ, is 0.

Note also that (13.88) has been obtained on the assumption that the x_i's are constant, but the expression for $p_R(r)$ does not depend on the actual values of the x_i's. Hence it follows that (13.88) gives the distribution of the sample correlation coefficient R, *whatever the distribution of the* x_i's, provided only that the conditional distribution of Y_j, given x_j, is normal with constant mean and variance.

It is worthwhile noting that x_j may be replaced by any explicit function of x_j, as pointed out in Section 12.5.2.

More generally, it can be shown that the *conditional* distribution of $\sqrt{F}$, given $x_1, \ldots, x_n$, is that of noncentral t with $(n - 2)$ degrees of freedom and noncentrality parameter $\sigma^{-1}(\beta - \beta_0)\sqrt{\sum (x_j - \bar{x})^2}$. Averaging over the distribution of $\sum (x_j - \bar{x})^2$, we can find the distribution of $\sqrt{F}$, and so of the sample correlation coefficient, R.

13.18   TEST FOR DEPARTURE FROM LINEARITY

The preceding discussion has been developed on the basis of the implied assumption that the linear model

$$Y_i = \alpha + \beta x_i + Z_i$$

is an adequate representation of the physical situation. However, it is not always that the assumption of *linear regression*

$$E(Y|x) = \alpha + \beta x$$

is justified. Sometimes, indeed, as we have seen in Chapter 12, we try to find curvilinear regression of certain specified forms. However, in many cases we do not have enough knowledge to specify a particular form; we would just like to know if a linear regression model suffices. This problem can be stated in terms of the general linear hypothesis in the following way.

We express the expected value of Y_t, given x_t in the form

$$E(Y_t|x_t) = \alpha + \beta x_t + \delta_t$$

where δ_t now represents a "departure from linearity." If there are k different values of x_t, then there are $k+2$ parameters $\alpha, \beta, \delta_1, \delta_2, \ldots, \delta_k$ in this model, and so we can impose 2 linear conditions on the parameters. The algebraic treatment is simplified if we use the conditions

$$\sum \delta_t = 0 = \sum \delta_t x_t$$

where $\sum$ denotes summation over all observations. The second of these conditions is sometimes felt to be a bit odd, because it implies that the values of the δ's depend on the values of the x's. Although this is true, the sole effect of the conditions is that $Y = \alpha + \beta x$ is the least squares fit of the regression of the actual expected values of Y (given x) on x. In other words we use the "best" values of α and β.

The model is

$$Y_t = \alpha + \beta x_t + \delta_t + Z_t$$

Since δ_t and Z_t share the same suffix, it proves impossible to obtain a Residual sum of squares (expected value proportional to σ^2) not affected by the δ_t's. However, if for at least *some* x's we have *more than one* observation of Y, then the model becomes

$$Y_{ti} = \alpha + \beta x_t + \delta_t + Z_{ti} \quad (t = 1, \ldots, k; \ i = 1, \ldots, n_t) \tag{13.89}$$

with at least one n_t greater than 1. Now the δ_t's and Z_{ti}'s have different suffixes and it becomes possible to test the linear hypothesis $H_0 : \delta_1 = \delta_2 = \cdots = \delta_k = 0$. In other words H_0 states "the regression of Y on x is linear."

The appropriate test can be developed by applying the method described in Section 13.16. Here, however, we will use a more heuristic method.

The model (13.89) is like that for a one-way classification. The Total sum of squares $(\Sigma_{t=1}^k \Sigma_{i=1}^{n_t}(Y_{ti} - \overline{Y}_{..})^2)$ can be split up into

$$\text{Between Groups}\left[\sum_{t=1}^{k} n_t \left(\overline{Y}_{t.} - \overline{Y}_{..} \right)^2 \right]$$

and

$$\text{Within Groups}\left[\sum_{t=1}^{k} \sum_{i=1}^{n_t} \left(Y_{ti} - \overline{Y}_{t.} \right)^2 \right] \text{sums of squares}$$

In turn the Between Groups sum of squares can be split up into sums of squares "Due to Linear Regression" and "Array Means about Linear Regression" according to the equation

$$\sum_{t=1}^{k} n_t \left(\overline{Y}_{t.} - \overline{Y}_{..} \right)^2 = B^2 \sum_{t=1}^{k} n_t \left(x_t - \overline{x} \right)^2 + \sum_{t=1}^{k} n_t \left[\overline{Y}_{t.} - \overline{Y}_{..} - B \left(x_t - \overline{x} \right) \right]^2$$

(Linear Regression) + (Array Means about Regression)

where

$$B = \frac{\displaystyle\sum_{t=1}^{k} n_t x_t \left(\overline{Y}_{t.} - \overline{Y}_{..} \right)}{\displaystyle\sum n_t \left(x_t - \overline{x} \right)^2} \quad \text{and} \quad \overline{x} = \frac{\displaystyle\sum_{t=1}^{k} n_t x_t}{\displaystyle\sum_{t=1}^{k} n_t}$$

Starting from (13.89), we find

$$\overline{Y}_{t.} = \alpha + \beta x_t + \delta_t + \overline{Z}_{t.} \tag{13.90}$$

therefore,

$$Y_{ti} - \overline{Y}_{t.} = Z_{ti} - \overline{Z}_{t.}$$

and $\Sigma_{t=1}^k \Sigma_{i=1}^t (Y_{ti} - \overline{Y}_{t.})^2$ is a true residual distributed as $\chi^2 \sigma^2$ with

$\sum_{t=1}^{k} n_t - k$ degrees of freedom. Also

$$\overline{Y}_{..} = \alpha + \beta\overline{x} + \overline{Z}_{..} \quad \left(\text{because } \sum_{t=1}^{k} n_t \delta_t = 0\right) \qquad (13.91)$$

and so

$$\overline{Y}_{t.} - \overline{Y}_{..} = \beta(x_t - \overline{x}) + \delta_t + \overline{Z}_{t.} - \overline{Z}_{..}$$

Finally,

$$B = \sum_{t=1}^{k} n_t x_t \left[\beta(x_t - \overline{x}) + \delta_t + \overline{Z}_{t.} - \overline{Z}_{..}\right] \left[\sum_{t=1}^{k} n_t (x_t - \overline{x})^2\right]^{-1}$$

$$= \beta + \frac{\sum_{t=1}^{k} n_t x_t (\overline{Z}_{t.} - \overline{Z}_{..})}{\sum_{t=1}^{k} n_t (x_t - \overline{x})^2} \quad \left(\text{because } \sum_{t=1}^{k} n_t \delta_t x_t = 0\right) \qquad (13.92)$$

Hence the "Array Means about Linear Regression" sum of squares is equal to

$$\sum_{t=1}^{k} n_t (\delta_t + Z_t)^2$$

where Z_t is a linear function of the Z_{ti}'s. It follows that the expected value of this statistic is equal to

$$(\text{multiple of } \sigma^2) + \sum_{t=1}^{k} n_t \delta_t^2$$

It can be shown (by a method similar to that used in Section 13.7) that the "multiple of σ^2" is in fact $(k-2)\sigma^2$. This sum of squares is increased by nonzero values of the δ_t's and is called "Departure from Linearity" sum of squares.

Finally, the "Due to Linear Regression" sum of squares has expected value

$$\sigma^2 + \beta^2 \sum_{t=1}^{k} n_t (x_t - \overline{x})^2$$

The analysis of variance table is shown in Table 13.28. Linearity of

TABLE 13.28
ANOVA FOR DEPARTURE FROM LINEARITY

SOURCE	DEGREES OF FREEDOM	SUM OF SQUARES	EXPECTED VALUE OF MEAN SQUARE
Linear Regression	1	$B^2 \sum_{t=1}^{k} n_t (x_t - \bar{x})^2$	$\sigma^2 + \beta^2 \sum_{t=1}^{k} n_t (x_t - \bar{x})^2$
Departure from Linearity	$k - 2$	$\sum_{t=1}^{k} n_t [\bar{Y}_{t.} - \bar{Y} - B(x_t - \bar{x})]^2$	$\sigma^2 + (k-2)^{-1} \sum_{t=1}^{k} n_t D_t^2$
Residual	$\sum_{t=1}^{k} n_t - k$	$\sum_{t=1}^{k} \sum_{i=1}^{n_t} (Y_{ti} - \bar{Y}_{t.})^2$	σ^2
Total	$\sum_{t=1}^{k} n_t - 1$	$\sum_{t=1}^{k} \sum_{i=1}^{n_t} (Y_{ti} - \bar{Y}_{..})^2$	

regression is tested by comparing the ratio

$$\frac{\text{"Departure from Linearity" mean square}}{\text{Residual mean square}}$$

with the F distribution with $k - 2$, $\sum_{t=1}^{k} n_t - k$ degrees of freedom.

It should be noted that *any* alternative to linearity of regression is included in the model (13.89). If a specific kind of alternative is suspected before carrying out the experiment, it should be allowed to influence the test to be used. Such a test is more sensitive (has greater power) with respect to this specific alternative than the analysis of variance test we have described; the latter test has to cater to a much wider range of possibilities.

If the alternative to linearity is suspected to be a polynomial regression of some (unknown) order, we can use the fact that (in the orthogonal polynomial notation of Section 12.5.3).

$$\sum_{i=1}^{N} (Y_i - \bar{Y})^2 = \sum_{i=1}^{N} \left[Y_i - \bar{Y} - B_1' \xi_1(x_i) - .. - B_s' \xi_s(x_i) \right]^2$$
$$+ \sum_{j=1}^{k} B_j'^2 \sum_{i=1}^{N} \left[\xi_j(x_i) \right]^2 \qquad (13.93)$$

The first term on the right side is the Residual sum of squares with $(N - s - 1)$ degrees of freedom. The quantities

$$B_j'^2 \sum_{i=1}^{N} \left[\xi_j(x_i) \right]^2$$

are mutually independent with 1 degree of freedom each and expected values

$$\sigma^2 + \beta_j'^2 \sum_{i=1}^{N} \left[\xi_j(x_i) \right]^2$$

Equation (13.93) is the basis of the polynomial regression analysis of variance table shown in Table 13.29. (See also Section 13.15.1)

TABLE 13.29
ANOVA USING ORTHOGONAL POLYNOMIALS

SOURCE	DEGREES OF FREEDOM	SUM OF SQUARES	EXPECTED VALUE OF MEAN SQUARE
Linear	1	$B_1'^2 \sum_{i=1}^{N} [\xi_1(x_i)]^2$	$\sigma^2 + \beta_1'^2 \sum_{i=1}^{N} [\xi_1(x_i)]^2$
Quadratic	1	$B_2'^2 \sum_{i=1}^{N} [\xi_2(x_i)]^2$	$\sigma^2 + \beta_2'^2 \sum_{i=1}^{N} [\xi_2(x_i)]^2$
Cubic	1	$B_3'^2 \sum_{i=1}^{N} [\xi_3(x_i)]^2$	$\sigma + \beta_3'^2 \sum_{i=1}^{N} [\xi_3(x_i)]^2$
⋮	⋮	⋮	⋮
s–ic	1	$B_s'^2 \sum_{i=1}^{N} \left[\xi_s(x_i) \right]^2$	$\sigma^2 + \beta_s'^2 \sum_{i=1}^{N} [\xi_s(x_i)]^2$
Residual	$N-s-1$	$\sum_{i=1}^{N} \left[Y_i - \bar{Y} - B_1'\xi_1(x_i) \cdots - B_s'\xi_s(x_i) \right]^2$	σ^2
Total	$N-1$	$\sum_{i=1}^{N} \left(Y_i - \bar{Y} \right)^2$	

Note that in this case we do not need to have repeated observations of Y for a given value of x in order to test for departure from linearity.

Usually each successive term in the polynomial regression is tested for significance until *two* successive terms are nonsignificant. This is a convenient practical rule, based on the fact that odd and even additional terms introduce different kinds of deviations from the previous polynomial regression. However, the rule is not applied automatically. If high-order terms are not considered of sufficient magnitude to be important, they are ignored, even though formally "significant"; on the other hand, analysis may be continued, even though two successive terms are nonsignificant, if prior considerations indicate the existence of higher-order terms.

Example 13.17 Consider the data of Example 12.10. These are shown in Table 13.30 and their graph is given in Figure 13.8. Table 13.31 gives the sums of squares for each term. The final column of the table is the mean square ratio for the significance test. Note that each Residual sum of squares is the result of subtracting the preceding number from that prior to it. This is a direct application of (12.65).

TABLE 13.30

BEHAVIOR OF LAMINATES CURED AT 225°F

LOG DAMPING DECREMENT (y) AS A FUNCTION OF TEMPERATURE (x)

Temperature $x(°C)$	Log damping decrement y
−30	0.053
−20	0.057
−10	0.061
0	0.068
10	0.072
20	0.081
30	0.093
40	0.105
50	0.115
60	0.130

The linear and quadratic terms are highly significant. The others are not significant, so we use the polynomial of second degree. We calculate the estimated regression coefficients, using (12.63)

$$A' = \frac{835 \times 10^{-3}}{10} = 83.5 \times 10^{-3}$$

$$\lambda_1 = 2 \qquad B_1' = \frac{2 \times 1403 \times 10^{-3}}{330} = 8.503 \times 10^{-3}$$

$$\lambda_2 = \tfrac{1}{2} \qquad B_2' = \frac{181 \times 10^{-3}}{2 \times 132} = \frac{181 \times 10^{-3}}{264} = 0.686 \times 10^{-3}$$

We fit the curve in x by putting $z_i = [x_i - (x_1 - w)]/w$. Then

$$z = \frac{x + 40}{10} = x' + 4 \qquad \text{where } x' = \frac{x}{10}$$

By (12.70) and (12.72)

$$\xi_1 = z - 5.5 = x' + 4 - 5.5 = x' - 1.5$$

$$\xi_2 = (z - 5.5)^2 - 8.25 = (x' - 1.5)^2 - 8.25$$

$$= x'^2 - 3x' + 2.25 - 8.25 = x'^2 - 3x' - 6$$

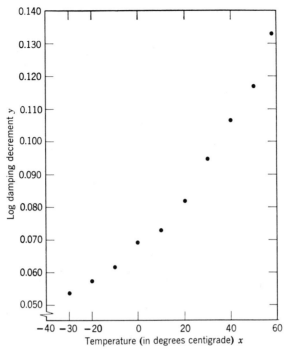

Figure 13.8 Log damping decrement (y) as a function of temperature (x).

TABLE 13.31

SOURCE	SUM OF SQUARES	D.F.	MEAN SQUARE	MEAN SQUARE RATIO
Total	6224.5×10^{-6}	9		
Linear term	5964.9×10^{-6}	1	5964.9×10^{-6}	184.1***
Residual (S_1)	259.6×10^{-6}	8	32.4×10^{-6}	
Quadratic term	248.2×10^{-6}	1	248.2×10^{-6}	152.3***
Residual (S_2)	11.4×10^{-6}	7	1.63×10^{-6}	
Cubic term	1.16×10^{-10}	1	1.16×10^{-10}	0.00006
Residual (S_3)	11.40×10^{-6}	6	1.90×10^{-6}	
4th order term	1.96×10^{-6}	1	1.96×10^{-6}	1.03
Residual (S_4)	9.44×10^{-6}	5	1.89×10^{-6}	
5th order term	0.68×10^{-6}	1	0.68×10^{-6}	0.31
Residual	8.76×10^{-6}	4	2.19×10^{-6}	

The estimated regression function is

$$E(Y|x') = \{83.5 + 8.503(x' - 1.5) + 0.686(x'^2 - 3x' - 6)\} \times 10^{-3}$$

$$= \{83.5 + 8.503x' - 12.754 + 0.686x'^2 - 2.058x' - 4.116\} \times 10^{-3}$$

$$= \{0.686x'^2 + 6.445x' + 66.63\} \times 10^{-3}$$

In terms of x as argument

$$E(Y|x) = \{0.00686x^2 + 0.645x + 66.63\} \times 10^{-3}$$

The analysis of variance for multiple linear regression follows similar lines.

First notice that the term called "quadratic" in Table 13.29 is actually quadratic *given linear* (regression)—it is the reduction in Residual sum of squares effected by introducing quadratic terms in addition to linear ones.

Similarly in multiple linear regression we can arrange the controllable variables in some order (for convenience we say $x_1, x_2, \ldots, x_s$) and consider the successive reduction in the Residual sum of squares effected by introducing these variables, one at a time, into the regression equation. There is, however, an important difference from the polynomial regression situation, in that there is no automatically natural order of choice for the X's. So with $s = 2$, for example, we could have either of the analyses:

SOURCE	DEGREES OF FREEDOM		SOURCE	DEGREES OF FREEDOM
X_1	1		X_2	1
X_2 given X_1	1	or	X_1 given X_2	1
Residual	$N - 3$		Residual	$N - 3$
Total	$N - 1$		Total	$N - 1$

The Residual and Total sum of squares have the same values in the 2 tables.

13.19 ONE-WAY CLASSIFICATION—BIVARIATE OBSERVATIONS

If *two* characters, x and Y, are measured on each individual in a one-way classification, we need to consider their mutual relationship as well as possible variation from group to group.

If we assume that there is linear regression of character Y on character x, possibly varying from group to group, a suitable model would be

$$Y_{ti} = \alpha_t + \beta_t x_{ti} + Z_{ti} \qquad (t = 1, \ldots, k; i = 1, \ldots, n_t) \qquad (13.94)$$

A natural problem to consider is whether it is possible for a *single* linear regression ($Y = \alpha + \beta X$) to represent the whole of the groups. This is a linear hypothesis $H_0: (\alpha_1 = \alpha_2 = \cdots = \alpha_k; \ \beta_1 = \beta_2 = \cdots = \beta_k)$ of order $2(k - 1)$. [Note that in the *linear model* (13.94) the x_{ti}'s are "known coefficients."]

It is convenient to split up this hypothesis into 3 parts, each of which may be tested separately.

$H_0^{(1)}$: The regression lines are parallel—that is, $\beta_1 = \beta_2 = \cdots = \beta_k$.

$H_0^{(2)}$: The group means lie on a straight line—that is the points $(\bar{x}_{t.}, \alpha_t + \beta_t \bar{x}_{t.})$ lie on a straight line.

$H_0^{(3)}$: The slope of this line equals the common value, β_c, say, of $\beta_1, \ldots, \beta_k$.

Firstly, we analyze the data for each group separately, obtaining fitted regression straight lines,

$$E[Y \mid x] = A_t + B_t x \qquad (t = 1, 2, \ldots, k)$$

The expected value of B_t is β_t, its variance is $\sigma^2 \, [\sum_{i=1}^{n_t} (x_{ti} - \bar{x}_{t.})^2]^{-1} = \sigma^2 / w_t$ where $w_t = \sum_{i=1}^{n_t} (x_{ti} - \bar{x}_{t.})^2$ can be thought of as the "weight" of B_t.

The analyses of variance for the linear regression in each group are based on the algebraic identities:

$$\sum_{i=1}^{n_t} \left(Y_{ti} - \bar{Y}_{t.} \right)^2 = w_t B_t^2 + \sum_{i=1}^{n_t} \left[Y_{ti} - \bar{Y}_{t.} - B_t \left(x_{ti} - \bar{X}_{t.} \right) \right]^2 \qquad (t = 1, 2, \ldots, k)$$

Adding these k identities we obtain

$$\sum_{t=1}^{k} \sum_{i=1}^{n_t} \left(Y_{ti} - \bar{Y}_{t.} \right)^2 = \sum_{t=1}^{k} w_t B_t^2 + \sum_{t=1}^{k} \sum_{i=1}^{n_t} \left[Y_{ti} - \bar{Y}_{t.} - B_t \left(x_{ti} - \bar{x}_{t.} \right) \right]^2$$

$$= \sum_{t=1}^{k} w_t B_t^2 + S_R$$

The last term is a Residual sum of squares with $\sum_{t=1}^{k} (n_t - 2) = N - 2k$ degrees of freedom, where $N = \sum_{t=1}^{k} n_t$. The computational formula for the weighted sum of squares of deviations from a weighted mean is

$$\sum_{t=1}^{k} w_t (B_t - B_c) = \sum_{t=1}^{k} w_t B_t^2 - w_c B_c^2$$

where

$$w_c = \sum_{t=1}^{k} w_t; \qquad B_c = \frac{\sum\limits_{t=1}^{k} w_t B_t}{w_c}$$

Therefore, we can split up $\Sigma_{t=1}^{k} w_t B_t^2$ into

$$w_c B_c^2 + \sum_{t=1}^{k} w_t (B_t - B_c)^2 = w_c B_c^2 + S_W$$

The expected value of the second quantity is $(k-1)\sigma^2 + \Sigma_{t=1}^{k} w_t (\beta_t - \beta_c)^2$ where $\beta_c = (\Sigma_{t=1}^{k} w_t \beta_t)/w_c$. (The subscript c can be thought of as meaning "common," referring to "Common (Average) Within Groups Slope" in this instance.)

The variance of B_c is σ^2/w_c.

Starting now from the standard algebraic identity, splitting up the Total sum of squares into Between Groups and Within Groups sums of squares:

$$\sum_{t=1}^{k} \sum_{i=1}^{n_t} \left(Y_{ti} - \overline{Y}_{..} \right)^2 = \sum_{t=1}^{k} n_t \left(\overline{Y}_{t.} - \overline{Y}_{..} \right)^2 + \sum_{t=1}^{k} \sum_{i=1}^{n_t} \left(Y_{ti} - \overline{Y}_{t.} \right)^2$$

and introducing the further breakup of the Within Groups sum of squares, which we have just developed, we have

$$(\text{Total sum of squares}) = \sum_{t=1}^{k} n_t \left(\overline{Y}_{t.} - \overline{Y}_{..} \right)^2 + w_c B_c^2$$

$$+ (\text{Between "Within Groups" Slopes sum of squares})$$

$$+ (\text{Residual sum of squares}) \qquad (13.95)$$

The "Between 'Within Groups' Slopes" sum of squares is, of course, $\Sigma_{t=1}^{k} w_t (B_t - B_c)^2$. Testing the corresponding mean square against the Residual mean square provides a test of the hypothesis $H_0^{(1)}$.

To obtain a test of $H_0^{(2)}$ we first fit c linear regression to the group mean points $(\overline{x}_{t.}, \overline{Y}_{t.})$ with weights n_t. This leads to the following splitting up of the "Between Groups" sum of squares

$$\sum_{t=1}^{k} n_t \left(\overline{Y}_{t.} - \overline{Y}_{..} \right)^2 = w_m B_m^2 + \sum_{t=1}^{k} n_t \left[\overline{Y}_{t.} - \overline{Y}_{..} - B_m \left(\overline{X}_{t.} - \overline{X}_{..} \right) \right]^2$$

$$= w_m B_m^2 + S_G \qquad (13.96)$$

where

$$B_m = \frac{\sum_{t=1}^{k} n_t (\bar{x}_{t.} - \bar{x}_{..})(\bar{Y}_{t.} - \bar{Y}_{..})}{\sum_{t=1}^{k} n_t (\bar{x}_{t.} - \bar{x}_{..})^2}$$

and

$$w_m = \sum_{t=1}^{k} n_t (\bar{x}_{t.} - \bar{x}_{..})^2$$

We note that $\text{var}(B_m) = \sigma^2 / w_m$. The expected value of the second term on the right side of (13.96) is

$$(k-2)\sigma^2 + \sum_{t=1}^{k} n_t \left[\alpha_t - \alpha_m - \beta_m \bar{x}_{..} \right]^2$$

where

$$\alpha_m = \frac{\sum_{t=1}^{k} n_t \alpha_t}{N}$$

$$\beta_m = \frac{\sum_{t=1}^{k} n_t (\bar{x}_{t.} - \bar{x}_{..})(\alpha_t + \beta_t \bar{x}_{t.})}{w_m}$$

Hypothesis $H_0^{(2)}$ is tested by comparing the mean square ratio

$$\frac{S_G/(k-2)}{S_R/(N-k)}$$

with the F distribution with $k-2$, $N-2k$ degrees of freedom.

$H_0^{(3)}$ is a linear hypothesis of first order: $\beta_m = \beta_c$. To test this we rearrange the sum of the 2 remaining unused quantities in the breakup of the Total sum of squares,

$$w_c B_c^2 + w_m B_m^2$$

into the form

$$w_o B_o^2 + \frac{w_c w_m}{w_o} (B_c - B_m)^2 = w_o B_o^2 + S_{WG}$$

where

$$w_o = w_c + w_m \left(= \sum_{t=1}^{k} \sum_{i=1}^{n_t} \left(x_{ti} - \bar{x}_{..} \right)^2 \right)$$

and

$$B_o = \frac{w_c B_c + w_m B_m}{w_o}$$

(Here, "o" is used to denote "overall," and B_o is the slope that would be obtained if a single linear regression were fitted to the whole of the data, ignoring subdivision into groups.)

Hypothesis $H_0^{(3)}$ is then tested by comparing the ratio

$$\frac{S_{WG}}{S_R / (N - k)}$$

with the F distribution with 1, $(N - 2k)$ degrees of freedom.

All the quantities needed for the separate tests of the 3 hypotheses which make up H_0 can be included in a single analysis of variance table as shown in Table 13.32.

If it is not desired to test $H_0^{(1)}$, $H_0^{(2)}$, and $H_0^{(3)}$ separately, the sums of squares for the second, third, and fourth sources in Table 13.32 are added and divided by $1 + (k - 2) + (k - 1) = 2(k - 1)$ to give a mean square; the ratio of this mean square to the Residual mean square is then compared with the F distribution with $2(k - 1)$, $N - 2k$ degrees of freedom.

13.20 ANALYSIS OF COVARIANCE

The preceding analysis is sometimes called "analysis of covariance." We prefer to reserve this term for analyses testing whether apparent differences in mean response (Y) from group to group can be explained by linear regression on a control ("concomitant") variable x.

Analysis of covariance takes different forms according to the model assumed. For a one-way classification by groups we might use the model

$$\text{(I)} \quad Y_{ti} = \alpha_t + \beta \left(x_{ti} - \bar{x}_{..} \right) + Z_{ti}$$

or the model

$$\text{(II)} \quad Y_{ti} = \alpha_t + \beta_t \left(x_{ti} - \bar{x}_{..} \right) + Z_{ti}$$

(There are other possibilities. These are the simplest and most often used.)

TABLE 13.32

ANOVA to Test $H_0^{(1)}$, $H_0^{(2)}$, and $H_0^{(3)}$

SOURCE	DEGREES OF FREEDOM	SUM OF SQUARES	EXPECTED VALUE OF MEAN SQUARE
Overall slope	1	$S_o = w_o B_o^2$	$\sigma^2 + w_o \beta_o^2$
Slope of group means vs. Average within group slope	1	$S_{WG} = \dfrac{w_c w_m}{w_o}(B_c - B_m)^2$	$\sigma^2 + \dfrac{w_c w_m}{w_o}(\beta_o - \beta_m)^2$
About linear regression of group means	$k-2$	$S_G = \sum_{t=1}^{k} n_t \left[\bar{Y}_{t\cdot} - \bar{Y}_{\cdot\cdot} - B_m(\bar{x}_{t\cdot} - \bar{x}_{\cdot\cdot}) \right]^2$	$\sigma^2 + (k-2)^{-1} \sum_{t=1}^{k} n_t[\alpha_t - \alpha_m - \beta_m \bar{x}_{t\cdot}]^2$
Between "within group" slopes	$k-1$	$S_W = \sum_{t=1}^{k} w_t (B_t - B_c)^2$	$\sigma^2 + (k-1)^{-1} \sum_{t=1}^{k} w_t (\beta_t - \beta_c)^2$
Residual (about "within group" linear regressions)	$N-2k$	$S_R = \sum_{t=1}^{k} \sum_{i=1}^{n_t} \left[Y_{ti} - \bar{Y}_{t\cdot} - B_t(x_{ti} - \bar{x}_{t\cdot}) \right]^2$	σ^2
Total	$N-1$	$\sum_{t=1}^{k} \sum_{i=1}^{n_t} (Y_{ti} - \bar{Y}_{\cdot\cdot})^2$	

697

For each model, we want to test the hypothesis

$$H_0: \alpha_1 = \alpha_2 = \cdots = \alpha_k$$

Appropriate tests can be derived using the method described in Section 13.16.

For case I we have (in the notation used earlier in the present section)

SOURCE	DEGREES OF FREEDOM	SUM OF SQUARES
Groups (adjusted for concomitant variable)	$k-1$	$S_{WG} + S_G$
Residual	$N-k-1$	$S_R + S_W$

For case II we have

SOURCE	DEGREES OF FREEDOM	SUM OF SQUARES
Groups (adjusted for concomitant variable)	$k-1$	$S_0 + S_{WG} + S_G + S_W$ $- w_0 \sum_{t=1}^{k} B_t'^2$
Residual	$N-2k$	S_R

where

$$B_t' = \frac{\displaystyle\sum_{i=1}^{n_t} (x_{ti} - \bar{x}_{..})(\bar{Y}_{ti} - \bar{Y}_{..})}{\displaystyle\sum_{i=1}^{n_t} (x_{ti} - \bar{x}_{..})^2}$$

Estimated values of α_t are

For case I: $\bar{Y}_{t.} - B_c (\bar{x}_{t.} - \bar{x}_{..})$

For case II: $\bar{Y}_{t.} - B_t (\bar{x}_{t.} - \bar{x}_{..})$

Provided one is really confident that $\beta_1 = \beta_2 = \cdots = \beta_k$ model I is better to use, as it gives more degrees of freedom in the residual. However, model II ensures against the possibility of an inflated residual due to unsuspected differences among the β's.

Further discussion of covariance techniques is given in Section 14.9.

REFERENCES

1. Brownlee, K. A., *Statistical Theory and Methodology in Science and Engineering*, 2nd ed., Wiley, New York, 1965.

2. Cochran, W. G. and Gertrude M. Cox, *Experimental Designs*, 2nd ed., Wiley, New York, 1957.

3. Cox, D. R., *Planning of Experiments*, Wiley, New York, 1958.

4. Davies, O. L., ed., *Statistical Methods in Research and Production*, 2nd ed., Griffin, London, Hafner, New York, 1967.

5. Duncan, D. B., "Multiple Range and Multiple F Tests," *Biometrics*, **11** (1955).

6. Fisher, R. A. and F. Yates, *Statistical Tables for Biological, Agricultural and Medical Research*, 6th ed., Oliver and Boyd, London, Hafner, New York, 1964.

7. Hartley, H. O., "The Use of Range in Analysis of Variance," *Biometrika*, **37** (1950).

8. Mandel, J., "The Partitioning of Interaction in the Analysis of Variance," *Journal of Research of the National Bureau of Standards*, **73B** (1969).

9. Neter, J. and W. Wasserman, *Applied Linear Statistical Models*, Irwin, Homewood, Ill., 1974.

10. Pearson, E. S. and H. O. Hartley, *Biometrika Tables for Statisticians*, Vol. 1, Cambridge University Press, 1968.

11. Tukey, J. W. "Comparing Individual Means in the Analysis of Variance," *Biometrics*, **5** (1949).

12. Tukey, J. W. "One Degree of Freedom for Non-Additivity," *Biometrics*, **5** (1949).

13. Williams, J. S., "A Confidence Interval for Variance Components," *Biometrika*, **49** (1962).

14. Yates, F., "Incomplete Latin Squares," *Journal of Agricultural Science*, **26** (1936).

15. Yates, F. and R. W. Hale, "The Analysis of Latin Squares when Two or More Rows, Columns or Treatments are Missing," *Journal of the Royal Statistical Society, Series B*, **6** (1939).

EXERCISES

1. Twenty sample specimens of the same type were taken at random from a day's production of bricks. These were assigned at random to 4 different storage conditions. Six of the bricks were lost owing to some error on the part of the experimenter. After being stored for 1 week the water content was determined for each. The data below represent the water content in percent.
(a) Construct a parametric model appropriate to these data.
(b) Test the hypothesis that the method of storage does not affect the water content.

METHOD OF STORING	WATER CONTENT				
1	7.3	8.3	7.6	8.4	8.3
2	5.4	7.4	7.1		
3	8.1	6.4			
4	7.9	9.5	10.0	7.1	

2. The following data came from a completely randomized experiment to compare the reflective properties of 4 kinds of paint. The 20 available test specimens were allotted to the paints at random. The response (in suitable units) was obtained by use of an optical instrument and coded into the data shown below.

PAINT 1	PAINT 2	PAINT 3	PAINT 4
195	45	230	110
150	40	115	55
205	195	235	120
120	65	225	50
160	145		80
	195		

(a) Construct an appropriate analysis of variance table.
(b) Are there differences among variances for the different paints?
(c) Determine a 95% confidence interval for the mean of the reflective characteristic of Paint 1.

3. In a study of Colombian molasses, one of the qualities of importance was Brix degrees. This is a measure of the quantity of solids in a molasses and is a key factor in the processing. The sources of the molasses were 3 different areas in the country. The data below present 8 sample values from each location.

	LOCATION	
I	II	III
81.6	81.8	82.1
81.3	84.7	79.6
82.0	82.0	83.1
79.6	85.6	80.7
78.4	79.9	81.8
81.8	83.2	79.9
80.2	84.1	82.6
80.7	85.0	81.9

(a) Are the 3 locations providing the same Brix degrees of molasses, on average?
(b) Obtain an estimate of variance within locations.
(c) Would you accept the hypothesis that the Brix degrees is 82
   (1) for all locations, or
   (2) for each location individually?
(d) Comment on the assumptions made in your answers to (a)-(c).

4. In a bombing contest between 4 wings, the following scores were recorded. The same type of aircraft and bombing equipment was used by the wings against the same target.

WING 1	WING 2	WING 3	WING 4
970	1960	3460	2060
4220	930	2480	1590
190	3400	2320	2840
650	4550	3830	2940
600	200	3660	1840
2690	530	3530	2110

(a) Describe the model used.

(b) Estimate the residual variance.

(c) Are there differences among the bombing capabilities of the wings?

(d) Is Wing 1 significantly different from the others?

(e) Discuss the data relative to the assumptions in the analysis of variance.

5. It is of importance to the astronomer to know the intrinsic brightness of a star. Owing to absorption of starlight by interstellar dust, the intrinsic brightness cannot be found without a correction for absorption. In the table below, a measurement of coefficient of absorption for 21 individual stars is given. Each star is randomly selected from 1 of 4 galactic longitudes.

LONGITUDE INTERVAL

43°–45°	45°–47°	100°–102°	102°–104°
0.757	0.744	0.655	0.750
0.793	0.811	0.679	0.741
0.681	0.833	0.681	0.717
0.791	0.826	0.762	0.706
0.852	0.752		0.736
0.852			
0.841			

(a) Test the hypothesis that the correction for absorption does not change with galactic longitude.

(b) Compare the first 2 longitude intervals with the last 2 longitude intervals for difference in absorption.

6. Three tread stocks are being tested for treadwear rating. The usual procedure is to compare these against a standard or control with performance designated as 100. Hence the accompanying figures are relative to the control.

(a) Stocks A and B are similar in composition except for a minor modification. Are these different in performance from C?

(b) Determine a 95% confidence interval for the average rating of each stock.

A	113	100	100	98	106	98	97	104	103
B	101	108	99	106	102	100	96	110	112
C	105	96	98	100	101	94	92	100	95

7. Five operators are chosen at random in order to determine the variability in the number of articles an operator is expected to produce in a fixed period of time. For each one, 4 unit intervals are selected at random and the output per unit interval recorded. On the basis of the data below, estimate the residual variance, supposing it to be the same for each operator, and determine a 95% confidence interval for the average output for an operator over 10 units of time. State the model you use.

OPERATORS

1	2	3	4	5
65	79	72	74	76
70	68	75	78	67
73	69	77	89	71
78	72	80	85	66

8. Alloys A, B, and C are identical except that a small percentage of an expensive component has been added in different proportions to alloys B and C to determine whether this affects the tensile strength of the uncast alloy. Six specimens of each were prepared by the investment casting method using molds which ranged in temperature to see if mold temperature affected the tensile strength. The data are coded.

TENSILE STRENGTH

MOLD TEMPERATURE	ALLOY A	B	C
1	170	211	180
2	188	142	162
3	150	110	120
4	183	244	216
5	40	85	112
6	196	214	179

(a) Construct a model.
(b) Carry out a complete analysis of variance.
(c) Draw conclusions.

9. In Exercise 8, suppose alloy B were considerably different from the others.
   (a) By appropriate linear comparisons, test whether its yield differs significantly from the others.
   (b) Compare C with A.
10. The percent pitch content in 5 batches of tar blocks for each of 4 days' production is recorded. The data below are the result of an experiment to determine the variation "between days" and within daily production.

DAY	PER CENT OF PITCH CONTENT				
I	7.5	7.2	7.3	7.3	7.5
II	7.4	7.6	7.4	7.4	7.5
III	7.0	6.9	6.7	6.9	6.8
IV	7.0	7.3	7.2	7.1	7.4

   (a) Construct a model.
   (b) Carry out an analysis of variance.
   (c) Estimate the residual variance and the "between days" variance.
11. The data below came from a particular operation, performed by 4 different machines on each of 5 different days. Consider the days as blocks and machines as treatments. Each entry represents the number of units produced by the machine in a day.

	MACHINE			
DAY	A	B	C	D
1	293	308	323	333
2	298	353	343	363
3	280	323	350	368
4	288	358	365	345
5	260	343	340	330

   (a) Construct an appropriate parametric model.
   (b) Test the differences among treatments for significance.
   (c) If Machine A is standard and Machines B and C have a feature different from D, test these data for meaningful comparison of effects.
12. In Exercises 10 and 11, apply separation of means techniques *where appropriate*.
13. As part of new car performance tests, mileage per gallon of gasoline was tested on new Zilch J-O's, using the same grade of gasoline. Five identical model cars were driven for the same course in the testing ground in Nevada on 10 different days. Consider days as block and cars as treatments. Each entry of the table following represents the mileage per gallon.

			CAR		
DAY	I	II	III	IV	V
1	19.0	19.2	21.2	18.5	21.4
2	20.1	20.8	21.7	19.4	21.7
3	20.4	21.5	22.3	20.2	22.2
4	19.0	20.3	21.4	19.5	21.6
5	20.2	19.9	22.0	18.7	23.0
6	18.9	20.4	21.4	19.2	21.4
7	21.2	20.7	21.4	20.4	22.0
8	19.8	21.2	20.2	18.4	20.8
9	20.3	19.4	19.9	20.2	21.6
10	18.6	18.9	21.2	20.5	20.8

(a) Construct a model and calculate the appropriate analysis of variance table.

(b) Estimate the residual variance and examine the residuals.

(c) Draw conclusions.

14. The following data are the result of preliminary tests. The data are coded weight loss on 2 competitive tires, A and B. If we can assume that the data were gathered in a properly controlled experiment, what conclusions can be drawn from this experiment? There are 2 types of test, fast wear and slow wear. These depend on the average speed of the vehicle used.

	FAST WEAR		SLOW WEAR	
TIRE A	151	101	118	95
	119	156	130	111
TIRE B	117	102	115	97
	133	100	109	92

(a) Construct a model.

(b) Construct an ANOVA table.

(c) Draw conclusions and present them in a concise report.

15. To determine the precision of a penetration test, 2 independent samples of 3 types of waxes were sent to the laboratory in each of 2 successive weeks. These were tested at 110°C. Wax A was in flake form and waxes B and C were in cake form. From the data below,

(a) test whether there is a week-to-week effect;

(b) test whether it is reasonable to suppose there is a constant residual variance, and, if this is so, estimate its value;

(c) test whether there is a difference among waxes.

(d) Is the difference between flake and cake wax significant?

WAX	FIRST WEEK		SECOND WEEK	
A	44	38	34	24
B	35	49	48	39
C	34	30	33	38

16. The product yields of an undesirable by-product were measured for 3 different catalysts, each at 2 different pressures. Measurements are expressed in percentages. Perform an analysis of variance on data below to determine whether the catalysts or pressure level have a significant effect on the yield. Is there an interaction?

	CATALYST I	CATALYST II	CATALYST III
HIGH PRESSURE	0.39	0.27	0.33
	0.45	0.45	0.68
	0.29	0.57	0.31
LOW PRESSURE	0.26	0.10	0.76
	0.54	0.38	0.78
	0.60	0.12	0.85

17. Four batches of crude rubber were sampled. From each batch 2 samples were taken. Three independent test specimens were prepared from each and analyzed. The data below give the modulus of elasticity as a percentage. Assuming that a component-of-variance model is applicable, construct the appropriate ANOVA table. Using this table obtain estimates of the variances of each component.

BATCH	1	2	3	4
	560	600	600	680
Sample 1	580	640	610	700
	600	620	640	730
	660	580	580	720
Sample 2	610	630	660	770
	600	670	620	740

18. An experiment on corrosion of zinc was conducted by determining the loss in weight over a full year. Three fixed exposure locations in Pennsylvania were selected. Two types of zinc, HG and EG, were tested. In all, 18 test plates were prepared as shown in the table below.

LOSS OF ZINC—OZ PER SQ FT PER YEAR

EXPOSURE LOCATION

TYPE	Pittsburgh	Sandy Hook	State College
HG	0.337	0.107	0.061
	0.385	0.116	0.061
	0.448	0.122	0.066
EG	0.332	0.098	0.050
	0.349	0.102	0.053
	0.402	0.112	0.055

(a) Construct an appropriate model.
(b) Is there a significant difference between types?
(c) Is there an interaction?

19. Three types of adhesive are tested in an adhesive assembly of glass specimens. A tensile test is performed to determine the bond strength of this glass to glass assembly. Three different types of assemblies, "cross-lap," "square-center," and "round-center" are tested. The data below present the results of 45 specimens.

(a) Are there differences among adhesives?
(b) Does the type of assembly influence the bond strength?
(c) Is there an interaction?
(d) Suppose adhesive "047" were considered in a different class from 00T and 001. By a linear comparison, determine whether it is significantly different from the other two.

BOND STRENGTH OF GLASS-GLASS ASSEMBLY

ADHESIVE	CROSS-LAP	SQUARE-CENTER	ROUND-CENTER
047	16	17	13
	14	23	19
	19	20	14
	18	16	17
	19	14	21
00T	23	24	24
	18	20	21
	21	12	25
	20	21	29
	21	17	24
001	27	14	17
	28	26	18
	14	14	13
	26	28	16
	17	27	18

20. An experiment is necessary to study the effects of etching compounds on wafer specimens after they have been etched and then heat-treated. There are 4 etching compounds to be compared and 3 furnaces available for heat-treating, but all furnaces are known to be different in their effects. Results from this experiment are shown in the accompanying table.

ETCHING COMPOUND	FURNACE NO.					
	1		2		3	
A	18	19	20	21	14	17
B	24	22	27	30	20	23
C	19	21	20	18	17	16
D	16	15	16	18	14	12

(a) Construct an appropriate parametric model.
(b) Since the furnaces are known to have a significant effect, why do we introduce them?
(c) Did the etching compounds have different effects on the wafers?

21. The data below represent test measurements with 4 densitometers chosen at random from a large number of available densitometers. Three specific types of film to be used in a photographic setup are tested. Two replicate samples are taken of each film for each densitometer.

FILMS	DENSITOMETERS			
	1	2	3	4
A	7	6	4	4
	8	4	3	6
B	8	5	3	7
	4	4	2	4
C	8	4	3	6
	7	6	4	3

(a) Construct an appropriate mixed model.
(b) Set up a complete analysis of variance table indicating the expected values of the mean squares.
(c) Estimate the residual variance and the other variance.
(d) Draw conclusions.

22. An experiment to determine the amount of warping of copper plates was conducted at 4 temperatures (50, 75, 100, and 125°C) and at 4 contents of

copper (40, 60, 80, and 100%). From the data below determine whether there are linear or quadratic effects owing to (*a*) temperature or (*b*) copper content.

	CONTENT			
TEMPERATURE	40	60	80	100
50°	17	16	24	28
75°	12	18	17	27
100°	16	18	25	20
125°	21	23	23	29

23. In an experiment similar to that of Exercise 22, 3 replicates were taken at each combination of temperature and copper content. From the data below answer the questions posed in the preceding exercise. What are the primary differences between the mathematical model for this and the preceding exercise?

	CONTENT			
TEMPERATURE	40	60	80	100
50°	17	16	24	28
	20	21	22	27
	18	12	22	30
75°	12	18	17	27
	9	13	12	31
	10	23	14	31
100°	16	18	25	30
	12	21	23	23
	15	16	26	25
125°	21	23	23	29
	17	21	22	31
	17	28	26	30

24. Four types of adhesives were tested for bond strength. A total of 48 specimens were prepared at a 300°F curing cycle. A second factor was tested within the experiment, namely curing pressure. The pressures were 100 psi, 200 psi, and 300 psi.

(*a*) Construct an appropriate mathematical model for the test results.

(*b*) Calculate a complete analysis of variance table and draw conclusions.

(c) If the pressure effect is significant, determine the linear and higher components.

(d) If the quantitative factor is not significant, is it reasonable to partition this into single degrees of freedom to test linearity, quadratic effect, etc.?

BOND STRENGTH AT THE RESPECTIVE CURING
PRESSURE (300°F CURING CYCLE)

ADHESIVE	100 PSI	200 PSI	300 PSI
031	14	14	10
	14	10	12
	14	19	13
	17	15	10
026	11	16	19
	15	15	10
	17	11	14
	30	10	8
047	15	16	18
	23	26	16
	24	25	6
	15	18	22
00T	3	20	23
	9	18	12
	6	21	17
	9	10	13

25. The data below [Adapted from "Statistics in Analytical Chemistry," by W. J. Youden, *Annals of the New York Academy of Sciences*, **52** (1950)] represent 16 determinations of the ratio of the reacting weights of iodine and silver. Two types of preparations of iodine are used with several types of preparations of silver.

RATIO OF IODINE TO SILVER

IODINE	SILVER				
	A	B	C	D	E
1	1.76422	1.76441	1.76429	1.76449	1.76455
	1.76425	1.76441	1.76437	1.76450	
2	1.76399	1.76423		1.76461	
	1.76440	1.76413		1.76448	
	1.76418				

(a) Treat these data as the result of 8 combinations. Estimate the error variance.

(b) Is there a significant difference among combinations? Use $\alpha = 0.05$.

(c) Determine a 95% confidence interval for the mean of the silver A–iodine 2 combination.

(d) Using the first 2 samples only of iodine 2 (silver) and disregarding the data for C and E silvers, run an analysis of the 2-way experiment. Determine the model equation, construct the ANOVA table, and estimate the residual variance.

26. Four paints were tested for resistance to wear. Two methods of abrasion which were supposed to yield similar results were compared using this test. In all, 24 test plates were prepared, but 4 had to be removed because of lack of control in the preparation. The data below are coded wear data in milligrams deviation from an arbitrary norm. Determine whether there are differences due to paint type, test method, or interaction.

	PAINT			
TEST METHOD	1	2	3	4
	40	53	12	27
1	32	43	21	57
	39	45		45
	64	60	24	45
2	48	42	11	39
	50			

27. Suppose for Exercise 15, some of the data were "lost" with the following results:

WAX	FIRST WEEK		SECOND WEEK	
A	44	38	34	
B	35		48	
C			33	38

Analyze this experiment with the factor "weeks" represented by parameters.

28. In Exercise 15 if the value 24 for wax A in the second week were missing, what estimate for this sample value would minimize the estimate of the residual variance?

29. For the data of Exercise 8 carry out the appropriate analysis of variance if the figure for mold temperature (4) and alloy B were missing.

30. Equimolar concentrations of adipic acid and ethylene glycol are reacted in a continuous reactor at 230°C. Batches of product are collected. The titration of the acid numbers of the product gives direct measure of the concentration of unreacted adipic acid and thus gives the conversion of the reactor (i.e., the higher the acid number, the lower the conversion).

In order to assess the relative magnitude of variation from batch to batch, and, within a batch, from sample to sample (taken for purposes of titration), and from titration to titration within a sample, the experimental results shown in the table were collected.

BATCH	SAMPLE	ACID NUMBER (REPLICATES) 1st	2nd
	A	71	73
1	B	74	72
	C	72	69
	A	71	69
2	B	70	73
	C	73	71
	A	74	70
3	B	71	73
	C	76	71
	A	69	72
4	B	72	75
	C	73	70
	A	72	66
5	B	68	73
	C	73	69

(a) Write down an appropriate component of variance model (assuming homoscedasticity of variances of similar nature).

(b) Estimate each of the 3 variances used in the model.

(c) Construct a 95% confidence interval for the residual variance.

31. The coating of steel pipes with plastic in a continuous extruding machine consists of heating plastic pellets, forcing the hot plastic through a tube forming die by means of a rotating screw, and then slitting the tube open and wrapping it over the steel tube. An experiment was conducted to study the effect of thermostat temperature profile (T), the speed of the screw (S), and the type of plastic (P) on the "output" of the machine. The output of the machine under different conditions are given in the table below.

(a) Assuming that the 3-factor interaction is 0, estimate the residual variance.

(b) Analyze each factor by separation of means using both the Tukey and Duncan techniques.

Speed	P_1 T_1	T_2	T_3	P_2 T_1	T_2	T_3	P_3 T_1	T_2	T_3
S_1	449	467	469	423	449	466	441	469	488
S_2	523	543	547	507	529	550	529	562	576
S_3	604	620	622	587	619	634	606	646	667

32. The data below represent the force in pounds per square inch required to break apart a bond used in fillings of teeth. A dental experiment was conducted with three factors, type of Adhesive (A), type of tooth (B), and configuration or shape (C) of the actual bond. There were 5 types of adhesives; 2 types of teeth, molars and bicuspids; and 5 shapes. For each of the 50 treatment levels ($5 \times 2 \times 5$) 3 die cast pairs (teeth and caps) were prepared and glued. The data below represent the average force for the set of three in each cell. Analyze the data and discuss your conclusions, using Tukey's and also Duncan's technique for separating means.

		Adhesive				
Teeth		1	2	3	4	5
Molars						
	C_1	67	263	184	185	122
	C_2	70	231	210	148	150
Shape	C_3	62	213	257	165	138
	C_4	70	183	201	150	130
	C_5	73	358	173	117	86
Biscuspids						
	C_1	69	352	164	146	102
	C_2	67	342	145	137	101
Shape	C_3	94	354	194	116	160
	C_4	79	288	153	164	136
	C_5	98	351	112	89	157

33. Three companies are bidding on a research and development project. All submit satisfactory technical proposals. A review of past performance by the companies is summarized below. It includes the ratio of final cost to bid and

	COMPANY A		COMPANY B		COMPANY C	
ESTIMATE	RATIO OF FINAL COST TO BID	TECH. PERFORM.	RATIO OF FINAL COST TO BID	TECH. PERFORM.	RATIO OF FINAL COST TO BID	TECH. PERFORM.
Under	1.50	1	0.78	1	1.10	1
$100,000	2.30	3	1.00	4	1.05	1
	2.80	2	3.70	2	2.60	1
$100,000	2.00	2	4.10	1	1.75	2
to	3.50	4	1.30	4	1.80	1
250,000	1.80	1	1.80	2	2.10	2
$250,000	3.50	2	1.00	2	5.00	5
to	2.75	4	2.50	3	1.40	1
500,000	1.90	2	2.10	2	1.10	2
$500,000	1.60	1	2.00	1	4.00	2
to	3.00	3	4.00	3	1.85	1
1,000,000	2.30	1	1.90	2	2.70	1
$1,000,000	1.80	2	2.30	3	1.40	2
to	2.00	2	2.50	2	1.60	1
5,000,000	1.45	2	2.70	1	4.50	2

an appraisal of technical performance of project, on a scale from 1 (best) to 5 (worst), for past completed contracts chosen at random from each of a number of ranges of estimated costs. Company A estimates cost at \$900,000, Company B at \$800,000, and Company C at \$1,200,000.

(a) To whom should the contract be awarded?

(b) What is the expected final cost?

34. In the situation described in Section 13.6.2, it is possible to construct a confidence interval for $\sigma^2 + n\sigma'^2$ using the Between Groups sum of squares, and for σ'^2/σ^2 using the mean square ratio. Show that the range of values for σ' for which $\sigma^2 + n\sigma'^2$ and σ'^2/σ^2 can both fall in their respective $100(1-\alpha/2)$ percent confidence intervals constitutes a confidence interval for σ', with confidence coefficient at least $100(1-\alpha)$ percent.

35. $S_0, S_1, S_2, \ldots, S_k$ are mutually independent random variables. S_j is distributed as χ^2 with ν_j degrees of freedom $(j=0, 1, \ldots, k)$. Show that the joint probability density function of $G_1, \ldots, G_k$, where $G_t = S_t/S_0$, is

$$p(G_1, \ldots, G_k) = \frac{\Gamma\left(\frac{1}{2}\sum_{j=0}^{k} \nu_j\right)}{\prod_{j=0}^{k} \Gamma\left(\frac{1}{2}\nu_j\right)} \cdot \frac{\prod_{t=1}^{k} G_t^{\frac{1}{2}\nu_t - 1}}{\left(1 + \sum_{t=1}^{k} G_t\right)^{\frac{1}{2}\sum_{1}^{k} \nu_j}} \qquad (G_t > 0)$$

Explain the relevance of this result to the calculation of probabilities of various combinations of results of significance tests applied to entries in the same analysis of variance table.

36. In the situation described in Exercise 35, show that the random variables

$$G_t = S_t \left(S_0 + \sum_{j=t+1}^{k} S_j\right)^{-1} \qquad \text{for} \quad t = 1, \ldots, (k-1)$$

and $G_k = S_k/S_0$ form a mutually independent set.

Explain the relevance of this result to the analyses discussed in Section 13.18 (Table 13.29).

CHAPTER 14

Analysis of Variance (II): Modified and Higher Order Designs

14.1 INTRODUCTION

In Chapter 13 we have considered models appropriate to the simpler forms of experimental design. We first considered the case where a single factor sufficed to classify the data, then extended our work to cases where 2 factors were present. The extension to more than 2 factors is simple, when the completely "balanced" features of the design are retained, and we confine our attention to "pure" designs where *only* cross-classification or *only* hierarchal (nested) classification is present.

In this chapter we consider a number of modifications of these pure completely balanced designs. Many of these modifications have been developed to suit frequently occurring practical problems. This is very clearly the case in the first set of designs we consider—balanced incomplete blocks (BIB) and other incomplete block designs. In these designs each block does not contain a representative of each level of the factor (often called "treatments") but a subset of these levels. By appropriate choice of these subsets considerable simplification of analysis is made possible, while the reduction of block size increases flexibility and may well reduce residual variation.

The second set of designs—Latin squares and hypersquares—is also aimed at reducing the size of experiments. In these cases the restriction is direct in that only a proportion of the possible combinations of factor levels actually appear in the experiment. In an $n \times n$ Latin square, for example, there are 3 factors, each with n levels, but only n^2 factor level combinations are used, instead of the n^3 possible combinations.

A similar, but technically different, property is possessed by *fractionally replicated* designs.

The technique of *confounding* enables us to reduce the block size by dividing the set of all possible factor level combinations into subsets and allocating each subset to a different block. In this way the block size is reduced, though all possible factor level combinations are used (possibly more than once).

The fractionally replicated designs referred to above are formed by omitting a proportion (one-half, three-quarters, etc.) of the blocks in a "confounded" design.

Fractionally replicated and confounded designs are discussed in Chapter 15. Split plot designs, which can be regarded as a special kind of confounded design, are also discussed in that chapter.

In this chapter we also consider mixed designs—for example, designs possessing, in part, the characteristics of both cross-classifications and nested designs.

For the most part, discussion is in terms of observations of a single variate. There is, however, some discussion of methods of analysis ("analysis of covariance") appropriate when 2 (or more) characters are measured on each individual.

14.2 INCOMPLETE RANDOMIZED BLOCKS

In a randomized block design each block contains every level of the "treatment" factor (that is, one complete replication). For this reason the design is sometimes called a *complete* randomized block design. Table 14.1 is an example of a complete randomized block design with 4 levels of "treatment," A, B, C, D, all of which appear in each Block (I, II, and III).

It can very well happen that the available blocks are not big enough to accommodate a complete replication. Recall that the factor represented by "blocks" is often a natural one, such as "days," "positions in a field," and "factory." If, for example, only 3 observations can be taken in 1 day, and the "treatments" factor has 4 levels, A, B, C, D, then a complete replicate

TABLE 14.1

RANDOMIZED BLOCKS FOR FOUR
TREATMENTS IN THREE BLOCKS

BLOCK	I	II	III
	D	A	C
	B	C	A
	A	B	D
	C	D	B

TABLE 14.2

BALANCED INCOMPLETE BLOCKS FOR FOUR
TREATMENTS IN BLOCKS OF THREE

BLOCK	I	II	III	IV
	B	A	C	B
	A	B	A	D
	C	D	D	C

cannot be observed in 1 day ("block"). However, a "balanced" design, in which there are 3 replicates arranged in 4 blocks, can be specified as in Table 14.2. The order in which the treatments are placed in each block has been randomized.

This is called an "incomplete" (randomized) block design. It is further described by the adjective "balanced," giving the name "balanced incomplete block," often abbreviated as BIB. The word "balanced" does not simply mean that the blocks are of the same size, and each treatment level appears the same number of times. By definition, in a BIB it must *also* be true that each pair of treatment levels appears together in the same block the same number of times (twice in Table 14.2).

Thus we describe a BIB in terms of certain standard parameters:

t = number of treatment levels
b = number of blocks
k = number of treatment levels ("*plots*") per block
r = number of replications of each treatment level
λ = number of blocks in which any given pair of treatment levels appear together

In Table 14.2, $t = 4$, $b = 4$, $k = 3$, $r = 3$, $\lambda = 2$. Two useful general relations between these parameters can be derived without difficulty. The total number of plots is (number of blocks) × (number of plots per block); it is also (number of treatment levels) × (number of replications). In symbols this is the equation

$$bk = tr \tag{14.1}$$

Similarly, expressing the total number of plots in those blocks containing a given treatment level in 2 ways we find

$$kr = r + \lambda(t - 1)$$

or

$$\lambda = r(k - 1)/(t - 1) \tag{14.2}$$

An alternative form of representation of Table 14.2 is Table 14.3.

TABLE 14.3

FOUR TREATMENTS IN FOUR BLOCKS OF
THREE PLOTS

TREATMENTS	BLOCKS			
	I	II	III	IV
A	×	×	×	
B	×	×		×
C	×		×	×
D		×	×	×

As further examples of BIB designs, consider Table 14.4. In design (*a*) there are 3 treatment levels ($t = 3$), 2 replicates ($r = 2$), 3 blocks ($b = 3$), 2 treatment levels per block ($k = 2$), and $\lambda = 1$. In (*b*) $t = 4$, $r = 3$, $b = 6$, $k = 2$, and $\lambda = 1$. In (*c*) $t = 11$, $r = 6$, $b = 11$, $k = 6$, and $\lambda = 3$.

Lists of BIB designs are given in Fisher and Yates [14.8] and Cochran and Cox [14.3]. When using a design selected from such a list it is a

TABLE 14.4

SOME BALANCED INCOMPLETE BLOCK DESIGNS

TREAT-MENT	BLOCK		
	I	II	III
A	×	×	
B	×		×
C		×	×

(*a*)

TREAT-MENT	BLOCK					
	I	II	III	IV	V	VI
A	×				×	×
B	×	×		×		
C		×	×		×	
D			×	×		×

(*b*)

TREATMENT	BLOCK										
	I	II	III	IV	V	VI	VII	VIII	IX	X	XI
A	×				×	×	×		×	×	
B		×				×	×	×		×	×
C	×		×				×	×	×		×
D	×	×		×				×	×	×	
E		×	×		×				×	×	×
F	×		×	×		×				×	×
G	×	×		×	×		×				×
H	×	×	×		×	×		×			
I		×	×	×		×	×		×		
J			×	×	×		×	×		×	
K				×	×	×		×	×		×

(*c*)

sensible precaution to label the blocks and treatment levels at random. Even if it is not felt necessary to invoke the use of randomization theory (see Section 13.7), for which such random assignment is essential, rearrangement of published designs is a good safeguard against giving unsuspected biases undue emphasis. For example, in Table 14.4, design (b), we could arrange the 6 blocks of the table in the order II, IV, I, VI, V, and III (the numbers 2, 4, 1, 6, 5, 3 might, for instance, occur in that order in a table of random numbers). By a similar kind of random selection the four treatments are assigned letters A, B, C, D.

We do not repeat this emphasis on the desirability of random assignment (if possible) on every relevant occasion, but its advisability should *always* be borne in mind.

14.2.1 Analysis of a BIB Design

The model for a complete randomized block design [equation (13.52)] will also apply to a BIB, since the fact that the blocks are smaller does not essentially alter the structure of the appropriate model (though it may be hoped that the magnitude of the standard deviation, σ, of the Residual terms may be reduced). This model is

$$X_{ij} = \alpha + \beta_i + \tau_j + Z_{ij} \tag{14.3}$$

where i denotes block and j denotes treatment level. This is a parametric model with

$$\sum_{i=1}^{b} \beta_i = 0 = \sum_{j=1}^{t} \tau_j$$

Just as in the randomized block, there is no term explicitly representing Block × Treatment interaction. We assume that this interaction is 0. If, however, there is such interaction, it will increase the size of the Residual variance [that is, $\sigma^2 = \text{var}(Z_{ij})$]. The *difference* from the complete randomized block model is simply that not all possible combinations of i and j in (14.3) are present in the experiment.

The analysis of variance for this design and model can be developed by applying the general method described in Section 13.16. Here, however, we use an alternative approach.

Consider now a particular treatment level—the qth, say. The sum of all observations for this treatment level is [from (14.3)]

$$X_{.q} = \sum_{i(q)} X_{iq} = r\alpha + \sum_{i(q)} \beta_i + r\tau_q + \sum_{i(q)} Z_{iq} \tag{14.4}$$

where $\sum_{i(q)}$ denotes summation over all blocks containing the qth treatment level. Similarly the sum of all the observations in the ith block is

$$X_{i.} = \sum_{j(i)} X_{ij} = k\alpha + k\beta_i + \sum_{j(i)} \tau_j + \sum_{j(i)} Z_{ij} \tag{14.5}$$

where $\sum_{j(i)}$ denotes summation over all treatment levels included in the ith block. Summing (14.5) over all blocks containing the qth treatment level, we find

$$\sum_{i(q)} \sum_{j(i)} X_{ij} = rk\alpha + k \sum_{i(q)} \beta_i + \sum_{i(q)} \sum_{j(i)} \tau_j + \sum_{i(q)} \sum_{j(i)} Z_{ij} \tag{14.6}$$

The double summation $\sum_{i(q)}\sum_{j(i)}$ looks complicated, but it simply means "sum over all plots in blocks which contain the qth treatment level." Therefore,

$$\sum_{i(q)} \sum_{j(i)} \tau_j = r\tau_q + \lambda \sum_{j \neq q} \tau_j$$

$$= (r - \lambda)\tau_q \qquad \text{since} \qquad \sum_{j=1}^{t} \tau_j = 0$$

Using this result and subtracting (14.6) from k times (14.4), we have

$$k \sum_{i(q)} X_{iq} - \sum_{i(q)} \sum_{j(i)} X_{ij} = (kr - r + \lambda)\tau_q + k \sum_{i(q)} Z_{iq} - \sum_{i(q)} \sum_{j(i)} Z_{ij} \tag{14.7}$$

Hence (since the expected value of each Z_{ij} is 0) using (4.2), an unbiased estimator of τ_q is

$$\hat{\tau}_q = \frac{1}{\lambda t} \left[k \sum_{i(q)} X_{iq} - \sum_{i(q)} \sum_{j(i)} X_{ij} \right] \tag{14.8}$$

This can be written

$$\hat{\tau}_q = \frac{k}{\lambda t} \sum_{i(q)} \left[X_{iq} - \overline{X}_{i.} \right] = \frac{k}{\lambda t} \left[X_{.q} - \sum_{i(q)} \overline{X}_{i.} \right]$$

where $\overline{X}_{i.} = k^{-1} \sum_j X_{ij} = $ arithmetic mean of observations in the ith block and

$$X_{.q} = \text{sum of all observations for the } q\text{th treatment}$$

Another form for this estimator of τ_q is

$$\hat{\tau}_q = \frac{kr}{\lambda t} \left[\overline{X}_{.q} - r^{-1} \sum_{i(q)} \overline{X}_{i.} \right]$$

where $\overline{X}_{.q} = r^{-1}\Sigma_{i(q)}X_{iq}$ = arithmetic mean of observations at the qth treatment level. (14.8) is perhaps the form of estimator in most common use.

Adding (14.3) over all observations we find that the overall mean

$$\overline{X}_{..} = \frac{\Sigma\Sigma X_{ij}}{kb}$$

is an unbiased estimator of α. Hence if we desire an unbiased estimator of the qth treatment level we add $\overline{X}_{..}$ to the appropriate expression (14.8). Treatment mean estimates and their variances are given in Table 14.5.

TABLE 14.5*

MEANS AND VARIANCE OF MEANS

	ESTIMATED MEAN	VARIANCE OF ESTIMATE
Single treatment	$\overline{X}_{..} + \hat{\tau}_q$	$\dfrac{\sigma^2}{r}\left[\dfrac{1}{t} + \dfrac{k(t-1)^2}{(k-1)t^2}\right]$
Difference between two treatments	$\hat{\tau}_q - \hat{\tau}_{q'}$	$\sigma^2\left[\dfrac{2k}{\lambda t}\right]$

*The multipliers of σ^2 in these formulas can be obtained as the sums of squares of coefficients of the X_{ij}'s in the corresponding expressions.

We now proceed to the calculation of the items appearing in the analysis of variance table. This is presented in Table 14.6. It will be noted that the same sources of variation appear as in the analysis of a complete randomized block experiment. The Blocks sum of squares is computed from the

TABLE 14.6

ANALYSIS OF VARIANCE TABLE OF A BIB DESIGN

SOURCE OF VARIATION	SUM OF SQUARES	DEGREES OF FREEDOM
Blocks	$S_1 = k^{-1}\displaystyle\sum_{i=1}^{b} X_{i.}^2 - (bk)^{-1}X_{..}^2$	$b-1$
Treatments (adjusted)	$S_2 = \dfrac{1}{k\lambda t}\displaystyle\sum_{j=1}^{t}\left[kX_{.j} - \sum_{i(j)}X_{i.}\right]^2$	$t-1$
Residual	S_e = difference	$bk-t-b+1$
Total	$S = \Sigma\Sigma X_{ij}^2 - (bk)^{-1}X_{..}^2$	$bk-1$

arithmetic means of observations in each block in precisely the same way as for the complete randomized block design. This is, however, only a crude sum of squares for Blocks; if the right side of (14.3) is used it will be found that this sum of squares depends on the parameters τ_j (representing differences between treatment levels), as well as the Block parameters β_i. In fact, this "crude" Blocks sum of squares is used only as an ancillary quantity in the calculation (by subtraction) of the Residual sum of squares.

The number of degrees of freedom of the Residual sum of squares should also be noted.

Example 14.1 In the production of a machine part, the inner diameter (ID) of a steel tube is a critical dimension. Samples of size 10 are taken, and their average calculated. These parts can be produced from 7 different compositions and on 7 different machines. A balanced incomplete block design is used with the machines constituting the "blocks," and the compositions the "treatments." The data of Table 14.7 represent the average deviations (in millimeters) of the ID from a specified standard value.

We have $b = t = 7$; $k = r = 3$; $\lambda = 1$. The (crude) sum of squares for Machines is calculated as (see Table 14.5):

$$S_1 = \tfrac{1}{3} \sum_{i=1}^{7} X_{i.}^2 - \tfrac{1}{21}X_{..}^2 = \tfrac{1}{3}(16^2 + 18^2 + \cdots + 11^2) - \tfrac{1}{21}(141)^2 = 72.96$$

The other sums of squares are calculated as follows:

$$S_2 = \frac{1}{3 \cdot 1 \cdot 7} \sum_{j=1}^{7} \left(3 \cdot X_j - \sum_{i(j)} X_{i.}\right)^2 = \tfrac{1}{21}\left\{[3(18) - (16 + 18 + 28)]^2 \right.$$

$$+ [3(30) - (28 + 19 + 27)]^2 + \cdots + [3(11) - (18 + 19 + 11)]^2\Big\}$$

$$= 75.90$$

$$S = \sum_{i}\sum_{j} X_{ij}^2 - (bk)^{-1}X_{..}^2 = 1103 - 946.71 = 156.29$$

$$S_e = S - S_1 - S_2 = 7.43$$

Hence we have the ANOVA table as given in Table 14.8. Note that the machines mean square (uncorrected) is not tested in a BIB design. In a symmetrical balanced incomplete design (as we shall see) we can test for blocks as well as treatments. However, the Block mean squares must also be corrected.

In Table 14.8 we see that the ratio of Compositions to Residual mean square is 13.62. This is significant at the 1 percent level ($F_{6,8,0.99} = 6.37$). We can say, then, that we have strong assurance that the ID is affected by the type of composition (and other factors which may go along with this) which is used in the machined part.

TABLE 14.7
AVERAGE DEVIATIONS IN I.D. (IN MM)

COMPOSITION	MACHINES							TOTALS
	1	2	3	4	5	6	7	
A	5	4	9					18
B			12	9	9			30
C	7			6		8		21
D			7			5	3	15
E	4				6		5	15
F		10			12	9		31
G		4		4			3	11
Total	16	18	28	19	27	22	11	141

TABLE 14.8
ANOVA ON AVERAGE DEVIATIONS IN I.D.

SQUARE	SUM OF SQUARES	D.F.	MEAN SQUARE	MEAN SQUARE RATIO
Machines (crude)	72.96	6		
Compositions (adjusted)	75.90	6	12.65	13.62**
Residual	7.43	8	0.929	
Total	156.29	20		

14.3 SYMMETRICAL BALANCED INCOMPLETE BLOCK DESIGNS

If the numbers of blocks and treatment levels are equal ($b = t$), then [by (14.1)] the number of replications equals the number of plots per block ($r = k$). In such cases the design is called a symmetrical balanced incomplete block design (SBIB). In these it is possible to test for block effects in a similar way to that in which we test for treatment effects.

Examples of SBIB designs are given in Tables 14.3, 14.4a, and 14.4c. The analysis of variance for an SBIB is given in Table 14.9. It should be noted that 2 separate analyses are needed. One gives the corrected Treatments sum of squares (and crude Blocks sum of squares). The other gives

TABLE 14.9

Analysis of Variance Table for SBIB Designs

SOURCE OF VARIATION	SUM OF SQUARES	DEGREES OF FREEDOM	MEAN SQUARE	MEAN SQUARE RATIO
Blocks	$S_1 = k^{-1}\sum_{i=1}^{b} X_{i.}^2 - (bk)^{-1}X_{..}^2$			
Treatments (adjusted)	$S_2 = \dfrac{1}{k\lambda t}\sum_{j=1}^{t}\left(kX_{.j} - \sum_{i(j)} X_{i.}\right)^2$	$t-1$	$\dfrac{S_2}{t-1}$	$\dfrac{S_2(bk - b - t + 1)}{S_e(t - 1)}$
Treatments	$S_3 = r^{-1}\sum_{j=1}^{t} X_{.j}^2 - (bk)^{-1}X_{..}^2$			
Blocks (adjusted)	$S_4 = \dfrac{1}{k\lambda t}\sum_{i=1}^{b}\left(rX_{i.} - \sum_{j(i)} X_{.j}\right)^2$	$b-1$	$\dfrac{S_4}{b-1}$	$\dfrac{S_4(bk - b - t + 1)}{S_e(b - 1)}$
Residual	$\begin{aligned}S_e &= S - (S_1 + S_2)\\ &= S - (S_3 + S_4)\end{aligned}$	$bk - b - t + 1$	$\dfrac{S_e}{bk - b - t + 1}$	
Total	$S = \sum\sum x_{ij}^2 - (bk)^{-1}X_{..}^2$	$bk - 1$		

(Note that $S_1 + S_2 = S_3 + S_4$.)

723

the corrected Blocks sum of squares (and crude Treatments sum of squares). These sums of squares are related by the formula

(corrected Blocks) + (crude Treatments)

$$= (\text{corrected Treatments}) + (\text{crude Blocks})$$

Example 14.2 The problem of Example 14.1 concerns both compositions (treatments) and machines (blocks). Since this is a SBIB design, it is also possible to test for differences among blocks, if we so desire. Suppose we also are interested in possible machine differences. We then use the formulas of Table 14.9 for the data as given in Table 14.7. We calculate then, in addition to S_1, S_2, and S (which we already have)

$$S_3 = \tfrac{1}{3} \sum_{j=1}^{7} X_j^2 - \tfrac{1}{21} X_{..}^2 = \tfrac{1}{3}(18^2 + 30^2 + \cdots + 11^2) - 946.71 = 118.96$$

$$S_4 = \frac{1}{3 \cdot 1 \cdot 7} \sum_{i=1}^{7} \left(3 \cdot X_{i.} - \sum_{j(i)} X_j \right)^2 = \tfrac{1}{21} \left\{ [3 \cdot (16) - (18 + 21 + 15)]^2 \right.$$

$$+ [3 \cdot (18) - (18 + 31 + 11)]^2 \cdots + [3 \cdot (11) - (15 + 15 + 11)]^2 \right\} = 29.90$$

$$S_e = S - S_3 - S_4 = 7.43$$

Note that $S_1 + S_2 = S_3 + S_4$. Hence S_4 could have been calculated by differencing. Carrying out the added computation, however, will give us a check on the numerical results. The ANOVA table is presented as Table 14.10.

TABLE 14.10

ANOVA on Average Deviation
(SBIB Design)

SOURCE	SUM OF SQUARES	D.F.	MEAN SQUARE	MEAN SQUARE RATIO
Machines (crude)	72.96			
Compositions (adjusted)	75.90	6	12.65	13.62**
Compositions (crude)	118.96			
Machines (adjusted)	29.90	6	4.98	5.36*
Residual	7.43	8	0.929	
Total	156.29	20		

The ratio of Machine (Block) mean square to Residual mean square is equal to 5.36. This is beyond the 5 percent critical value $F_{6,8,0.95}(=3.58)$. Hence we can say that there is some evidence of machine effect.

Tables 14.11 (*a* and *b*) present the "estimated means" for machines (Table 14.11*a*) and for compositions (Table 14.11*b*) as well as the standard deviations of each mean and of the difference between means. Note that the "estimated means" differ from the simple arithmetic means of observed values for the corresponding machine or composition. The latter "crude means" for machines $1, 2, 3, \ldots$ would be $5.3, 6.0, 9.3, \ldots$, for example.

TABLE 14.11(*a*)

ESTIMATED MACHINE MEANS (MM)

MACHINE

	1	2	3	4	5	6	7
Estimated mean	$5\frac{6}{7}$	$5\frac{6}{7}$	$9\frac{5}{7}$	6	$7\frac{3}{7}$	$6\frac{4}{7}$	$5\frac{4}{7}$

Estimated standard deviation (mm) of each estimated mean: 0.62; of difference between means: 0.89.

TABLE 14.11(*b*)

ESTIMATED COMPOSITION MEANS (MM)

COMPOSITION

	A	B	C	D	E	F	G
Estimated mean	$5\frac{4}{7}$	9	$7\frac{4}{7}$	$4\frac{3}{7}$	$5\frac{3}{7}$	$10\frac{3}{7}$	$4\frac{4}{7}$

Estimated standard deviation (mm) of each estimated mean: 0.62; of difference between means: 0.89.

14.4   LATIN SQUARES

A design which is used in many problems of applied statistics is the Latin square. This has been referred to, briefly, in Section 14.1. Three factors can be examined in the same experiment. The existence of differences between levels can be tested separately for each of the 3 factors. By the very thorough balance of the design, the 3 sums of squares for factor level differences, and the Residual sum of squares against which each can be tested in turn, can be formed into a single analysis of variance table with the 4 sums of squares adding up to the Total sum of squares of the whole set of observations.

Latin squares first found extensive use in agricultural experimentation. Two of the 3 factors simply defined position in a 2-dimensional coordinate system representing location on a field. The third would be a "treatment" factor (representing actual treatments, or perhaps different varieties). The experimental design enabled variation of average yield (for example) with changes in treatment level to be examined, substantially reducing the unwanted effects of variation in natural fertility from one position to another in the field (represented by the other 2 factors).

This type of design soon found application in other kinds of work, and particularly in industrial investigations. In such applications it frequently occurs that 2 or sometimes all 3 factors are of interest in themselves, and not introduced merely to allow for effects which we would be well pleased to have eliminated altogether.

The Latin square design may be represented as in Table 14.12. Here 2 of the factors are represented by the rows and columns of a square checkerboard. Each row or column corresponds to one level of the appropriate factor. The third factor is indicated by the letters appearing in the squares of the checkerboard. These are so arranged that, in each row, each letter appears just once, and also in each column, each letter appears just once. If there are n rows and n columns, then there are n^2 squares and each square indicates by the combination row, column, and letter the levels of the first, second, and third factors to be combined. There are thus n^2 different factor level combinations to be used, as compared with n^3 possible such combinations—a very substantial saving.

TABLE 14.12
A 4 × 4 LATIN SQUARE DESIGN

	FACTOR II			
FACTOR I	1	2	3	4
1	A	B	C	D
2	D	A	B	C
3	C	D	A	B
4	B	C	D	A

It seems that we have reduced our experimental effort to $(1/n)$th without losing the possibility of separate evaluation of the effects of changes in level of each factor. When we come to construct the model to be used as a basis for the analysis of the data, however, we can see the price that we do, in fact, pay.

Let $X_{(ijk)}$ denote the observed value corresponding to the ith row, jth column, and kth letter. Evidently i, j, and k each take one of the values $1, 2, \ldots, n$. The subscripts are placed in brackets—thus (ijk)—because only 2 of them are needed to identify the observation. For example, in Table 14.12 if $i = 2$ and $k = 3$ (corresponding to C) then $j = 4$ because C appears in the fourth column of row 2.

The mathematical model for the Latin square is

$$X_{(ijk)} = \alpha + \rho_i + \kappa_j + \tau_k + Z_{(ijk)} \tag{14.9}$$

where

$$\sum_{i=1}^{n} \rho_i = \sum_{j=1}^{n} \kappa_j = \sum_{k=1}^{n} \tau_k = 0$$

These conditions signify that $(\alpha + \rho_i)$, $(\alpha + \kappa_j)$, and $(\alpha + \tau_k)$ represent the mean of the expected values for the ith row, the jth column, and the kth letter, respectively.

This is a parametric model. In this design the analysis is the same for a component-of-variance (or mixed) model as for a parametric model.

It will be noted that no interaction terms appear in the model (14.9). This is because any general interaction term—for example, $(\rho\kappa)_{ij}$—could not be separated from $Z_{(ijk)}$, and so no Residual sum of squares, not depending on any parameter values, could be obtained. (The reader will appreciate that the pair of subscripts ij is equivalent to the triplet (ijk), as explained above.)

So the price that we have to pay for the reduction (to $1/n$th) of experimental effort is ignorance of interactions. If real interactions are present they swell the size of σ (the standard deviation of each $Z_{(ijk)}$), but the analysis remains valid in the sense that

(i) levels of significance can be assigned accurately, and

(ii) each test really does test for effects of changes in the appropriate factor level.

But the increased σ means a loss of sensitivity (or power). Some chemical engineers are somewhat disturbed about the use of the Latin square because they feel that there are very few, if any, problems in their cognizance which do not involve interactions. However, it should be recognized that average effects of factor level changes (the ρ_i's, κ_j's, and τ_k's of model (14.9)) can be estimated and compared, by using a Latin square design, even if some features of the problem must perforce be neglected. (See also Exercise 33.)

There is a further special feature of the Latin square design which is sometimes felt to be a drawback. This is that all 3 factors must appear at the *same* number of levels (*n* in our general notation). This effect can be reduced by using "dummy" levels which are in fact repetitions of other levels. Table 14.13 shows one such design with 2 factors (rows and columns) each at 4 levels, and the third (letters) at 3 levels. The level *A* is repeated. (The design is formed from Table 14.12 by replacing *D* by *A*.) Usually the factor level (or levels) to be repeated will be chosen so that levels of more interest are repeated more frequently. The analysis of such modified designs presents some special features which will be discussed later.

TABLE 14.13

A 4 × 4 LATIN SQUARE WITH
ONE DUMMY LEVEL

	FACTOR II			
FACTOR I	1	2	3	4
1	*A*	*B*	*C*	*A*
2	*A*	*A*	*B*	*C*
3	*C*	*A*	*A*	*B*
4	*B*	*C*	*A*	*A*

We see, in the next chapter, that a Latin square design can be a part (or "unit") of a larger experiment.

14.4.1   Construction and Analysis of Latin Square Designs

A number of Latin square designs are listed by Cochran and Cox [14.3] and by Fisher and Yates [14.8]. Usually a "standard" square is selected and the correspondence of levels to rows, columns, and letters assigned at random.

Table 14.14 gives another example of a Latin square. This is a 6×6 Latin square in which the letters originally followed a cyclic pattern as in Table 14.12, but subsequently rows and columns have been randomly permuted.

The analysis of variance for an *n* × *n* Latin square is shown in Table 14.15. It is almost self-explanatory. The calculations for the Factor I, Factor II, and Factor III sums of squares are exactly parallel to each other. (In fact, the 3 factors enter symmetrically into the design: in any Latin square design the factors represented by rows, columns, and letters can be permuted in any order, still producing Latin squares.) The Residual sum of squares is obtained by subtraction.

TABLE 14.14
A 6 × 6 Latin Square Design

FACTOR II

FACTOR I	1	2	3	4	5	6
1	F	D	B	A	C	E
2	A	E	C	B	D	F
3	B	F	D	C	E	A
4	D	B	F	E	A	C
5	C	A	E	D	F	B
6	E	C	A	F	B	D

TABLE 14.15
Analysis of Variance Table for a Latin Square

SOURCE	SUM OF SQUARES	D.F.	MEAN SQUARE	MEAN SQUARE RATIO
Factor I	$S_1 = n^{-1} \sum_i X^2_{(i..)} - X^2_{(...)}/n^2$	$n-1$	$S_1/(n-1)$	$\dfrac{S_1(n-2)}{S_e}$
Factor II	$S_2 = n^{-1} \sum_j X^2_{(.j.)} - X^2_{(...)}/n^2$	$n-1$	$S_2/(n-1)$	$\dfrac{S_2(n-2)}{S_e}$
Factor III	$S_3 = n^{-1} \sum_k X^2_{(..k)} - X^2_{(...)}/n^2$	$n-1$	$S_3/(n-1)$	$\dfrac{S_3(n-2)}{S_e}$
Residual	$S_e = \text{Difference}$	$(n-1)(n-2)$	$S_e/(n-1)(n-2)$	
Total	$S = \sum_i \sum_j x^2_{(ijk)} - X^2_{(...)}/n^2$	n^2-1		

Example 14.3 A particular missile alternator design is made up of 3 separate power generating sections, each considered independent of the other. The alternator is driven by a turbine which is powered by hot gas supplied from a solid grain gas generator. The parasitic section of the alternator supplies power to a dummy electrical load as required in order to maintain alternator speed at a constant value of 24,000 rpm. The parasitic section of the alternator is comprised of a 4-pole stator, 6-pole rotor, and a shaft. The rotor turns concentrically within the stator bore, while the stator is held fixed within the housing. The stator is wound with both DC and AC turns of fixed wire size. AC output voltage is a function of DC input current and AC turns. The rotor is stacked from individual laminations punched from 0.004-in.-thick stock. Laminations are coated for insulation purposes.

The purpose of the experiment was to determine which factors were most closely associated with performance and what levels of these factors gave best performances. A 5×5 Latin square experiment was designed with the factors and levels as follows:

1. Number of AC turns for the stators. The levels were at 145, 150, 155, 160, and 165 AC turns.
2. Number of laminations per stack for the rotors. The levels were 230, 240, 250, 260, and 270.
3. Quality (visual) of lamination coatings. The five levels were on an arbitrary scale with A the best and E the worst.

A conventional alternator was built up for test purposes. The unit was assembled and disassembled as necessary to test components and follow the Latin square design. A random testing order was established. The background of the test conditions were controlled as rigidly as possible. The character observed was the maximum parasitic AC output voltage. The data are given in Table 14.16. These data were coded by subtracting 300 from each yield; the coded data are presented in Table 14.17.

TABLE 14.16
OUTPUT VOLTAGES OF MISSILE ALTERNATORS

	STATORS				
ROTORS	145	150	155	160	165
230	310 C	312 B	320 A	306 D	300 E
240	309 D	310 C	324 B	300 E	305 A
250	312 B	303 E	325 C	307 A	302 D
260	316 A	306 D	318 E	304 C	294 B
270	314 E	308 A	323 D	309 B	303 C

TABLE 14.17
CODED DATA OF OUTPUT VOLTAGES

	STATORS					
ROTORS	145	150	155	160	165	TOTALS
230	10 C	12 B	20 A	6 D	0 E	48
240	9 D	10 C	24 B	0 E	5 A	48
250	12 B	3 E	25 C	7 A	2 D	49
260	16 A	6 D	18 E	4 C	−6 B	38
270	14 E	8 A	23 D	9 B	3 C	57
Totals	61	39	110	26	4	240
Coating Quality	A	B	C	D	E	
Totals	56	51	52	46	35	

As an example of the calculation, consider the sum of squares for stators. This is

$$\frac{\sum\limits_{i} X^2_{(i..)}}{n} - \frac{X^2_{(...)}}{n^2} = \tfrac{1}{5}[61^2 + 39^2 + 110^2 + 26^2 + 4^2] - \tfrac{240^2}{25}$$

$$= 3606.8 - 2304.0 = 1302.8$$

[Note that we have interchanged Factors I and II of (14.9). This does not affect the structure of the design, which is symmetrical in all 3 factors.]

As a result of the analysis we can say that the number of turns on the stator is a highly significant factor. *We could see this quite clearly once the data were coded.* However, upon looking at the total of Table 14.17 one might suspect a significant effect due to the quality level of the coatings. Its mean square ratio is 13.10/12.03 $= 1.09 < F_{4,12,0.95}(= 3.26)$. Hence we do not consider the coating quality, within this range, to be a factor affecting performance.

Although one can test this independently by a significance test, it is clear that the maximum of the 5 preselected number of turns for the stators is achieved at about 155 turns. The estimate of the mean output voltage at this level is 322. Further, the estimate of the residual variance is 12.03. Hence $S = 3.47$ V.

One further point of interest is partitioning the factor "stators" into linear, quadratic, cubic, and quartic effects. Applying orthogonal polynomials in the way described in Section 13.15 the terms in the analysis of variance are obtained as shown below Table 14.18.

TABLE 14.18

ANALYSIS OF VARIANCE TABLE FOR CODED OUTPUT VOLTAGES

SOURCE	SUM OF SQUARES	DEGREES OF FREEDOM	MEAN SQUARE	MEAN SQUARE RATIO
Stators	1302.8	4	325.7	27.07***
Rotors	36.4	4	9.10	0.76
Coating Quality	52.4	4	13.10	1.09
Residual	144.4	12	12.03	
Total	1536.0	24		

SOURCE	SUM OF SQUARES	D.F.
Linear	322.58	1
Quadratic	343.21	1
Cubic	19.22	1
Quartic	617.79	1
Total stator effect	1302.80	4

If we were to insert these in Table 14.18, we would find that each of the linear, quadratic, and quartic components are significant. The purely quartic effect is somewhat meaningless, since it really includes "quartic and higher." The "higher" cannot be tested since there are only 4 degrees of freedom with which to work. The conclusion then is that a curvilinear regression on number of turns for the stators could be used in this range. However, if it is desired to use a polynomial regression the degree of the polynomial is not well fixed. For the present it may well be better just to aim at 155 turns for the maximum output voltage.

A table of means and 95 percent confidence intervals on each level of the stator effect is given as Table 14.19. These confidence limits were calculated from the formula

$$\overline{X}_{(i..)} \pm t_{12,0.975} \sqrt{\tfrac{1}{5}(\text{Residual mean square})}$$

where $\overline{X}_{(i..)}$ is the mean of the 5 observations for the ith stator level.

TABLE 14.19

MEANS AND 95 PER CENT CONFIDENCE INTERVALS ON OUTPUT VOLTAGES

STATORS	LOWER CONFIDENCE LIMIT	MEAN	UPPER CONFIDENCE LIMIT
145	308.8	312.2	315.6
150	304.4	307.8	311.2
155	318.6	322.0	325.4
160	301.8	305.2	308.6
165	297.4	300.8	304.2

14.4.2 Graeco-Latin Squares and Hypersquares

With the Latin square design, 3 factors can be studied simultaneously and separately. In most cases a fourth factor can be included in the study, symmetrically with the other 3 factors, by superposing on the Latin square a further Latin square of the same size ($n \times n$) which is "orthogonal" to the original Latin square. "Orthogonal" in this context means that each letter in one Latin square appears in the same position as each letter of the other Latin square *just once*. Usually Greek letters are used in the second Latin square. (This can be regarded as the basis for the name "Graeco-Latin.") Three examples of Graeco-Latin squares appear in Table 14.20. These include 3×3, 4×4, and 7×7 squares.

It is not always possible to construct a Graeco-Latin square. For a long time it was thought that no $(4n+2) \times (4n+2)$ Graeco-Latin squares existed, but in 1959, 10×10 and 22×22 Graeco-Latin squares were constructed, and it has been shown that $n \times n$ Graeco-Latin squares can be constructed provided n is not equal to 2 or 6 [14.1].

The analysis of variance for a Graeco-Latin square design (see Table 14.21) is very similar to that for a Latin square. It is, in fact, the latter analysis with a further "Factor IV" sum of squares, and, of course, with

TABLE 14.20
GRAECO-LATIN SQUARE DESIGNS

FACTOR I	FACTOR II		
	1	2	3
1	$A\ \alpha$	$B\ \gamma$	$C\ \beta$
2	$C\ \gamma$	$A\ \beta$	$B\ \alpha$
3	$B\ \beta$	$C\ \alpha$	$A\ \gamma$

(a) 3 × 3 square

FACTOR I	FACTOR II			
	1	2	3	4
1	$A\ \delta$	$D\ \beta$	$B\ \alpha$	$C\ \gamma$
2	$B\ \gamma$	$C\ \alpha$	$A\ \beta$	$D\ \delta$
3	$C\ \beta$	$B\ \delta$	$D\ \gamma$	$A\ \alpha$
4	$D\ \alpha$	$A\ \gamma$	$C\ \delta$	$B\ \beta$

(b) 4 × 4 square

FACTOR I	FACTOR II						
	1	2	3	4	5	6	7
1	$C\ \zeta$	$A\ \alpha$	$F\ \eta$	$B\ \gamma$	$G\ \delta$	$E\ \epsilon$	$D\ \beta$
2	$A\ \gamma$	$C\ \beta$	$G\ \alpha$	$E\ \eta$	$D\ \epsilon$	$B\ \zeta$	$F\ \delta$
3	$B\ \delta$	$D\ \gamma$	$E\ \zeta$	$G\ \epsilon$	$C\ \alpha$	$F\ \beta$	$A\ \eta$
4	$D\ \eta$	$E\ \delta$	$B\ \beta$	$F\ \alpha$	$A\ \zeta$	$G\ \gamma$	$C\ \epsilon$
5	$E\ \alpha$	$B\ \epsilon$	$D\ \delta$	$A\ \beta$	$F\ \gamma$	$C\ \eta$	$G\ \zeta$
6	$G\ \beta$	$F\ \zeta$	$A\ \epsilon$	$C\ \delta$	$B\ \eta$	$D\ \alpha$	$E\ \gamma$
7	$F\ \epsilon$	$G\ \eta$	$C\ \gamma$	$D\ \zeta$	$E\ \beta$	$A\ \delta$	$B\ \alpha$

(c) 7 × 7 square

the degrees of freedom of the Residual sum of squares reduced by $(n-1)$. It should be noted that if $n=3$, the number of degrees of freedom for the Residual sum of squares is *zero*. (In fact the sum of squares can be seen, by algebraic analysis, to be exactly equal to 0.) Therefore, although 3×3 Graeco-Latin squares can be constructed, they are not very useful, except perhaps as parts of a larger experiment or for a qualitative comparison of the four factors.

The mathematical model is a natural extension of the model for a Latin square design. It is

$$X_{(ijkl)} = \alpha + \rho_i + \kappa_j + \lambda_k + \gamma_l + Z_{(ijkl)} \qquad (14.10)$$

TABLE 14.21
ANALYSIS OF VARIANCE TABLE FOR A GRAECO-LATIN SQUARE

SOURCE	SUM OF SQUARES	DEGREES OF FREEDOM	MEAN SQUARE	MEAN SQUARE RATIO
Factor I (Rows)	$S_1 = \dfrac{\sum_i X^2_{(i\ldots)}}{n} - \dfrac{X^2_{(\ldots)}}{n^2}$	$n-1$	$\dfrac{S_1}{n-1}$	$\dfrac{S_1(n-3)}{S_e}$
Factor II (Columns)	$S_2 = \dfrac{\sum_j X^2_{(.j..)}}{n} - \dfrac{X^2_{(\ldots)}}{n^2}$	$n-1$	$\dfrac{S_2}{n-1}$	$\dfrac{S_2(n-3)}{S_e}$
Factor III (Latin letters)	$S_3 = \dfrac{\sum_k X^2_{(..k.)}}{n} - \dfrac{X^2_{(\ldots)}}{n^2}$	$n-1$	$\dfrac{S_3}{n-1}$	$\dfrac{S_3(n-3)}{S_e}$
Factor IV (Greek letters)	$S_4 = \dfrac{\sum_l X^2_{(...l)}}{n} - \dfrac{X^2_{(\ldots)}}{n^2}$	$n-1$	$\dfrac{S_4}{n-1}$	$\dfrac{S_4(n-3)}{S_e}$
Residual	$S_e = $ difference	$(n-1)(n-3)$	$\dfrac{S_e}{(n-1)(n-3)}$	
Total	$S = \sum_i \sum_j x^2_{(ijkl)} - \dfrac{X^2_{(\ldots)}}{n^2}$	$n^2 - 1$		

where λ_k and γ_l are now used for the factors represented by Latin and Greek letters, respectively, and

$$\sum_{i=1}^{n} \rho_i = \sum_{j=1}^{n} \kappa_j = \sum_{k=1}^{n} \lambda_k = \sum_{l=1}^{n} \gamma_l = 0 \qquad (i,j,k,l = 1,2,\ldots,n)$$

In this design any 2 of the 4 subscripts ($ijkl$) suffice to define an observation. There are many more interactions neglected than in the Latin square design and the effects of this (increasing the residual standard deviation) are more often to be expected with Graeco-Latin than with Latin square designs.

Example 14.4 As an illustration of a Graeco-Latin square, consider the following application in the semiconductor industry. [Taken from "Applications of Analytical Techniques and Experimental Designs to the Semiconductor Industry," R. W. Anderson, *Fourteenth Midwest Quality Control Conference* (1959).]

"The purpose of this experiment was to determine what effect the following variables had on etch rate: (1) discolored nitric acid, (2) volume of etch, (3) size of bars, and (4) time in the etchant. It was feared that discoloration was a function of decomposition of the nitric acid and that the more discolored the acid the more dilute the concentration of nitric became in the etch solution. If the etch were diluted as a result of the discolored nitric, then the length of time necessary to remove a given amount of material would be an important variable. Another method of controlling etch rate might be to control the bar size very closely. Therefore, the effect of bar size on etch rate was of interest. Lastly, we wanted to determine if the volume of etch used had any influence on etch rate. These were the variables of prime interest.

The procedure was as follows: five bottles of fresh colorless nitric acid were exposed to varying degrees of bright sunlight until a range in discoloration from clear to brilliant yellow was obtained. These acids were then used in making up the necessary etch solution. Five groups of silicon bars were then carefully sorted by weight. Five different etch volumes and 5 different etch times were used. The order of testing was randomized.

As an index of etch strength, weight loss in each silicon bar was recorded following the etching, rinsing, and drying operations. If the discoloration of acid or any of the other variables had an effect on the etch rate, then it was felt that weight loss would be a good index of the influence attributable to each variable."

The data are given in Table 14.22.

TABLE 14.22
WEIGHT LOSS OF SILICON BARS (IN MILLIGRAMS × 10⁻²)

COLOR (ACID)

VOLUME	1		2		3		4		5	
1	65		82		108		101		126	
	A	α	B	γ	C	ϵ	D	β	E	δ
2	84		109		73		97		83	
	B	β	C	δ	D	α	E	γ	A	ϵ
3	105		129		89		89		52	
	C	γ	D	ϵ	E	β	A	δ	B	α
4	119		72		76		117		84	
	D	δ	E	α	A	γ	B	ϵ	C	β
5	97		59		94		78		106	
	E	ϵ	A	β	B	δ	C	α	D	γ

The model equation is

$$X_{(ijkl)} = \alpha + \kappa_i + \nu_j + \lambda_k + \tau_l + Z_{(ijkl)} \text{ with } i,j,k,l = 1,2,\ldots,5$$

κ_i designates the ith color (or level of discolored nitric acid), ν_j, the jth volume of etch, λ_k (at levels A, B, C, D, E), the kth size bar, and τ_l (at levels $\alpha, \beta, \gamma, \delta, \epsilon$), the lth time in the etchant.

Following the formulas of the analysis of variance table, we obtain Table 14.23.

TABLE 14.23

ANALYSIS OF VARIANCE TABLE WEIGHT LOSS OF SILICON BARS

SOURCE	SUM OF SQUARES	D.F.	MEAN SQUARE	MEAN SQUARE RATIO
Color (Acid)(C)	227.76	4	56.94	0.47
Volume (V)	285.76	4	71.44	0.59
Size (S)	2867.76	4	716.94	5.96*
Time (T)	5536.56	4	1384.14	11.50**
Residual	962.72	8	120.34	
Total	9880.56	24		

As a typical calculation, consider the sum of squares due to size:

$$S_3 = \left(\frac{372^2 + 429^2 + 484^2 + 528^2 + 481^2}{5} \right) - \left(\frac{2294^2}{25} \right)$$

$$= 2867.76$$

If in Table 14.23 we compute the ratios of the mean squares for Size and for Time to that of the Residual, we obtain 5.96 and 11.50, respectively. Critical values are $F_{4,8,0.95} = 3.84$ and $F_{4,8,0.99} = 7.01$.

We conclude then that the discoloration (C) and volume (V) of the acid do not have a significant effect on the weight loss. However, the size (L) of the bar and the time (T) in the etchant do affect loss of weight. The fact that the 2 latter factors had an important effect was expected. But the failure of the decomposition (discoloration) and the volume factors to show up significantly were important discoveries. With respect to discoloration, further experiments were conducted to ensure that harmful effects in processing were not introduced by using acids that were discolored. As a result of the complete experiment (4 replicates were taken in the original problem), several lots of acid were accepted at times when their rejection in incoming inspection would have meant shutting down production lines. Since volume was not important, smaller amounts of etchant could be used, resulting in sizable savings in cost of etchant.

The concept of orthogonality (described earlier in this section) can be further extended, leading to the construction of hyper-squares—or, more precisely, hyper-Graeco-Latin squares. For squares of sizes for which hyper-squares can be formed, the maximum number of factors that can be handled is $n + 1$ (where $n \times n$ is the size of the square). If $n + 1$ different factors are considered, the total number of degrees of freedom, $n^2 - 1 = (n + 1)(n - 1)$ is completely utilized in the Factors I to $(n + 1)$ sums of squares, and there is no Residual sum of squares. (The case $n = 3$ has already been mentioned.) In such a case it is not possible to carry out formal tests of significance, but relative magnitudes of each factor's effect can be judged from the appropriate mean squares (or, more simply, from tables of means). It is useful to notice that a verdict of significance on comparing mean squares for 2 factors *can* be relied upon as indicating a real effect for the factor with the larger mean square.

Example 14.5 gives an illustration of a hyper-square utilized in a problem in structural plastics testing.

Example 14.5   In the field of structural plastics a great deal of testing is involved in an attempt to produce the strongest laminates from the known resins. An experimental design was constructed at the General Electric Company to test 5 different factors. [T. J. Gair, P. E. Thompson, and M. H. Monahan, "Statistics Shortens Structural Plastics Testing," *Plastics World* (May, 1956).] The character observed was flexural strength. The test pieces were strips of 4 by 1 in. These were to be tested on a Tinius Olsen Testing Machine using the 1000-lb scale. Using information available prior to the formulation of the experimental design, it was decided to consider the 4 levels of the 5 variables as given in Table 14.24. The design is shown in Table 14.25. Note that all the degrees of freedom are utilized. To carry out the analysis of variance an independent estimate of the residual variance (σ^2) must be determined. Though this design makes it possible to investigate the

TABLE 14.24

FACTORS IN STRUCTURAL PLASTICS TESTING

VARIABLE	LEVELS	CODE DESIGNATOR
Number of plies of laminates	11	Factor I 1
	12	2
	13	3
	14	4
Orientation of glass	Parallel lamination	Factor II 1
	Face to face	2
	Alt. 0°—90°	3
	Face to face and 90°	4
Press time	1 hr	A
	$\frac{3}{4}$ hr	B
	$\frac{1}{2}$ hr	C
	$\frac{1}{4}$ hr	D
Press temperature	140°	α
	150°	β
	160°	γ
	180°	δ
After-bake time	8 hrs	a
	16 hrs	b
	24 hrs	c
	32 hrs	d

TABLE 14.25
4 × 4 HYPERSQUARE WITH 5 FACTORS

FACTOR II

FACTOR I	1	2	3	4
1	A α a	B β b	C γ c	D δ d
2	D γ b	C δ a	B α d	A β c
3	C β d	D α c	A δ b	B γ a
4	B δ c	A γ d	D β a	C α b

effects of variation in many factors, it is not possible (from the experimental data alone) to estimate the accuracy with which these effects are measured. Also it is unlikely that all the ignored interactions will, in fact, be negligible.

14.5   YOUDEN SQUARES

If each of 3 factors has the same number of levels it is possible to use a Latin square design. If this is not so, it may still be possible to use the Latin square design by repeating certain levels, treating the repeated levels *formally* as different levels. Very often, however, the number of "dummy" levels so introduced may be excessive or it may be actually impossible (for lack of space and time) to use a full Latin square design. In such cases certain special kinds of incomplete Latin squares, suggested by W. J. Youden, and called *Youden squares*, may be useful. These are essentially Latin squares with a number of rows missing. However, they are further specialized by the requirement that, regarding the columns as blocks, we have a BIB design. Thus although

A	B	C	D		A	B	C	D
B	C	D	A	is part of the Latin square	B	C	D	A
					C	D	A	B
					D	A	B	C

it is not a Youden square because, for example, A appears once in the same column as B and D, but not at all with C.

Lists of Youden squares are given in Fisher and Yates [14.8] and Cochran and Cox [14.3].

Three examples of Youden squares are given in Table 14.26. In (a) we have a 7 (rows or blocks)×3 (columns) square; (b) is 7×4, and (c) is 11×5. We will suppose Factor I to be represented by rows, Factor II by columns, and Factor III by letters. Then, necessarily, Factors I and III have the same number of levels ($t = b$ in the BIB notation). In (a), Factors I and III have 7 levels each, but Factor II has only 3 levels. If (as in many agricultural applications) "rows" and "columns" represent real "location" factors, then the number of levels of the "treatment" factor must equal the larger of the number of levels of the 2 location factors.

Youden squares can be regarded as incomplete Latin squares (of a special kind) or as BIB designs with the special property that a third factor (Factor II in our notation) can be allowed for in the analysis.

TABLE 14.26
YOUDEN SQUARES

FACTOR I	FACTOR II		
	1	2	3
1	G	A	C
2	A	B	D
3	B	C	E
4	C	D	F
5	D	E	G
6	E	F	A
7	F	G	B

(a) 7 × 3 Square

FACTOR I	FACTOR II				
	1	2	3	4	5
1	A	B	C	D	E
2	G	A	F	J	C
3	I	H	A	F	B
4	K	I	G	A	D
5	J	K	E	H	A
6	H	G	B	C	K
7	B	F	D	K	J
8	F	C	K	E	I
9	C	D	J	I	H
10	E	J	I	B	G
11	D	E	H	G	F

(c) 11 × 5 Square

FACTOR I	FACTOR II			
	1	2	3	4
1	D	F	G	A
2	E	G	A	B
3	F	A	B	C
4	G	B	C	D
5	A	C	D	E
6	B	D	E	F
7	C	E	F	G

(b) 7 × 4 Square

TABLE 14.27
SBIB DESIGN AND YOUDEN SQUARE (7 × 4)
[FACTOR II IN PARENTHESES]

	FACTOR I						
TREATMENTS	1	2	3	4	5	6	7
A	(4)	(3)	(2)		(1)		
B		(4)	(3)	(2)		(1)	
C			(4)	(3)	(2)		(1)
D	(1)			(4)	(3)	(2)	
E		(1)			(4)	(3)	(2)
F	(2)		(1)			(4)	(3)
G	(3)	(2)		(1)			(4)

In Table 14.27 the Youden square (b) of Table 14.26 is set out as a BIB design. In this BIB design $t = 7$, $r = 4$, $b = 7$, $k = 4$, and $\lambda = 2$. (Factor II thus has 4 levels.) The analysis of variance is just the same as that for the BIB design, with the modification that a sum of squares "(Between levels of) Factor II" is calculated in a straightforward fashion, and the Residual sum of squares is reduced by this amount. The number of degrees of freedom for Factor II is, of course, $k - 1$, and the number of degrees of freedom for the Residual is reduced by $(k - 1)$, as compared with standard BIB analysis. The analysis of variance for a Youden square, using the standard symbols of a BIB design, is given in Table 14.28. (Recall that in a Youden square we must have $b = t$ and $k = r$.) The model is

$$X_{(ijl)} = \alpha + \beta_i + \tau_j + \gamma_l + Z_{(ijl)} \qquad (14.11)$$

where $i = 1, 2, \ldots, b$; $j = 1, 2, \ldots, t (= b)$; $l = 1, 2, \ldots, k$ $(< t)$ and $\Sigma_{i=1}^{b} \beta_i = \Sigma_{j=1}^{t} \tau_j = \Sigma_{l=1}^{k} \gamma_l = 0$. Note that as in the Latin square the triple subscript (ijl) is enclosed in brackets, to remind us that the appropriate observation is, in fact, uniquely defined by any 2 out of the 3 subscripts. Note that this model is, in fact, of exactly the same structure as the model (14.9) for the Latin square.

Example 14.6 In the determination of octane number of gasoline a particular method concerns the use of a base gasoline and 6 additives as candidates for inclusion in a new blend. The test is run in a Youden square of size 7×3. Each fuel is given 2 min in the engine and the result is recorded on the "knockmeter." The knockmeter is read at 60, 90, and 120 sec to check the stability. A marked shift between 90 and 120 sec is cause for alarm. The data for 7 gasolines are given in Table 14.29. The order of running the test is left to right on the top row (Block 1), then left to right on the second row (Block 2), etc. "Blocks" are groups of 3 2-min

TABLE 14.28

ANALYSIS OF VARIANCE TABLE FOR A YOUDEN SQUARE

SOURCE OF VARIATION	SUM OF SQUARES	DEGREES OF FREEDOM	MEAN SQUARE	MEAN SQUARE RATIO
Blocks (crude)	$S_1 = k^{-1} \sum\limits_{i=1}^{b} X_{(i..)}^2 - (bk)^{-1} X_{(...)}^2$			
Treatments (adjusted)	$S_2 = \dfrac{1}{k\lambda t} \sum\limits_{j=1}^{t} \left(k X_{(j.)} - \sum\limits_{i(j)} X_{(i..)} \right)^2$	$t - 1$	$\dfrac{S_2}{t-1}$	$\dfrac{S_2(bk - t - b - k + 2)}{S_e(t-1)}$
Treatments (crude)	$S_3 = r^{-1} \sum\limits_{j=1}^{t} X_{(.j.)}^2 - (tr)^{-1} X_{(...)}^2$			
Blocks (adjusted)	$S_4 = \dfrac{1}{k\lambda t} \sum\limits_{i=1}^{b} \left(r X_{(i..)} - \sum\limits_{j(i)} X_{(.j.)} \right)^2$	$b - 1$	$\dfrac{S_4}{b-1}$	$\dfrac{S_4(bk - t - b - k + 2)}{S_e(b-1)}$
Factor II	$S_5 = b^{-1} \sum\limits_{l=1}^{k} X_{(..l)}^2 - (bk)^{-1} X_{(...)}^2$	$k - 1$	$\dfrac{S_5}{k-1}$	$\dfrac{S_5(bk - t - b - k + 2)}{S_e(k-1)}$
Residual	$\begin{aligned} S_e &= S - (S_1 + S_2 + S_5) \\ &= S - (S_3 + S_4 + S_5) \end{aligned}$	$bk - t - b - k + 2$	$\dfrac{S_e}{N - t - b - k + 2}$	
Total	$S = \sum\sum\sum X_{(ijl)}^2 - (bk)^{-1} X_{(...)}^2$	$bk - 1$		

(Note that $S_1 + S_2 = S_3 + S_4$.)

TABLE 14.29
OCTANE NUMBER OF SEVEN GASOLINES

COLUMNS

BLOCKS	1		2		3		TOTALS
1	43	A	34	B	47	D	124
2	36	B	32	C	46	E	114
3	33	C	47	D	43	F	123
4	44	D	40	E	33	G	117
5	41	E	35	F	44	A	120
6	36	F	32	G	32	B	100
7	33	G	41	A	27	C	101
Totals	266		261		272		799

TABLE 14.30
ANALYSIS OF VARIANCE OF OCTANE NUMBERS OF SEVEN GASOLINES

SOURCE OF VARIATION	SUM OF SQUARES	DEGREES OF FREEDOM	MEAN SQUARE	MEAN SQUARE RATIO
Order (blocks)	$S_1 = 196.95$			
Gasoline (adjusted)	$S_2 = 493.62$	6	82.27	64.27***
Gasoline	$S_3 = 608.29$			
Order (adjusted)	$S_4 = 82.29$	6	13.72	10.72**
Time	$S_5 = 8.67$	2	4.34	3.39
Residual	$S_e = 7.71$†	6	1.28	
Total	$S = 706.95$	20		

† $S - S_3 - S_4 - S_5 = 7.70$

intervals. The entire design takes 42 min plus the time of testing reference gasolines. These reference gasolines are used for bracketing results for screening purposes and a decision as to whether to further test particular blends.

The standard analysis for Treatment effects gives the corrected knockmeter readings for each of the 7 fuels. The standard analysis for Block effects gives the time trend over the 42 min, in terms of 3 2-min intervals per block. Column totals also indicate time trend. If columns vary significantly, the engine is obviously unstable. If blocks show a smooth trend, all is well. If the Block effect is significant but does not indicate a steady trend, the engine is suspected of being unstable, and the treatment evaluations are suspect.

The analysis of variance table is calculated as shown in Table 14.30. As examples of the calculations, consider the sums of squares due to Blocks (unadjusted) and Blocks (adjusted):

$$S_1 = k^{-1} \sum_{i=1}^{b} X^2_{(i..)} - (bk)^{-1} X^2_{(...)}$$

$$= \tfrac{1}{3}(124^2 + 114^2 + \cdots + 101^2) - \tfrac{1}{21}(799)^2$$

$$= 30597 - 30400.05 = 196.95$$

$$S_4 = \frac{1}{k\lambda t} \sum_{i=1}^{b} \left(rX_{(i..)} - \sum_{j(i)} X_{(.j.)} \right)^2$$

$$= \frac{1}{3 \cdot 1 \cdot 7} \left[(3 \cdot 124 - 128 - 102 - 138)^2 + \cdots + (3 \cdot 101 - 128 - 92 - 98)^2 \right]$$

$$= \tfrac{1}{21} \left[4^2 + 21^2 + \cdots + (-15)^2 \right] = 82.29$$

Noting the results in the analysis of variance (Table 14.30), the variation due to gasolines is very highly significant, since $64.27 > F_{6,6,0.999}(= 20.0)$. The effect due to blocks is also highly significant since $10.72 > F_{6,6,0.99}$ ($= 8.47$). Hence we can say that some of the gasolines have means much different from the others and that there is a strong suspicion of instability of the engine. Further analysis can now be carried out, such as tests for differences among particular treatments and estimates of variances. Note that the estimate of the residual variance is 1.28. Hence $S = \sqrt{1.28} = 1.13$.

14.6 COMPLETELY HIERARCHAL DESIGNS

In this section we present a brief exposition of the completely hierarchal design for more than 2 factors. In a completely hierarchal design each factor is nested within the preceding factor as described in Section 13.8. We may have either random or parametric terms or a mixture of these. A general, rather rapid procedure for determining the expectations of the mean squares (EMS) is described in Section 14.8.

In this section we consider a purely hierarchal design with all the levels of each factor random variables from an infinite population. The mathematical model for a purely hierarchal design with 4 factors at h, m, p, and q levels for the main, sub, sub², and sub³ factors, respectively, is

$$X_{tijkl} = \alpha + U_t + V_{ti} + W_{tij} + Y_{tijk} + Z_{tijkl} \qquad (14.12)$$

with $t=1, 2,\ldots,h$; $i=1, 2,\ldots,m$; $j=1, 2,\ldots,p$; $k=1, 2,\ldots,q$, and $l=1, 2,\ldots,n$. The ANOVA table is given in Table 14.31. Note that in the EMS column of the table each variance is tested against the next "higher" variance. For example, the EMS associated with W_{tij} is $\sigma^2 + n\sigma^2_{tijk} + qn\sigma^2_{tij}$, whereas that associated with Y_{tijk} is $\sigma^2 + n\sigma^2_{tijk}$.

Figure 14.1 is a diagrammatic representation of the design corresponding to model (14.12) with $h=4$, $m=2$, $p=2$, $q=3$, and $n=3$.

In the next section we describe the analysis appropriate to a purely crossed design with a completely parametric model. It should be understood that component-of-variance models can apply to crossed designs and parametric models to hierarchal designs. The cases described here have been selected for discussion because they are rather more common, in our experience, than the other designs.

Example 14.7 In a study of the determination of the viscosity of a polymeric material, the original sample was subdivided (for ease of handling) into 2 random subsamples which can be represented by random variables U_1 and U_2. Each of these was then divided into 10 aliquots $(V_1, V_2,\ldots, V_{10})$. The first step in the analysis of the polymer was then carried out. Each aliquot was again divided into 2 $(W_1$ and $W_2)$ and the second step of the analysis was performed. Finally, each of the 20 aliquots from the preceding step was again split into 2 parts $(Y_1$ and $Y_2)$ and the last step of the analysis was performed.

The experimental results of the viscosity determinations are given in Table 14.32. Each of the factors is nested in the preceding factor. Further, each factor is

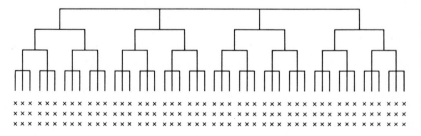

Figure 14.1 Diagrammatic representation of a purely hierarchal (nested) design.

TABLE 14.31
Analysis of Variance of a Completely Hierarchal Experiment (All Random Factors)

SOURCE	SUM OF SQUARES	D.F.	MEAN SQUARE	EXPECTED VALUE OF MEAN SQUARE
u	$\sum_t X_{t....}^2/mpqn - X_{.....}^2/hmpqn$	$h-1$	—	$\sigma^2 + n\sigma_{tijk}^2 + qn\sigma_{tij}^2 + pqn\sigma_{ti}^2 + mpqn\sigma_t^2$
v	$\sum_t\sum_i X_{ti...}^2/pqn - \sum_t X_{t....}^2/mpqn$	$h(m-1)$	—	$\sigma^2 + n\sigma_{tijk}^2 + qn\sigma_{tij}^2 + pqn\sigma_{ti}^2$
w	$\sum_t\sum_i\sum_j X_{tij..}^2/qn - \sum_t\sum_i X_{ti...}^2/pqn$	$hm(p-1)$	—	$\sigma^2 + n\sigma_{tijk}^2 + qn\sigma_{tij}^2$
y	$\sum_t\sum_i\sum_j\sum_k X_{tijk.}^2/n - \sum_t\sum_i\sum_j X_{tij..}^2/qn$	$hmp(q-1)$	—	$\sigma^2 + n\sigma_{tijk}^2$
Residual	$\sum_t\sum_i\sum_j\sum_k\sum_l X_{tijkl}^2 - \sum_t\sum_i\sum_j\sum_k X_{tijk.}^2/n$	$hmpq(n-1)$	—	σ^2
Total	$\sum_t\sum_i\sum_j\sum_k\sum_l X_{tijkl}^2 - X_{.....}^2/hmpqn$	$hmpqn-1$		

TABLE 14.32
VISCOSITY DETERMINATIONS

			v_1	v_2	v_3	v_4	v_5	v_6	v_7	v_8	v_9	v_{10}
u_1	w_1	y_1	59.8	66.6	64.9	62.7	59.5	69.0	64.5	61.6	64.5	65.2
		y_2	59.4	63.9	68.8	62.2	61.0	69.0	66.8	56.6	61.3	63.9
	w_2	y_1	58.2	61.8	66.3	62.9	54.6	60.6	60.2	64.5	72.7	60.8
		y_2	63.5	62.0	63.5	62.8	61.5	61.8	57.4	62.3	72.4	61.2
u_2	w_1	y_1	59.8	65.0	65.0	62.5	59.8	68.8	65.2	59.6	61.0	65.0
		y_2	61.2	65.8	65.2	61.9	60.9	69.0	65.6	58.5	64.0	64.0
	w_2	y_1	60.0	64.5	65.5	60.9	56.0	62.5	61.0	62.3	73.0	62.0
		y_2	65.0	64.5	63.5	61.5	57.2	62.0	59.3	61.5	71.7	63.0

represented by a random variable, so we have a pure component of variance model. Model (14.12) is appropriate with $h = 2$, $m = 10$, $p = 2$, $q = 2$, and $m = 1$. (The last term is included to represent variability in measurement.)

The usual assumptions for a component-of-variance model are made. The hypotheses we would like to test are $\sigma_U^2 = 0$, $\sigma_V^2 = 0$, $\sigma_W^2 = 0$, and $\sigma_Y^2 = 0$. However, since there is only a single replicate at the final stage, in order to carry out tests of significance we must assume that $\sigma_Y^2 = 0$ or equivalently decide to use $\sigma_Y^2 + \sigma^2$ as residual variance of Y. We cannot test the hypothesis $\sigma_Y^2 = 0$.

Table 14.33 presents the analysis of variance. Column 4 contains the mean squares. In this problem we note that the test of the hypothesis $\sigma_W^2 = 0$ gives significance at the 0.001 level. However, our major concern is to *estimate* these

TABLE 14.33
ANOVA OF VISCOSITY DETERMINATIONS

SOURCE	S.S.	D.F.	MEAN SQUARE	MEAN SQUARE RATIO	EXPECTED VALUE OF MEAN SQUARE
Sample (u)	0.11	1	0.11	< 1	$\sigma^2 + 2\sigma_W^2 + 4\sigma_V^2 + 40\sigma_U^2$
Step 1 (v) (within Samples)	503.50	18	27.97	1.23*	$\sigma^2 + 2\sigma_W^2 + 4\sigma_V^2$
Step 2 (w) (within Step 1)	454.72	20	22.74	8.33***	$\sigma^2 + 2\sigma_W^2$
Step 3 (y) (within Step 2)	109.13	40	2.73		σ^2
Total	1067.43	79			

variances. These estimates are obtained from the algebraic manipulation of the formulas obtained by equating columns 4 and 6. We have then

$$\widehat{\sigma^2} = 2.73$$

$$\widehat{\sigma_W^2} = \frac{22.74 - 2.73}{2} = 10.01$$

$$\widehat{\sigma_V^2} = \frac{27.97 - 22.74}{4} = 1.31$$

$$\widehat{\sigma_U^2} = 0 \quad \text{(since the variance cannot be negative)}$$

Note that since the sample sum of squares has only 1 degree of freedom it is very variable. $\widehat{\sigma_U^2}$ will consequently also be variable.

14.7 COMPLETELY CROSSED DESIGNS

In a completely crossed design all the levels of each factor appear in all combinations with all the other factors. The number of terms in the mathematical model (besides the grand mean and residual) is a total of all of the factors, taken 1 at a time, 2 at a time, and so forth. For 4 factors,* for example, there are $4[=\binom{4}{1}]$ main effects, $6[=\binom{4}{2}]$ 2-factor interactions, $4[=\binom{4}{3}]$ 3-factor interactions, and $1[=\binom{4}{4}]$ 4-factor interaction. Its mathematical model (for all factors parametric) is

$$X_{tijkl} = \alpha + \beta_t + \gamma_i + \delta_j + \epsilon_k + (\beta\gamma)_{ti} + (\beta\delta)_{tj}$$
$$+ (\beta\epsilon)_{tk} + (\gamma\delta)_{ij} + (\gamma\epsilon)_{ik} + (\delta\epsilon)_{jk}$$
$$+ (\beta\gamma\delta)_{tij} + (\beta\gamma\epsilon)_{tik} + (\beta\delta\epsilon)_{tjk}$$
$$+ (\gamma\delta\epsilon)_{ijk} + (\beta\gamma\delta\epsilon)_{tijk} + Z_{tijkl} \qquad (14.13)$$

There are sets of relations like $\Sigma_t (\beta\gamma\epsilon)_{tik} = \Sigma_i (\beta\gamma\epsilon)_{tik} = \Sigma_k (\beta\gamma\epsilon)_{tik} = 0$ for each main effect and interaction.

The model is sometimes not presented with the terms in this order, but in a "standard" order, with each factor in turn combined with previous factors. For example, we might have $\alpha + \beta_t + \gamma_i + (\beta\gamma)_{ti} + \delta_j + (\beta\delta)_{tj} + (\gamma\delta)_{tj} + (\beta\gamma\delta)_{tij} + \cdots$.

We present only the ANOVA table (Table 14.34) for this model, but do not take a specific example. In the next section a rather large example of a mixed model is considered. This covers more general cases. Suppose the

*This is sometimes called a 4-way classification.

SOURCE	SUM OF SQUARES

B

$$\frac{\sum_t X^2_{t....}}{mpqn} - \frac{X^2_{.....}}{hmpqn}$$

C

$$\frac{\sum_i X^2_{.i...}}{hpqn} - \frac{X^2_{.....}}{hmpqn}$$

D

$$\frac{\sum_j X^2_{..j..}}{hmqn} - \frac{X^2_{.....}}{hmpqn}$$

E

$$\frac{\sum_k X^2_{...k.}}{hmpn} - \frac{X^2_{.....}}{hmpqn}$$

BC

$$\frac{\sum_t \sum_i X^2_{ti...}}{pqn} - \frac{\sum_t X^2_{t....}}{mpqn} - \frac{\sum_i X^2_{.i...}}{hpqn} + \frac{X^2_{.....}}{hmpqn}$$

BD

$$\frac{\sum_t \sum_j X^2_{t.j..}}{mqn} - \frac{\sum_t X^2_{t....}}{mpqn} - \frac{\sum_j X^2_{..j..}}{hmqn} + \frac{X^2_{.....}}{hmpqn}$$

BE

$$\frac{\sum_t \sum_k X^2_{t..k.}}{mpn} - \frac{\sum_t X^2_{t....}}{mpqn} - \frac{\sum_k X^2_{...k.}}{hmpn} + \frac{X^2_{.....}}{hmpqn}$$

CD

$$\frac{\sum_i \sum_j X^2_{.ij..}}{hqn} - \frac{\sum_i X^2_{.i...}}{hpqn} - \frac{\sum_j X^2_{..j..}}{hmqn} + \frac{X^2_{.....}}{hmpqn}$$

CE

$$\frac{\sum_i \sum_k X^2_{.i.k.}}{hpn} - \frac{\sum_i X^2_{.i...}}{hpqn} - \frac{\sum_k X^2_{...k.}}{hmpn} + \frac{X^2_{.....}}{hmpqn}$$

DE

$$\frac{\sum_j \sum_k X^2_{..jk.}}{hmn} - \frac{\sum_j X^2_{..j..}}{hmqn} - \frac{\sum_k X^2_{...k.}}{hmpn} + \frac{X^2_{.....}}{hmpqn}$$

BCD

$$\sum_t \sum_i \sum_j X^2_{tij..}/qn - \sum_t \sum_i - \sum_t \sum_j - \sum_i \sum_j + \sum_t + \sum_i + \sum_j - \frac{X^2_{.....}}{hmpqn}$$

BCE

$$\sum_t \sum_i \sum_k X^2_{ti.k.}/pn - \sum_t \sum_i - \sum_t \sum_k - \sum_i \sum_k + \sum_t + \sum_i + \sum_k - \frac{X^2_{.....}}{hmpqn}$$

CROSSED PARAMETRIC MODEL

D.F.	M.S.	E.M.S.
-1	—	$\sigma^2 + (h-1)^{-1}mpqn \sum_{t=1}^{h} \beta_t^{\,2}$
$1-1$	—	$\sigma^2 + (m-1)^{-1}hpqn \sum_{i=1}^{m} \gamma_i^{\,2}$
-1	—	$\sigma^2 + (p-1)^{-1}hmqn \sum_{j=1}^{p} \delta_j^{\,2}$
-1	—	$\sigma^2 + (q-1)^{-1}hmpn \sum_{k=1}^{q} \epsilon_k^{\,2}$
$h-1)(m-1)$	—	$\sigma^2 + [(h-1)(m-1)]^{-1}pqn \sum_{t=1}^{h} \sum_{i=1}^{m} (\beta\gamma)_{ti}^2$
$h-1)(p-1)$	—	$\sigma^2 + [(h-1)(p-1)]^{-1}mqn \sum_{t=1}^{h} \sum_{j=1}^{p} (\beta\delta)_{tj}^2$
$h-1)(q-1)$	—	$\sigma^2 + [(h-1)(q-1)]^{-1}mpn \sum_{t=1}^{h} \sum_{k=1}^{q} (\beta\epsilon)_{tk}^2$
$m-1)(p-1)$	—	$\sigma^2 + [(m-1)(p-1)]^{-1}hqn \sum_{i=1}^{m} \sum_{j=1}^{p} (\gamma\delta)_{ij}^2$
$m-1)(q-1)$	—	$\sigma^2 + [(m-1)(q-1)]^{-1}hpn \sum_{i=1}^{m} \sum_{k=1}^{q} (\gamma\epsilon)_{ik}^2$
$p-1)(q-1)$	—	$\sigma^2 + [(p-1)(q-1)]^{-1}hmn \sum_{j=1}^{p} \sum_{k=1}^{q} (\delta\epsilon)_{jk}^2$
$h-1)(m-1)(p-1)$	—	$\sigma^2 + [(h-1)(m-1)(p-1)]^{-1}qn \sum_{t=1}^{h} \sum_{i=1}^{m} \sum_{j=1}^{p} (\beta\gamma\delta)_{tij}^2$
$h-1)(m-1)(q-1)$	—	$\sigma^2 + [(h-1)(m-1)(q-1)]^{-1}pn \sum_{t=1}^{h} \sum_{i=1}^{m} \sum_{k=1}^{q} (\beta\gamma\epsilon)_{tik}^2$

SOURCE	SUM OF SQUARES
BDE	$\sum_t \sum_j \sum_k X^2_{t \cdot jk \cdot}/mn - \sum_t \sum_j - \sum_t \sum_k - \sum_j \sum_k + \sum_t + \sum_j + \sum_k - \dfrac{X^2_{\cdots\cdots}}{hmpqn}$
CDE	$\sum_i \sum_j \sum_k X^2_{\cdot ijk \cdot}/hn - \sum_i \sum_j - \sum_i \sum_k - \sum_j \sum_k + \sum_i + \sum_j + \sum_k - \dfrac{X^2_{\cdots\cdots}}{hmpqn}$
BCDE	$\sum_t \sum_i \sum_j \sum_k X^2_{tijk \cdot}/n - \sum_t \sum_i \sum_j - \cdots + \sum_t \sum_i + \cdots - \sum_t - \cdots + \dfrac{X^2_{\cdots\cdots}}{hmpqn}$
Residual	Difference
Total	$\sum_t \sum_i \sum_j \sum_k \sum_l x^2_{tijkl} - X^2_{\cdots\cdots}/hmpqn$

number of levels of each factor, β through ϵ, are h, m, p, and q, respectively, and there are n replicates. In Table 14.34, the latter part of the sum of squares column is abbreviated. The subscripts indicate the quantity to appear with each summation.

Note that from the Expected Value of mean square column it is clear that each of the ratios of the mean squares is taken with the same Residual mean square in the denominator. This applies only for purely parametric models. More is said about this in the next section.

14.8   MIXED MODELS AND DESIGNS

In the preceding 2 sections we were concerned with completely nested (hierarchal) or completely crossed experiments. There are many situations where these are the correct designs. However, in any general, fairly extensive problem, the design may very well be a combination of nested and crossed classifications. Also some factors may be represented by parametric terms and others by component-of-variance terms in the model. In this section we consider the general situation of mixed models and mixed experimental designs. Some general rules for calculating the sums of squares and the expectations of the mean squares are evolved. The reader should note that in obtaining expectations of mean squares we are not always considering variances alone. A quick look at the different models of 2-factor experiments in Chapter 13 shows this.

14.34 (*Continued*)
CROSSED PARAMETRIC MODEL

D.F.	M.S.	E.M.S.
$(h-1)(p-1)(q-1)$	—	$\sigma^2 + [(h-1)(p-1)(q-1)]^{-1}mn \sum\limits_{t=1}^{h} \sum\limits_{j=1}^{p} \sum\limits_{k=1}^{q} (\beta\delta\epsilon)_{tjk}^2$
$(m-1)(p-1)(q-1)$	—	$\sigma^2 + [(m-1)(p-1)(q-1)]^{-1}hn \sum\limits_{i=1}^{m} \sum\limits_{j=1}^{p} \sum\limits_{k=1}^{q} (\gamma\delta\epsilon)_{ijk}^2$
$(h-1)(m-1)$ $\times (p-1)(q-1)$	—	$\sigma^2 + [(h-1)(m-1)(p-1)(q-1)]^{-1}n \sum\limits_{t=1}^{h} \sum\limits_{i=1}^{m} \sum\limits_{j=1}^{p} \sum\limits_{k=1}^{q} (\beta\gamma\delta\epsilon)_{tijk}^2$
$hmpq(n-1)$	—	σ^2
$hmpq\ n-1$		

It cannot be too strongly stated that the model must be formulated following very closely the actual problem under study. Unless this is done the expectations of the mean squares are very likely to be incorrect. Since these expressions indicate what ratios to take, inappropriate ratios of mean squares might be used, yielding subsequent wrong conclusions.

We consider now mixed designs with 3 or more factors where at least 1 of these is crossed with another and 1 is nested within 1 or more other factors.

14.8.1 Calculation of the Sums of Squares and Degrees of Freedom for all Complete Balanced Designs (Factorials)

It is now convenient to introduce some new features of notation, which facilitate the calculation of expected values in some of the more complicated cases.

We consider, for the sake of simplicity, only factorial designs that have the same number of elements per cell and for which each level of each factor is completely replicated. For example, in a 2-factor hierarchal classification as presented in the preceding chapter in (13.22) we require $j = 1, 2, \ldots, n$ for all i, and $i = 1, 2, \ldots, m$ for all t. Let us first concentrate on the subscripts in the mathematical model. Consider 3 simple cases

(*i*) $X_{tij} = \alpha + U_t + V_{i(t)} + Z_{j(ti)}$
(*ii*) $X_{tij} = \alpha + \beta_t + V_{i(t)} + Z_{j(ti)}$
(*iii*) $X_{tij} = \alpha + \beta_t + \gamma_i + (\beta\gamma)_{ti} + Z_{j(ti)}$

with $t = 1, 2, \ldots, h$; $i = 1, 2, \ldots, m$; $j = 1, 2, \ldots, n$. In (i) and (ii) note the symbol $i(t)$, indicating that i is nested in t. For example, factor V is nested in factor β in case (ii). Case (i) is a component-of-variance model of a nested experiment. Each of the levels of i is nested in t. Case (ii) is again a nested experiment, but factor β is represented by parametric terms whereas V is represented by random variables. This is a mixed model. Case (iii) is a parametric model for a 2-factor, crossed classification. It is, in fact, the same model as (13.25) with the subscript of Z written in a special form to indicate the design structure.

Consider first the sum of squares (SS) for each source of variation. In case (i) we shall call these SS[t], SS[$i(t)$], SS[e], and SS(total). Recall that

$$\mathrm{SS}[\,t\,] = (mn)^{-1} \sum_t X_{t..}^2 - (mnh)^{-1} X_{...}^2$$

$$\mathrm{SS}[\,i(t)\,] = n^{-1} \sum_t \sum_i X_{ti.}^2 - (mn)^{-1} \sum_t X_{t..}^2$$

$$\mathrm{SS}[\,e\,] = \mathrm{SS}[\,j(ti)\,] = \sum_t \sum_i \sum_j X_{tij}^2 - n^{-1} \sum_t \sum_i X_{ti.}^2.$$

$$\mathrm{SS}[\,\mathrm{Total}\,] = \sum_t \sum_i \sum_j X_{tij}^2 - (mnh)^{-1} X_{...}^2$$

We now introduce the symbols

$$G_t = (mn)^{-1} \sum_t X_{t..}^2 \qquad G_{tij} = \sum_t \sum_i \sum_j X_{tij}^2$$

$$G_{ti} = n^{-1} \sum_t \sum_i X_{ti.}^2 \qquad G = (mnh)^{-1} X_{...}^2$$

Then

$$\mathrm{SS}[\,t\,] = G_t - G \qquad \mathrm{SS}[\,j(ti)\,] = G_{tij} - G_{ti}$$

$$\mathrm{SS}[\,i(t)\,] = G_{ti} - G_t \qquad \mathrm{SS}[\,\mathrm{Total}\,] = G_{tij} - G$$

We could have obtained the specific G values by noting the subscripts in the model and making use of a simple algebraic rule. We will now present the subscripts in tabular form, together with the algebraic rule and the G values. The rule is that whenever the subscript is *not* in parentheses we substitute that subscript minus 1; if it is in parentheses it remains unchanged. The algebraic signs resulting from the product precede the G

values. Thus in case (i)

FACTOR OR VARIABLE	SUBSCRIPT	RULE			SUM OF SQUARES
u	t	$\rightarrow$	$t - 1$	$\rightarrow$	$G_t - G$
v	$i(t)$	$\rightarrow$	$(i - 1)t$	$\rightarrow$	$G_{ti} - G_t$
z	$j(ti)$	$\rightarrow$	$(j - 1)ti$	$\rightarrow$	$G_{tij} - G_{ti}$

Following the above algebraic notation we obtain the same sums of squares for case (ii). In case (iii) we obtain, from the subscripts,

FACTOR OR VARIABLE	SUBSCRIPT	RULE			SUM OF SQUARES
B	t	$\rightarrow$	$t - 1$	$\rightarrow$	$G_t - G$
C	i	$\rightarrow$	$i - 1$	$\rightarrow$	$G_i - G$
BC	ti	$\rightarrow$	$(t - 1)(i - 1)$	$\rightarrow$	$G_{ti} - G_t - G_i + G$
z	$j(ti)$	$\rightarrow$	$ti(j - 1)$	$\rightarrow$	$G_{tij} - G_{ti}$

The Total sum of squares is $G_{tij} - G$ which is the total of the individual sum of squares of deviations from the overall mean.

In order to calculate the number of degrees of freedom, we use the same algebraic rule, except that t is replaced by h, i is replaced by m, and j by n. Hence the number of degrees of freedom for the last of the above models is

FACTOR OR VARIABLE	SUBSCRIPT:		DEGREES OF FREEDOM
B	t	$\rightarrow$	$h - 1$
C	i	$\rightarrow$	$m - 1$
BC	ti	$\rightarrow$	$(h - 1)(m - 1)$
z (Residual)	$j(ti)$	$\rightarrow$	$hm(n - 1)$
Total			$hmn - 1$

Example 14.8   For the following mathematical model let us determine the sums of squares and number of degrees of freedom for $(\gamma\epsilon)_{ik(t)}$.

$$X_{tijkl} = \alpha + \beta_t + \gamma_{i(t)} + \delta_{j(ti)} + \epsilon_k$$

$$+ (\beta\epsilon)_{tk} + (\gamma\epsilon)_{ik(t)} + (\delta\epsilon)_{jk(ti)} + Z_{l(tijk)}$$

with $t = 1,\ldots,h$; $i = 1,\ldots,m$; $j = 1,\ldots,p$; $k = 1,\ldots,q$; $l = 1,\ldots,n$. The sum of squares for $(\gamma\epsilon)_{ik(t)}$ has the subscripts $ik(t)$, yielding the product $(i - 1)(k - 1)t$. Hence

$$SS[ik(t)] = G_{tik} - G_{ti} - G_{tk} + G_t$$

The number of degrees of freedom is $(m-1)(q-1)h$. The determination of the other sums of squares and number of degrees of freedom follow in a similar manner. Note that the determination of the sums of squares of any factor is *not* affected by whether this or any other factor is represented by parameters or by random variables. This does, however, enter into the determination of the expected value of the mean square.

14.8.2 Auxiliary Tables for Determining the Expected Values of the Mean Squares

In this section, formulas are introduced which apply when the levels of some factors are chosen at random from a *finite* population of possible levels. The number of levels in the population is denoted by the capital letter corresponding to the lowercase letter which represents the number of levels appearing in the experiment. Thus in (14.14) h represents the number of possible levels of the factor represented by U_t. If h is finite, we have a parametric term; if h is infinite we have a random term.

Consider a model which involves a complete factorial and both parametric and random terms. We shall take a particular case and evolve the general rule, using this as an example. Consider the model*

$$X_{tijkl} = \alpha + U_t + V_{i(t)} + W_j + (UW)_{tj} + (VW)_{ij(t)} + \beta_k$$

$$+ (U\beta)_{tk} + (V\beta)_{ik(t)} + (W\beta)_{jk} + (UW\beta)_{tjk}$$

$$+ (VW\beta)_{ijk(t)} + Z_{l(tijk)} \qquad (14.14)$$

with $t=1,\dots,h$; $i=1,\dots,m$; $j=1,\dots,p$; $k=1,\dots,q$ and $l=1,\dots,n$. According to the model there are 12 sources of variation. We list these, in terms of their subscripts, in tabular form. Table 14.35 has as its rows the subscript designation of each effect or source of variation. The column designators are the individual letters in the subscript. The model number (I or II) is also designated.

The rules for completing the table are as follows:

1. In each row under each column write the total number of levels of each letter not represented in the row designator. Hence in row $ij(t)$ we place in the last 2 columns q and n, respectively, because k and l do not appear in the row designator.

2. In each row under each column in which the column designator is bracketed in the row designator, write *unity*.

3. In the remainder of each column write $1 - \lambda/\Lambda$. Thus if both the column and the row designators are represented by parameters, $1 - \lambda/\Lambda =$

---

*If any factor in an interaction is represented by a random variable, it will be assumed that the interaction should be represented by a random variable. For example, $(U\beta)_{tk}$ is a random variable. Though it is mathematically possible for this rule to fail, it is very unlikely to be inapplicable in any model of a real situation.

TABLE 14.35
AUXILIARY TABLE FOR EXPECTATIONS OF MEAN SQUARES
(INDEPENDENT MIXED INTERACTIONS)

COLUMN DESIGNATORS*

ROW DESIGNATORS	t(II)	i(II)	j(II)	k(I)	l(II)
t	$1 - h/H$	m	p	q	n
$i(t)$	1	$1 - m/M$	p	q	n
j	h	m	$1 - p/P$	q	n
tj	$1 - h/H$	m	$1 - p/P$	q	n
$ij(t)$	1	$1 - m/M$	$1 - p/P$	q	n
k	h	m	p	0	n
tk	$1 - h/H$	m	p	$1 - q/Q$	n
$ik(t)$	1	$1 - m/M$	p	$1 - q/Q$	n
jk	h	m	$1 - p/P$	$1 - q/Q$	n
tjk	$1 - h/H$	m	$1 - p/P$	$1 - q/Q$	n
$ijk(t)$	1	$1 - m/M$	$1 - p/P$	$1 - q/Q$	n
$l(tijk)$	1	1	1	1	$1 - n/N$

0, if, by random variables with Λ infinite, $1-\lambda/\Lambda=1$. Otherwise $0<1-\lambda/\Lambda<1$. λ is the number of levels of the sample and Λ is the number of levels of the population of the factor in question.

We are assuming here that the interactions are represented by mutually *independent* random variables as long as one of its elements is a random variable. If this were not so then step 3 will produce a 1 if both row and column are random variables (or $1-\lambda/\Lambda$ if the column is random but finite) and 0 otherwise.

In order to evaluate the expected value of the mean square in the analysis of variance table one now proceeds as follows:

1. For each factor or effect in the analysis of variance table, the column (or columns) containing the subscript in question is (are) deleted.

2. For each row which contains in its designator the subscript of the "variance in question" the product of the entries in each of the columns is obtained. This is the coefficient of the "variance term" for each of these. This is done for each of the rows containing the subscripts being considered.

TABLE 14.36
EXPECTATIONS OF MEAN SQUARES

FACTOR OR VARIABLE	EXPECTATION OF MEAN SQUARE FOR MODEL (14.14)
U_t	$\sigma^2 + \left(1-\dfrac{m}{M}\right)\left(1-\dfrac{p}{P}\right)n\sigma^2_{ijk(t)} + m\left(1-\dfrac{p}{P}\right)n\sigma^2_{tjk}$ $+ \left(1-\dfrac{m}{M}\right)pn\sigma^2_{ik(t)} + mpn\sigma^2_{tk} + \left(1-\dfrac{m}{M}\right)\left(1-\dfrac{p}{P}\right)qn\sigma^2_{ij(t)}$ $+ m\left(1-\dfrac{p}{P}\right)qn\sigma^2_{tj} + \left(1-\dfrac{m}{M}\right)pqn\sigma^2_{i(t)} + mpqn\sigma^2_t$
$V_{i(t)}$	$\sigma^2 + \left(1-\dfrac{p}{P}\right)n\sigma^2_{ijk(t)} + pn\sigma^2_{ik(t)} + \left(1-\dfrac{p}{P}\right)qn\sigma^2_{ij(t)} + pqn\sigma^2_{i(t)}$
W_j	$\sigma^2 + \left(1-\dfrac{m}{M}\right)n\sigma^2_{ijk(t)} + \left(1-\dfrac{h}{H}\right)mn\sigma^2_{tjk} + hmn\sigma^2_{jk}$ $+ \left(1-\dfrac{m}{M}\right)qn\sigma^2_{ij(t)} + \left(1-\dfrac{h}{H}\right)mqn\sigma^2_{tj} + hmqn\sigma^2_j$
$(UW)_{tj}$	$\sigma^2 + \left(1-\dfrac{m}{M}\right)n\sigma^2_{ijk(t)} + mn\sigma^2_{tjk} + \left(1-\dfrac{m}{M}\right)qn\sigma^2_{ij(t)} + mqn\sigma^2_{tj}$
$(VW)_{ij(t)}$	$\sigma^2 + n\sigma^2_{ijk(t)} + qn\sigma^2_{ij(t)}$

β_k $\sigma^2 + \left(1 - \dfrac{m}{M}\right)\left(1 - \dfrac{p}{P}\right) n\sigma^2_{ijk(t)} + \left(1 - \dfrac{h}{H}\right) m\left(1 - \dfrac{p}{P}\right) n\sigma^2_{ijk} + \left(1 - \dfrac{p}{P}\right) n\sigma^2_{jk} + hm\left(1 - \dfrac{p}{P}\right) n\sigma^2_{ik(t)} + \left(1 - \dfrac{m}{M}\right) pn\sigma^2_{ik(t)}$

$\qquad\qquad + \left(1 - \dfrac{h}{H}\right) mpn\sigma^2_{ik} + \langle hmpn\sigma^2_k \rangle$

$(U\beta)_{tk}$ $\sigma^2 + \left(1 - \dfrac{m}{M}\right)\left(1 - \dfrac{p}{P}\right) n\sigma^2_{ijk(t)} + m\left(1 - \dfrac{p}{P}\right) n\sigma^2_{ijk} + \left(1 - \dfrac{m}{M}\right) pn\sigma^2_{ik(t)} + mpn\sigma^2_{ik}$

$(V\beta)_{ik(t)}$ $\sigma^2 + \left(1 - \dfrac{p}{P}\right) n\sigma^2_{ijk(t)} + pn\sigma^2_{ik(t)}$

$(W\beta)_{jk}$ $\sigma^2 + \left(1 - \dfrac{m}{M}\right) n\sigma^2_{ijk(t)} + \left(1 - \dfrac{h}{H}\right) mn\sigma^2_{ijk} + hmn\sigma^2_{jk}$

$(UW\beta)_{ijk}$ $\sigma^2 + \left(1 - \dfrac{m}{M}\right) n\sigma^2_{ijk(t)} + mn\sigma^2_{ijk}$

$(UV\beta)_{ijk(t)}$ $\sigma^2 + n\sigma^2_{ijk(t)}$

Residual

$Z_{l(tijk)}$ σ^2

For example, the terms in the expectation for $V_{i(t)}$ are (working up):

$$\sigma^2_{l(tijk)}(\text{or } \sigma^2), \quad \sigma^2_{ijk(t)}, \quad \sigma^2_{ik(t)}, \quad \sigma^2_{ij(t)}, \quad \text{and} \quad \sigma^2_{i(t)}$$

After eliminating columns "t" and "i," we take the products (from Table 14.35)

$$1 \text{ for } \sigma^2, \quad \left(1 - \frac{p}{P}\right)n \text{ for } \sigma^2_{ijk(t)}, \quad pn \text{ for } \sigma^2_{ik(t)},$$

$$\left(1 - \frac{p}{P}\right)qn \text{ for } \sigma^2_{ij(t)}, \quad \text{and } pqn \text{ for } \sigma^2_{i(t)}$$

One must bear in mind that there is no such thing as a "variance" for a set of parameters. Nevertheless, we use this general technique to indicate where a nonzero component of expectation of mean squares does exist. We shall put a *bracket* ($\langle \ \rangle$) around any such term to indicate that this is not a true variance. The expectation of the mean square for the preceding model is given in Table 14.36. We limit ourselves to the case when the number of individuals available for measurement is effectively infinite, so that $n/N = 0$. We also let $1 - q/Q = 1$ for purposes of illustration.

Example 14.9   An investigation was performed to evaluate some of the factors that affect the tensile properties of a particular titanium alloy and to determine the relative magnitude of these effects. Since yield strength is the most important tensile property when considering the application of the part being studied (a jet engine compressor rotor blade) the following factors which were thought to contribute to yield strength were studied:

(1) *Vendors*. Bar stock was purchased from 3 vendors. Any systematic difference in processing used by these vendors that will influence yield strength should show up in this effect.

(2) *Bar Size*. To make the various size blades used in the different stages of a jet engine compressor rotor, bar stock of 3 difference diameters (1, $1\frac{1}{4}$, and $1\frac{3}{4}$ in.) was used. Since processing of different size bar stock from a common ingot size represents different amount of forging and probably different rolling techniques, this bar size factor has been investigated.

(3) *Heats*. Different heats, or melts, of material represent slightly different chemical compositions as well as usually representing one specific lot of material processed at one time. Therefore this factor includes the effect of varying composition and also any variation of processing technique with respect to time. Since only one vendor can make any one heat of material and usually processes the entire heat into one size of bar stock in one production lot, this heat effect is nested within the combinations of bar stock size and vendor.

(4) *Test Specimens*.   Standard 0.250-in.-diameter round tensile specimens with 1-in. gauge length were used in this investigation. These tensile specimens were made from 3 types of test "coupons" which are: raw bar stock (B.S.), forgedowns

(F.D.), and finish forged blades (B.). Since each of these 3 types of test coupons were prepared for each heat of material, this effect is crossed with both the vendor and bar-stock size, and it is also crossed with the heat effect within each vendor-bar size combination.

The mathematical model is then

$$Y_{ijklm} = \alpha + \nu_i + \beta_j + (\nu\beta)_{ij} + H_{k(ij)}$$

$$+ \tau_l + (\nu\tau)_{il} + (\beta\tau)_{jl} + (\nu\beta\tau)_{ijl}$$

$$+ (H\tau)_{kl(ij)} + Z_{m(ijkl)}$$

where i ($= 1,2,3$) denotes vendor; j ($= 1,2,3$) denotes bar size; k ($= 1,2,3$) denotes heat nested in vendor $\times$ bar size; l ($= 1,2,3$) denotes test specimen; and m ($= 1,2,3$) denotes replicate. This model is parametric in vendors, bar size, and test specimen. Random variables represent the effect of heats and the heat $\times$ test specimen interaction. The data are given in Table 14.37. The data are recorded in 100,000 psi.

In presenting an illustration of this size, namely, $3^4 \times 3$ replicates, one may wonder whether it is necessary to have such a large experiment to gain the information that is required. We consider, in the next chapter, methods for reducing the size of the experiment. For the present, let us consider this as a legitimate experiment, in which different effects are to be studied in further detail after the analysis of variance has been completed.

Another reason for presenting this example is to show the importance of having a clear concept of the physical procedure. The mathematical model must be consistent with this. Note that the heats are nested within the vendor $\times$ bar size 2-factor effect. Finally, the types of test specimens are crossed with all of the other factors.

The data of Table 14.38 are coded as follows:

$$Y_{ijklm} = 1000 \text{ (observed value} - 1.260)$$

In order to determine the necessary sums of squares and mean squares, we now construct a supplementary table (Table 14.39) presenting the various totals. One may wish to omit this table in carrying out the calculation. An advantage of this table, however, is a visual presentation of the effects of the different factors. A table of means of the original data rather than these totals will be more effective especially after the analysis of variance has been carried out.

Following the procedure for calculating the G values and finally the sums of squares and degrees of freedom we have the analysis of variance table for the coded data as given in Table 14.41. Note that the column headed "expected value of mean square" has only 3 *true* variances, namely σ^2, $\sigma_{k(ij)}^2$, and $\sigma_{kl(ij)}^2$. The other values, which are in brackets, are in fact sums of squares of deviations (times a constant) from some mean.

In order to determine the coefficients of the terms in column (5) we carry out the technique shown in the preceding section and in Table 14.35. This is shown in

TABLE 14.37
YIELD STRENGTH OF TITANIUM ALLOY
(RESULTS IN HUNDREDS OF THOUSANDS PSI)

HEAT	VENDOR A			VENDOR B			VENDOR C		
	1	2	3	1	2	3	1	2	3
1″ Bar Stock									
B.S.	1.430	1.410	1.325	1.407	1.490	1.400	1.423	1.401	1.398
	1.422	1.400	1.320	1.450	1.478	1.380	1.409	1.396	1.382
	1.431	1.402	1.331	1.432	1.479	1.390	1.413	1.397	1.380
F.D.	1.450	1.480	1.431	1.490	1.467	1.489	1.442	1.490	1.427
	1.470	1.462	1.442	1.460	1.449	1.482	1.449	1.501	1.438
	1.463	1.467	1.442	1.485	1.452	1.481	1.447	1.497	1.434
B.	1.470	1.481	1.419	1.421	1.497	1.400	1.500	1.409	1.446
	1.472	1.460	1.427	1.421	1.489	1.402	1.512	1.423	1.429
	1.481	1.459	1.426	1.430	1.489	1.396	1.499	1.417	1.439
$1\frac{1}{4}″$ Bar Stock									
B.S.	1.356	1.320	1.382	1.361	1.320	1.339	1.300	1.298	1.362
	1.349	1.305	1.373	1.349	1.318	1.351	1.302	1.287	1.373
	1.352	1.308	1.373	1.348	1.321	1.341	1.311	1.301	1.373
F.D.	1.456	1.420	1.441	1.460	1.400	1.421	1.412	1.400	1.429
	1.470	1.410	1.432	1.452	1.412	1.420	1.409	1.429	1.436
	1.466	1.399	1.430	1.451	1.402	1.409	1.411	1.417	1.437
B.	1.470	1.400	1.400	1.462	1.459	1.442	1.431	1.449	1.401
	1.462	1.390	1.390	1.473	1.467	1.436	1.432	1.457	1.400
	1.480	1.415	1.402	1.470	1.462	1.432	1.436	1.460	1.398
$1\frac{3}{4}″$ Bar Stock									
B.S.	1.320	1.270	1.280	1.269	1.300	1.279	1.254	1.347	1.312
	1.312	1.292	1.290	1.271	1.291	1.280	1.249	1.361	1.300
	1.313	1.283	1.302	1.273	1.291	1.280	1.242	1.350	1.289
F.D.	1.420	1.423	1.430	1.407	1.438	1.401	1.421	1.399	1.400
	1.401	1.419	1.420	1.407	1.440	1.411	1.420	1.400	1.410
	1.412	1.418	1.442	1.402	1.437	1.405	1.409	1.430	1.407
B.	1.430	1.400	1.493	1.429	1.391	1.400	1.379	1.381	1.432
	1.450	1.389	1.472	1.428	1.402	1.400	1.392	1.391	1.419
	1.442	1.393	1.473	1.435	1.392	1.398	1.392	1.393	1.426

TABLE 14.38
YIELD STRENGTH OF TITANIUM ALLOY (CODED)

HEAT	VENDOR A			VENDOR B			VENDOR C		
	1	2	3	1	2	3	1	2	3
1″									
Bar Stock									
	170	150	65	147	230	140	163	141	138
B.S.	162	140	60	190	218	120	149	136	122
	171	142	71	172	219	130	153	137	120
	190	220	171	230	207	229	182	230	167
F.D.	210	202	182	200	189	222	189	241	178
	203	207	182	225	192	221	187	237	174
	210	221	159	161	237	140	240	149	186
B.	212	200	167	161	229	142	252	163	169
	221	199	166	170	229	136	239	157	179
$1\frac{1}{4}″$									
Bar Stock									
	96	60	122	101	60	79	40	38	102
B.S.	89	45	113	89	58	91	42	27	113
	92	48	113	88	61	81	51	41	113
	196	160	181	200	140	161	152	140	169
F.D.	210	150	172	192	152	160	149	169	176
	206	139	170	191	142	149	151	157	177
	210	140	140	202	199	182	171	189	141
B.	202	130	130	213	207	176	172	197	140
	220	155	142	210	202	172	176	200	138
$1\frac{3}{4}″$									
Bar Stock									
	60	10	20	9	40	19	−6	87	52
B.S.	52	32	30	11	31	20	−11	101	40
	53	23	42	13	31	20	−18	90	29
	160	163	170	147	178	141	161	139	140
F.D.	141	159	160	147	180	151	160	140	150
	152	158	182	142	177	145	149	170	147
	170	140	233	169	131	140	119	121	172
B.	190	129	212	168	142	140	132	131	159
	182	133	213	175	132	138	132	133	166

TABLE 14.39
TABLE OF TOTALS

VENDORS

BAR SIZES	TEST SPECIMEN	A — HEATS 1	2	3	$Y_{1j;l.}$	B — HEATS 1	2	3	$Y_{2j;l.}$	C — HEATS 1	2	3	$Y_{3j;l.}$	$Y_{.j;l.}$
1″	B.S.	503	432	196	1131	509	667	390	1566	465	414	380	1259	3956
	F.D.	603	629	535	1767	655	588	672	1915	558	708	519	1785	5467
	B.	643	620	492	1755	492	695	418	1605	731	469	534	1734	5094
	$Y_{i1k..}$	1749	1681	1223	4653	1656	1950	1480	5086	1754	1591	1433	4778	14517 = $Y_{.1..}$
		$Y_{111..}$	$Y_{112..}$	$Y_{113..}$	$Y_{11...}$	$Y_{211..}$	$Y_{212..}$	$Y_{213..}$	$Y_{21...}$	$Y_{311..}$	$Y_{312..}$	$Y_{313..}$	$Y_{31...}$	
1¼″	B.S.	277	153	348	778	278	179	251	708	133	106	328	567	2053
	F.D.	612	449	523	1584	583	434	470	1487	452	466	522	1440	4511
	B.	632	425	412	1469	625	608	530	1763	519	586	419	1524	4756
	$Y_{i2k..}$	1521	1027	1283	3831	1486	1221	1251	3958	1104	1158	1269	3531	11320 = $Y_{.2..}$
		$Y_{121..}$	$Y_{122..}$	$Y_{123..}$	$Y_{12...}$	$Y_{221..}$	$Y_{222..}$	$Y_{223..}$	$Y_{22...}$	$Y_{321..}$	$Y_{322..}$	$Y_{323..}$	$Y_{32...}$	
1¾″	B.S.	165	65	92	322	33	102	59	194	−35	278	121	364	880
	F.D.	453	480	512	1445	436	535	437	1408	470	449	437	1356	4209
	B.	542	402	658	1602	512	405	418	1335	383	385	497	1265	4202
	$Y_{i3k..}$	1160	947	1262	3369	981	1042	914	2937	818	1112	1055	2985	9291 = $Y_{.3..}$
		$Y_{131..}$	$Y_{132..}$	$Y_{133..}$	$Y_{13...}$	$Y_{231..}$	$Y_{232..}$	$Y_{233..}$	$Y_{23...}$	$Y_{331..}$	$Y_{332..}$	$Y_{333..}$	$Y_{33...}$	
	$Y_{i....}$	$Y_{1....}$			11853				11981				11294	35128
		6889	14187	14052	$Y_{1....}$				$Y_{2....}$				$Y_{3....}$	$(G) = Y_{.....}$

TEST SPECIMEN $Y_{....l.}$: 1 = 6889, 2 = 14187, 3 = 14052

$Y_{i..l.}$ — VENDORS

TEST SPECIMEN	A	B	C
B.S.	2231	2468	2190
F.D.	4796	4810	4581
B.	4826	4703	4523

TABLE 14.40
COEFFICIENT TABLE

ROW DESIGNATORS AND FACTOR LETTERS	i(I)	j(I)	k(II)	l(I)	m(II)
ν_i	0	3	3	3	3
β_j	3	0	3	3	3
$(\nu\beta)_{ij}$	0	0	3	3	3
$H_{k(ij)}$	1	1	1	3	3
τ_l	3	3	3	0	3
$(\nu\tau)_{il}$	0	3	3	0	3
$(\beta\tau)_{jl}$	3	0	3	0	3
$(\nu\beta\tau)_{ijl}$	0	0	3	0	3
$(H\tau)_{kl(ij)}$	1	1	1	1	3
$Z_{m(ijkl)}$	1	1	1	1	1

Table 14.40. In this table the row designators are accompanied by their corresponding factor-indicating letters to help identify the effect represented in each row.

Let us return to the analysis of variance table. We first test the hypothesis that the interaction between Heats and Test Specimens (within the Vendor × Bar Size interaction) is 0. We calculate the mean square ratio = 2360/60 = 39.43 which is highly significant judged against the distribution of $F_{36,162}$. (Because of the large numbers of degrees of freedom, this test is very sensitive, and even small differences are quite likely to be detected as significant.)

We proceed to test the 3-factor interaction of Vendor, Bar Size, and Test Specimen. Noting the last column of Table 14.41, we take the ratio of this mean square to that of Test Specimen × Heat interaction. We obtain

$$\frac{2465}{2360} = 1.04$$

which is not significant. Working up Table 14.41 we test each of the $B \times T$, $V \times T$, and T mean squares against the $h \times T$ interaction mean square. $B \times T$ and T yield mean square ratios of 13542/2357 = 5.73** and 215201/2360 = 91.2***, respectively.

If we note the expectations of the mean squares, the Heats mean square is to be tested against the $h \times T$ mean square. Hence the mean square ratio is 3795/2360 = 1.61. This is not significant judged against the distribution of $F_{18,36}$. The $V \times B$ interaction is tested against Heats mean square, giving ratio of 2024/3795 = 0.53, which is not significant. Finally, the Bar Size and Vendor mean squares, tested against Heats mean square yield values of 85697/3795 = 22.6***, and 1648/3795 = 0.43.

We proceed now to a discussion of the results. Since at the time of the experiment only one known commercial process was available for extracting titanium from its ore and since all titanium was processed to its ingot form in an

TABLE 14.41
ANALYSIS OF VARIANCE TABLE YIELD STRENGTH OF ROTOR BLADES

SOURCE	SUM OF SQUARES	DEGREES OF FREEDOM	MEAN SQUARE	MEAN SQUARE RATIO	EXPECTED VALUE OF MEAN SQUARE
Vendors (V)	3296	2	1648	0.43	$\sigma^2 + 3\sigma^2_{kl(ij)} + 9\sigma^2_{k(ij)} + \langle 81\sigma^2_i \rangle$
Bar Sizes (B)	171394	2	85697	22.6***	$\sigma^2 + 3\sigma^2_{kl(ij)} + 9\sigma^2_{k(ij)} + \langle 81\sigma^2_j \rangle$
$V \times B$	8097	4	2024	0.53	$\sigma^2 + 3\sigma^2_{kl(ij)} + 9\sigma^2_{k(ij)} + \langle 27\sigma^2_{ij} \rangle$
Heats (h) with ($V \times B$)	68312	18	3795	1.61	$\sigma^2 + 3\sigma^2_{kl(ij)} + 9\sigma^2_{k(ij)}$
Test Specimens (T)	430402	2	215201	91.2***	$\sigma^2 + 3\sigma^2_{kl(ij)} + \langle 81\sigma^2_T \rangle$
$V \times T$	1313	4	328	0.14	$\sigma^2 + 3\sigma^2_{kl(ij)} + \langle 27\sigma^2_{il} \rangle$
$B \times T$	54083	4	13521	5.73**	$\sigma^2 + 3\sigma^2_{kl(ij)} + \langle 27\sigma^2_{jl} \rangle$
$V \times B \times T$	19724	8	2465	1.04	$\sigma^2 + 3\sigma^2_{kl(ij)} + \langle 9\sigma^2_{ijl} \rangle$
$h \times T$ within ($V \times B$)	84954	36	2360	39.43***	$\sigma^2 + 3\sigma^2_{kl(ij)}$
Residual	9695	162	60		σ^2
Total	851270	242			

almost identical manner, it is not surprising that no vendor effect was apparent. The different suppliers do not really represent different processes.

On the other hand, a difference in bar stock diameter should cause a difference in the yield strength of the material. Since all 3 bar stock sizes are rolled from the same size ingots, the smaller diameter receives more working and thus will have higher tensile properties. As expected, this bar size effect is seen to be highly significant.

A significant effect noted was that of test specimens. Test specimens taken from forgedowns and blades are significantly higher than those taken from bar stock. Again, this effect is caused by the added forging operations used to produce a blade or forgedown from bar stock.

The interaction of Bar Size and Test Specimen may be explained by the fact that the larger bar stock is forged a greater amount in producing blades than is the smaller bar stock. Therefore, the increase in yield strength from forging is not as great for the small bar stock as for the larger bar stock.

The conclusions may be summarized as follows:

1. No real differences in yield strength are evident among the vendors.

2. Yield strength is significantly affected by bar size with the smaller bar size rendering the higher yield strengths. This effect holds true for bar stock tests, forgedown tests, and forged blade tests.

3. The effect of *changing* test specimens has an influence on yield strength of this titanium alloy. The yield strengths of forgedowns and blades are approximately equal, while the yield strength of the raw bar stock is markedly less than that of forgedowns and blades.

4. There is a definite interaction between Heats and Test Specimens, but not Vendor and Bar Stock nor Vendor and Test Specimens.

Since the effects of changes in Bar Sizes and Test Specimens are of particular interest we proceed to give tables of estimated means $\hat{\alpha} + \hat{\beta}_j$, and $\hat{\alpha} + \hat{\tau}_l$) for the different levels of these factors. The value of $(\xi + \beta_j)$ is estimated by $\overline{Y}_{j\ldots}$ and the variance of this estimator is $\frac{1}{9}\sigma^2_{k(ij)} + \frac{1}{27}\sigma^2_{kl(ij)} + \frac{1}{81}\sigma^2$. An unbiased estimate of the variance of $\overline{Y}_{j\ldots}$ is

$$\frac{1}{81} \times [\text{Heats within Vendor–Bar Size combinations mean square}]$$

$$= 46.85 \times 10^4 \text{ psi}^2$$

and this is based on 18 degrees of freedom. Individual $100(1-\alpha)$ percent confidence limits (in units of 100,000 psi) are then calculated from the formula $\overline{Y}_{j\ldots} \pm t_{18, 1-\alpha/2} \cdot 10^{-3}\sqrt{46.85}$. If $\alpha = 0.05$, the limits are $\overline{Y}_{j\ldots} \pm 0.0144$. The results of these calculations (with $\alpha = 0.05$) are shown below. If simultaneous confidence

BAR SIZE	LOWER LIMIT	ESTIMATE OF EXPECTED VALUE	UPPER LIMIT
1"	1.425	1.439	1.454
$1\frac{1}{4}$"	1.385	1.400	1.414
$1\frac{3}{4}$"	1.360	1.375	1.389

intervals are required, then (from Table 15.14) the limits would be $(\overline{Y}_{j\ldots} \pm 0.0144 \times$
$(2.617/2.101) \times 10^5 = (\overline{Y}_{j\ldots} \pm 0.0179) \times 10^5$ psi.

Turning now to the factor Test Specimen, we see that an unbiased estimator of
$\alpha + \tau_l$ is $\overline{Y}_{\ldots l}$. The variance of this estimator is $\frac{1}{9}(\sigma^2_{k(ij)} + \sigma^2_{kl(ij)}) + \frac{1}{81}\sigma^2$. Unfortunately, it is not possible to estimate this quantity unbiasedly as a simple multiple
of a single mean square in the analysis of variance table. An unbiased estimate is
provided by

$\frac{1}{81}$[Heats within Vendor-Bar Size combinations mean square]

$+ \frac{2}{81}$[(Heat $\times$ Test Specimen interaction within Vendor-Bar Size

combinations mean square)—(Residual mean square)]

$$= \tfrac{1}{81}(3795) + \tfrac{2}{81}[2360 - 60] = 103.64$$

It is not possible to regard this as the observed value of a multiple of a χ^2 random
variable. It is, rather, to be regarded as the observed value of a random variable
distributed as

$$\frac{\chi^2_{18}\left(\sigma^2 + \sigma^2_{kl(ij)} + 9\sigma^2_{k(ij)}\right)}{81 \times 18} + \frac{2\chi^2_{36}\left(\sigma^2 + 3\sigma^2_{kl(ij)}\right)}{81 \times 36} - \frac{2\chi^2_{162}\sigma^2}{81 \times 162}$$

where the three χ^2's are mutually independent. For purposes of approximation this
is represented, using Satterthwaite's method [F. E. Satterthwaite, "An approximate
Distribution of Estimates of Variance Components," *Biometrics*, **2** (1946)], by

$$\nu^{-1}\chi^2_\nu\left[\tfrac{1}{9}\left(\sigma^2_{k(ij)} + \sigma^2_{kl(ij)}\right) + \tfrac{1}{81}\sigma^2\right]$$

where (making variances agree)

$$\frac{2}{\nu}\left[\tfrac{1}{9}\left(\sigma^2_{k(ij)} + \sigma^2_{kl(ij)}\right) + \tfrac{1}{81}\sigma^2\right]^2$$

$$= 2\left[\frac{\left(\tfrac{1}{9}\sigma^2_{k(ij)} + \tfrac{1}{27}\sigma^2_{kl(ij)} + \tfrac{1}{81}\sigma^2\right)^2}{18} + \frac{\left(\tfrac{2}{27}\sigma^2_{kl(ij)} + \tfrac{2}{81}\sigma^2\right)^2}{36} + \frac{\left(\tfrac{2}{81}\sigma^2\right)^2}{162}\right]$$

Inserting estimated values of the variances we find

$$\nu \doteq \frac{(103.64)^2}{\dfrac{1}{18}\left(\dfrac{3795}{81}\right)^2 + \dfrac{1}{36}\left(\dfrac{4720}{81}\right)^2 + \dfrac{1}{162}\left(\dfrac{120}{81}\right)^2}$$

$$\doteq \frac{10741.25}{216.28} = 49.7$$

(Note that ν depends only on *ratios* of estimates of variances.)

Individual 95 percent confidence limits for the value of the "Test-Specimen
mean" $(\alpha + \tau_l)$ are then calculated from the formula

$$\overline{Y}_{\ldots l} \pm t_{49.7, 0.975} \cdot 10^{-3}\sqrt{103.64} = \overline{Y}_{\ldots l} \pm 0.02045$$

The resulting values for the limits are shown below. If simultaneous confidence limits (with joint confidence coefficient 95 percent) are required they would be obtained by using the formula

$$\left(\bar{Y}_{....l.} \pm 0.02045 \times \left(\frac{2.468}{2.009} \right)^{-1} \right) \times 10^5 = (\bar{Y}_{....l.} \pm 0.02512) \times 10^5 \, \text{psi}$$

TEST SPECIMEN	LOWER LIMIT	ESTIMATE OF EXPECTED VALUE	UPPER LIMIT
Raw bar stock	1.325	1.345	1.365
Forgedowns	1.415	1.435	1.456
Finished forged blades	1.413	1.433	1.454

Since the variance among heats is significantly different from 0, it is well to estimate this variance. An unbiased estimate is

$$\hat{\sigma}^2_{k(ij)} = \tfrac{1}{9}[(\text{Heats mean square}) - (\text{Heats} \times \text{Treatments mean square})]$$

$$= \tfrac{1}{9}[3795 - 2360] = 159.44$$

This gives $\hat{\sigma}_{k(ij)} = 1260$ psi (since the coded unit is 100 psi).

Figure 14.2 illustrates how some of the results may be presented graphically. Note how clearly various interactions can be picked up. For simplicity the totals of Table 14.39 are plotted. Of course, these can all be translated to the original units of yield strength in 100,000 psi.

*14.8.3 Recovery of Interblock Information

In the analysis of a balanced incomplete block design developed in Section 14.2.1 the model (14.3) was used, in which both block and treatment effects are represented by parameters. As a consequence the estimators $\hat{\tau}_q$ [see (14.8)] of treatment direction depend only on differences among observed values within the same block.

If the block effects are represented by random variables, giving the model

$$X_{ij} = \xi + \tau_j + Y_i + Z_{ij} \qquad (14.15)$$

with $E[Y_i] = 0$, $\text{var}(Y_i) = \sigma'^2$, and the Y's and Z's all mutually uncorrelated then a different analysis results, in which some information from differences *between* blocks is used in estimating the τ's. The model (14.15) can be rewritten

$$X_{ij} = \xi + \tau_j + W_{ij} \qquad (14.15')$$

with $W_{ij} = Y_i + Z_{ij}$. It is easy to see that $E[W_{ij}] = 0$; $\text{var}(W_{ij}) = \sigma'^2 + \sigma^2$; $\text{cov}(W_{ij}, W_{i'j'}) = 0$ if $i \neq i'$, $\text{cov}(W_{ij}, W_{ij'}) = \sigma'^2$ (for $j \neq j'$).

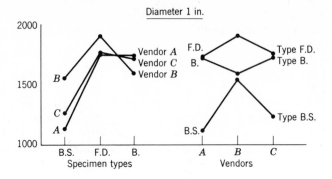

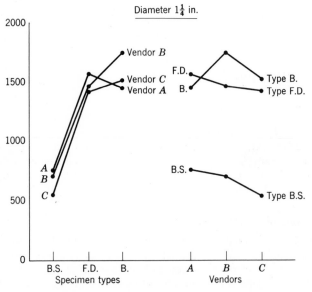

Figure 14.2 Totals of coded yield strengths for different diameters, specimen types, and vendors.

The variance–covariance matrix of the W's is

$$
V = \begin{bmatrix}
V_1 & 0 & \cdot & 0 \\
0 & V_1 & \cdot & 0 \\
\cdot & \cdot & \cdot & \cdot \\
\cdot & \cdot & \cdot & \cdot \\
\cdot & \cdot & \cdot & \cdot \\
0 & 0 & \cdot & V_1
\end{bmatrix}
$$

where V_1 is a $k \times k$ matrix with diagonal elements each equal to $(\sigma^2 + \sigma'^2)$ and off-diagonal elements each equal to σ'^2.

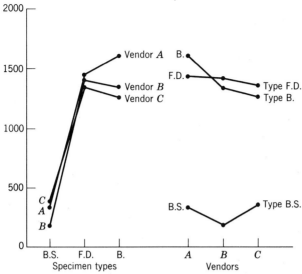

Figure 14.2   (*Continued*)

The best linear unbiased estimators of the parameters are obtained by minimizing

$$(X - E[X|\xi,\tau])'V^{-1}(X - E[X|\xi,\tau])$$

with respect to ξ and τ, subject of course to the condition $\sum_{j=1}^{\tau}\tau_j = 0$.
The estimator of ξ is, as before, $\hat{\xi} = \overline{X}_{..}$. The estimator of τ_q, however, is now

$$\tau_q^* = (\lambda t\omega + r)^{-1}\left[\lambda t\omega\hat{\tau}_q + r\left(\overline{X}_{.q} - \overline{X}_{..}\right)\right] \tag{14.16}$$

where $\hat{\tau}_q$ is as given in (14.8) and $\omega = \sigma'^2/\sigma^2$. This is a weighted mean of the "intrablock estimator" $\hat{\tau}_q$ and of a new "estimator" $(\overline{X}_{.q} - \overline{X}_{..})$. The latter is the estimator which would be obtained if we put $\beta_i = 0$ in (14.3), or $Y_i = 0$ in (14.15). It may therefore be called the "block-ignored estimator."
The variance of (14.15) is

$$\text{var}(\tau_q^*) = \frac{1+k\omega}{r+\lambda t\omega} \cdot \frac{t-1}{t}\sigma^2 \tag{14.17}$$

and the variance of the intrablock estimator [if model (14.15) *is valid*] is

$$\text{var}(\hat{\tau}_q) = \frac{k(t-1)}{\lambda t^2}\sigma^2 \tag{14.18}$$

The excess of $\text{var}(\hat{\tau}_q)$ over $\text{var}(\tau_q^*)$ is

$$\frac{(kr - \lambda t)(t-1)}{\lambda(r + \lambda t\omega)t^2}\sigma^2 = \frac{(r-\lambda)(t-1)}{\lambda(r + \lambda t\omega)t^2}\sigma^2$$

Since $r > \lambda$ (for a BIB) this is always positive, as indeed it must be since τ_q^* is the *best* linear unbiased estimator of τ_q.

Unfortunately, the value of $\omega(= \sigma'^2 / \sigma^2)$, which is needed to compute τ_q^*, is usally not known. If no previous information on this ratio is available, estimates of it must be obtained from the data. In Example 14.10 we describe two methods that may be used. Neither will be very accurate, but the estimator τ_q^* is unbiased, even if an incorrect value of ω is used and there are some results (see, in particular [14.12]) which indicate that loss of accuracy in τ_q^* is not great for quite large errors in ω.

Example 4.10   We use the data of Example 14.1 as an illustration. Values of the two kinds of estimators of the τ's are set out:

q	1	2	3	4	5	6	7
Composition	A	B	C	D	E	F	G
$\hat{\tau}_q$ (intrablock)	-1.14	2.29	0.86	-2.29	-1.29	3.71	-2.14
$\bar{X}_q - \bar{X}$ (blocks ignored)	-0.71	3.29	0.29	-1.71	-1.71	3.62	-3.05

From (14.16) (with $\lambda = 1$, $r = k = 3$, $b = t = 7$)

$$\tau_q^* = (7\omega + 3)^{-1}\left[7\omega\hat{\tau}_q + 3\left(\bar{X}_{.q} - \bar{X}_{..}\right)\right]$$

We do not know ω, and so must estimate it.

We may use the formula (see [14.3], p. 150)

$$\hat{\omega} = \frac{b - 1}{bk - t}\left[\frac{\text{(adjusted) Blocks mean square}}{\text{Residual mean square}} - 1 \right]$$

which, in this case, gives

$$\hat{\omega} = \frac{6}{14}\left[\frac{4.98}{0.929} - 1 \right] = 1.87$$

(The mean squares are obtained from Table 14.8.)

An alternative method of estimating ω is based on the fact that the model (14.15) implies that the correlation between true residuals $(X_{ij} - \xi - \tau_j = Y_i + Z_{ij})$ in the same block (i.e., with common i) is

$$\rho = \frac{\sigma'^2}{\sigma^2 + \sigma'^2} = \frac{\omega}{1 + \omega} \qquad (14.19)$$

We use the estimated residuals

$$\hat{W}_{ij} = X_{ij} - \bar{X}_{..} - \tau_j^*(\omega_1)$$

where $\tau_j^*(\omega_1)$ denotes that we have taken $\omega = \omega_1$ in the calculation. A new value for ω, ω_2 say, is then obtained by making

$$\frac{\omega_2}{1 + \omega_2} \quad = \quad \text{product-moment correlation between all possible pairs } \hat{W}_{ij}, \hat{W}_{ij'}$$

[There are $bk(k-1)$ such pairs in all, taking account of order.]
This is equivalent to making

$$\frac{\omega_2}{1+\omega_2} = \frac{1}{k-1}\left[\frac{\sum\limits_{i=1}^{k}\left(\sum\limits_{j(i)}\hat{W}_{ij}\right)^2}{\sum\limits_{i=1}^{k}\sum\limits_{j(i)}\hat{W}_{ij}^2} - 1\right]$$

The process is continued until the value of ω settles down. Alternatively we may proceed as below.

With the present data, taking

$$\omega_1 = 0 \quad \text{gives} \quad \omega_2 = 0.77$$

$$\omega_1 = 1 \quad \text{gives} \quad \omega_2 = 2.51$$

$$\omega_1 = \infty \quad \text{gives} \quad \omega_2 = 3.12$$

We find that $\omega_1 = 3$ gives $\omega_2 = 2.88$ and so a value $\hat{\omega}$ of about 3.0 is indicated.

Although this differs from the value 1.87 obtained from the analysis of variance mean squares it does not lead to greatly different values of τ_q^* as can be seen from the following table.

Values of τ_q^*

Composition	A	B	C	D	E	F	G
$\hat{\omega} = 1.87$	-1.00	2.48	0.75	-2.18	-1.37	3.69	-2.31
$\hat{\omega} = 3$	-1.09	2.40	0.79	-2.22	-1.34	3.70	-2.25

14.9   COVARIANCE TECHNIQUES

The *analysis of covariance* is a term applied, often rather loosely, to a wide variety of methods employed in the analysis of experimental data when there is not only a "response variable," X, of primary importance but also one or more "concomitant variables," $Y_1, Y_2, \ldots, Y_m$, say. We have already encountered some methods of this kind in Section 13.20.

In this section we are concerned with methods applicable to a restricted class of models. These models are all of form

(observed value of X) = (model of a kind described in Chapters 13, 14 and the present chapter)

+ [linear function of concomitant variable(s)]

$$= \mathfrak{M} + \mathfrak{L}, \text{ say} \tag{14.20}$$

In the particular case when

$$\mathfrak{M} = \xi + Z$$

(Z being a random residual as described in Section 13.2), we have a (simple or multiple) regression of the kind described in Sections 13.17 and 13.18. If $\mathfrak{M}$ is a one-way classification model (see 13.4) with

$$\mathfrak{M} = \xi + \gamma_t + Z_{ti}$$

corresponding to X_{ti}, and for $\mathfrak{L}$ we take $\beta_t(Y_{ti} - \overline{Y}_t)$, in an obvious notation we have the model

$$X_{ti} = \xi + \gamma_t + \beta_t\left(Y_{ti} - \overline{Y}_t\right) + Z_{ti}$$

discussed in Section 13.19.

The analysis of covariance methods that we will describe are mostly concerned with making inferences about the parameters in $\mathfrak{M}$, rather than the regression coefficients in $\mathfrak{L}$. The $\mathfrak{L}$ terms are included in the analysis in order to reduce the effect of the concomitant variable(s) on the inferences about the parameters in $\mathfrak{M}$. This should be clearly understood: sometimes we may be more interested in crude comparisons, in that differences among the Y values may, in fact, reflect intrinsic differences among the experimental categories.

For example, suppose we are comparing tensile strength of rubber bands made by two processes, and use average cross-sectional area as a concomitant variable. Although the fact that an apparent difference between the average tensile strengths for the two processes may be "explained" by differences in average cross-sectional area may be of interest, it is quite likely that we will be more interested in the fact that the difference does exist, however caused.

Nevertheless, provided proper care in interpretation is exercised, "elimination" [to an extent depending on adequacy of the model] of concomitant variable(s) can lead to valuable insights into the structure of variation of experimental material.

We give here a detailed discussion in terms of a randomized block experiment, as in (13.52) of Section 13.11. Methods appropriate to other designs may be developed by similar arguments.

If a single variate X were measured on each "plot," the standard parametric model would be

$$X_{ij} = \xi + \delta_i + \tau_j + Z_{ij} \qquad (i = 1, \ldots, b; j = 1, \ldots, t)$$

with $\Sigma_{i=1}^{b} \delta_i = 0 = \Sigma_{j=1}^{t} \tau_j$; Z_{ij}'s mutually independent; $E[Z_{ij}] = 0$; and $\mathrm{var}(Z_{ij}) = \sigma^2$. This is the $\mathfrak{M}$ part of the model of (14.20).

If a further (concomitant) variable Y is measured on each plot, we can introduce the $\mathfrak{L}$ part of the model as (i) $\beta(Y_{ij} - \overline{Y}_{..})$ or (ii) $\beta_i(Y_{ij} - \overline{Y}_{..})$. (There are many other possibilities. These are two of the simplest.) In (i) a constant regression slope is assumed, in (ii) the regression slope is allowed to vary from block to block. In both cases, simple linear regression is assumed.

We proceed to consider case (i). The analysis is, in fact, based on the general method described in Section 13.16, but we couch it in language which, we believe, makes methods more comprehensible and easier to remember.

We first recall the standard analysis for the X values:

SOURCE	SUM OF SQUARES	DEGREES OF FREEDOM
Blocks	$t \sum_{i=1}^{b} (\bar{X}_{i.} - \bar{X}_{..})^2$	$b-1$
Treatments	$b \sum_{j=1}^{t} (\bar{X}_{.j} - \bar{X}_{..})^2$	$t-1$
Residual	$\sum_{i=1}^{b} \sum_{j=1}^{t} (X_{ij} - \bar{X}_{i.} - \bar{X}_{.j} + \bar{X}_{..})^2$	$(b-1)(t-1)$

The least squares estimator of β is

$$\hat{\beta} = \frac{\sum_i \sum_j X'_{ij} Y'_{ij}}{\sum_i \sum_j Y'^{2}_{ij}}$$

where $X'_{ij} = X_{ij} - \bar{X}_{i.} - \bar{X}_{.j} + \bar{X}_{..}$ and Y'_{ij} is a similar function of the Y_{ij}'s.

The adjusted residual sum of squares is obtained by subtracting $\hat{\beta}\sum_i\sum_j X'_{ij} Y'_{ij} = \hat{\beta}^2 \sum\sum Y'^{2}_{ij}$ from the standard residual $\sum\sum X'^{2}_{ij}$. It has $(b-1) \times (t-1) - 1 = bt - b - t$ degrees of freedom (1 less than in the standard case).

It is tempting to suppose that the other sums of squares could be adjusted in a similar fashion (though without changing the degrees of freedom). This would give

Adjusted Treatment sum of squares $= b \sum_{j=1}^{t} \left(\bar{X}_{.j} - \bar{X}_{..} \right)^2 - \hat{\beta}^2 b \sum_{j=1}^{t} \left(\bar{Y}_{.j} - \bar{Y}_{..} \right)^2$

In fact, however, the formula should be

Adjusted Treatment sum of squares

$$= b \sum_{j=1}^{t} \left(\bar{X}_{.j} - \bar{X}_{..} \right)^2 - \hat{\beta}^2 \sum_{i=1}^{b} \sum_{j=1}^{t} \left(Y_{ij} - \bar{Y}_{i.} \right)^2$$

$$+ \hat{\beta}^2 \sum_{i=1}^{b} \sum_{j=1}^{t} \left(Y_{ij} - \bar{Y}_{i.} - \bar{Y}_{.j} + \bar{Y}_{..} \right)^2$$

where

$$\hat{\beta} = \frac{\sum\limits_{i=1}^{b} \sum\limits_{j=1}^{t} \left(X_{ij} - \bar{X}_{i.}\right)\left(Y_{ij} - \bar{Y}_{i.}\right)}{\sum\limits_{i=1}^{b} \sum\limits_{j=1}^{t} \left(Y_{ij} - \bar{Y}_{i.}\right)^2}$$

To see why this is so, note that we must have

(Adjusted Treatment sum of squares) + (Residual sum of squares)

$$= \min \sum_{i=1}^{b} \sum_{j=1}^{t} \left[X_{ij} - \xi - \delta_i - \beta\left(Y_{ij} - \bar{Y}_{..}\right) \right]^2$$

$$= \sum_{i=1}^{b} \sum_{j=1}^{t} \left(X_{ij} - \bar{X}_{i.}\right)^2 - \hat{\beta}^2 \sum_{i=1}^{b} \sum_{j=1}^{t} \left(Y_{ij} - Y_{i.}\right)^2$$

The calculations for this analysis may be set out as follows (the notation is self-explanatory):

(1)	(2)	(3)	(4)	(5)	(6)
SOURCE	SUMS OF SQUARES		SUMS OF PRODUCTS	$=(4)/(2)$	$=(5)\times(4)$
Blocks	$S_{XX}^{(B)}$	$S_{YY}^{(B)}$	$S_{XY}^{(B)}$	—	—
Treatments	$S_{XX}^{(T)}$	$S_{YY}^{(T)}$	$S_{XY}^{(T)}$	—	—
Residual	$S_{XX}^{(R)}$	$S_{YY}^{(R)}$	$S_{XY}^{(R)}$	$\hat{\beta}$	$S^{(R)}$
Blocks and Residual	$S_{XX}^{(C)}$	$S_{YY}^{(C)}$	$S_{XY}^{(C)}$	$\hat{\beta}$	$S^{(C)}$

[Note: $S^{(C)} \equiv S^{(B)} + S^{(T)}$]
Adjusted Treatment sum of squares: $S_{XX}^{(T)} - (S^{(C)} - S^{(R)})$
Residual sum of squares: $S_{XX}^{(R)} - S^{(R)}$

*14.10 CHOICE OF VARIABLES

In all the preceding models it was assumed that the observations conform to some additive model. Suppose, however, that the "correct" model were multiplicative. Then an additive model could be applied to the logarithm of the observed characteristics. For example,

$$W_{tij} = \alpha\beta_t\gamma_i(\beta\gamma)_{ti}Z_{tij}$$

Then

$$Y_{tij} = \log W_{tij} = \log\alpha + \log\beta_t + \log\gamma_i + \log(\beta\gamma)_{ti} + \log Z_{tij}$$

For such variates as time to failure, corrosion, decay, and the like, it is not uncommon for a logarithmic transformation of the yield to be represented by a linear model. As an illustration of this we shall present an example by Bowen, Groot, and Jaech ["Variable Interaction: A Statistical Solution," *Corrosion*, **15** (1959)] in which they carry out an analysis of variance on the linear model and follow with an analysis of the transformed data.

Example 14.10 In the design of a power-producing nuclear reactor, a high purity, light water coolant and aluminum jacketed fuel elements were used. Even after the reactor is shut down it continues to generate heat by radioactive decay. In case the main cooling system fails, it is necessary to have available some supplementary cooling system. This can be done most easily by running ordinary potable water through the reactor from the fire or sanitary water systems. The purpose of the experiment was to determine whether low purity water at low temperature would produce excessive corrosion in the system. Two matched autoclaves were used, one with deionized water and one with specified alternating waters. Three different aluminum alloys were selected at random: 1245, M-388, and X-2219. The test was carried out for periods of 1 and 3 months.

Samples were sized 0.5 by 1 in. They were hung in the autoclaves and protected from each other to prevent dependence of results. In this corrosion test, the observations were in milli-inches penetration of the samples. The mathematical model used in the *first analysis* was

$$X_{ijkl} = \beta + A_i + \omega_j + (A\omega)_{ij} + \tau_k + (A\tau)_{ik} + (\omega\tau)_{jk} + (A\omega\tau)_{ijk} + Z_{l(ijk)}$$

where $i(=1,2,3)$ denotes alloy, $j(=1,2)$ denotes water (deionized and alternating), $k(=1,2)$ denotes times (1 month and 3 months), and $l(=1,2)$ denotes replicates. The data are given in Table 14.42.

TABLE 14.42
CORROSION OF ALUMINUM IN DEIONIZED AND ALTERNATING WATER

TIME	ALLOY	DEIONIZED WATER		ALTERNATING WATER	
1 month	1245	0.09	0.07	0.22	0.21
	M-388	0.06	0.06	0.21	0.19
	X-2219	0.07	0.09	0.29	0.29
3 months	1245	0.09	0.07	0.30	0.28
	M-388	0.11	0.08	0.29	0.31
	X-2219	0.14	0.11	0.45	0.48

By the computational techniques described in Section 14.7, the values in the ANOVA table can be calculated. These appear in Table 14.43.

Each of the mean square ratios indicated by the EMS column of Table 14.43 is significant except the triple interaction.

TABLE 14.43
ANOVA OF RAW DATA ($\times 100$)

SOURCE	SUM OF SQUARES	D.F.	MEAN SQUARE	EXPECTED VALUE OF MEAN SQUARE [FROM TABLE 14.35]
Alloys (a)	300.2	2	150.1	$\sigma^2 + 2\sigma_{ijk}^2 + 4\sigma_{ik}^2 + 4\sigma_{ij}^2 + 8\sigma_i^2$
Water (W)	2562.7	1	2562.7	$\sigma^2 + 2\sigma_{ijk}^2 + 4\sigma_{ij}^2 + \langle 12\sigma_j^2 \rangle$
Time (T)	308.2	1	308.2	$\sigma^2 + 2\sigma_{ijk}^2 + 4\sigma_{ik}^2 + \langle 12\sigma_k^2 \rangle$
$a \times W$	140.1	2	70.0	$\sigma^2 + 2\sigma_{ijk}^2 + 4\sigma_{ij}^2$
$a \times T$	53.1	2	26.5	$\sigma^2 + 2\sigma_{ijk}^2 + 4\sigma_{ik}^2$
$W \times T$	121.5	1	121.5	$\sigma^2 + 2\sigma_{ijk}^2 + 6\sigma_{jk}^2$
$a \times W \times T$	12.3	2	6.15	$\sigma^2 + 2\sigma_{ijk}^2$
Residual	25.9	12	2.16	σ^2
Total	3524.0	23		

Since it was recognized that the appropriate model is not necessarily linear, the model

$$X'_{ijkl} = \beta' + A'_i + \omega'_j + (A'\omega')_{ij} + \tau'_k + (A'\tau')_{ik}$$

$$+ (\omega'\tau')_{jk} + (A'\omega'\tau')_{ijk} + Z'_{ijkl}$$

was used, where $X' = \log X$, $B' = \log B$, $A'_i = \log A_i$, and so forth.

The ANOVA table of the transformed values is presented in Table 14.44.

In the analysis of the transformed data we see that the variation among alloys is no longer significant. The low-purity water (alternating cycle in autoclave) does produce a markedly significant effect. The random effect of the alloy–time interaction is not significant. It is very nearly at the 5 percent level. The remaining effects are not now significant.

TABLE 14.44
ANALYSIS OF VARIANCE OF TRANSFORMED DATA

SOURCE	DEGREES OF FREEDOM	MEAN SQUARE	MEAN SQUARE RATIO
Alloy	2	0.0532	< 19.7
Water	1	1.6622	615.6***
Time	1	0.1347	12.4
$a' \times W'$	2	0.0027	0.87
$a' \times T'$	2	0.0109	3.52
$W' \times T'$	1	0.0025	0.81
$a' \times W' \times T'$	2	0.0031	1.07
Residual	12	0.0029	
Total	23		

The change in model (from a linear model for the original observations to a linear model for the logarithms of these observations) has reduced the number of significant interactions. This may well be regarded as evidence justifying the change, in that it indicates that a simpler model, containing fewer terms, will be adequate to describe the results. Further, in the present case, physical considerations indicate that the second model should be preferable to the original model.

REFERENCES

1. Bose, R. C., S. S. Shrikande, and E. T. Parker, "Further Results on the Construction of Mutually Orthogonal Latin Squares and the Falsity of Euler's Conjecture," *Canadian Journal of Mathematics*, **12**, 189–203 (1960).

2. Brownlee, K. A., *Statistical Theory and Methodology in Science and Engineering*, 2nd ed., Wiley, New York, 1965, Chapters 14, 15.

3. Cochran, W. G. and G. M. Cox, *Experimental Designs*, 2nd ed., Wiley, New York, 1957.

4. Cooper, B. E., *Statistics for Experimentalists*, Pergamon Press, London and New York, 1969, Chapters 9, 10.

5. Cox, D. R., *Planning of Experiments*, Wiley, New York, 1958.

6. Davies, O. L., ed., *Statistical Methods in Research and Production*, 3rd ed., Hafner, New York, Oliver and Boyd, London, 1957.

7. Federer, W. T., *Experimental Design: Theory and Application*, Macmillan, New York, 1955.

8. Fisher, R. A. and F. Yates, *Statistical Tables for Biological, Agricultural and Medical Research*, 6th ed., Hafner, New York; Oliver and Boyd, London, 1963.

9. Hicks, C. R., *Fundamental Concepts in the Design of Experiments*, 2nd ed., Holt, Rinehart and Winston, New York, 1973, Chapters 5–7, 9.

10. John, P. W. M., *Statistical Design and Analysis of Experiments*, Macmillan, New York, 1971.

11. Peng, K. C., *The Design and Analysis of Scientific Experiments*, Addison-Wesley, Reading, Mass., 1967, Chapters 5–6.

12. Siskind, V., "On Using an Incorrect Value of σ_B^2/σ^2 in Incomplete Block Designs," *Biometrika*, **55** (1968).

13. Yates, F., *Experimental Design: Selected Papers of Frank Yates*, Griffin, London, 1970.

EXERCISES

1. Four paints are being tested for weight loss in an abrasion machine. These paints differ in the proportion of one basic ingredient. The abrasion machine can handle 3 paint panels at one time. Yield is measured in terms of weight loss in a fixed period of time and at a fixed speed of the machine. Four runs are made. Consider each run as a block. The data are stated in coded units.

   (a) What is the estimated average weight loss of each paint, independent of blocks?

	RUN			
PAINT	1	2	3	4
A	13	10	15	
B	15	11		21
C	18		23	29
D		20	26	28

(b) Are the paints different with respect to abrasion loss?

(c) Are the runs significantly different?

2. An experiment was conducted to determine the effect of 4 treatments applied to the coils of TV tube filaments on the current flow through these coils. Because of the amount of time needed to perform this experiment, only 3 sample values could be run each day. Four treatments, A, B, C, and D were tested. The data are given, in coded form, in the table below.

DAY	TREATMENT			
(BLOCKS)	A	B	C	D
1	12		30	17
2		42	24	13
3	14	23	41	
4	10	33		21

(a) Is there a significant effect among treatments?

(b) Is there a significant day (block) effect?

3. Ten specimens of rubber are sent to the laboratory for a test of flexural strength. There are 5 curing times. However, each specimen is sufficient for only 2 samples. Hence a balanced incomplete block is proposed. Specimens are considered as blocks and curing times as treatments. Investigate the effect of curing time on flexural strength, using the coded data below.

	CURING TIMES				
SPECIMENS	1	2	3	4	5
1	25				6
2	10		3		
3	3			16	
4	15	11			
5			0		6
6				14	11
7		6			17
8			10	27	
9		10	5		
10		7		21	

4. Six manufacturers of critical material claim that their product can withstand a severe tensile test criterion. Each manufacturer supplies enough material for 5 tests. However, it is planned to carry out these tests under 10 different conditions. The total of 30 test pieces are arranged in a balanced incomplete block in order to test the effect of manufacturers. (The data (shown below) are coded.)

	CONDITION									
MANUFACTURER	1	2	3	4	5	6	7	8	9	10
1	1.62			2.10		1.50	2.30	2.23		
2		1.93			1.90		1.77	1.64	1.75	
3			2.22			1.56		2.29	1.92	2.46
4	2.14	1.58	1.88				1.81			1.86
5	1.65		1.61	1.65	2.03				2.15	
6		1.99		1.64	1.74	2.46				2.62

(*a*) What assumptions are made in the model?
(*b*) Carry out a complete analysis of variance.
(*c*) Draw conclusions.
5. The following data present the results of an exploratory experiment on the first stage of a purification process which involves absorbing a substance on carbon. Two factors, pH and the quantity of carbon as a percentage of the volume of solution, were to be studied at 5 levels each. The "treatments" consist of 5 different grades of carbon (*A–E*). Evaluate the effect of each factor.

	PER CENT CARBON					
pH	0.05	0.10	0.20	0.40	0.80	Totals
4.0	17 *A*	39 *D*	65 *B*	19 *C*	12 *E*	152
5.0	32 *E*	33 *C*	61 *A*	71 *B*	94 *D*	291
6.0	56 *C*	49 *A*	84 *D*	90 *E*	100 *B*	379
7.0	76 *D*	81 *B*	97 *E*	98 *A*	100 *C*	452
8.0	93 *B*	90 *E*	97 *C*	100 *D*	100 *A*	480
Totals	274	292	404	378	406	1754

Grade	*A*	*B*	*C*	*D*	*E*
Totals	325	410	305	393	321

6. An experiment to determine the amount of warping of copper plates was conducted in 4 different laboratories (1,2,3,4). Four temperatures were used (50°, 75°, 100°, 125°C) and 4 percentage compositions of copper were considered (40, 60, 80, and 100 ≡ *A*, *B*, *C*, and *D*, respectively). The results are shown below.

LABORATORIES

TEMPERATURE	1	2	3	4
50°	C 24	A 17	B 16	D 28
75°	D 27	B 18	A 12	C 17
100°	A 16	C 25	D 20	B 18
125°	B 23	D 23	C 29	A 21

Estimate the effect of each factor. Is there a linear or quadratic effect of temperature and of composition?

7. Analyze the following 4×4 Graeco-Latin square:

COLUMN

ROW	1	2	3	4
a	$A_1 = 6$	$B_3 = 4$	$C_4 = 7$	$D_2 = 5$
b	$B_2 = 5$	$A_4 = 6$	$D_3 = 3$	$C_1 = 4$
c	$C_3 = 4$	$D_1 = 5$	$A_2 = 8$	$B_4 = 4$
d	$D_4 = 3$	$C_2 = 2$	$B_1 = 8$	$A_3 = 6$

Letters $(A, B, C, D) \equiv 4$ known sources of variation
Subscripts $\equiv 4$ processes being investigated

8. The data below represent a series of 25 runs made at 5 temperatures and 5 durations to find the effect of these variables on the extent of conversion. The values are in percent conversion. To test for the possible effects of different reactors and different operators, the 25 runs were made in 5 reactors by 5 operators in a balanced arrangement so that each operator used each reactor only once at each temperature and duration. Latin letters in the table identify reactors and subscripts identify operators. Make an appropriate analysis of these data.

TEMP. °F	TIME (MINUTES)					TOTALS
	30	60	90	120	150	
100	16 A_1	40 B_3	50 C_5	20 D_2	15 E_4	141
125	30 B_2	25 C_4	62 D_1	67 E_3	30 A_5	214
150	50 C_3	50 D_5	83 E_2	85 A_4	45 B_1	313
175	80 D_4	80 E_1	95 A_3	98 B_5	70 C_2	423
200	90 E_5	92 A_2	98 B_4	100 C_1	88 D_4	468
TOTALS	266	287	388	370	248	1559

9. In Exercise 1, let the order of entry in each column refer to position on the machine, thus forming a Youden square. Test each of the factors for significance.

POSITION	RUN 1	2	3	4
1	A 13	D 20	C 23	B 21
2	B 15	A 10	D 26	C 29
3	C 18	B 11	A 15	D 28

10. The data below give the average deviations in millimeters of samples of size 10 on an inner diameter of a tube. The tubes were manufactured on 7 different machines, from 7 compositions of material, and on 3 different shifts.

COMPOSITION	MACHINES 1	2	3	4	5	6	7
A	5 α	4 β	9 γ				
B		12 α	9 β	9 γ			
C	7 β			6 α		8 γ	
D			7 β		5 α		3 γ
E	4 γ				6 β		5 α
F		10 γ			12 α	9 β	
G		4 α		4 γ			3 β

Determine whether there are significant differences due to machines, compositions or shifts. The shifts are designated as α, β, and γ.

11. The octane number is measured for 7 different gasolines. Gasoline F is the base, while the other 6 are the result of 6 different additives. The design is a 7×3 Youden square. Each fuel is given 2 minutes in the engine. The knockmeter is read at 60, 90, and 120 seconds to check the stability of the fuel. The data are given in the accompanying table. Consider the engines as blocks. (*a*) Are there any differences among blocks? (*b*) Are there any differences among gasolines?

TIME	ENGINE 1	2	3	4	5	6	7
60	A 43	B 36	C 33	D 44	E 41	F 36	G 33
90	B 34	C 32	D 47	E 40	F 35	G 32	A 41
120	D 47	E 46	F 43	G 33	A 44	B 32	C 27

12. The data in the table below are taken from David Frazier, "A Statistical Intercomparison of Thirteen Company and Competitive Motor Oils," presented at the Gordon Research Conferences, Statistics in Chemistry (1952). Thirteen oils were tested in 13 cars. Each car was run for 4 months with a different oil each month. The design was balanced in a 13×4 Youden square. The data below are grouped according to the 13 cars $(1, 2, \ldots, 13)$, the 13 oils $(A, B, \ldots, M)$ and the 4 months (I, II, III, IV). The 2 figures are (1) the $\log_{10}$ (oil consumption in coded units) and (2) the $\log_{10}$ (mileage on car at start of test period in units of 1000 mi.). Using (1) only, construct a suitable model and test for

(a) average differences among oils;

(b) a trend over the period of the experiment.

| | | MONTH | | | |
CAR		I	II	III	IV
1	E	0.82 1.10	D 1.07 1.13	G 0.98 1.18	M 1.04 1.21
2	F	0.82 0.96	E 0.79 1.02	H 0.78 1.08	A 0.91 1.13
3	D	1.05 1.09	C 0.71 1.15	F 0.71 1.12	L 1.09 1.17
4	I	1.23 0.80	H 1.14 0.95	K 1.16 1.05	D 1.16 1.14
5	G	1.17 1.05	F 1.13 1.11	I 1.05 1.14	B 1.02 1.18
6	M	0.92 1.02	L 0.78 1.06	B 0.86 1.10	H 0.80 1.13
7	L	1.07 1.50	K 0.57 1.52	A 1.00 1.54	G 1.03 1.55
8	C	0.88 1.17	B 0.98 1.10	E 0.63 0.90	K 0.89 1.01
9	H	0.88 1.17	G 0.84 1.38	J 0.92 1.43	C 0.83 1.48
10	J	1.09 1.34	I 0.82 1.30	L 0.88 1.35	E 0.81 1.38
11	K	0.69 1.24	J 0.46 1.51	M 1.06 1.54	F 0.42 1.56
12	A	0.43 1.51	M 0.80 1.54	C 0.56 1.56	I 0.36 1.59
13	B	0.75 1.27	A 0.80 1.22	D 0.68 1.34	J 0.61 1.31

13. Joe Zilch, in studying the effect of different brands of flashbulbs (denoted by A through E) on photographic density, made use of 5 varieties of cameras (ranging from box type to expensive) and 5 types of film (slow to fast ASA ratings). Data are coded by the formula $100(x - 0.65)$.

			CAMERAS		
FILM	1	2	3	4	5
1	(C) 3	(D) 3	(A) −1	(B) 3	(E) 4
2	(E) 12	(A) 4	(B) 7	(C) 5	(D) 7
3	(A) −3	(E) 3	(C) 1	(D) 1	(B) −5
4	(D) 4	(B) 5	(E) 9	(A) 2	(C) 3
5	(B) 0	(C) 0	(D) 5	(E) 7	(A) −2

(a) Construct a mathematical model.
(b) Estimate the standard deviation of any one measurement assuming this to be the same for each measurement (in coded form).
(c) What effects are significant?
(d) Determine a 95% confidence interval for the mean photographic density of film No. 4.

14. Three factors are being studied to determine their effect on percentage conversion. These are catalyst type (A), catalyst concentration (B), and reaction temperature (C). The data are given below.

	CONCENTRATION			
REACTION	CATALYST 1		CATALYST 2	
TEMPERATURE	0.1%	0.5%	0.1%	0.5%
120°C	75	85	60	67
160°C	88	90	83	94

(a) State the mathematical model.
(b) Are any of the main effects or interactions significant?
(c) Discuss the design and size of the experiment.

15. A study was made of the variations involved in sampling and testing a tar-based item used in lining a basic oxygen furnace. This was part of a more extensive study aimed at increasing the life of the lining. The manufacturing company was interested in comparing variations between production runs, between bags within each run, between samples from the same bag, and between the chemical determinations for the same sample. The measurements

pertain to a ratio of 2 prime ingredients in the tar mix and are stated in coded form.

	RUN A				RUN B		
BAG 1		BAG 2		BAG 1		BAG 2	
SAMPLE 1	SAMPLE 2	SAMPLE 1	SAMPLE 2	SAMPLE 1	SAMPLE 2	SAMPLE 1	SAMPLE 2
0.46	0.49	0.50	0.52	0.58	0.60	0.54	0.51
0.48	0.51	0.49	0.50	0.55	0.57	0.54	0.54

(a) Construct an appropriate model.
(b) What are the assumptions?
(c) Estimate the variances of the different factors.

16. In an investigation to test the fired modulus of rupture of 3 plastic pressed body materials, A 380, A 384, and A 386, 3 batches of raw material were taken from each composition. These, in turn, were each divided into 2 samples, from each of which 3 test specimens were prepared. The data below give the rupture modulus.

BATCH	A 380		A 384		A 386	
	SAMPLE 1	SAMPLE 2	SAMPLE 1	SAMPLE 2	SAMPLE 1	SAMPLE 2
	1450	1215	1228	1113	1147	1276
1	1390	1300	1006	1207	1084	1537
	1176	1171	1140	1188	1359	1364
	1080	1218	1031	1325	1293	1566
2	1510	1092	1209	1043	1112	1218
	1373	1163	1017	1264	1276	1231
	1334	1318	1214	1209	1255	1128
3	1219	1353	1070	1192	1417	1481
	1464	1277	1117	1137	1221	1463

(a) State the model equations and the assumptions.
(b) Estimate the variances involved.
(c) Do the 3 materials differ significantly?

17. The product yields of an undesirable by-product were measured for 3 different catalysts, each at 2 different pressures. Measurements are expressed in percentages. The experiment was carried out at 2 laboratories; 3 replicates were taken at each laboratory.

(a) Does either the catalyst or pressure level have a significant effect on the yield?
(b) Are the laboratories consistent with each other?

CATALYST

	I		II		III	
	Lab. A	Lab. B	Lab. A	Lab. B	Lab. A	Lab. B
HIGH PRESSURE	0.40	0.39	0.53	0.27	0.32	0.33
	0.32	0.45	0.43	0.45	0.12	0.68
	0.29	0.12	0.45	0.57	0.21	0.31
LOW PRESSURE	0.61	0.26	0.60	0.10	0.48	0.76
	0.24	0.54	0.42	0.98	0.67	0.78
	0.11	0.60	0.95	0.12	0.58	0.85

18. A laboratory procedure consists of the exposure of zinc in dilute aqueous solutions for a given period of time. The solutions are buffered so that they maintain a constant pH (hydrogen ion concentration, or acidity). There are 4 pH levels. The solution is exposed to 3 different levels of oxygen pressure, giving rise to 3 levels of oxygen concentration in the solution. The electrical potential (emf) of the zinc is observed hourly for 6 hr. In general, the emf starts at an active potential and changes to a less active potential. It is this change in emf that is analyzed. The levels of pH, pressure, and time were chosen arbitrarily and should be considered fixed quantitative levels. For the data below assume that the three factor interaction is 0.

(a) Specify the model and construct a complete analysis of variance table.

(b) Obtain an estimate of the residual variance.

(c) What further inferences can be made on each of the factors?

(T)	(P)	(pH)			
TIME	PRESSURE	3.90	4.35	5.25	6.73
1	1	0.020	0.095	0.118	0.145
	3	0.252	0.271	0.203	0.262
	5	0.268	0.307	0.237	0.268
2	1	0.095	0.219	0.196	0.266
	3	0.243	0.263	0.212	0.250
	5	0.250	0.298	0.245	0.281
3	1	0.100	0.241	0.208	0.272
	3	0.228	0.261	0.210	0.228
	5	0.238	0.292	0.246	0.286
4	1	0.116	0.247	0.189	0.264
	3	0.211	0.260	0.209	0.218
	5	0.233	0.289	0.246	0.286

(TABLE *Continues*)

(T)	(P)	(pH)			
TIME	PRESSURE	3.90	4.35	5.25	6.73
	1	0.135	0.246	0.172	0.262
	1	0.135	0.246	0.172	0.262
5	3	0.197	0.260	0.209	0.213
	5	0.228	0.288	0.246	0.287
	1	0.149	0.243	0.159	0.260
6	3	0.193	0.259	0.211	0.212
	5	0.201	0.286	0.247	0.288

(T) time in hours (P) oxygen pressure in atmosphere
(pH) negative $\log_{10}$ of hydrogen ion concentration
(Each entry represents the change in Emf (volts) from
time zero up to a particular time, for the stated pressure,
and pH.)

19. An analysis was conducted to investigate the effects due to resins, batch-to-batch differences, and curing on the hardness of varnish cake as determined by a modified Universal Precision Penetrometer. Specifically, 3 hardness readings were taken on cakes selected from 4 batches of each of 2 resins after a cure of 16 hrs at 110° and also after an additional cure of 6 hr at 135°C. The data are penetrations in tenths of a millimeter.

	RESIN A		RESIN B			RESIN A		RESIN B	
	CURE		CURE			CURE		CURE	
BATCH	1	2	1	2	BATCH	1	2	1	2
	32	16	34	16		30	14	40	17
I	33	17	35	17	III	29	15	41	17
	34	16	35	18		31	14	44	18
	26	11	37	17		22	13	33	17
II	29	10	37	18	IV	23	12	35	16
	29	11	36	17		20	11	33	16

Cure 1 represents a cure of 16 hours at 110°.
Cure 2 represents an additional cure of 6 hours at 135°.

(a) Are the 2 resins different in their residual variances?
(b) If not, test each of the factors involved. If so, treat each resin as a separate problem and test the factors, Cure and Batch in each case.
(c) What is the expected average change between the 2 cures?
(d) Determine a 95% confidence interval for this expected change.

20. Carbon black was added to latex to prepare an oil-black master. A control was prepared by dry-mixing the black into the oil master polymer in a Banbury mill. The steps of mixing, curing and testing were carried out for each of 2 different centers (A and B). The data recorded are tensile strength in psi.

MIXING

		A		B	
CURING	TESTING	DRY MIX	OIL-BLACK MASTER	DRY MIX	OIL-BLACK MASTER
A	A	3450	3700	3000	2900
	B	3100	3300	3200	3400
B	A	3500	3050	3050	2750
	B	3250	3550	3050	3200

(a) Construct a mathematical model.

(b) Test the significance of each of the main effects and interactions.

21. In testing the operation of voltage regulators, 4 setting stations were available. From each of these, 3 regulators were selected at random and each was tested at 4 test stations. The data are given below [adapted from "Quality Control on the Setting of Voltage Regulators," by D. J. Desmond, *Applied Statistics*, **3** (1954)].

(a) Construct an appropriate model.

(b) Perform an analysis of variance, giving a table of the means of each setting station, with estimated standard error of each mean.

(c) Draw conclusions.

(Assume a parametric model for the setting stations and test stations.)

SETTING STATION	REGULATOR NO.	TEST STATION			
		A	B	C	D
I	1	16.5	16.1	16.2	16.0
	2	15.9	15.4	15.8	15.5
	3	16.9	15.9	16.0	15.8
II	4	16.7	16.1	15.7	16.2
	5	17.0	16.4	16.4	16.4
	6	16.3	16.1	16.1	15.8
III	7	17.0	16.1	15.8	16.0
	8	16.6	16.3	15.9	15.7
	9	16.3	15.9	16.2	15.3
IV	10	16.8	16.7	16.3	16.2
	11	16.1	16.0	16.0	15.6
	12	16.2	16.1	16.1	16.0

22. An experiment was performed to determine if a capacitor, used in the tuned circuit of a device which indicates the frequency of a transmitter or receiver in a communications system, was satisfying the specifications. Several factors were introduced into the experiment to allow a broad applicability of the results. These were

(i) 3 temperatures: ambient (room), $-65°F$, and $+257°F$
(ii) 2 different days of testing
(iii) 2 bridges
(iv) 3 units to be tested within each day

The data shown below (frequency values in microfarads) are taken from a larger experiment.

		DAY 1		DAY 2	
TEMP.	UNIT	BRIDGE A	BRIDGE B	BRIDGE A	BRIDGE B
	1	0.892	0.925	0.855	0.882
$-65°$	2	0.853	0.885	0.871	0.900
	3	0.818	0.842	0.900	0.930
	1	0.933	0.945	0.896	0.908
Room	2	0.894	0.900	0.910	0.920
	3	0.855	0.865	0.938	0.950
	1	1.020	0.900	0.974	0.978
$+257°$	2	0.947	0.940	0.991	1.000
	3	0.910	0.890	1.212	1.350

(a) Construct an appropriate model.
(b) Construct a complete analysis of variance table.
(c) Obtain formulas for the expected values of mean squares in this table.
(d) Draw appropriate conclusions.

23. It was suspected that there was considerable variation of purity of product. An experiment was designed to test the purity of material as follows: at the final product packaging, just prior to shipping, 3 production batches (each containing a large number of bags) were randomly chosen. From each of these batches 3 bags were randomly selected and 2 samples were randomly taken from each. These samples were each further divided into 3 portions and a portion sent to each of 3 different laboratories. The data, in percentage purity, are given below.

(a) Construct an appropriate mathematical model.
(b) Estimate the residual variances, the within-batch variance, and among-batch variance.
(c) Are there any differences among laboratories?
(N.B. There were only 3 laboratories—the same 3 for each bag.)

	SAMPLE	BATCH 1		BATCH 2		BATCH 3	
		1	2	1	2	1	2
BAG	LAB.						
	1	31.4	30.9	31.1	30.8	31.3	30.7
1	2	31.1	31.0	31.6	31.0	31.0	30.8
	3	31.0	31.5	30.8	30.9	31.1	30.8
	1	30.8	30.3	30.6	30.5	30.5	30.3
2	2	30.3	30.5	30.5	30.3	30.5	30.5
	3	30.5	31.4	31.0	30.3	30.3	30.4
	1	32.1	31.2	30.6	30.9	30.1	30.4
3	2	31.9	30.6	30.5	30.9	30.3	30.6
	3	31.6	31.4	30.7	31.0	30.5	30.6

24. The inventory levels of 9 different privately owned electric utility companies were compared annually, using the second and fourth quarter figures. Thus the 4 factors in the analysis are:

    (*i*)   Size, as determined by book value of transmission and distribution facilities, a prime factor in determining size of inventory [(1) less than $150 million, (2) between $150 million and $200 million, (3) greater than $200 million].

    (*ii*)   Company within size category (3 companies in each size category).

    (*iii*)   Years (1954, 1955, 1956).

    (*iv*)   Quarters (second and fourth).

(*a*) Give a formal description of the structure of the experiment.

(*b*) Construct an analysis of variance appropriate to this structure.

(*c*) Give a detailed report on changes associated with the 2 time factors (Years and Quarters).

		YEAR					
		1954		1955		1956	
	QUARTER	2	4	2	4	2	4
SIZE	COMPANY						
	1	0.92	0.88	0.84	0.87	0.97	1.06
(1)	2	1.53	1.31	1.46	1.42	1.64	1.63
	3	1.97	1.89	1.90	2.06	2.19	2.27
	7	2.78	3.14	2.96	3.19	3.72	4.01
(2)	8	3.46	3.18	3.45	4.11	4.24	4.06
	9	4.56	4.32	4.47	4.73	4.74	5.00
	13	3.96	4.04	4.27	4.38	4.36	4.61
(3)	14	4.97	4.70	4.94	5.14	5.42	5.59
	15	5.30	5.29	5.06	5.43	5.91	6.27

25. Consider an experiment designed to test the relative accuracy attainable with 3 different types of bombing systems B_1, B_2, B_3. Eighteen planes are available, 6 with B_1, 6 with B_2, and 6 with B_3. Since a comparison of bombing systems may depend on type of target, 3 different targets are selected: T_1, T_2, T_3. The crews of the 18 planes represent 6 different bombing groups G_{11}, G_{12}, G_{21}, G_{22}, G_{31}, G_{32}, where G_{11} and G_{12} are all equipped with B_1, G_{21} and G_{22} possess B_2, etc. The 18 crews are designated by $C_{111}, C_{112}, C_{113}, C_{121}, \ldots, C_{323}$.

Bombing accuracy then depends on the bombing system, the target, the group within the system, and the crew within the group. Interaction effects are possible between system and target, and between group and target within a system. No other interaction effects are possible, because of the hierarchal breakdown of the bombing system factor. The table below presents the bombing scores (in coded form) together with the necessary totals and subtotals:

(a) construct a mathematical model;

(b) carry out a complete analysis of variance;

(c) draw conclusions.

BOMBING SYSTEM	GROUP	CREW NUMBER	T_1	TOTALS	T_2	TOTALS	T_3	TOTALS	TOTALS
B_1	G_{11}	C_{111}	1.446		1.340		1.895		4.681
		C_{112}	1.622		1.250		1.955		4.827
		C_{113}	1.417	4.485	2.009	4.599	1.880	5.730	5.306	14.814
	G_{12}	C_{121}	1.290		1.665		1.362		4.317
		C_{122}	1.283		1.710		1.799		4.792
		C_{123}	1.511	4.084	1.679	5.054	1.539	4.700	4.729	13.838
				8.569		9.653		10.430		28.652
B_2	G_{21}	C_{211}	0.875		1.783		1.742		4.400
		C_{212}	1.673		1.474		2.044		5.191
		C_{213}	1.004	3.552	0.940	4.197	1.501	5.287	3.445	13.036
	G_{22}	C_{221}	1.765		1.644		1.860		5.269
		C_{222}	1.233		1.167		1.228		3.628
		C_{223}	1.464	4.462	1.605	4.416	1.444	4.532	4.513	13.410
				8.014		8.613		9.819		26.446
B_3	G_{31}	C_{311}	1.111		1.021		1.508		3.640
		C_{312}	0.892		0.918		0.991		2.801
		C_{313}	1.270	3.273	1.279	3.218	1.334	3.833	3.883	10.324
	G_{32}	C_{321}	0.982		0.968		1.104		3.054
		C_{322}	1.057		1.369		1.698		4.124
		C_{323}	1.009	3.048	1.511	3.848	1.456	4.258	3.976	11.154
				6.321		7.066		8.091		21.478
		Totals		22.904		25.332		28.340		76.576

26. In Exercise 13, flashbulbs were compared for 5 cameras and 5 film types. It is now desired to add another factor, namely, 5 filter types (denoted by α through ϵ). Two duplicates are taken for each combination of the 4 factors. The data are given (in uncoded form) below.

				CAMERA					
FILM	1		2		3		4		5
1	$(A\alpha)$	0.64 0.66	$(B\gamma)$	0.70 0.74	$(C\epsilon)$	0.73 0.69	$(D\beta)$	0.66 0.66	$(E\delta)$ 0.66 0.64
2	$(B\beta)$	0.62 0.64	$(C\delta)$	0.63 0.61	$(D\alpha)$	0.69 0.67	$(E\gamma)$	0.70 0.72	$(A\epsilon)$ 0.78 0.76
3	$(C\gamma)$	0.65 0.64	$(D\epsilon)$	0.72 0.73	$(E\beta)$	0.68 0.68	$(A\delta)$	0.64 0.65	$(B\alpha)$ 0.74 0.70
4	$(D\delta)$	0.64 0.63	$(E\alpha)$	0.73 0.72	$(A\gamma)$	0.68 0.70	$(B\epsilon)$	0.74 0.74	$(C\beta)$ 0.72 0.75
5	$(E\epsilon)$	0.74 0.74	$(A\beta)$	0.73 0.71	$(B\delta)$	0.67 0.66	$(C\alpha)$	0.74 0.75	$(D\gamma)$ 0.78 0.78

(a) Determine the residual variance.

(b) What effects are significant? (NOTE: the duplicates were run at the same time. Hence this may not be a true measurement of error.)

(c) Determine a 99% confidence interval for the mean density for camera No. 5.

(d) Write a report for Joe Zilch describing differences between the results of the two experiments and trying to account for them. (Try to guess the system of coding used in Exercise 13.)

27. In Exercise 12 the second figure in each cell can be considered as a covariate. Determine whether oil consumption is independent of this covariate.

28. Explain, with appropriate numerical illustrations, the effects of the following additional pieces of information on the analysis of the corresponding data.

(a) In Exercise 15, Bag 1 was filled before Bag 2 in each run.

(b) In Exercise 22, 9 different units were used each day, three at each of the three temperatures.

(c) In Exercise 21, Regulators 1,2,3,4,5,6 were identical with 7,8,9,10,11,12, respectively (but different from each other).

29. A consulting firm collected data on dustfall and sulfation rate for a large electric power company. This was done in an attempt to reconcile the allegation that this company was not contributing substantially to the overall pollution. The data below present the results in coded form for 15 fixed locations in a 3-mi radius.

(a) Analyze the data and give conclusions.

(b) Criticize the overall experiment for the data as seen from this brief exposure.

| | DUSTFALL | | | | | | | |
| | WINTER | | SPRING | | SUMMER | | FALL | |
SAMPLE LOCATION	1971	1972	1971	1972	1971	1972	1971	1972
1	2	9	9	5	3	4	6	25
2	55	10	4	5	11	7	9	6
3	6	10	8	11	10	9	8	5
4	2	6	5	4	5	4	4	4
5	12	5	12	11	3	6	7	4
6	3	5	4	6	9	3	7	3
7	3	4	3	4	5	3	9	4
8	4	5	7	8	6	8	9	6
9	3	3	4	5	4	2	5	2
10	12	12	15	14	16	7	16	19
11	4	5	4	5	5	4	5	4
12	3	5	14	12	16	13	13	13
13	3	4	14	7	1	4	23	3
14	3	2	5	3	7	4	4	3
15	2	2	5	5	2	4	6	2

30. A particular foundry produces cylindrical steel magnets approximately 16 mm high. One complete casting operation produces 1080 cylinders, arranged as follows: one box contains 9 circles in a 3 by 3 pattern. Around each circle are 5 cylindrical molds. Hence each box has 45 individual molds. Six of these boxes are stacked one above the other. Four of these stacks are filled with one complete pour from the ladle of steel. Suppose the variance components include those parts due to stacks, position of boxes within stacks, circles within boxes, and molds within circles.

(a) Construct an appropriate model.

(b) Set up the outline for an ANOVA table including sums of squares, degrees of freedom, mean squares, and expected values of mean squares.

(c) Obtain an estimate of the total variance which accompanies the size of any particular molded cylinder.

[See "Infusing Statistics into Industry," H. C. Hamaker, *Bulletin of the International Statistics Institute*, **35** (1957).]

31. A *modified Latin square* design is formed by arranging rg "treatments" in r rows and r columns, each of the r^2 cells containing g different treatments in such a way that each treatment appears once in each row and once in each column. For example, a design with $r = 3$ and $g = 2$ is

$$
\begin{array}{ccc}
1,2 & 3,4 & 5,6 \\
3,5 & 1,6 & 2,4 \\
4,6 & 2,5 & 1,3
\end{array}
$$

Using the model

$$X_{ijl} = \xi + \rho_i + \gamma_j + U_{ij} + \tau_l + Z_{ijl}, \quad (i = 1,\dots,r; j = 1,\dots,r;$$
$$l = 1,\dots,rg)$$

where ξ, ρ_i, γ_j, τ_l are parameters and U_{ij} and Z_{ijl} are mutually independent random variables, construct an analysis of variance for the above design and give the expected values of each mean square in the analysis of variance.

32. A special kind of modified Latin square design (see Exercise 31) can be constructed by (i) dividing the rg treatments into g groups of r and (ii) superposing g orthogonal Latin squares, one for each of the groups.
   An example of such a design with $r = 3$, $g = 2$, and groups $(1,2,3)(4,5,6)$ is

1,4	2,5	3,6
2,6	3,4	1,5
3,5	1,6	2,4

(a) Why is the design in Exercise 31 not of this class?
(b) Give an appropriate analysis for general designs of this class. (Assume that the g orthogonal Latin squares exist.)
   [See L. A. Derby and N. Gilbert, "The Trojan Square," *Euphytica*, **7**, 183–188 (1958). The authors suggest the name "Trojan square" for designs with $g > 2$.]

33. Show that if the model (14.9) is modified by addition of a term $\theta \rho_i \kappa_j \tau_k$ then Tukey's method (see Section 13.14.7) leads to an estimator

$$\hat{\theta} = \frac{\sum\sum\hat{Z}_{(ijk)}\hat{\rho}_i\hat{\kappa}_j\hat{\tau}_k}{\sum\sum\hat{\rho}_i^2\hat{\kappa}_j^2\hat{\tau}_k^2}$$

of θ, where $\hat{\rho}_i = \bar{X}_{(i\cdot\cdot)} - \bar{X}_{(\cdots)}$, $\hat{\kappa}_j = \bar{X}_{(\cdot j\cdot)} - \bar{X}_{(\cdots)}$, $\hat{\tau}_k = \bar{X}_{(\cdot\cdot k)} - \bar{X}_{(\cdots)}$, $\hat{Z}_{(ijk)} = X_{ijk} - \bar{X}_{(\cdots)} - \hat{\rho}_i - \hat{\kappa}_j - \hat{\tau}_k$, and summation is over all n^2 combinations of i,j,k in the experiment.
   Show also that if $\theta = 0$ then

$$\frac{S_\theta}{(S_e - S)(n^2 - 3n + 1)^{-1}}$$

where $S_\theta = \hat{\theta}^2\sum\sum\hat{\rho}_i^2\hat{\kappa}_j^2\hat{\tau}_k^2$ is distributed as F with 1, $n^2 - 3n + 1$ degrees of freedom.

CHAPTER 15

Analysis of Variance (III): Factorial Designs

15.1 INTRODUCTION

In the two preceding chapters we have been concerned with some of the mathematical bases for analysis of variance, at least for the simpler cases. We have considered parametric, component-of-variance, and randomization models, randomization theory, crossed and nested experiments, and mixed experiments. Some incomplete designs were analyzed, as well as the Latin square and hypersquares. Some elements of covariance analysis were also presented in Sections 13.20 and 14.9.

We now consider more particularly the experimental design aspect of this area of statistical methodology. To analyze an experiment of 5 factors, each at 2 levels, may seem anticlimactic in comparison with some of the problems of Chapter 14. But looking at an experiment in detail from the design point of view, investigating various ways of "blocking" the experiment, and interpreting consequent confounding (or mixing up) of effects with blocks are of primary importance. An investigation of the overall results of taking a balanced fraction of the entire experiment is also worthwhile. This chapter could easily be entitled "experimental design." Rather than separate analysis of variance from experimental design we feel that it is preferable to handle them together, recalling that the model is first postulated, the experiment is then designed, and finally the results are submitted to analysis.

In this chapter we shall first discuss a 2^k experiment, that is, k factors, each at 2 levels. The straightforward analyses we have so far considered are appropriate to a k factor *completely crossed* experiment. We now reintroduce the concept of blocking and investigate the result of not having a complete replicate in each block. A fraction of the entire experiment, say, $2^{-r} \times 2^k$, is studied in each block. There is a similar discussion of 3^k

experiments with only a fraction 3^{-r} of possible factor level combinations in each block. These involve the introduction of appropriate fresh algebraic techniques. Finally, we consider some mixed factorial designs and note how these can be handled completely or in part by the methods introduced.

Several additional designs including split plot experiments in particular, are introduced in this chapter. Many specialized designs peculiar to fields of application other than the physical and engineering sciences are mentioned rather briefly but are not studied in detail. There still remains the area of response surface methodology, which is covered in Chapter 17.

15.2 k FACTORS—EACH AT TWO LEVELS

Consider a simple experiment with 2 factors, A and B, each at 2 levels. We let capital letters stand for "effects" and lower case letters for actual combinations of treatment levels. "A" then refers to the effect of factor A and "a" refers to the "high" level of A appearing in a treatment level combination. We arbitrarily refer to the 2 levels of each factor as the "low" and "high" levels. (These, in fact, may be low and high on some scale or simply an arbitrary distinction.) The 4 treatment combinations for this simple 2^2 experiment are, as shown in Table 15.1: $(1), a, b, ab$. Our method of designating these treatments is to include a particular lower case letter if the factor is at the high level and to exclude the letter if the factor is at the low level. If all factors are at the "low" level, the symbol (1) is used. For convenience, we call the lower and upper levels of A, A_0 and A_1, respectively (likewise for the other factors). These subscripts 0 and 1 are of advantage in later discussions. [The symbols ab, b, etc., are now understood to represent the observation (or sum of observations) for the corresponding treatment level combinations.]

The average effect of A can be estimated from this 2^2 experiment as

$$A: \tfrac{1}{2}\left\{(ab-b)+\left[a-(1)\right]\right\}$$

TABLE 15.1

TREATMENT LEVEL COMBINATIONS
IN A 2^2 EXPERIMENT

	A_0	A_1
B_0	(1)	a
B_1	b	ab

This is the average difference of the upper and lower level of A (taken at the upper, then at the lower level of B). Occasionally the coefficient $\frac{1}{2}$ is omitted and the *total effect* of A estimated. Likewise, for the average effect of B, we have

$$B: \tfrac{1}{2}\{(ab-a)+[b-(1)]\}$$

If we define the interaction AB as the average *difference*, that is, the effect of A at the upper level of B minus the effect of A at the lower level of B, we have the following estimate

$$AB: \tfrac{1}{2}\{(ab-b)-[a-(1)]\}$$

It is easy and instructive to write the appropriate parametric model and check that these estimators do, in fact, estimate unbiasedly the magnitudes of the corresponding parameters.

The preceding formulas could have been generated as follows: (Coefficients are omitted, and we consider total effects.)

$$A: (a-1)(b+1) = ab - b + a - (1)$$

$$B: (a+1)(b-1) = ab - a + b - (1) \tag{15.1}$$

$$AB: (a-1)(b-1) = ab - a - b + (1)$$

The left sides are formally expanded, using ordinary algebra, and "1" is replaced by (1) in the resulting expression. To determine whether the yield of a particular treatment is to be added or subtracted, we form the product of each of the letters minus 1, if this factor is included in the effect, or plus 1 if the factor is not included. As a further example, in a 3-factor (2^3) problem with the factors A, B, and C, the expressions for the effects and interactions (apart from a multiplying factor) are as follows:

$$A: (a-1)(b+1)(c+1) = abc + ab + ac - bc + a - b - c - (1)$$

$$B: (a+1)(b-1)(c+1) = abc + ab - ac + bc - a + b - c - (1)$$

$$C: (a+1)(b+1)(c-1) = abc - ab + ac + bc - a - b + c - (1)$$

$$AB: (a-1)(b-1)(c+1) = abc + ab - ac - bc - a - b + c + (1) \tag{15.2}$$

$$AC: (a-1)(b+1)(c-1) = abc - ab + ac - bc - a + b - c + (1)$$

$$BC: (a+1)(b-1)(c-1) = abc - ab - ac + bc + a - b - c + (1)$$

$$ABC: (a-1)(b-1)(c-1) = abc - ab - ac - bc + a + b + c - (1)$$

Example 15.1 An experiment was conducted at the National Bureau of Standards [Zelen and Connor, "Multifactor experiments," *Industrial Quality Control*, **15** (March 1959). (The levels of B were $400°F$ (B_0) and $600°F$ (B_1); details of A and C levels were not given.)] on the evaluation of the strength of steel. Three factors were considered

A: Carbon content
B: Tempering temperature
C: Method of cooling

Call the levels A_0, A_1, B_0, B_1, and C_0, C_1. The data are the results of a strength test on 8 steel specimens and are recorded in pounds per square inch divided by 10^3. These data are given in Table 15.2. Recall first the treatment combinations for this 2^3 experiment. These are given in Table 15.3. It should be emphasized at this point that although these treatment combinations are usually stated in some fixed order, the actual order in which the test is taken is randomized within the limitations of the experimental procedure.

Following the notation of (15.2), we can obtain the total main effects and interaction. For example,

$$AB: (a-1)(b-1)(c+1) = abc + ab - ac - bc - a - b + c + (1)$$

$$= 134 + 135 - 165 - 143 - 167 - 145 + 173 + 169$$

$$= -9$$

TABLE 15.2

TENSILE STRENGTH
(IN UNITS OF 10^3 LBS/IN^2)

	A_0		A_1	
	B_0	B_1	B_0	B_1
C_0	169	145	167	135
C_1	173	143	165	134

TABLE 15.3

TREATMENT COMBINATIONS OF A 2^3
EXPERIMENT

	A_0		A_1	
	B_0	B_1	B_0	B_1
C_0	(1)	b	a	ab
C_1	c	bc	ac	abc

TABLE 15.4

ANOVA Table of a 2^3 Experiment

SOURCE	SUM OF SQUARES	DEGREES OF FREEDOM	MEAN SQUARE	MEAN SQUARE RATIO	AVERAGE EFFECT
A	105.125	1	105.125	18.7*	−7.25
B	1711.125	1	1711.125	304.2***	−29.25
C	0.125	1	0.125	0.02	−0.25
$A \times B$	10.125 ⎫	1 ⎫			
$A \times C$	3.125 ⎬ 22.500	1 ⎬ 4	5.625		
$B \times C$	3.125 ⎪	1 ⎪			
$A \times B \times C$	6.125 ⎭	1 ⎭			
Total	1838.875	7			

$$\text{Critical Values} \begin{cases} F_{1,4,0.95} = 7.71 \\ F_{1,4,0.99} = 21.20 \end{cases}$$

We can now set out the average effect of each of the main effects and interactions in tabular form. In constructing the sums of squares in Table 15.4 we can use the conventional methods shown in the 2 preceding chapters or simply take the square of the total effect divided by the size of the experiment. For example, for the sum of squares for A, we have $(X_{1..} - X_{2..})^2/8$, which is identical with $(X_{1..}^2 + X_{2..}^2)/4 - X_{...}^2/8$ as given by the earlier formulas.

In Table 15.4 we have arbitrarily assumed that all the interactions were really 0 and used the 4 corresponding sums of squares to estimate the residual variance. The reader should be cautioned against doing this indiscriminately. In general we certainly cannot assume that all the 2-factor interactions are 0. Even if our assumptions were correct, the number of degrees of freedom is so small that we cannot expect much in the way of accuracy or sensitivity for our estimating and testing procedures. However, even in an experiment of this size, with so few degrees of freedom, an inspection of the size of the mean squares is enough to tell us that factors A and B are significant while the others are not. We might possibly suspect some interaction between A and B, but even this sum of squares is of a different order of magnitude from that of A.

According to the preceding chapter we note that the mathematical model used in this example is

$$X_{tij} = \xi + \alpha_t + \beta_i + \gamma_j + Z_{tij} \quad \left(\text{with } \sum_t \alpha_t = \sum_i \beta_i = \sum_t \gamma_j = 0 \right)$$

That is, we have a 3-factor, parametric model and completely crossed classification. For this analysis all the interactions are assumed to be 0.

15.2.1 Notation for Calculating Effects

We now consider a way of further representing the total effects of each of the factors. Its purpose is to evolve a rather rapid method of estimating any individual effect as well as to show the orthogonality of each of these effects. Later we use this notation as one way of deciding upon a particular "confounding" interaction. We set up Table 15.5 to calculate the effects of each factor. Let us use as column designators the main effects and interactions (and I—the total of the entire experiment). The row designators are the treatment combinations. The body of the table is made up of "$+$" and "$-$" symbols. For each effect the plus and minus signs indicate how each treatment yield is to be combined. For example, under I there are all plus values. This states that the grand total is the sum of all of the yields. The effect A has in its 8 rows a *plus* wherever the treatment includes the letter "a" (that is, its upper level), and a minus in the rows where "a" is not included. Once the signs for the main effects have been established, the signs for the remaining columns are obtained by proper multiplication of some of the preceding columns. For example, the signs of AB are the "products" of the A and B signs, row by row. Table 15.5 is thus a concise way of summarizing (15.2).

TABLE 15.5

ALGEBRAIC SIGNS FOR CALCULATING EFFECTS

	I	A	B	AB	C	AC	BC	ABC
(1)	$+$	$-$	$-$	$+$	$-$	$+$	$+$	$-$
a	$+$	$+$	$-$	$-$	$-$	$-$	$+$	$+$
b	$+$	$-$	$+$	$-$	$-$	$+$	$-$	$+$
ab	$+$	$+$	$+$	$+$	$-$	$-$	$-$	$-$
c	$+$	$-$	$-$	$+$	$+$	$-$	$-$	$+$
ac	$+$	$+$	$-$	$-$	$+$	$+$	$-$	$-$
bc	$+$	$-$	$+$	$-$	$+$	$-$	$+$	$-$
abc	$+$	$+$	$+$	$+$	$+$	$+$	$+$	$+$

Note the following properties of the table: (1) except for column I, the number of plus and minus signs are equal in each column. (2) The sum of products of signs of any 2 columns (pairwise) is 0. That is, the "product" has an equal number of plus and minus signs. (3) The product of any 2 columns yields a column included in the table. For example, $A \times B = AB$, $AB \times B = A$, $ABC \times AB = C$, etc. All these properties are implied by orthogonality, a concept that was first discussed in Chapter 12 and again in the

2 following chapters. Note also the way in which the products

$$AB \times B = AB^2 = A$$

$$ABC \times AB = A^2B^2C = C$$

$$ABC \times C = ABC^2 = AB$$

are formed. We have here a product, *modulo* 2. That is, the exponent can be only 0 or 1. If the power is greater than 1, we reduce it by multiples of 2 until it is 0 or 1.

15.2.2 Yates' Algorithm

A simplified technique devised by Frank Yates [15.8], given in Table 15.6, is a mechanical method for obtaining the total effects of each of the factors (and their interactions) of a 2^k experiment. Later we shall modify this technique for a 3^k experiment. The data used in Table 15.6 are those of Example 15.1. In this table, column (1) indicates the yields of the treatment combinations given in the preceding column. This order of the treatment combinations is always maintained. A fourth factor D would be added below this as d, ad, bd, abd, cd, etc. Column (2) is obtained by adding the yields in adjacent pairs to make up the first half of the column and subtracting in adjacent pairs for the second half. For example, $336 = 169 + 167$, $280 = 145 + 135, \ldots$; $-2 = 167 - 169$, $-10 = 135 - 145$, etc. The differences are always taken in exactly this same order: second minus the first, fourth minus the third, etc.

Column (3) is obtained from column (2) in exactly the same manner as (2) is obtained from (1). So also is column (4) obtained from (3). This

TABLE 15.6

Yates' Algorithm for a 2^3 Experiment

TREATMENT COMBINATION	(1) YIELD	(2)	(3)	(4)	EFFECT	(5) AVERAGE EFFECT (4) ÷ 4	(6) SUM OF SQUARES (4)² ÷ 8
(1)	169	336	616	1231	I		
a	167	280	615	−29	A	−7.25	105.125
b	145	338	−12	−117	B	−29.25	1711.125
ab	135	277	−17	−9	AB	−2.25	10.125
c	173	−2	−56	−1	C	−0.25	0.125
ac	165	−10	−61	−5	AC	−1.25	3.125
bc	143	−8	−8	−5	BC	−1.25	3.125
abc	134	−9	−1	7	ABC	1.75	6.125
Total	1231						1838.875

process is carried out 3 times. For a 2^k experiment, there are k steps of this type. Column (4) is the total effect of the factor (or interaction) designated (in lower case letters) at the beginning of the row. To obtain the average effect we divide by 4, which is the number of differences in each total effect. Finally, column (6) is the sum of squares for each of the factors of the experiments. Note that the denominator of the squared "total effect" is 2^3 or the size of the experiment. Another way of considering the sum of squares in column (6) in each case is that it is a linear comparison (of which there are 7 in this example) with a single degree of freedom. We may use equation (13.66), namely

$$SS(A) = \frac{\left(\sum_i a_i X_{i..} \right)^2}{n \sum a_i^2}$$

where the a_i are the coefficients in a linear comparison and are either $+1$ or -1. So the divisor for each of the sums of squares is $n2^k$.

15.2.3 Analysis of a 2^5 Experiment using Yates' Algorithm

The data in the following example have been considerably simplified. They are presented to show how Yates' algorithm expedites the necessary calculations when little or no mechanical aid to computation is available.

Example 15.2 A factorial experiment was carried out on a pilot plant scale. A product was being purified by a form of steam distillation process. The 5 factors, each at 2 levels, were concentration of material (A), rate of distillation (B), volume of solution (C), stirring rate (D), and solvent-to-water ratio (E). The residual acidity of material from one run on each of the 32 experimental treatment combinations was determined. The results (in coded form) are given in Table 15.7.

The data of Table 15.7 are analyzed by means of the Yates algorithm as shown in Table 15.8. Column (7) presents the average effect of each factor and the

TABLE 15.7

RESIDUAL ACIDITY IN PILOT PLANT EXPERIMENT
(CODED DATA)

		A_0				A_1			
		D_0		D_1		D_0		D_1	
		E_0	E_1	E_0	E_1	E_0	E_1	E_0	E_1
B_0	C_0	9	3	11	8	10	9	13	7
	C_1	3	5	7	7	5	6	10	7
B_1	C_0	8	4	9	8	6	6	16	6
	C_1	6	4	7	5	10	10	13	6

TABLE 15.8

YATES' ALGORITHM—PILOT PLANT DATA

TREAT-MENT COMBI-NATION	(1) YIELD	(2)	(3)	(4)	(5)	(6)	EFFECT	(7) AVERAGE EFFECT (6) ÷ 16	(8) SUM OF SQUARES (6)² ÷ 32
(1)	9	19	33	57	143	244	I		
a	10	14	24	86	101	36	A	2.250	40.500
b	8	8	49	47	23	4	B	0.250	0.500
ab	6	16	37	54	13	8	AB	0.500	2.000
c	3	24	22	5	7	−22	C	−1.375	15.125
ac	5	25	25	18	−3	10	AC	0.625	3.125
bc	6	17	29	15	7	18	BC	1.125	10.125
abc	10	20	25	−2	1	14	ABC	0.875	6.125
d	11	12	−1	3	−21	36	D	2.250	40.500
ad	13	10	6	4	−1	−4	AD	−0.250	0.500
bd	9	11	9	1	7	−4	BD	−0.250	0.500
abd	16	14	9	−4	3	8	ABD	0.500	2.000
cd	7	15	8	−1	15	−10	CD	−0.625	3.125
acd	10	14	7	8	3	−2	ACD	−0.125	0.125
bcd	7	14	−3	1	3	−18	BCD	−1.125	10.125
$abcd$	13	11	1	0	11	−14	$ABCD$	−0.875	6.125†
e	3	1	−5	−9	29	−42	E	−2.625	55.125
ae	9	−2	8	−12	7	−10	AE	−0.625	3.125
be	4	2	1	3	13	−10	BE	−0.625	3.125
abe	6	4	3	−4	−17	−6	ABE	−0.375	1.125
ce	5	2	−2	7	1	20	CE	1.250	12.500
ace	6	7	3	0	−5	−4	ACE	−0.250	0.500
bce	4	3	−1	−1	9	−12	BCE	−0.750	4.500
$abce$	10	6	−3	4	−1	8	$ABCE$	0.500	2.000†
de	8	6	−3	13	−3	−22	DE	−1.375	15.125
ade	7	2	2	2	−7	−30	ADE	−1.875	28.125
bde	8	1	5	5	−7	−6	BDE	−0.375	1.125
$abde$	6	6	3	−2	5	−10	$ABDE$	−0.625	3.125†
cde	7	−1	−4	5	−11	−4	CDE	−0.250	0.500
$acde$	7	−2	5	−2	−7	12	$ACDE$	0.750	4.500†
$bcde$	5	0	−1	9	−7	4	$BCDE$	0.250	0.500†
$abcde$	6	1	1	2	−7	0	$ABCDE$	0	0 †
Total	244								

TABLE 15.9

ANALYSIS OF VARIANCE—PILOT PLANT DATA

SOURCE	SUM OF SQUARES	DEGREES OF FREEDOM	MEAN SQUARE	MEAN SQUARE RATIO
A	40.500	1	40.500	14.96**
B	0.500	1	0.500	—
C	15.125	1	15.125	5.58
D	40.500	1	40.500	14.96**
E	55.125	1	55.125	20.36**
AB	2.000	1		
AC	3.125	1		
AD	0.500	1		
AE	3.125	1		
BC	10.125	1		
BD	0.500	1		
BE	3.125	1		
CD	3.125	1		
CE	12.500	1	12.500	4.62
DE	15.125	1	15.125	5.58
ABC	6.125	1		
ABD	2.000	1		
ABE	1.125	1		
ACD	0.125	1		
ACE	0.500	1		
ADE	28.125	1	28.125	10.4*
BCD	10.125	1		
BCE	4.500	1		
BDE	1.125	1		
CDE	0.500	1		
4− and 5− Factor Interactions	16.250	6	2.708	
Total	275.500	31		

Significance limits for F: $F_{1,6,0.90} = 3.78$, $F_{1,6,0.95} = 5.99$, $F_{1,6,0.99} = 13.75$.

interactions. Finally, the sums of squares [in column (8)] are presented in Table 15.9. We can simply look at the sums of squares and note which factors stand out. More formally, if we can assume that the 4- and 5-factor interactions are 0, then the subtotal sum of squares calculated from these 6 effects (indicated in Table 15.8 by the sign †) can be used for the residual sum of squares. The pooled estimate of σ^2 is then 2.708 in coded units. The critical F ratios are given at the bottom of Table 15.9.

Three of the main effects, namely, concentration of material (A), stirring rate (D), and solvent-to-water ratio (E) are strongly significant. No 2-factor interactions are significant at the 5 percent level. However, CE and DE show some slight evidence of a significant effect. A disturbing result is the significance of the 3-factor interaction, ADE. If this cannot be explained from an engineering or physical viewpoint, then some further investigation is in order. A more definitive experiment may be called for. Of course, it may be just a chance effect. It should be remembered that no fewer than 25 effects are separately being tested for significance. This is discussed further in Section 15.2.4.

Note that factor B did not show up as significant either as a main effect or as part of an interaction. Does this mean that the rate of distillation is not important? Not necessarily—the results of the experiment state that *within the range of rates of distillation chosen*, the data did not show up a real difference due to this factor. If there is a true difference in this range, it has not shown up strongly enough.

Several additional points should be noted here. If a new experiment were to be planned, a single level of factor B should be chosen or a new range of levels of this factor should be studied. We have been considering average effects. If we are concerned with a quantitative factor where a nonlinear regression is in question, more than 2 levels should be chosen for this factor.

Example 15.2 is concerned with a "singly replicated" experiment. In the following example there is a 2^4 experiment with 2 replicates. Column (2) of Table 15.11 is made up of the sums of the 2 replicates for each treatment combination. The remainder of the table is calculated as in Table 15.8 with the exception of the divisors for columns (7) and (8) and the calculation of the Residual mean square with 16 degrees of freedom.

Example 15.3   The development of a practical industrial fermentation process usually begins with a laboratory study of the physiological requirements of the microorganism concerned. In one such study it was found that a useful substance is secreted by a species of mold when grown in a liquid culture medium. It was desired to increase the yield. The formation of the substance was already known to depend mainly upon the levels of 2 culture medium ingredients, provided certain broad requirements for temperature, aeration, pH, and culture age are met.

It was suspected that 4 of these 6 factors might be interdependent. To test this, a 2^4 factorial experiment was performed. Two culture medium ingredients (X_1, X_2) and 2 environmental factors (X_4, X_5) were examined; duplicate cultures were

TABLE 15.10

DATA FROM A 2^4 FERMENTATION EXPERIMENT

x_4	x_5	x_2	x_1 -1		$+1$	
			-1	$+1$	-1	$+1$
-1	-1		32.7	90.4	70.6	115.0
			19.3	89.8	84.5	108.6
	$+1$		20.2	94.1	76.1	133.6
			29.9	96.5	73.3	131.6
$+1$	-1		50.0	72.6	104.2	81.3
			52.1	76.9	103.4	88.2
	$+1$		50.5	91.8	78.6	108.3
			49.1	86.9	74.1	108.3

prepared for each treatment.* The data, presented in Table 15.10, are coded. Effects are reported as "yield units" (Y.U.) per "design unit" (D.U.). There are 2 replicates of each of the 16 combinations of factors.

Note that the values for the high levels of X_1 and X_4 and the low level of X_5 seem to be reversed. It is essential to investigate these values before proceeding with the analysis. (The authors did check this out with the original experiment and learned that these values were, in fact, correct.)

The data are analyzed by means of the Yates' algorithm as shown in Table 15.11. The treatment combinations are given as (1), $x_1, x_2, x_1x_2, \ldots$ instead of the usual (1), $a, b, ab, \ldots$. Column (2) presents the sum of the duplicates.

The main effects of X_1 and X_2 account for most of the differences among the preparations. The rather large negative interaction is plausible; certain of the nutritional requirements of the mold can be supplied by either of the ingredients. It was surprising to find that the factors X_4 and X_5 have little direct effect, but exert their influences mainly by interaction with X_2 and, to a lesser extent, with X_1. Clearly, none of the 4 factors is independent of the others, in the sense of affecting the yield in a purely additive manner.

*The variables X_1, X_2, X_4, and X_5 are used, since X_3 will be used in Example 17.13, first with X_1 and X_2, then with X_4 and X_5.

TABLE 15.11

Yates' Algorithm—Fermentation Experiment

TREATMENTS (1)	TOTAL YIELD (2)	(3)	(4)	(5)	(6)	(6) $\div$ 16 AVERAGE EFFECT (7)	(6)^2 $\div$ 32 MEAN SQUARE (8)
(1)	52.0	207.1	610.9	1239.6	2542.5		
x_1	155.1	403.8	628.7	1302.9	536.9	33.56	9008.2***
x_2	180.2	309.7	655.3	272.0	605.3	37.83	11449.6***
x_1x_2	223.6	319.0	647.6	264.9	−185.1	−11.57	1070.7***
x_4	102.1	199.5	146.5	206.0	10.1	0.63	3.2
x_1x_4	207.6	455.8	125.5	399.3	−103.9	−6.49	337.4***
x_2x_4	149.5	252.3	173.9	−145.2	−300.7	−18.79	2825.6***
$x_1x_2x_4$	169.5	395.3	91.0	−39.9	−16.3	−1.02	8.3
x_5	50.1	103.1	196.7	17.8	63.3	3.96	125.2*
x_1x_5	149.4	43.4	9.3	−7.7	−7.7	−0.44	1.6
x_2x_5	190.6	105.5	256.3	−21.0	193.3	12.08	1167.7***
$x_1x_2x_5$	265.2	20.0	143.0	−82.9	105.3	6.58	346.5***
x_4x_5	99.6	99.3	−59.7	−187.4	−25.5	−1.59	20.3
$x_1x_4x_5$	152.7	74.6	−85.5	−113.3	−61.9	−3.87	119.7*
$x_2x_4x_5$	178.7	53.1	−24.7	−25.8	74.1	4.63	171.6**
$x_1x_2x_4x_5$	216.6	37.9	−15.2	+9.5	35.3	2.21	38.9
Residual						(321.6/16) = 20.1	

Total Sum of Squares = 27016.1

Significance limits for F: $F_{1,16,0.95} = 4.49$, $F_{1,16,0.99} = 8.53$, $F_{1,16,0.999} = 16.12$

15.2.4 Half-Normal Plots

It is often advantageous to study the magnitude of the effects of a 2^k experiment without formally considering the ANOVA table. Sometimes, also, there is a question as to which effects can reasonably be pooled together into the Residual sum of squares, that is, which interactions will be assumed to be 0 (or at least negligible) in future experiments. An approach proposed by C. Daniel which is helpful in connection with this kind of work is the use of the *half-normal plot* [15.3]. In this the effects are arranged in order of estimated absolute magnitude. They are then plotted on half-normal probability paper in which the vertical scale is that of normal probability paper modified to

$$P' = 2P - 100$$

where P is the cumulative probability of the normal distribution and P', the cumulative probability used in the half-normal paper. It can be seen that P' represents the cumulative distribution of $U' = |U|$ where U has a standard normal distribution (see Section 5.10.1).

Rather than have a cumulative probability for the vertical scale one may prefer an ordered rank, $1, 2, 3, \ldots, 2^k - 1$, that is, the cumulative rank (with $U_1 < U_2 < \cdots < U_n$) of the empirical data. A general conversion formula for the vertical scale is

$$P' = \frac{i - \frac{1}{2}}{n} \qquad (i = 1, 2, \ldots, n)$$

An illustration of this is shown in Figure 15.1 in the example that follows.

If one wishes, further, to place significance limits on the chart, it would be preferable to first standardize the data. This can be done by using as an approximation to σ the value U_α, the absolute value of the rank order statistic (see Section 6.4) from the n observations (effects) that is nearest to $(0.683n + 0.5)$. Values of α are given in Table 15.12 for different values of $n = 2^k - 1$. To determine the "center line" and "significance limits" for the standard chart one uses the constants in Table 15.12. These lines are determined as follows:

Center line: through the points $(0,0)$ and $\tau, n)$

Significance limit: through the point $(\tau + d, n)$ and the intersection of the center line and the horizontal line with ordinate equal to the rank of U_α, i.e., [1, rank (U_α)]

An illustration of this is given in Example 15.4 and in Figure 15.2.

TABLE 15.12
CONSTANTS FOR CENTER LINE AND LIMITS OF A HALF-NORMAL PLOT

n	τ	rank of U_α	d $\alpha = 0.05$	0.20
15	2.09	11	0.94	0.26
31	2.42	22	0.94	0.33
63	2.69	44	0.66	0.23
127	2.88	88	0.66	0.25

Example 15.4 The average effects of Example 15.2 (as shown in column (7) of Table 15.8) are arranged in order of absolute magnitude, as shown in Table 15.13. These are then plotted in Figure 15.1 where the ordinate is the ascending order number (or rank) and the abscissa is the value of the effects. Note that the effects E, A, D, and ADE show up quite strongly as in Table 15.9. Of course, this simply indicates that these effects are larger than would be expected of the 4 greatest values out of 31 mutually independent half-normal variables. Individual significance can still be judged by the actual values of estimated individual effects (or their squares).

In order to construct the standardized-normal plot for $n = 31$, we first divide each value by an estimate of σ, that is, $U_\alpha = U_{22} = 0.875$. The standardized observations

TABLE 15.13
RANKING EFFECTS FOR A HALF-NORMAL PLOT

RANK	AVERAGE EFFECT	$\dfrac{\lvert\text{AVERAGE EFFECT}\rvert}{\lvert U_{22}\rvert}$	RANK	AVERAGE EFFECT	$\dfrac{\lvert\text{AVERAGE EFFECT}\rvert}{\lvert U_{22}\rvert}$
1	0	0	16	0.625	0.714
2	0.125	0.143	17	0.625	0.714
3	0.250	0.286	18	0.625	0.714
4	0.250	0.286	19	0.750	0.857
5	0.250	0.286	20	0.750	0.857
6	0.250	0.286	21	0.875	1.000
7	0.250	0.286	22	0.875	1.000
8	0.250	0.286	23	1.125	1.286
9	0.375	0.429	24	1.125	1.286
10	0.375	0.429	25	1.250	1.429
11	0.500	0.571	26	1.375	1.571
12	0.500	0.571	27	1.375	1.571
13	0.500	0.571	28	1.875	2.143
14	0.625	0.714	29	2.250	2.571
15	0.625	0.714	30	2.250	2.571
			31	2.625	3.000

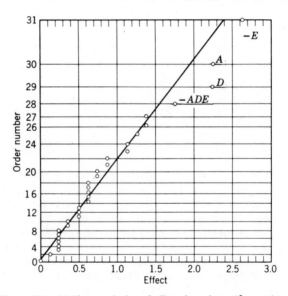

Figure 15.1 Half-normal plot of pilot plant data (2^5 experiment).

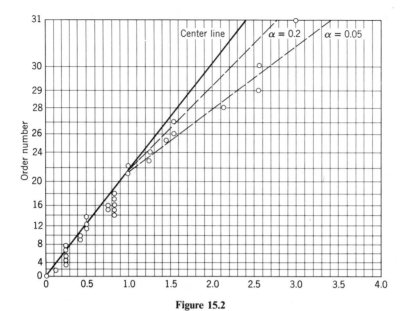

Figure 15.2

are plotted in Figure 15.2. The $\alpha = 0.2$ and $\alpha = 0.05$ limit lines as well as the "center line" are also plotted. These go through the points:

$$
\begin{array}{lll}
\text{Center line:} & (0,0) & \text{and} \ (2.42, 31) \\
\alpha = 0.2: & (1,22) & \text{and} \ (2.75, 31) \\
\alpha = 0.05: & (1,22) & \text{and} \ (3.36, 31)
\end{array}
$$

Note that in the standard chart the 2 largest values are inside the limit line for $\alpha = 0.05$, but the next 2 are outside. This should not be construed as an indication that the 2 former are not significant.

15.2.5 The Maximum Variance Ratio Test

When there are several mean squares (each with ν degrees of freedom) in an analysis of variance table, and the *largest* of these mean squares is selected for comparison with a Residual mean square (with ν_0 degrees of freedom), we are, in fact, using a critical region of the following kind. Let $S_0^2, S_1^2, \ldots, S_k^2$ be mutually independent random variables, with S_j^2 distributed as $\chi_{\nu_j}^2 \sigma^2 / \nu_j$, where $\nu_1 = \nu_2 = \cdots = \nu_k = \nu$. Then we are using a critical region of the form

$$
\max_j \left(S_j^2 / S_0^2 \right) > K
$$

In order to determine K so as to make the significance level equal to α it is necessary to solve the equation

$$\frac{1}{2^{\nu_0/2}\Gamma(\nu_0/2)}\int_0^\infty x_0^{\nu_0/2-1}e^{-x_0/2}$$

$$\times \left[\int_0^{Kx_0\nu/\nu_0}\frac{1}{2^{\nu/2}\Gamma(\nu/2)}x^{\nu/2-1}e^{-x/2}dx\right]^k dx_0 = 1-\alpha$$

The following short table (based, with permission, on Table 19 of Pearson and Hartley [13.10].) gives some values of K for $\alpha = 0.05$ and $\nu = 1$. Recalling the notation of Section 13.14.2, we have $K = q'^2_{k,\nu_0,1-\alpha}$.

$$K = q'^2_{k,\nu_0,1-\alpha}$$

	k	2	3	4	6	8	10
	10	6.79	8.00	8.96	10.52	11.79	12.87
	12	6.44	7.53	8.37	9.68	10.68	11.53
	15	6.12	7.11	7.86	8.98	9.82	10.52
ν_0	20	5.81	6.72	7.40	8.39	9.13	9.71
	30	5.52	6.36	6.97	7.87	8.51	9.03
	60	5.25	6.02	6.58	7.38	7.96	8.41
	∞	5.00	5.70	6.21	6.92	7.44	7.84

In Table 15.9, for example, the largest mean square ratio (that for E) out of $k = 25$ such ratios, is 20.36, while $\nu_0 = 6$. The values shown above do not include $k = 25$, $\nu_0 = 6$, but very rough extrapolation shows that the maximum variance ratio would give a significant result at 5 percent level for these data.

15.3 CONFOUNDING IN A 2^k EXPERIMENT

We introduce now a concept in experimental design called *confounding*. Suppose, for example, there are 2 methods (1 and 2) for determining the reflectance of test panels of paint. We are interested in 2 different paint compositions (A and B). If the data are analyzed in such a manner that composition A is analyzed by method 1, and composition B by method 2, then any difference observed could be due to either composition, or method, or both. Composition and Method differences have been confounded. It is not possible to distinguish whether any observed difference is due to one or the other factor.

In the 2 preceding chapters we have sometimes been concerned with the inclusion of "blocks" as a factor in the model equation. Suppose it is not possible to have a complete replicate of the experiment within one block. For example, in a 2^3 experiment, suppose that it is possible to handle only 4 experimental units at one time. This may be necessitated by the equipment available or may be a self-imposed limitation to keep the size of the block quite small. If this experiment were planned as in Figure 15.3 (i), then factor C would be confounded with blocks. For the effect of factor C is

$$C: \left[c - (1) \right] + (ac - a) + (bc - b) + (abc - ab)$$

which is the same as the difference between the totals for block 1 and block 2. In Figure 15.3 (ii) factor AB is confounded with blocks. For the effect of AB is

$$AB: (a-1)(b-1)(c+1) = abc + ab + c + (1)$$

$$-a - b - ac - bc$$

which is also the block difference. In Figure 15.3 (iii) the interaction ABC is confounded with blocks.

Block 1	Block 2	Block 1	Block 2	Block 1	Block 2
(1)	c	(1)	a	(1)	a
a	ac	ab	b	ab	b
b	bc	c	ac	ac	c
ab	abc	abc	bc	bc	abc
(i)		(ii)		(iii)	

Figure 15.3 2^3 experiments in 2 blocks.

In order to avoid having blocks confounded with main effects or some particular higher-order interactions, the interaction to be confounded can be chosen in advance. A higher-order interaction is often selected. This should not be construed as an inviolable rule of selection. We may want to pick (for confounding) 2-factor interactions which we feel are not different from 0 in their effect on the experiment in question, or which although known to be important, are not of special interest in the proposed investigation.

The block containing the treatment (1) is called the *principal block*. The remaining block can be generated from this. The rule for construction of blocks is as follows: the treatments included in the principal block are *all* those with an *even* or *zero* number of letters in common with the "*confounding interaction.*" Thus in Figure 15.3(*iii*), (1) has 0 letters in common with ABC; ab has 2 letters in common with ABC, as have ac and bc. There are no others. The principal block is then made up of (1), ab, ac, and bc. To construct the other block we choose a treatment combination not in the principal block, say, a. (This is in block 2.) All the other treatments can be obtained by taking the products, modulo 2. Therefore,

$$a \times (1) = a$$

$$a \times ab = a^2 b = b$$

$$a \times ac = a^2 c = c$$

$$a \times bc = abc$$

A particular property of the principal block should be noted. This is the fact that it is a closed set, modulo 2. We may "multiply" any 2 (or more) of the treatment combinations in this block with the resulting treatment always being in this block. This property is not true of any other blocks.

Example 15.5   To construct a 2^4 experiment in 2 blocks of 8 treatments each, suppose we choose a confounding interaction, say, ACD. The principal block is made up of the treatments with an even (or zero) number of letters in common with ACD. These are $(1), ac, ad, cd, b, abc, abd, bcd$. Using the property of a closed set (under multiplication modulo 2) we could have selected treatments $(1), ac, ad$. Then cd could be the result of $ac \times ad (= a^2 cd = cd)$. Selecting b, we can finally take its product with the preceding treatment combinations to evolve the remainder of the block. To construct the second block, choose a new treatment, say a. Multiply this (modulo 2) by each of the treatments in the principal block. This gives us the 2 blocks in Figure 15.4.

$$\begin{bmatrix} (1) & b \\ ac & abc \\ ad & abd \\ cd & bcd \end{bmatrix} \quad \begin{bmatrix} a & ab \\ c & bc \\ d & bd \\ acd & abcd \end{bmatrix}$$

Figure 15.4   A 2^4 experiment in 2 blocks
(ACD confounded with blocks).

We may wish to divide the entire experiment into more than 2 blocks. Suppose, for example, we want a 2^4 experiment (without replication) in 4 blocks of 4 treatments. Since there are 4 blocks, 3 degrees of freedom of the total 15 degrees of freedom must be confounded with blocks. These 3 are not independent of each other. We may arbitrarily choose 2 of them.

The third is the resulting product (modulo 2) sometimes called their *generalized interaction*. For example, if we choose $ABCD$ and ABC, the other effect confounded with blocks is $D(=A^2B^2C^2D)$. Noting this, we must choose the confounding interactions carefully. Since we do not, in general, wish to confound a main effect, we cannot confound both a 3- and 4-factor interaction (for a 2^4 in 4 blocks). We choose then a 2-factor interaction which we can safely assume equal to 0. If we confound this with $ABCD$, then another 2-factor interaction is confounded; if, with a 3-factor interaction, then another 3-factor interaction can be the third confounding interaction.

Suppose that we can safely assume that all the 3-factor interactions are 0 and so also is the interaction AB. We choose AB and BCD. Then $ACD(=AB^2CD)$ is also a confounding interaction. Let us use our earlier rule for the principal block, but add that each treatment combination must have an even (or 0) number of elements in common with *all* the confounding interactions (really just 2 of them in this case, since the third follows from these). Let us choose the combinations (1), cd, and abd, which give us also $abc(=abcd^2)$.

For each additional block choose the treatments a, b, and c, in order, and generate each block by multiplication (modulo 2) with the principal block (Figure 15.5).

$$
\begin{bmatrix} (1) \\ cd \\ abd \\ abc \end{bmatrix}
\begin{bmatrix} a \\ acd \\ bd \\ bc \end{bmatrix}
\begin{bmatrix} b \\ bcd \\ ad \\ ac \end{bmatrix}
\begin{bmatrix} c \\ d \\ abcd \\ ab \end{bmatrix}
$$

Figure 15.5 A 2^4 experiment in 4 blocks confounding AB, BCD, and ACD.

Example 15.6 We wish to construct a 2^6 experimental plan arranged in 8 blocks, each containing 8 treatment combinations. The factors are A, B, C, D, E, and F. Seven effects or interactions will be confounded; 3 of these (none being the generalized interaction of the other 2) can be chosen arbitrarily. In order to determine the selection of the treatment combinations for each block we first chose 2 confounding interactions. Note that any one of $\binom{57}{2} = 1596$ possible selections can be made. In general, we wish to avoid confounding too many lower-order interactions. We choose $ABCD$ and AEF. This also gives us $BCDEF(=A^2BCDEF)$ as a confounded interaction. We then choose a fourth interaction, say $ACDE$. The other confounded interaction are then $BE(=ABCD \times ACDE = A^2BC^2D^2E)$, $CDF(=A^2CDE^2F)$, and $ABF(=ABC^2D^2E^2F)$. We now determine the principal block by selecting the treatments with an even (or zero) number of letters in common with each of the independently chosen interactions, $ABCD$, AEF, and $ACDE$. No other interactions need be used, since all the others were generated from these. The treatments are (1), abe, acf, cd, and their products (modulo 2). The principal block, one other block, and the "generators" for the remaining blocks are given in Figure 15.6.

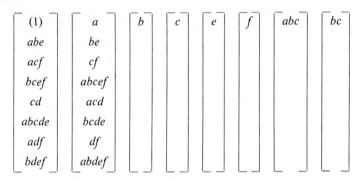

Figure 15.6   A 2^6 experiment in 2^3 blocks of 8.

Generating the remaining blocks is left as an exercise for the reader. Note that the treatment combinations used in generating the additional blocks must not have appeared in earlier blocks. Otherwise some block(s) will be repeated.

From this example we can note 2 important points, bearing in mind that the number of blocks to be selected is 2^3: (1) We must confound $2^3 - 1$ interactions. (2) Only 3 independent choices of interactions can be made. All the other confounding interactions are generated from these. In general, for a 2^k experiment in 2^r blocks: (1) $2^r - 1$ interactions are confounded and (2) r of these interactions are chosen independently. Note also the fact that in the principal block we need only $k - r$ independent choices [besides (1)] and can generate the remaining ones from these.

15.3.1   A Confounded Experiment

The following example illustrates a 2^4 experiment arranged in 2 blocks of 8 treatment combinations each. It includes a discussion of the random order of testing (within the limitations of the experiment) as well as an analysis using Yates' algorithm.

Example 15.7   A new rifle was being tested for performance. The development had reached the stage where the design of rifle to be used had already been decided upon. Some of the characteristics of the weapon, namely, the propellant charge, the weight of the projectile, and the propellant web, were still being investigated. The propellant web concerned the geometric configuration of the propellant. A 2^4 experiment was planned as an initial examination of the factors. These factors, all quantitative, were:

A: propellant charge in pounds

B: projectile weight in pounds

C: propellant web

D: weapon

Two weapons were selected at random to provide some check on whether there was any sizable variance due to variation between weapons. A character of primary interest was velocity of the projectile. Because of the limitations of testing equipment and the time it took to carry out this test (and make other required measurements), only 8 tests could reasonably be taken on any single day. It was logical to consider each day as a block. The experiment was, then, to be 2^4 in 2 blocks of 2^3 units each.

In order that any extraneous factors not be allowed to bias the results, a balanced order of firing was planned for each day. This is shown in Table 15.14 where the numbers in the table indicate the order. In this the underlined numbers were those tests taken on the second day. It was arbitrarily decided to confound the highest order interaction, namely $ABCD$, with the day effect. The 2 blocks are shown in Figure 15.7. Rather than choose the firing program as indicated in Table 15.14 one could have made a random order of firing within each day. The choice of the present order was made in an attempt to pick up trends in firing order if possible. The data are presented in Table 15.15. The velocity has been coded. The Yates' algorithm is used in Table 15.16 to obtain the effect of the various factors and their interactions.

TABLE 15.14

FIRING PROGRAM OF A BALLISTICS TEST

CHARGE WEIGHT		A_0		A_1	
PROJECTILE WEIGHT		B_0	B_1	B_0	B_1
PROPELLANT WEB	WEAPON				
C_0	D_0	1	9	10	2
	D_1	13	5	6	14
C_1	D_0	11	3	4	12
	D_1	7	15	16	8

	DAY 1		DAY 2	
(1)	ad		a	d
	ab	bd	b	abd
	ac	cd	c	acd
	bc	$abcd$	abc	bcd

Figure 15.7 Daily schedule for firing ($ABCD$ interaction confounded with days).

TABLE 15.15

VELOCITY OF PROJECTILE (CODED)

CHARGE WEIGHT		A_0		A_1	
PROJECTILE WEIGHT		B_0	B_1	B_0	B_1
PROPELLANT WEB	WEAPON				
C_0	D_0	97	68	151	150
	D_1	75	53	145	141
C_1	D_0	39	15	100	66
	D_1	26	-16	97	54

TABLE 15.16

YATES' ALGORITHM—VELOCITY DATA

TREATMENT COMBINATION	(1) VELOCITY	(2)	(3)	(4)	(5)		(6) (5) ÷ 8	(7) $(5)^2 ÷ 16$
(1)	97	248	466	686	1261	I		
a	151	218	220	575	547	A	68.375	18700.5625
b	68	139	414	248	-199	B	-24.875	2475.0625
ab	150	81	161	299	35	AB	4.375	76.5625
c	39	220	136	-88	-499	C	-62.375	15562.5625
ac	100	194	112	-111	-41	AC	-5.125	105.0625
bc	15	123	158	18	-87	BC	-10.875	473.0625
abc	66	38	141	17	-57	ABC	-7.125	203.0625†
d	75	54	-30	-246	-111	D	-13.875	770.0625
ad	145	82	-58	-253	51	AD	6.375	162.5625
bd	53	61	-26	-24	-23	BD	-2.875	33.0625
abd	141	51	-85	-17	-1	ABD	-0.125	0.0625†
cd	26	70	28	-28	-7	CD	-0.875	3.0625
acd	97	88	-10	-59	$+7$	ACD	0.875	3.0625†
bcd	-16	71	18	-38	-31	BCD	-3.875	60.0625†
$abcd$	54	70	-1	-19	$+19$	Days	2.375	(22.5625)
Total	1261							

Column (6) of Table 15.16 presents the average effect on velocity for each of the factors. These results are stated (for the main effects) with both 95 and 99 percent confidence intervals in Table 15.18. Increase in propellant charge increases the velocity, while increased projectile weight and size of propellant web reduce the velocity in this range under study. The final column of Table 15.16 gives the sums of squares for each factor and interaction. The sum of squares due to *days* is that confounded with $ABCD$ (which we assume is 0).

The analysis of variance table is given as Table 15.17. Without any additional assumptions (on interactions equaling 0) we can easily see that charge weight (A) and propellant web (C) as well as propellant weight (B) are very strongly significant. The next strongest effect is that due to weapons (D). Only one interaction (BC) seems significant. If, further, we can assume that the 3-factor interactions (marked † in Table 15.16) are 0 and that their sum of squares is, in fact, the Residual sum of squares, then we can apply a formal test. The estimate of the residual variance is then 66.56 with 4 degrees of freedom, and of the standard deviation is 8.16. As already noted, A, B, and C are strongly significant, while D is significant at the 5 percent level. The BC interaction is just short of this 5 percent level. Hence it should not be completely disregarded.

Let us reconsider the effect of factor D. If we assume that the weapons were chosen independently and at random, what we are measuring is a variance among weapons. The ratio of 770.06 to 66.56 is appropriate to test whether this variance exists. Our major interest, however (as far as weapons variation is concerned), is in

TABLE 15.17
ANALYSIS OF VARIANCE TABLE—VELOCITY DATA

SOURCE	SUM OF SQUARES	DEGRESS OF FREEDOM	MEAN SQUARE	MEAN SQUARE RATIO
Propellant Charge (A)	18700.56	1		280.96***
Projectile Weight (B)	2475.06	1		37.19**
Propellant Web (C)	15562.56	1		233.81***
Weapon (D)	770.06	1		11.57*
Days (confounded with ABCD)	22.56	1		0.34
AB	76.56	1		1.15
AC	105.06	1		1.58
AD	162.56	1		2.44
BC	473.06	1		7.11
BD	33.06	1		0.50
CD	3.06	1		0.05
ABC	⎡ 203.06	1 ⎤		
ABD	266.24 ⎰ 0.06	1 ⎰ 4	66.56	
ACD	⎱ 3.06	1 ⎱		
BCD	⎣ 60.06	1 ⎦		
Total	38650.40†	15		

Critical Values $\begin{cases} F_{1,4,0.95} = 7.71 \\ F_{1,4,0.99} = 21.20 \end{cases}$

---

†Exact value is 38650.4375. Round off error appears since 0.0025 is omitted from each sum of squares.

TABLE 15.18
ESTIMATED EFFECT OF CHANGES IN FACTOR LEVELS ("HIGH" — "LOW")
AND CONFIDENCE INTERVALS*

	AVERAGE EFFECT	95% CONFIDENCE LIMITS		99% CONFIDENCE LIMITS	
Propellant Charge	68.375	57.05	79.70	49.59	87.16
Projectile Weight	−24.875	−36.20	−13.55	−43.66	−6.19
Propellant Web	−62.375	−73.70	−51.05	−81.16	−43.59

* There is a rounding error, since $\widehat{\sigma^2} = 66.56$ is carried to only 2 decimal places.

the actual magnitude of this variance σ_w^2. An unbiased estimate of σ_w^2 is

$$\hat{\sigma}_w^2 = \frac{(770.06 - 66.56)}{8} = 87.94$$

In other words, in any test result, the estimated variance in velocity of a shot fired from a weapon chosen at random is $154.50 (= 66.56 + 87.94)$. That is, a standard deviation estimated at 12.4 accompanies each test result. Table 15.18 presents the expected difference between lower and upper levels of factors A, B, and C, as well as the 95 and 99 percent confidence limits on this difference. Note, however, that the average of, say, A_1 (or A_0) is equal to the grand mean plus (or minus) one-half the average effect of factor A. Hence

Average of $A_1 = 78.8125 + \frac{1}{2}(68.375) = 113.0000$

Average of $A_0 = 78.8125 - \frac{1}{2}(68.375) = 44.6250$

Average effect of $A = 68.375$

Since these confidence intervals are based on *independent* linear functions of the observations, the factors in Table I cannot be used to obtain a set of simultaneous confidence intervals. In this case the method described in the *latter* part of Section 13.14.2 has to be used. We need to find values of $q'_{k,\nu_0,1-\alpha}$ for $k=3$ and $\nu_0=4$. The values of $q'^2_{k,\nu_0,1-\alpha}$ given in Section 15.2.5 do not include such low values of ν. However, it can be seen that $q'_{3,4,0.95}$ would be of the order of $\sqrt{16} = 4$, as compared with $t_{4,0.975} = 2.766$, the multiplier for the individual 95 percent confidence limits. A few values of the ratio $t_{\nu_0,0.975}/q'_{k,\nu_0,0.95}$ are given below.

VALUES OF RATIO $t_{\nu_0,0.975}/q'_{k,\nu_0,0.95}$

			k			
ν_0	2	3	4	6	8	10
10	0.86	0.79	0.75	0.69	0.65	0.62
20	0.87	0.80	0.77	0.72	0.69	0.67
∞	0.88	0.82	0.79	0.75	0.72	0.70

TABLE 15.19 MAIN EFFECT AND INTERACTIONS IN 2^2, 2^3, 2^4, AND 2^5 FACTORIAL DESIGNS*

EFFECTS

TREATMENT COMBINATIONS	2^2				2^3				2^4								2^5															
	I	A	B	AB	C	AC	BC	ABC	D	AD	BD	ABD	CD	ACD	BCD	ABCD	E	AE	BE	ABE	CE	ACE	BCE	ABCE	DE	ADE	BDE	ABDE	CDE	ACDE	BCDE	ABCDE
(1)	+	−	−	+	−	+	+	−	−	+	+	−	+	−	−	+	−	+	+	−	+	−	−	+	+	−	−	+	−	+	+	−
a	+	+	−	−	−	−	+	+	−	−	+	+	+	+	−	−	−	−	+	+	+	+	−	−	+	+	−	−	−	−	+	+
b	+	−	+	−	−	+	−	+	−	+	−	+	+	−	+	−	−	+	−	+	+	−	+	−	+	−	+	−	−	+	−	+
ab	+	+	+	+	−	−	−	−	−	−	−	−	+	+	+	+	−	−	−	−	+	+	+	+	+	+	+	+	−	−	−	−
c	+	−	−	+	+	−	−	+	−	+	+	−	−	+	+	−	−	+	+	−	−	+	+	−	+	−	−	+	+	−	−	+
ac	+	+	−	−	+	+	−	−	−	−	+	+	−	−	+	+	−	−	+	+	−	−	+	+	+	+	−	−	+	+	−	−
bc	+	−	+	−	+	−	+	−	−	+	−	+	−	+	−	+	−	+	−	+	−	+	−	+	+	−	+	−	+	−	+	−
abc	+	+	+	+	+	+	+	+	−	−	−	−	−	−	−	−	−	−	−	−	−	−	−	−	+	+	+	+	+	+	+	+
d	+	−	−	+	−	+	+	−	+	−	−	+	−	+	+	−	−	+	+	−	+	−	−	+	−	+	+	−	+	−	−	+
ad	+	+	−	−	−	−	+	+	+	+	−	−	−	−	+	+	−	−	+	+	+	+	−	−	−	−	+	+	+	+	−	−
bd	+	−	+	−	−	+	−	+	+	−	+	−	−	+	−	+	−	+	−	+	+	−	+	−	−	+	−	+	+	−	+	−
abd	+	+	+	+	−	−	−	−	+	+	+	+	−	−	−	−	−	−	−	−	+	+	+	+	−	−	−	−	+	+	+	+
cd	+	−	−	+	+	−	−	+	+	−	−	+	+	−	−	+	−	+	+	−	−	+	+	−	−	+	+	−	−	+	+	−
acd	+	+	−	−	+	+	−	−	+	+	−	−	+	+	−	−	−	−	+	+	−	−	+	+	−	−	+	+	−	−	+	+
bcd	+	−	+	−	+	−	+	−	+	−	+	−	+	−	+	−	−	+	−	+	−	+	−	+	−	+	−	+	−	+	−	+
abcd	+	+	+	+	+	+	+	+	+	+	+	+	+	+	+	+	−	−	−	−	−	−	−	−	−	−	−	−	−	−	−	−
e	+	−	−	+	−	+	+	−	−	+	+	−	+	−	−	+	+	−	−	+	−	+	+	−	−	+	+	−	+	−	−	+
ae	+	+	−	−	−	−	+	+	−	−	+	+	+	+	−	−	+	+	−	−	−	−	+	+	−	−	+	+	+	+	−	−
be	+	−	+	−	−	+	−	+	−	+	−	+	+	−	+	−	+	−	+	−	−	+	−	+	−	+	−	+	+	−	+	−
abe	+	+	+	+	−	−	−	−	−	−	−	−	+	+	+	+	+	+	+	+	−	−	−	−	−	−	−	−	+	+	+	+
ce	+	−	−	+	+	−	−	+	−	+	+	−	−	+	+	−	+	−	−	+	+	−	−	+	−	+	+	−	−	+	+	−
ace	+	+	−	−	+	+	−	−	−	−	+	+	−	−	+	+	+	+	−	−	+	+	−	−	−	−	+	+	−	−	+	+
bce	+	−	+	−	+	−	+	−	−	+	−	+	−	+	−	+	+	−	+	−	+	−	+	−	−	+	−	+	−	+	−	+
abce	+	+	+	+	+	+	+	+	−	−	−	−	−	−	−	−	+	+	+	+	+	+	+	+	−	−	−	−	−	−	−	−
de	+	−	−	+	−	+	+	−	+	−	−	+	−	+	+	−	+	−	−	+	−	+	+	−	+	−	−	+	−	+	+	−
ade	+	+	−	−	−	−	+	+	+	+	−	−	−	−	+	+	+	+	−	−	−	−	+	+	+	+	−	−	−	−	+	+
bde	+	−	+	−	−	+	−	+	+	−	+	−	−	+	−	+	+	−	+	−	−	+	−	+	+	−	+	−	−	+	−	+
abde	+	+	+	+	−	−	−	−	+	+	+	+	−	−	−	−	+	+	+	+	−	−	−	−	+	+	+	+	−	−	−	−
cde	+	−	−	+	+	−	−	+	+	−	−	+	+	−	−	+	+	−	−	+	+	−	−	+	+	−	−	+	+	−	−	+
acde	+	+	−	−	+	+	−	−	+	+	−	−	+	+	−	−	+	+	−	−	+	+	−	−	+	+	−	−	+	+	−	−
bcde	+	−	+	−	+	−	+	−	+	−	+	−	+	−	+	−	+	−	+	−	+	−	+	−	+	−	+	−	+	−	+	−
abcde	+	+	+	+	+	+	+	+	+	+	+	+	+	+	+	+	+	+	+	+	+	+	+	+	+	+	+	+	+	+	+	+

* A larger table will be found in [15.4].

In the preceding examples on confounding, the 2 (or 4) blocks could have been generated by reference to Table 15.19. This is an expansion of Table 15.5. In Example 15.7, for instance, we could consult Table 15.19 and in the $ABCD$ column, call the treatment combinations with the same sign as that of (1) the principal block. The others constitute the second block. If 3 interactions are confounded with blocks, then we carry this out in stages. For the first stage, divide the 2^k treatment combinations into 2 groups of 2^{k-1} each. Retain in one group those combinations with the same sign as that of (1). At each successive stage, repeat this procedure, always retaining those with the same sign as that of (1). Keep in mind, further, that if there are 2^r blocks, then there are only r independently chosen confounding interactions.

15.3.2 Partial Confounding

If an experiment can be arranged in a number of separate complete replicates it is not necessary to confound the same interaction(s) in each replicate. For example, Figure 15.8 shows a 2^3 experiment with 4 replicates, each arranged in 2 blocks of 4 plots each. The 3-factor interaction ABC is confounded in the first replicate; in the other 3 replicates the three 2-factor interactions are confounded.

REPLICATION	I		II		III		IV	
	(1)	a	(1)	b	(1)	a	(1)	a
	bc	b	bc	c	b	ab	ab	b
	ac	c	a	ab	ac	c	c	ac
	ab	abc	abc	ac	abc	bc	abc	bc
CONFOUNDING	ABC		BC		AC		AB	

Figure 15.8 Design with partially confounded interactions.

For convenience the principal block has been placed on the left in each case. In practice the arrangement could be randomized, if desired and practicable.

Such an experimental design is said to be *partially confounded*. The particular arrangement shown in Figure 15.8 is also *balanced*, in that each member of *any given order of interactions* is equally confounded. Omitting replication I still leaves a balanced partially confounded experiment; omission of II, III, or IV destroys the balance.

The value of partially confounded designs lies in the fact that *some* information is available for each of the interactions. In the analysis the sum of squares for a partially confounded interaction is based only on observations in those replicates where it is *not* confounded. If a Yates table is used, the calculations can be made in terms of the complete totals of observations for each treatment combination. Then the confounded effects should be corrected by subtracting the "block" difference (confounded with this interaction) from the appropriate replicate(s). The divisor of the square of the "effect" should be appropriately modified. Thus for the design shown in Figure 15.8 the divisor for each of the main effects is 32, but for each of ABC, BC, AC, and AB it is 24.

In the analysis of variance table, partially confounded and unconfounded effects have the same number of degrees of freedom as in an unconfounded experiment. The effect of confounding appears in the expected values of the mean squares. For the design of Figure 15.8 the "(effect)^2" part of this expected value of the mean square for a partially confounded effect has a multiplier only three-quarters of the multiplier for unconfounded effects. This shows that the mean square ratio test is less sensitive for the partially confounded effects. Sometimes this feature of a partially confounded design is expressed by saying that "one-quarter of the information on each of ABC, BC, AC, and AB is lost."

Example 15.8   In a preliminary laboratory test of ceramic ware 3 factors were under consideration, namely, firing time (A), firing temperature (B), and formulation of basic ingredients (C). In an early study, 2 levels of each factor were chosen. The test consisted of an abrasion of the surface, and lasted 20 hr. Two machines were available for the test, each of which held 4 samples. Yield was measured in weight loss, and the data were coded.

Each of the 3 factors, (A), (B), and (C) were at 2 levels. Machines are to be regarded as blocks each containing 4 "plots." The 2 replicates used were equivalent to replicated I and IV of Figure 15.8 (here called replicates I and II, respectively). In the first replicate ABC is confounded with blocks; in the second, AB is confounded. The positions of the test pieces on the abrasion machine were randomized (within the design allocation). The data are given in Table 15.20.

TABLE 15.20

A PARTIALLY CONFOUNDED EXPERIMENT ON WEIGHT
LOSS DUE TO ABRASION

REPLICATE I		REPLICATE II	
Block 1	Block 2	Block 1	Block 2
$(1) = 9$	$a = 8$	$(1) = 0$	$a = 9$
$bc = 13$	$b = 3$	$ab = 8$	$b = 2$
$ac = 5$	$c = 15$	$c = 14$	$ac = 10$
$ab = 11$	$abc = 11$	$abc = 13$	$bc = 12$

TABLE 15.21

YATES' CALCULATION FOR A PARTIALLY CONFOUNDED
EXPERIMENT ON WEIGHT LOSS DUE TO ABRASION

TREATMENT COMBINATIONS (1)	YIELD (2)	(3)	(4)	(5)	EFFECT (6)	AVERAGE EFFECT (5)/8	MEAN SQUARE $(5)^2/16$
(1)	9	26	50	143	I		
a	17	24	93	7	A	0.875	3.0625
b	5	44	22	3	B	0.375	0.5625
ab	19	49	-15	19	AB	(not applicable)	
c	29	8	-2	43	C	5.375	115.5625
ac	15	14	5	-37	AC	-4.625	85.5625
bc	25	-14	6	7	BC	0.875	3.0625
abc	24	-1	13	7	ABC	(not applicable)	

The complete data are initially analyzed by Yates' algorithm as shown in Table 15.21. Note that column (2) contains the total of 2 replicates for each treatment combination. In columns (6), (7), and (8), the average effect and mean square of AB and ABC are not applicable. The mean square of AB can be calculated directly from the 8 treatments of replicate I; the ABC mean square is calculated from replicate II. Alternatively the difference between blocks in replicate I(II) can be subtracted from the $ABC(AB)$ total in the Yates table to give corrected $ABC(AB)$ totals. For the ABC sum of squares this would be $[7-(37-38)]^2/8 = 8^2/8 = 8$. This method is preferable when more than 2 interactions are partially confounded.

The ANOVA table is given as Table 15.22. The Replicate sum of squares and the Blocks (within Replicate) sum of squares can be calculated directly from the data of Table 15.20.

From Table 15.22 one notes that factor C (Formulations) is significant, as is the interaction between Firing Time (A) and the Formulations (C). Firing Time (A) and Firing Temperature (B), in the range being studied (and with just 2 levels considered), do not show any significant effect on weight loss. Their interaction (AB), although larger than either main effect, is also not significant.

Table 15.23 presents a two-way table of means for Formulations and Firing Time and the estimate of the standard deviations of means. This is given in order to exhibit the Formulation effect and the (Formulation)×(Firing Time) interaction.

Example 15.9 As a second example on the use of partial confounding, we consider some experiments on the formulation of emulsified pesticides. (Terao and Mitamura, "Formulation of New Pesticides," *Reports of Statistical Application Research, Union of Japanese Scientists and Engineers*, **7** (2) (1960).) The factors considered in the experiment were Concentration of Emulsifier (A) at 2 levels, Solvent (B) at 3 levels, and Emulsifier (C) at 4 levels. There were thus $2 \times 3 \times 4 = 24$

TABLE 15.22

ANOVA OF WEIGHT LOSS DUE TO ABRASION

SOURCE	SUM OF SQUARES	DEGREES OF FREEDOM	MEAN SQUARE	MEAN SQUARE RATIO
A	3.0625	1		
B	0.5625	1		
C	115.5625	1	115.5625	14.7*
AB	36.1250	1	36.1250	4.59
AC	85.5625	1	85.5625	10.9*
BC	3.0625	1		
ABC	8.0000	1	8.0000	1.02
Replications	3.0625	1		
Blocks within				
Replications	0.6250	2		
Residual	39.3125	5	7.8625	
Total	294.9375	15		

Critical F values: $F_{1,5,0.95} = 6.61$;

$F_{1,5,0.99} = 16.26$.

TABLE 15.23

TWO-WAY TABLE: AVERAGE LOSS FOR COMBINATIONS OF
FORMULATIONS AND FIRING TIME

FIRING TIME (A)	FORMULATION (C) 1	2	MEANS
1	3.5	13.5	8.50
2	9.0	9.75	9.375
Means	6.25	11.625	

Estimated Standard Deviation of each mean: 0.99

factor-level combinations. Three replications of these 24 combinations were used. As only 12 samples could be tested at a time it was necessary to divide each replication into 2 blocks of 12 factor-level combinations each. A design given by Kitagawa and Mitome (p. II-136) (see Section 15.6) was used. The design is shown in Table 15.24 [together with the observed "infestation indices" (I.I.)]; the 3 numbers represent the levels of factors A, B, C, respectively, in each treatment combination. One degree of freedom of the $A \times C$ interaction is confounded in one replication, and 2 different degrees of freedom of the $A \times B \times C$ interaction are confounded in the other replications (one in each replication). The analysis of variance calculated from the experimental I.I. values is shown in Table 15.25. It should be noted that the Residual includes unconfounded $A \times C$ and $A \times B \times C$

TABLE 15.24
PARTIALLY CONFOUNDED $2 \times 3 \times 4$ EXPERIMENT ON PESTICIDE FORMULATIONS

(T.C. ≡ Treatment Combinations; I.I. ≡ Infestation Index)

REPLICATE I				REPLICATE II				REPLICATE III			
BLOCK 1		BLOCK 2		BLOCK 1		BLOCK 2		BLOCK 1		BLOCK 2	
T.C.	I.I.	T.C.	I.I.	T.C.	I.I.	T.C.	I.I.	T.C.	I.I.	T.C.	I.I.
000	34	100	26	000	28	100	30	000	43	100	44
110	96	010	52	110	48	010	64	010	65	110	71
120	36	020	47	020	63	120	92	120	58	020	64
101	55	001	20	101	13	001	41	101	40	001	51
011	72	111	45	011	46	111	29	111	74	011	36
021	78	121	34	121	70	021	20	021	81	121	63
002	83	102	90	002	62	102	41	002	56	102	52
112	31	012	47	112	77	012	69	012	46	112	61
122	25	022	70	022	18	122	54	122	11	022	52
103	88	003	64	103	62	003	101	103	72	003	46
013	39	113	65	013	73	113	60	113	84	013	36
023	45	123	49	123	70	023	72	023	98	123	56

TABLE 15.25
ANOVA OF PESTICIDE FORMULATION EXPERIMENT

SOURCE	DEGREES OF FREEDOM	SUM OF SQUARES	MEAN SQUARE	MEAN SQUARE RATIO*
Concentration (A)	1	1.4	1.4	—
Solvent (B)	2	332.5	166.3	—
Emulsifier (C)	3	2,991.2	997.1	2.62
A × B	2	834.2	417.1	1.10
B × C	6	6,998.5	1,164.7	3.06*
Replicates	2	113.2	56.6	—
Blocks within replicates	3	683.1	227.7	—
Residual	52	19,787.4	380.5	
Total	71	31,731.5		

* The Residual Mean Square was used as denominator in each mean square ratio corresponding to an assumed purely parametric model.

effects. These were presumably felt to be of negligible amount. The only significant mean square is the interaction $(B \times C)$ between Solvent and Emulsifier. It must be borne in mind that real $A \times C$ and (or) $A \times B \times C$ effects would tend to reduce the apparent significance derived form formal analysis of variance procedures. As well as the Solvent $\times$ Emulsifier interaction effect, which appears fairly clearly established, it is possible that there is also a sizable Emulsifier (C) effect masked by a swollen Residual mean square.

It would be useful to provide tables giving estimates of mean I.I. for each Solvent-Emulsifier combination (and, perhaps. also overall averages for each Emulsifier). The best (lowest I.I.) combinations of Solvent and Emulsifier are

(*i*) both at level 0, and

(*ii*) both at level 2.

Concentration of Emulsifier (A) appears to have little effect on the I.I.

15.4 FRACTIONAL REPLICATION

Consider a k factor experiment with each factor at 2 levels. If k is large then the size of a complete factorial experiment could easily be beyond the reach of our experimental effort. Even an experiment of the size of 2^5 may be too great an effort owing to limitations of cost, time, and other restrictions. If we can plan a suitably balanced fraction of a very large experiment, however, we may still be able to get the information desired.

We have already encountered the idea of fractional replication in the use of the Latin square. We noted that this could be considered an $n^{-1} \times n^3$ experiment, giving n^2 treatment combinations. We also noted that its model is

$$X_{(tij)} = \xi + \rho_t + \kappa_i + \tau_j + Z_{(tij)}$$

We sacrifice some information on interaction effects in order to reduce the size of the experiment.

Now let us consider one-half of a 2^4 experiment. Since there are only 8 experimental units (hence only 7 degrees of freedom), only 7 distinct effects are measurable. In fact, the original 15 effects (4 main effects, 6 two-factor, 4 three-factor and 1 four-factor interaction) are now intermixed. There now are 7 pairs of inseparable effects plus 1 more that cannot be measured at all. For example, consider a $\frac{1}{2} \times 2^4$ experiment with the 8 treatment combinations of the factors A, B, C, and D shown in Figure 15.9. This is, in fact, one block of a 2^4 experiment in which the interaction $ABCD$ is confounded.

Consult Table 15.19 for the estimation of main effects and interactions from these 8 combinations. We note that the effect of A is measured by

$$-(1) + ab + ac - bc + ad - bd - cd + abcd$$

		A_0		A_1	
		B_0	B_1	B_0	B_1
C_0	D_0	(1)			ab
	D_1		bd	ad	
C_1	D_0		bc	ac	
	D_1	cd			$abcd$

Figure 15.9   Design of a $\frac{1}{2} \times 2^4$ experiment.

The effect of BCD is also measured by

$$-(1) + ab + ac - bc + ad - bd - cd + abcd$$

which is identical to that of A. Further investigation of the table shows that the effects in each of the pairs below are represented by the same function of the available observations. That is, for each pair we cannot distinguish between the 2 effects in the pair.

I	$ABCD$	D	ABC
A	BCD	AB	CD
B	ACD	AC	BD
C	ABD	AD	BC

(I denotes the total of all observations).

Two effects or interactions are said to be an *alias pair* if we cannot distinguish between them in the analysis. There are 7 alias pairs above (I and $ABCD$ are not usually regarded as an alias pair). Their common estimate can measure one or the other of the 2, or both of them. A $\frac{1}{2} \times 2^k$ experiment is just 1 block of a 2^k experiment confounded in 2 blocks. The confounded interaction in the latter is the *defining contrast* in the former. We can find the alias of any effect by finding its generalized interaction with the defining contrast by "multiplication" (modulo 2). For example, in the above illustration the defining contrast is $ABCD$ and A is aliased with $BCD (\equiv A \times ABCD \equiv A^2BCD)$, AB is aliased with $CD (\equiv AB \times ABCD \equiv A^2B^2CD)$, and so on.

The procedure for selecting the $\frac{1}{2}$ replicate is as follows: (1) choose the defining contrast; (2) use this contrast to split the original 2^k experiment into 2 blocks; (3) choose either of the 2 blocks for the treatment combinations of the experiment. Each treatment combination in any one of the blocks will have the same algebraic signs for calculating the effects for each factor in an alias-pair, while those in the other block will have opposite signs. For example, suppose we had chosen the other treatment combinations for the illustration, namely,

$$a \quad c \quad abc \quad acd$$
$$b \quad d \quad abd \quad bcd$$

Then effect A would be measured by

$$a - b - c - d + abc + abd + acd - bcd$$

and BCD would be measured by

$$-a + b + c + d - abc - abd - acd + bcd$$

This change of signs holds true for the other pairs of aliases.

Example 15.10 We wish to plan a $\frac{1}{2} \times 2^5$ experiment. First we choose the defining contrast, say, $ABCDE$. The alias pairs are given in Table 15.26 and are obtained by taking the product (modulo 2) of the effect and $ABCDE$.

There are 2 choices for the $\frac{1}{2}$ replicate, either the principal or the other block, using $ABCDE$ as if it were the confounding interaction. This principal block is given in Table 15.27.

TABLE 15.26

ALIASES IN A $\frac{1}{2}$ × 2^5 EXPERIMENT WITH THE
DEFINING CONTRAST $ABCDE$

EFFECT	ALIAS	EFFECT	ALIAS
I	$ABCDE$	AD	BCE
A	$BCDE$	AE	BCD
B	$ACDE$	BC	ADE
C	$ABDE$	BD	ACE
D	$ABCE$	BE	ACD
E	$ABCD$	CD	ABE
AB	CDE	CE	ABD
AC	BDE	DE	ABC

TABLE 15.27

$\frac{1}{2}$ REPLICATE OF A 2^5 EXPERIMENT
($ABCDE$ is the defining contrast)

(1)	ae	cd	$abce$
ab	bc	ce	$abde$
ac	bd	de	$acde$
ad	be	$abcd$	$bcde$

Of course, no particular significance attaches to the order in which these 16 treatment combinations are written down. In an experiment this should be randomized. Note that the effect A has the opposite algebraic signs from that of $BCDE$. This is true also of all of the effects and their aliases. In general, for the principal block the algebraic signs in computing effects of pairs of aliases are the same if the defining contrast has an even number of letters. Otherwise the signs are opposite. If now we can assume that the 3-factor (and higher) interactions are 0, then the 15 degrees of freedom allow us to estimate the main effects and 2-factor interactions separately. Unless some of the 2-factor interactions are 0, however, it will be unlikely that a Residual sum of squares can be formed.

15.4.1   A 2^{-r} Fraction of a 2^k Experiment in One Block

Rather than a $\frac{1}{2}$ replicate of a 2^5 experiment let us suppose that we want to use only 8 experimental units. We obtain such a design by taking just 1 block of a 2^5 experiment arranged in 4 blocks of 8 plots each. This means that we use a $\frac{1}{4}$ replicate, and every effect has 3 aliases. That is, there are only 7 degrees of freedom accounting for 28 effects. The remaining 3 effects are defining contrasts and are, in fact, the confounding interactions in the complete design from which our block was chosen. Just as in the case of confounding interactions, 2 of these contrasts can be chosen independently, while the third is their product (modulo 2). Suppose that we arbitrarily choose ABC and CDE as 2 of the defining contrasts. Then $ABDE$ is the third defining contrast. The effects and their aliases are given in Table 15.28. The three aliases of each effect are obtained by "multiplying" the effect by ABC, CDE, and $ABDE$, respectively.

In order to find the 8-treatment combinations for the experiment, we use the procedure already given for selecting the 4 blocks confounded with the interactions ABC, CDE, and $ABDE$. We then choose any one of these.

TABLE 15.28

ALIASES IN A $\frac{1}{4} \times 2^5$ EXPERIMENT WITH
ABC, CDE, AND $ABDE$, AS DEFINING CONTRASTS

EFFECT	ALIASES		
I	ABC	CDE	$ABDE$
A	BC	$ACDE$	BDE
B	AC	$BCDE$	ADE
C	AB	DE	$ABCDE$
D	$ABCD$	CE	ABE
E	$ABCE$	CD	ABD
AD	BCD	ACE	BE
AE	BCE	ACD	BD

The principal block is composed of the treatments

$$(1) \qquad de \qquad acd \qquad ace$$

$$ab \qquad abde \qquad bcd \qquad bce$$

Example 15.11 We wish to show the design of a $\frac{1}{4} \times 2^5$ experiment as a fraction of a 2^5 experiment. This is given in Figure 15.10. In this the defining contrasts are those just given, namely, ABC, CDE, and $ABDE$.

		A_0				A_1		
		B_0		B_1		B_0		B_1	
		C_0	C_1	C_0	C_1	C_0	C_1	C_0	C_1
D_0	E_0	(1)						ab	
	E_1				bce		ace		
D_1	E_0				bcd		acd		
	E_1	de							$abde$

Figure 15.10 Design of a $\frac{1}{4} \times 2^5$ experiment.

Note that each main effect has an equal number of treatments at the lower and upper levels. Although this $\frac{1}{4} \times 2^5$ experiment is used as an illustration, it must be kept in mind that this experiment is quite small. Only the main effects appear in alias groups in which the other 3 members are of higher order. We usually try to arrange for each (or as many as possible) low-order interactions to be aliased only with higher-order interactions.

15.4.2 A 2^{-r} Fraction of a 2^k Experiment in 2^b Blocks each of 2^{k-r-b} Experimental Units

We now proceed one further step beyond a 2^{k-r} experiment. For example, suppose the size of the blocks in the experiment has been determined, and it is such that we must divide the 2^{k-r} experiment into more than 1 block, say, 2^b blocks. The procedure is as follows:

(1) Choose r independent defining contrasts. The remaining $2^r - r - 1$ defining contrasts are developed from these as generalized interactions.

(2) Determine a suitable set of confounding interaction(s), guarding against the selection of a main effect and its alias, or any of the defining contrasts.

(3) Select that block which has an even (or zero) number of letters in common with the independently chosen defining contrasts and the independently chosen confounding interactions. Use this or a block generated from this. We illustrate this by Example 15.12.

Example 15.12   For a 6-factor (each at 2 levels) experiment we wish to use a design of size 16 arranged in 2 blocks. Suppose we choose as the defining contrasts $ABCD$, $ABEF$, and their "product" $CDEF$. The alias groups are given in Table 15.29.

TABLE 15.29

ALIAS GROUPS IN A $\frac{1}{4} \times 2^6$ EXPERIMENT WITH DEFINING
CONTRASTS $ABCD$, $ABEF$, AND $CDEF$

EFFECT	ALIASES		
I	$ABCD$	$ABEF$	$CDEF$
A	BCD	BEF	$ACDEF$
B	ACD	AEF	$BCDEF$
C	ABD	$ABCEF$	DEF
D	ABC	$ABDEF$	CEF
E	$ABCDE$	ABF	CDF
F	$ABCDF$	ABE	CDE
AB	CD	EF	$ABCDEF$
AC	BD	$BCEF$	$ADEF$
AD	BC	$BDEF$	$ACEF$
AE	$BCDE$	BF	$ACDF$
AF	$BCDF$	BE	$ACDE$
CE	$ABDE$	$ABCF$	DF
CF	$ABDF$	$ABCE$	DE
ACE	BDE	BCF	ADF
ACF	BDF	BCE	ADE

Note that a large number of 2-factor interactions appear together in alias groups. To select 2 blocks of 8 treatment combinations, we first choose a single block of 16 units. We select the principal block (the choice of this block as the fractional replicate is arbitrary) on the basis of the 3 defining contrasts. The treatments in this block are

(1)	bce	$abef$	acf
ab	ace	ef	bcf
cd	bde	$abcdef$	adf
$abcd$	ade	$cdef$	bdf

We now confound this design, confounding AD and its aliases BC, $BDEF$, and $ACEF$. With AD as the confounding interaction, the 2 blocks are given in Table 15.30.

TABLE 15.30

$\frac{1}{4} \times 2^6$ Design in 2 Blocks, Confounding the Alias-Group
AD, BC, $BDEF$, $ACEF$

BLOCK 1		BLOCK 2	
(1)	ef	ab	abef
abcd	abcdef	cd	cdef
bce	bcf	ace	acf
ade	adf	bde	bdf

Usually we would try to arrange that none of the interactions in the alias-groups AD, BC, $BDEF$, and $ACEF$ were of particular interest.

Note that if we wished to have 4 blocks of 4 treatments, another interaction, independently chosen and not aliased with the preceding interactions, would be necessary. To generate these 4 blocks let us arbitrarily choose AF, realizing that its aliases BE, $BCDF$, and $ACDE$ will also be confounded with blocks, and still use AD and its alias group. Then DF (the product, modulo 2, of AD and AF) and its aliases will also be confounded. The new principal block contains (1), bce, $abcdef$, adf. By our preceding rules we can now generate the 4 blocks shown in Table 15.31. We now have a $\frac{1}{4} \times 2^6$ experiment in 2^2 blocks of 2^2 units. The defining contrasts are $ABCD$, $ABEF$, and their product, $CDEF$. The confounded interactions are AD, AF, and their product, DF, and the alias groups to which these 3 interactions belong.

TABLE 15.31

A $\frac{1}{4} \times 2^6$ Experiment in 4 Blocks of 4 Units

BLOCK 1	BLOCK 2	BLOCK 3	BLOCK 4
(1)	ef	ab	cd
bce	bcf	ace	bde
abcdef	abcd	cdef	abef
adf	ade	bdf	acf

15.4.3   Analysis of a Fractional Experiment using Yates' Algorithm

Consider a $\frac{1}{2}$ replicate of a 2^5 experiment with $ABCDE$ the defining contrast. The principal block contains the treatments

(1)	ae	cd	abce
ab	bc	ce	acde
ac	bd	de	abde
ad	be	abcd	bcde

For the first step, we set up as the first column in the Yates' algorithm the usual order of treatment combinations for 4 factors, namely,

(1)	c	d	cd
a	ac	ad	acd
b	bc	bd	bcd
ab	abc	abd	abcd

Second, we add (in parentheses) the letter "e" to the 1 and 3 letter combinations. This gives us the first column of Table 15.32. These are the treatment combinations of the desired fractional design. Third, we carry out the usual process of summing and differencing in pairs, and label the

TABLE 15.32

YATES' ALGORITHM FOR A $\frac{1}{2} \times 2^5$ EXPERIMENT

TREATMENT COMBINATIONS (1)	(2)	(3)	(4)	(5)	(6)	(7)
(1)	(1) + a(e)	(1) + a(e) + b(e) + ab	—	—	I	ABCDE
a(e)	b(e) + ab	—	—	—	A	BCDE
b(e)	—	—	—	—	B	ACDE
ab	—	—	—	—	AB	CDE
c(e)	—	—	—	—	C	ABDE
ac	—	—	—	—	AC	BDE
bc	—	—	—	—	BC	ADE
abc(e)	—	—	—	—	ABC	DE
d(e)	a(e) − (1)	b(e) + ab − (1) − a(e)	—	—	D	ABCE
ad	ab − b(e)	—	—	—	AD	BCE
bd	—	—	—	—	BD	ACE
abd(e)	—	—	—	—	ABD	CE
cd	—	—	—	—	CD	ABE
acd(e)	—	—	—	—	ACD	BE
bcd(e)	—	—	—	—	BCD	AE
abcd	—	—	—	—	ABCD	E

resulting effects as in Table 15.32. Columns (6) and (7) present the pairs of aliased effects. (If the defining contrast were $ABCD$, the first step could be to present the 16 combinations of A, B, C, and E. The second step would add "d" values to 8 of the combinations giving the desired 16 treatments which, in fact, constitute the principal block. In general, the finally inserted letter must be one of those appearing in the defining contrast.)

Example 15.13   In the preparation and curing of rubber in the manufacture of tires, a basic investigation was planned to determine the variability of the mixing, curing, and testing stages of a compound. In the exploratory stages, a $\frac{1}{2}$ replicate of a 2^4 experimental design was used. Carbon black was added to the latex to prepare an oil-black master. A control was prepared by dry-mixing the black into the oil-master polymer in a Banbury mill. The 3 stages of mixing, curing, and testing of the compounds were carried out in 2 different locations, A_0 and A_1. After each stage, the resulting material was distributed equally to each location for the next stage. The data recorded were measurements of tensile strength in psi. These are given in Table 15.33. A diagram representing the stages is presented in Figure 15.11.

TABLE 15.33

$\frac{1}{2} \times 2^4$ EXPERIMENT—TENSILE STRENGTH
IN PSI OF RUBBER COMPOUNDS

		A_0		A_1	
		B_0	B_1	B_0	B_1
C_0	D_0	3400			2850
	D_1		3350	3200	
C_1	D_0		3150	3050	
	D_1	3250			3200

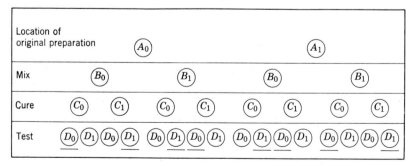

Figure 15.11   Mixing, testing, and curing at 2 locations.

The factors in the experiment are then (1) location of the original mixing (A_0 and A_1), (2) dry mix and oil-black master (B_0 and B_1), (3) location of the curing (C_0 and C_1), and (4) location of the testing (D_0 and D_1). The $\frac{1}{2}$ replicate which was used is indicated in Figure 15.11 by the underlined circles in the last stage.

The defining contrast is $ABCD$. This aliases the main effects with 3-factor interactions, and 2-factor interactions with other 2-factor interactions. The data are coded by subtracting 3000 from each value and dividing by 50. These coded yields are given in Table 15.34.

TABLE 15.34

CODED VALUES OF TENSILE STRENGTH IN $\frac{1}{2} \times 2^4$ EXPERIMENT

		A_0		A_1	
		B_0	B_1	B_0	B_1
C_0	D_0	8			-3
	D_1		7	4	
C_1	D_0		3	1	
	D_1	5			4

We now use Yates' algorithm in Table 15.35 inserting (d) where necessary. Suppose we could make the assumption that there are no interaction effects. Then the sums of squares (marked †) for the alias pairs $AB \equiv CD$, $AC \equiv BD$, and $BC \equiv AD$ could be combined to give a Residual mean square equal to 8.46 with 3 degrees of freedom. We need only examine the last column of Table 15.35 to conclude our analysis. This indicates that beyond some of the 2-factor interactions ($AC \equiv BD$ and $BC \equiv AD$) and the main effect (D), the greatest source of variation

TABLE 15.35

YATES' ALGORITHM FOR A $\frac{1}{2} \times 2^4$ DESIGN

TREATMENT COMBINATIONS (1)	YIELDS (2)	(3)	(4)	(5)	EFFECT	ALIAS	AVERAGE EFFECT (5) ÷ 4	SUM OF SQ. (5)² ÷ 8
(1)	8	12	16	29	I	$ABCD$		
$a(d)$	4	4	13	-17	A	BCD	-4.25	36.125
$b(d)$	7	6	-14	-7	B	ACD	-1.75	6.125
ab	-3	7	-3	-1	AB	CD	-0.25	0.125†
$c(d)$	5	-4	-8	-3	C	ABD	-0.75	1.125
ac	1	-10	1	11	AC	BD	2.75	15.125†
bc	3	-4	-6	9	BC	AD	2.25	10.125†
$abc(d)$	4	1	5	11	ABC	D	2.75	15.125
Totals	29							83.875

is in the location of the original mixing stage (A). The greatest difference is not between dry mix and oil-black master but rather between locations at this first preparation of the mix. Further, for an experiment of this size we need not go through a formal analysis of variance table. The one effect A stands out strongly enough, even though it is not formally significant. Any further examination should include a reexamination of the preparation of the mix. This could have been a chance result, but since it is large relative to the other main effects, it is well worth including in future experiments.

*15.4.4 A 4×2^k Experiment

If an experiment contains $k + 1$ factors, where k of the factors are each at 2 levels while 1 is at 4 levels, the techniques of this chapter can still be applied. This is done by splitting the 4-level factor (say A) into 2 "dummy" factors, each at 2 levels. If the 4 levels of treatment A are denoted by a_0, a_1, a_2, and a_3, the 2 factors A' and A'' can be arranged as in Table 15.36.

TABLE 15.36

Four-Level Factor Reduced to 2 Two-Level Factors

	A'_0	A'_1
A''_0	a_1	a_2
A''_1	a_0	a_3

The effects of A', A'', and $A' \times A''$ are 3 orthogonal effects, each with a single degree of freedom. If the factor is quantitative we can determine the linear quadratic, and higher-order effects. If the experiment is to be fractionally replicated or confounded, it could be handled as shown previously in this chapter, but it would be considered as a 2^{k+2} experiment.

Similar methods can be applied to $4^l 2^k$, $8^m 4^l 2^k$, etc., experiments.

*15.4.5 Modulo Sums Technique

In the preceding notational representations of treatments and effects we could have resorted to the numbers 0, 1, modulo 2. For any treatment combinations, instead of using the lowercase letters, we could have substituted "1" where a letter exists and "0" otherwise. For example, the treatments for a 2^3 experiment would be stated as in Table 15.37.

TABLE 15.37

2^3 Experiment in 0, 1 Notation

A B C	A B C
$(1) = 0 \ 0 \ 0$	$c = 0 \ 0 \ 1$
$a = 1 \ 0 \ 0$	$ac = 1 \ 0 \ 1$
$b = 0 \ 1 \ 0$	$bc = 0 \ 1 \ 1$
$ab = 1 \ 1 \ 0$	$abc = 1 \ 1 \ 1$

In generating 2 blocks of a 2^4 experiment with the confounding interaction $ABCD$ we originally had the treatment combinations

(1)	ad	a	d
ab	bd	b	abd
ac	cd	c	acd
bc	abcd	abc	bcd

where the second block was generated by the product, modulo 2 of a, and the treatment combinations of the first block. We can now restate these as Table 15.38.

TABLE 15.38

2^4 EXPERIMENT IN 2 BLOCKS USING 0, 1 NOTATION
($ABCD$ Is the Confounding Interaction)

BLOCK 1		BLOCK 2	
0000	1001	1000	0001
1100	0101	0100	1101
1010	0011	0010	1011
0110	1111	1110	0111

The second block is generated by adding (modulo 2) $1000(=a)$ to each of the elements of the principal block. For example, $1000(=a) + 1100(=ab) = 0100(=b)$.

Example 15.14   Let us represent the treatments of Table 15.30 in 0, 1 notation. These are for the factors A, B, C, D, E, F.

FACTORS	A	B	C	D	E	F	A	B	C	D	E	F
	0	0	0	0	0	0	1	1	0	0	0	0
	1	1	1	1	0	0	0	0	1	1	0	0
	0	1	1	0	1	0	1	0	1	0	1	0
	1	0	0	1	1	0	0	1	0	1	1	0
	0	0	0	0	1	1	1	1	0	0	1	1
	1	1	1	1	1	1	0	0	1	1	1	1
	0	1	1	0	0	1	1	0	1	0	0	1
	1	0	0	1	0	1	0	1	0	1	0	1

*15.4.6   Construction of Blocks using 0, 1 Notation

Let us construct 2 blocks of a 2^4 experiment using the 0, 1 notation. We write the family of equations

$$k_A x_1 + k_B x_2 + k_C x_3 + k_D x_4 = 0 \bmod 2$$

$$k_A x_1 + k_B x_2 + k_C x_3 + k_D x_4 = 1 \bmod 2$$

(15.3)

where k_A, k_B, k_C, and k_D are *zero* or *one* depending upon whether the letters A, B, C, and D, respectively are included in the confounding interaction. If the confounding interaction is $ABCD$, then the equations

$$x_1 + x_2 + x_3 + x_4 = 0 \bmod 2 \tag{15.4}$$

and

$$x_1 + x_2 + x_3 + x_4 = 1 \bmod 2 \tag{15.5}$$

generate the principal and other block, respectively. We choose those treatments whose values (0 or 1 for each factor, in order), when substituted in the equations, yield the equality. For example, $1100(x_1 = 1, x_2 = 1, x_3 = 0, x_4 = 0)$ satisfies (15.4). That is, $1 + 1 + 0 + 0 = 0 \bmod 2$. $0100(x_1 = 0, x_2 = 1, x_3 = 0, x_4 = 0)$ satisfies (15.5). The former is in the principal block whereas the latter is not. The principal block and the other block are shown in Table 15.38. This should be compared with Figure 15.7. Note that the treatments of the second block of Table 15.38 could be generated by the sum (modulo 2) of 1000 and each treatment of the principal block.

Example 15.15   We wish to construct a $\frac{1}{4} \times 2^6$ design in 2 blocks of 8 treatments each. As in Example 15.12 we use $ABCD$, $ABEF$ (and $CDEF$) as the defining contrasts and AD as the confounding interaction. The treatment combinations we select must satisfy the equations

$$x_1 + x_2 + x_3 + x_4 = 0 \bmod 2$$

$$x_1 + x_2 + x_5 + x_6 = 0 \bmod 2$$

The 16 treatments must be further subdivided into 2 blocks satisfying

$$x_1 + x_4 = 0 \bmod 2 \quad \text{for the principal block}$$

and

$$x_1 + x_4 = 1 \bmod 2 \quad \text{for the other block}$$

One method of generating the 16 treatment combinations is to first choose 000000, 111100, 011010 which satisfy the first 2 equations. This generates (by addition, modulo 2) 100110. We further choose 000011. This generates (by addition to the previous values) 111111, 011001, and 100101. Finally choose 110000 and add this to each of the preceding numbers. Once the 16 values are obtained, then we separate them according to the equations $x_1 + x_4 = i \bmod 2$ where $i = 0$ yields the principal block and $i = 1$ yields the other block. The result is given in Example 15.14.

15.5   FACTORS AT THREE LEVELS

In the consideration of factors at 3 levels not only can we investigate the main effect of each factor (and the interactions) but we can also study the linear and quadratic effects of quantitative factors (as has been described in Section 13.18). Three level factors can be analyzed by the methods learned earlier. We propose further a modification of Yates' algorithm for the analysis of 3^k experiments. Our primary interest in this section is similar to that for the 2^k experiments, namely, the design of the experiment. Again we ask, "How can we arrange a 3^k experiment into more than one block and how can we construct a proper fractional replicate?"

15.5.1 Notation for a 3^k Experiment

We use the notation of Section 15.4.6 to designate the levels of each factor. The 3 levels of each factor are designated by 0, 1, and 2. We make use of multiplication and addition, modulo 3. For a 3^3 experiment, the treatment combinations are as shown in Table 15.39. The first number indicates the level of factor A, the second, the level of factor B, and the third, the level of factor C.

TABLE 15.39

TREATMENT COMBINATIONS IN A 3^3 EXPERIMENT

	A_0			A_1			A_2		
	B_0	B_1	B_2	B_0	B_1	B_2	B_0	B_1	B_2
C_0	000	010	020	100	110	120	200	210	220
C_1	001	011	021	101	111	121	201	211	221
C_2	002	012	022	102	112	122	202	212	222

15.5.2 Extension of Yates' Algorithm to the Analysis of 3^k Experiments

Recall that in the Yates algorithm for a 2^k experiment we repeatedly calculated the sums and differences in order to evolve the total effects and finally sums of squares. We now consider each factor as if it were quantitative, evolve the linear and quadratic effects and their interactions, and finally regroup these for total effects and linear contrasts where this is appropriate.

Example 15.16 Consider a 3^2 experiment where the factor A is quantitative and B is qualitative. An experiment was designed to determine the amount of warping of copper plates. Three temperatures (A_0, A_1, A_2) were used, namely, 50, 75, and 100°C. The experiments were conducted in 3 different laboratories (B_0, B_1, B_2). This was, in fact, part of a much larger experiment. For the sake of illustration we consider only these 9 treatments. The data are given (in coded form) in Table

TABLE 15.40

AMOUNT OF WARPING OF COPPER PLATES

| | TEMPERATURE | | |
LABORATORIES	50°	75°	100°
1	24	17	16
2	27	18	12
3	18	25	20

15.40. We wish to determine whether there are significant differences among either laboratories or temperatures and if there is a linear or quadratic effect due to temperature.

We further code these data by subtracting 15 from each value. The coded figures are given in Table 15.41.

TABLE 15.41

CODED WARP VALUES OF COPPER PLATES

| | TEMPERATURE | | |
LABORATORIES	50°	75°	100°
1	9	2	1
2	12	3	−3
3	3	10	5

The modified Yates' algorithm for a 3^k experiment with factors all quantitative is given (for the data of Table 15.41) in Table 15.42. The first column presents the treatment combinations, while column (2) gives the actual observations. The treatments are "introduced" one at a time, each level being combined in turn with every combination of levels above it. (Level "0" is used until the treatment is "introduced.") Column (3) is obtained as follows. (a) The first third of the column is made up of the sums of each of the 3 groups of 3 observed values from column (2), (b) the second third is the third minus the first observed values in the same groups of 3 (this estimates the linear component), and (c) the last third is the sum of the first and third observed values minus twice the second value (this estimates the quadratic component). For example, in column (3) the first, fourth, and seventh values are $12 = 9 + 2 + 1$, $-8 = 1 - 9$, and $6 = 9 + 1 - 2 \times 2$. Column (4) is obtained from column (3) just as column (3) was obtained from column (2). Column (5) designates the effects. In order to obtain the sum of squares for each of these effects (all with 1 degree of freedom), we must use the proper divisor. Since these are linear functions of response the square of the entries in column (4) must be divided by the sum of squares of coefficients contributing to these effects. A

TABLE 15.42

YATES' TABLE FOR ANALYSIS OF 3^2 EXPERIMENT
ON COPPER PLATES

(1)	(2)	(3)	(4)	EFFECT (5)	DIVISOR (6)	SUM OF SQUARES (7)
00	9	12	42			
10	2	12	−21	A_L	$2\times3\times1$	73.5
20	1	18	−3	A_Q	$6\times3\times1$	0.5
01	12	−8	6	B_L	$3\times2\times1$	6.0
11	3	−15	10	$A_L B_L$	$2\times2\times1$	25.0
21	−3	2	−18	$A_Q B_L$	$6\times2\times1$	27.0
02	3	6	6	B_Q	$3\times6\times1$	2.0
12	10	3	24	$A_L B_Q$	$2\times6\times1$	48.0
22	5	−12	−12	$A_Q B_Q$	$6\times6\times1$	4.0

formula for calculating the divisor is

$$2^p 3^q n$$

where p is the number of factors in the interaction considered, q is the number of factors examined in the entire experiment minus the number of linear terms in this interaction and n is the number of replicates. For example, A_L (linear effect of A) has the divisor $2^1 \times 3^1 \times 1$, while $A_Q B_Q$ has the divisor $2^2 \times 3^2 \times 1$. The divisors can also be calculated as follows:

(sum of squares for coefficients of A effect)

$\times$ (sum of squares for coefficients of B effect)

$\times$ (number of replications)

If 1 factor is absent we have to regard its "effect" as being the sum, with "sum of squares of coefficients" equal to the number of levels of the factor. This method is more general and can be used for any number of levels of a factor and any number of factors. The divisors in Table 15.42 are calculated in this manner.

In Table 15.42, column (7) would give all of the required sums of squares if both A and B were quantitative factors. Since factor A is quantitative while B is not, only the effects shown in Table 15.43 are relevant. Assuming there is no interaction between A and B we can use the interaction mean square

$$\frac{73.0+31.0}{4} = 26.0$$

as a residual mean square with 4 degrees of freedom. None of the effects in Table 15.43 is then significant, but in the case of the linear A effect it appears possible that significance might be established with further observations. Looking at the figures from the point of view of the physical problem, one would be inclined to expect nearly linear dependence of warp on temperature and fit a linear rather than a quadratic regression equation to these data.

TABLE 15.43

ANALYSIS OF VARIANCE TABLE FOR CODED WARP DATA

SOURCE		SUM OF SQUARES	DEGREES OF FREEDOM	MEAN SQUARE
Temperature	Linear	73.5	1	73.5
(A)	Quadratic	0.5	1	0.5
Laboratories				
$B (= \text{``}B_L\text{''} + \text{``}B_Q\text{''})$		8.0	2	4.0
$A_L B (= \text{``}A_L B_L\text{''} + \text{``}A_L B_Q\text{''})$		73.0	2	36.5
$A_Q B (= \text{``}A_Q B_L\text{''} + \text{``}A_Q B_Q\text{''})$		31.0	2	15.5

*15.5.3 Confounding in a 3^k Experiment

We introduce as a mathematical convenience the powers "1" and "2" for factors included in an interaction. Each of these "interactions" has 2 degrees of freedom. For example, the interaction AB with 4 degrees of freedom can now be split into "AB" and "AB^2," each with 2 degrees of freedom. By convention we do not include "A^2B^2" or "A^2B." In general, the first factor of an interaction is assumed to include *all* the appropriate degrees of freedom.

In order to design an experiment in 3 blocks, for example, we choose a 2-(or more) factor interaction (the confounding interaction) and state the family of equations

$$\sum a_j x_j = i \bmod 3 \qquad (15.6)$$

where a_1 is the power of A, a_2 the power of B, etc; $i = 0, 1, 2$ and j extends over the number of factors. For each value of i, $\frac{1}{3}$ of the treatment combinations will satisfy the equation. For example, consider a 3^3 experiment in 3 blocks of 9 treatments. Let AB^2C^2 be the confounding interaction (with 2 degrees of freedom). Then those combinations satisfying each equation

$$x_1 + 2x_2 + 2x_3 = i \bmod 3 \quad (i = 0, 1, 2)$$

will be included within each of the 3 blocks.

The equation

$$x_1 + 2x_2 + 2x_3 = 0 \bmod 3$$

will be satisfied by the elements of block 1 of Table 15.44. These can be generated by first choosing 000, 012, and 101, which satisfy the last equation. That is, $x_1, x_2, x_3 = 0, 0, 0$; $0, 1, 2$ and $1, 0, 1$ all satisfy the last equation. The remaining treatments are generated as indicated below. New treatments are obtained by addition, modulo 3.

(1)	000	(4)	101	(7)	202(= 101 + 101)
(2)	012	(5)	110(= 101 + 012)	(8)	211(= 202 + 012)
(3)	021(= 012 + 012)	(6)	122(= 101 + 021)	(9)	220(= 202 + 021)

TABLE 15.44

3^3 Experiment in 3 Blocks, Each of 9 Treatments

$(AB^2C^2 = $ Confounding Interaction$)$

BLOCK 1	BLOCK 2	BLOCK 3
000	100	200
012	112	212
021	121	221
101	201	001
110	210	010
122	222	022
202	002	102
211	011	111
220	020	120

This is the principal block. The 2 remaining blocks can be obtained by placing all those treatment combinations which satisfy $x_1 + 2x_2 + 2x_3 = 1 \bmod 3$ in one block and those satisfying $x_1 + 2x_2 + 2x_3 = 2 \bmod 3$ in the other block. It is easier to find one treatment in each of the other blocks and generate the remainder of the block by addition, modulo 3, of this treatment combination with those of the principal block. The resulting 3 blocks are given in Table 15.44. Note again that the principal block is a closed set. Any treatment combination in this block "added" to itself or another treatment combination in the block gives a result still in this block.

In Table 15.44, the interaction AB^2C^2 is confounded with blocks. Realizing that the notation AB^2C^2 is simply a mathematical convenience, we say that a component of the 3-factor interaction of A, B, and C is confounded with blocks. Since only 2 degrees of freedom are utilized for blocks, the remaining sums of squares and degrees of freedom (6) for the triple interaction are then assigned to the residual, if it is felt that the interaction itself is really negligible.

Example 15.17 Consider a 3^4 experiment in 9 blocks of 9 treatments each. The factors A, B, C, and D are each at 3 levels. We may choose 2 confounding interactions, and generate the other 2 from these. Suppose the confounding interactions are ABC and AC^2D. The other 2 are determined as follows:

(1) $ABC \times AC^2D = A^2BC^3D = (A^2BD)^2 = A^4B^2D^2 = AB^2D^2$

(2) $ABC \times (AC^2D)^2 = A^3BC^5D^2 = BC^2D^2$

Note that we can square these interactions in order to obtain the first power for the first term in the resulting interactions. The resulting interactions are unique, although the specific products are not. For example, BC^2D^2 could have been obtained from $(ABC)^2 \times AC^2D = A^3B^2C^4D = (B^2CD)^2 = BC^2D^2$.

With the confounding interactions equal to ABC, AC^2D, and their resulting products, we need use only these 2 to generate the contents of each of the 9 blocks.

With these interactions we have the 9 simultaneous equations

$$x_1 + x_2 + x_3 = i \bmod 3 \quad i = 0, 1, 2$$

$$x_1 + 2x_3 + x_4 = k \bmod 3 \quad k = 0, 1, 2$$

Each of the 9 pairs of values of i, k gives us a pair of equations which is satisfied by each treatment within a given block. The equations

$$x_1 + x_2 + x_3 = 0 \bmod 3$$

$$x_1 + 2x_3 + x_4 = 0 \bmod 3$$

provide the principal block. Table 15.45 presents the 9 blocks. Note that once the principal block is established each succeeding block is obtained by one "new" treatment added modulo 3 to each treatment in the principal block. The column headings of Table 15.45 are the values of i and k which determine the treatment combinations within the particular block.

TABLE 15.45

A 3^4 EXPERIMENT IN NINE BLOCKS OF NINE TREATMENTS
(Confounded Interactions $= ABC$, AC^2D, AB^2D^2, BC^2D^2)

BLOCK								
0,0	0,1	0,2	1,0	1,1	1,2	2,0	2,1	2,2
0000	0001	0002	0100	1000	0010	0200	0020	2000
0122	0120	0121	0222	1122	0102	0022	0112	2122
0211	0212	0210	0011	1211	0221	0111	0201	2211
1021	1022	1020	1121	2021	1001	1221	1011	0021
1110	1111	1112	1210	2110	1120	1010	1100	0110
1202	1200	1201	1002	2202	1212	1102	1222	0202
2012	2010	2011	2112	0012	2022	2212	2002	1012
2101	2102	2100	2201	0101	2111	2001	2121	1101
2220	2221	2222	2020	0220	2200	2120	2210	1220

*15.5.4 Fractional Replicates of a 3^k Design

Just as in the case of a fraction of a 2^k experiment, we also can choose defining contrasts to select the particular treatment combinations we use.

Suppose, for example, we wish to select a $\frac{1}{9}$ replicate of a 3^4 experiment. If we choose a defining contrasts ABC and AC^2D, then AB^2D^2 and BC^2D^2 are necessarily defining contrasts. We can then select any 1 of the 9 blocks of Table 15.45. There remain to be shown the aliases of each of the factors and interactions. These are given in Table 15.46.

TABLE 15.46

ALIAS-GROUPS IN A $\frac{1}{9} \times 3^4$ EXPERIMENT

(Defining contrasts are ABC, AC^2D, AB^2D^2, BC^2D^2.)

EFFECT			ALIASES					
A	AB^2C^2	BC	ACD^2	CD^2	ABD	BD	ABC^2D^2	AB^2CD
B	AB^2C	AC	ABC^2D	AB^2C^2D	AD^2	ABD^2	BCD	CD
C	ABC^2	AB	AD	ACD	AB^2CD^2	$AB^2C^2D^2$	BD^2	BCD^2
D	$ABCD$	$ABCD^2$	AC^2D^2	AC^2	AB^2	AB^2D	BC^2	BC^2D

In order to obtain the aliases we must "multiply" the main effect or interaction by the defining contrasts and their squares.

From the table of aliases we note that each main effect is aliased with 3 2-factor, 3 3-factor and 2 4-factor interaction components. These main effects take up the total 8 degrees of freedom of the experiment.

Example 15.18 In a study of the dispersion stability of paints, 7 factors were under consideration in the initial phases of the investigation. Each factor could be maintained at 3 levels. This led to a 3^7 factorial experiment, which was necessarily fractionalized due to the immense number of possible factor level combinations. The factors (and their levels) being studied are given in Table 15.47. It was decided to take a $\frac{1}{27}$ replication of this 3^7 experiment. The defining contrasts are $ACDEF^2G$,

TABLE 15.47

FACTORS IN A $\frac{1}{27} \times 3^7$ FACTORIAL EXPERIMENT

A	Resin (Model Alkyd)	E	Dispersing Agent Type
	0 Low Mol. Weight		0 Tetramethyl Ammonium Hydroxide
	1 Med. Mol. Weight		1 Ammonium Hydroxide
	2 High Mol. Weight		2 Morpholine
B	Solvent Type	F	Dispersing Agent Level
	0 50:50 Xylol-Butanol		(Ratio of acid to base functionality)
	1 Xylol		0 1:1
	2 50:50 Xylol-Ethanol		1 1:2
			2 1:3
C	Solvent Level		
	0 40% Solvent	G	Equipment Type
	1 50% Solvent		0 Ultra-sonics
	2 60% Solvent		1 Waring-Blender
			2 Manton Gaulin Colloid Mill
D	Water Level		
	0 40% Water		
	1 50% Water		
	2 60% Water		

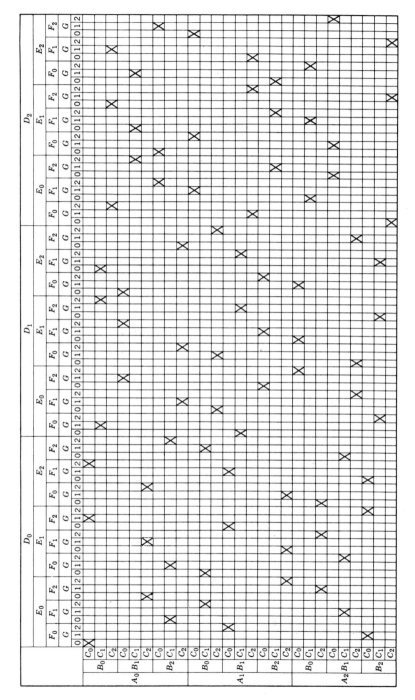

BC^2EF^2G, and $ABCEG^2$ and the 10 others generated from these. The equations for determining the correct treatment combinations for the principal block (of the 27 possible blocks which could have been chosen) are as follows:

$$x_1 + x_3 + x_4 + x_5 + 2x_6 + x_7 = 0 \bmod 3$$

$$x_2 + 2x_3 + x_5 + 2x_6 + x_7 = 0 \bmod 3$$

$$x_1 + x_2 + x_3 + x_5 + 2x_7 = 0 \bmod 3$$

Rather than enumerate the treatment combinations as was done in the preceding example, we present as Table 15.48 a diagrammatic representation of the experiment, in which the particular treatment combinations used in the actual experiment are designated by $\times$.

15.6 TABLES FOR FRACTIONAL REPLICATES OF 2^k, 3^k, AND $2^k \times 3^l$ EXPERIMENTS

Very useful sets of tables have been constructed at the National Bureau of Standards for fractional factorial experiments. These tables are very valuable when seeking to design an experiment of many factors each at 2 (or 4 or 8) levels, or at 3 (or 9) levels, or a combination of these. The first of these publications, *Fractional Factorial Experiment Designs for Factors at Two Levels* (N.B.S. Applied Math. Series No. 48), considers fractions of experiments from 5 through 16 factors. Each table has 3 designators: (*a*) the size of the replicate, (*b*) the number of factors, and (*c*) the number of units per block. For each design the principal block is given. The other blocks are either given or a method for generating them is stated. For each fraction, 3 possible confoundings are given: (*a*) none at all, (*b*) with blocks, and (*c*) with rows and blocks. Finally, the interactions which are measurable are enumerated.

The second publication, *Fractional Factorial Experiment Designs for Factors at Three Levels* (N.B.S. Applied Math. Series No. 54), appeared in 1959, 2 years after the first. The third of the series is entitled *Fractional Factorial Designs for Experiments with Factors at Two and Three Levels* (N.B.S. Applied Math. Series No. 58). This series is available from the U.S. Department of Commerce.

A number of useful designs are given by Kitagawa and Mitome in *Tables for the Design of Factorial Experiments* (Baifukan Co. Ltd., Tokyo; and Charles E. Tuttle Co., Rutland, Vt., 1953). Fractional replicates can be formed by taking parts of confounded designs, of which many examples are given in this last reference.

15.7 SPLIT-PLOT EXPERIMENTS

It is sometimes convenient to confound a main effect. This is the case, for instance, when we are willing to sacrifice some accuracy in estimating effects of changes in levels of certain factors in order to obtain increased accuracy in estimating other effects or interactions.

Consider, for example, the construction of an experiment comparing 3 methods of work organization and 4 sources of raw material, which is to be carried out at 3 factories. This is illustrated in Figure 15.12. If this is arranged as a complete $3 \times 4 \times 3$ cross-classification, then each factory could be divided into 3 areas, one for each type of organization of work, and each area subdivided into 4 parts—one for each source of raw material. By this arrangement we might hope to get specially accurate comparison of results for different sources of raw material because they will come from (we hope) closely similar types of working conditions.

WORK ORGANIZATION

(1)	(2)	(3)		(3)	(1)	(2)		(1)	(3)	(2)
A	A	D		D	A	A		C	A	C
C	D	A		C	B	C		A	B	B
B	C	B		B	C	B		D	D	A
D	B	C		A	D	D		B	C	D

| FACTORY 1 | FACTORY 2 | FACTORY 3 |

(Sources of raw material denoted by $A,\, B,\, C,\, D$)

Figure 15.12   Split-plot design.

It will be realized that this split-plot experiment is just a complete $3^2 \times 4$ cross-classification. Its distinguishing feature is (in this case) the geographical arrangement of the various factor level combinations.

The sums of squares appearing in the analysis are exactly those appearing in the analysis of a standard $3^2 \times 4$ cross-classification with a single replicate, as shown below.

SOURCE	DEGREES OF FREEDOM
Organization	2
Factories	2
Organization $\times$ Factories	4
Sources	3
Sources $\times$ Organization	6
Sources $\times$ Factories	6
Sources $\times$ Organization $\times$ Factories	12
Total	35

It will be noted that the entries in the table have been arranged in an unusual order. This is because the first 2 sums of squares ("Organization" and "Factories") are to be compared against the third ("Organization × Factories"), while the next 3 are to be compared against the "Sources × Organization × Factories" sum of squares. "Organization" and "Factories," are called *main plot factors*, "Sources" is a *sub-plot factor*. For testing main-plot factors for significance we use their (formal) interaction, usually termed "Residual (i)" while the sub-plot factor *and its interaction with main-plot factors* are tested against the interaction of the sub-plot factor with "Residual (i)." This last interaction is called "Residual (ii)."

The model underlying this analysis is the standard cross-classification model in which the interaction between main-plot factors is represented by a random variable. Thus for the particular case described above the model would be of form

$$X_{tij} = \xi + \omega_t + \phi_i + (OF)_{ti} + \gamma_j + (\omega\gamma)_{tj} + (\phi\gamma)_{ij} + Z_{tij}$$

where $t = 1, 2, 3$ denotes "work organization"
$\quad\quad\ \ i = 1, 2, 3$ denotes "factory"
$\quad\quad\ \ j = 1, 2, 3, 4$ denotes "source of raw material"

Z_{tij} represents the triple interaction; this is used as the Residual (ii), testing for existence of nonzero γ_j, $(\omega\gamma)_{tj}$, or $(\phi\gamma)_{ti}$ terms. The random variable $(OF)_{ti}$ contributes to the Residual (i) used in testing for nonzero ω_t or ϕ_i terms. The expected value of Residual (ii) mean square is σ_Z^2; the expected value of the Residual (i) mean square is $\sigma_Z^2 + 4\sigma_{OF}^2$. If (as is often hoped for) σ_Z^2 is considerably less than σ_{OF}^2, it can be seen that the tests for effects involving the factor Sources are likely to be more sensitive than those for the main-plot factors, Organization, and Factories.

This type of experimental design can be extended in a natural manner by subdividing each sub-plot into sub-sub-plots (or sub²-plots) according to levels of a *sub²-plot factor*, and so on to further subdivisions, if desired and practicable. Each further subdivision leads to a further Residual term in the analysis. For example, a sub²-plot design will require the use of 3 Residual sums of squares in the analysis.

Although the attainment of special accuracy in regard to certain comparisons is a common reason for using a split-plot design, this type of design may sometimes be used by force of circumstances. In the case we have described, for example, it may well have been impracticable to have methods of work organization changing too much from one small area of the factory to another.

15.7.1 Examples of Split-Plot Experiments

These examples are intended to exhibit different ways in which split-plot designs may be found useful. In each case we essentially have a straightforward crossed design, but there are circumstances peculiar to each example calling for special ways of grouping the factor level combinations.

Example 15.19 In the study of strength properties of polymers the intent of a particular experiment was to distinguish between several polymers and to evaluate the particular type of test. Five different polymers were chosen. It was expected that 2 were superior, 1 intermediate, and the other 2 quite inferior. Still these 5 polymers were chosen in order to see how strongly they were distinguished and what were the errors involved.

An experimental paper was used, which was assumed to have no strength properties when wet (that is, in comparison to that of the polymers). The paper was dipped into a solution of the polymer, the excess solution was removed to achieve a constant pickup. Then the specimens were dried. Two cure times were chosen—4 and 10 mins. The oven was maintained at a constant temperature, and the positions within the ovens were assumed to have homogeneous temperatures. This was ascertained by some earlier tests. The test specimens were measured according to Standard Test Method 61–1962 of the American Association of Textile Chemists and Colorists, in the "Launder-Ometer." These specimens, 2×2 in., were then placed in steel cylindrical containers, each container having 10 *small* steel balls, a fixed amount of water, and detergent. One specimen from each of the polymers was placed in each of 5 containers for the 4-min group and similarly for the 10-min group. The containers were fastened on a drum which rotated at a fixed frequency in a constant temperature bath. The data are represented in Figure 15.13 below. After washing for 60 min, the specimens were removed and examined along some arbitrary scales.

This is a split-plot experiment with the 10 cylinders as "main plots." (Cylinders are, also, in fact, "sub-plots," with Times as main plots, in a hierarchal classification.) Each cylinder is then split into 5 sub-plots, 1 for each polymer.

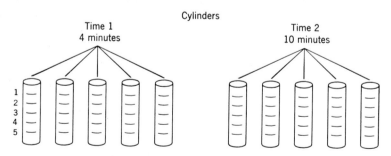

Figure 15.13 Split-plot design of a Launder-Ometer test.

The model equation is

$$X_{ijk} = \xi + \tau_i + U_{j(i)} + \pi_k + (\tau\pi)_{ik} + Z_{ijk}$$

where $i = 1$, 2 times; $j = 1, 2, \ldots, 5$ containers within time, and $k = 1, 2, \ldots, 5$ polymers. Note that Z_{ijk} is used as the error term. The corresponding "Residual (ii)" sum of squares is in fact the "Polymers $\times$ (Cylinders within Times)" interaction sum of squares. We assume that this interaction is, in itself, negligible.

Given the model equation just presented we can determine the expectations of the mean squares, which suggest the appropriate significance tests. Using the method of Section 14.8.2 we obtain Table 15.49 in which the last column is the expectation of the mean square.

TABLE 15.49

EXPECTED VALUES OF MEAN SQUARES (EMS) IN THE
ANOVA OF A SPLIT-PLOT EXPERIMENT (FIGURE 15.13)

	$i(I)$	$j(II)$	$k(I)$	EMS	
i	0	5	5	$\sigma^2 + 5\sigma_{ij}^2 + \langle 25\sigma_i^2 \rangle$	
$j(i)$	1	1	5	$\sigma^2 + 5\sigma_{ij}^2$	(Residual (i))
k	2	5	0	$\sigma^2 + \langle 10\sigma_k^2 \rangle$	
ik	0	5	0	$\sigma^2 + \langle 5\sigma_{ik}^2 \rangle$	
$kj(i)$	1	1	1	σ^2	(Residual (ii))

Hence the factor T (time) is tested against whole plot error of cylinders within time periods Factors P (polymers) and TP (polymer $\times$ time interaction) are tested against sub-plot error. With the proper model, not only do we have estimates of the whole plot and sub-plot variance, but we can test such things as individual linear comparisons among polymers, we can obtain confidence intervals on means, and extract a great deal of additional information.

Split-plot analyses are easy to construct if it is realized that split-plot designs are a special type of complete crossed designs. The main point is to recognize the main and sub-plot factors and express the situation correctly in the model equation. The remaining tests and consequent analysis follow lines similar to those of Chapter 14.

Example 15.20 An experiment [taken from Bobalek, Leone, and Hawkins, "Design of an Experiment to Evaluate Alternative Processes for Pigment Dispersion in Aqueous Vehicles," *Official Digest* (*Journal of Paint Technology and Engineering*) (September, 1963)] was designed to evaluate alternative processes for pigment dispersion in aqueous vehicles. Two mills were chosen, a roller mill and a Manton-Gaulin mill. Three different pigments were selected. (For this illustration we concern ourselves with only 1 of these—phthalocyanine blue—B-4790, Harmon Color Co.). Two different dispersion solutions were used. A schematic diagram of the experiment is given in Figure 15.14.

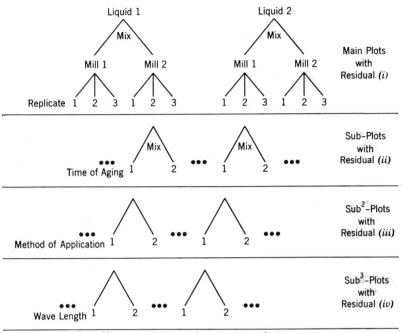

Measurement: $X_{ml\alpha ij}$ in percentage reflectance

Figure 15.14 A split-plot experiment to evaluate alternative processes for pigment dispersion in aqueous vehicles.

Note the 4 stages of the process:

(*i*) The mixing and dispersion stages with factor L, 2 specific dispersion solutions (liquid 1 and liquid 2) and factor M, 2 mills (mill 1, a roller mill, and mill 2, the Manton-Gaulin mill). Three replicates were provided. That is, each pigment was prepared 3 times and each individually dispersed. These (mills × replications) constituted the main plot with Residual (*i*).

(*ii*) Each single liquid dispersion was mixed by weight with a titanium dioxide-pigmented water–dispersed polystyrene paint. This was done within $\frac{1}{2}$ hr for part of the solution, and after an aging of 1 week for the remainder of the solution. This constitutes, then, the sub-plot (aging) with Residual (*ii*).

(*iii*) Two types of panel were prepared—a spray panel and a flow coat panel. This constitutes the sub²-plot with Residual (*iii*).

(*iv*) In order to obtain critical values of percent reflectance, 2 wavelengths were chosen after some spectrophotometric studies. This last selection of the levels of the factor constitutes the sub³-plot with Residual (*iv*).

The factors are as follows:
Mill (M): $m = 1, 2$ (roller and Manton-Gaulin, respectively)
Liquid (L): $l = 1, 2$ (ammoniated Lustrex 820 and ammoniated casein, respectively)

Replicates: $\alpha = 1, 2, 3$ (NOTE: there were 12 replications in all)

Aging time (T): $t = 1, 2$ ($\frac{1}{2}$ hr and 1 week, respectively)

Method of application (S): $i = 1, 2$ (spray and flow, respectively)

Wave length (W): $j = 1, 2$ (470 and 660 mμ, respectively)

The data are given in Table 15.50. The yield is in percentage reflectance. We should note that in the mathematical model, the factors M, L, T, S, and W are all parametric. The model may be stated

$$X_{ml\alpha tij} = \xi + \mu_m + \lambda_l + (\mu\lambda)_{ml} + U_{\alpha(ml)} \qquad \left.\begin{array}{l} \\ \\ \end{array}\right\} \begin{array}{l} \text{Main} \\ \text{Plots} \end{array}$$

$$+ \tau_t + (\mu\tau)_{mt} + (\lambda\tau)_{lt} + (\mu\lambda\tau)_{mlt} + V_{\alpha(mlt)} \qquad \left.\begin{array}{l} \\ \\ \end{array}\right\} \begin{array}{l} \text{Sub-} \\ \text{plots} \end{array}$$

$$\begin{array}{l} + \gamma_i + (\mu\gamma)_{mi} + (\lambda\gamma)_{li} + (\mu\lambda\gamma)_{mli} + (\tau\gamma)_{ti} \\ + (\mu\tau\gamma)_{mti} + (\lambda\tau\gamma)_{lti} + (\mu\lambda\tau\gamma)_{mlti} + (U'_{\alpha(mli)} + V'_{\alpha(mlti)}) \end{array} \quad \left.\begin{array}{l} \\ \\ \end{array}\right\} \begin{array}{l} \text{Sub}^2\text{-} \\ \text{plots} \end{array}$$

$$\begin{array}{l} + \omega_j + (\mu\omega)_{mj} + (\lambda\omega)_{lj} + (\mu\lambda\omega)_{mlj} + (\tau\omega)_{tj} \\ + (\mu\tau\omega)_{mtj} + (\lambda\tau\omega)_{ltj} + (\mu\lambda\tau\omega)_{mltj} + (\gamma\omega)_{ij} \\ + (\mu\gamma\omega)_{mij} + (\lambda\gamma\omega)_{lij} + (\mu\lambda\gamma\omega)_{mlij} + (\tau\gamma\omega)_{tij} \\ + (\mu\tau\gamma\omega)_{mtij} + (\lambda\tau\gamma\omega)_{ltij} + (\mu\lambda\tau\gamma\omega)_{mltij} \\ + (U''_{\alpha(mlj)} + V''_{\alpha(mltj)} + U'''_{\alpha(mlij)} + V'''_{\alpha(mltij)}) \end{array} \quad \left.\begin{array}{l} \\ \\ \\ \\ \\ \end{array}\right\} \begin{array}{l} \text{Sub}^3\text{-} \\ \text{plots} \end{array}$$

The residuals are presented in a rather clumsy fashion (algebraically) for greater ease in calculation. That is, one can use the general factorial methods presented in Chapter 14, remembering that "Replicates" are not crossed with the other factors but nested within Mill-Liquid combinations. One precaution is necessary—each stage (main plot, sub-plot, ...) must be handled as a separate part of the auxiliary

TABLE 15.50

PERCENTAGE REFLECTANCE OF PHTHALOCYANINE BLUE PIGMENT

| | | | L_1 | | | | | | L_2 | | | | |
| | | | M_1 | | | M_2 | | | M_1 | | | M_2 | | |
			R_1	R_2	R_3	R_1	R_2	R_3	R_1	R_2	R_3	R_1	R_2	R_3
T_1	S_1	W_1	44.5	44.5	45.0	47.0	50.5	48.0	44.0	44.5	44.0	47.5	48.5	46.5
		W_2	7.5	8.0	8.5	8.5	8.5	8.5	7.5	7.5	7.5	9.0	9.0	8.5
	S_2	W_1	43.5	42.5	44.0	45.5	44.5	45.0	43.0	44.0	43.0	44.5	43.0	43.0
		W_2	10.0	9.5	10.0	12.0	12.5	12.5	9.0	10.5	9.0	10.5	9.5	10.0
T_2	S_1	W_1	45.0	44.0	45.5	47.0	46.5	45.5	44.0	44.5	43.5	48.5	46.5	46.5
		W_2	7.5	7.5	8.0	8.0	8.0	8.0	7.5	7.5	7.0	9.5	8.5	8.0
	S_2	W_1	44.0	43.0	44.5	45.0	44.0	45.5	43.5	44.0	43.5	43.5	43.5	43.0
		W_2	10.5	10.0	11.0	12.0	11.0	11.0	9.0	9.0	9.0	9.5	10.5	10.0

table of the EMS. The random terms in the mathematical model contributing to the various Residuals are as follows:

$$\text{Residual } (iv) = U''_{\alpha\,(mlj)} + V''_{\alpha\,(mltj)} + U'''_{\alpha(mlij)} + V'''_{\alpha(mltij)}$$

$$\text{Residual } (iii) = \text{Above} + U'_{\alpha(mli)} + V'_{\alpha(mlti)}$$

$$\text{Residual } (ii) = \text{Above} + V_{\alpha(mlt)}$$

$$\text{Residual } (i) = \text{Above} + U_{\alpha(ml)}$$

The extra terms in the last 2 lines could have been called $Y_{\alpha(mlti)}$ and $Z_{\alpha(mltij)}$. However, by using 2 and 4 terms with appropriate subscripts, the proper sum of squares, degrees of freedom, and EMS can be determined automatically by use of tables presented in Chapter 14.

Results of the analysis are shown in the ANOVA table (Table 15.51). Note that Mill and Liquid mean squares are tested against Residual (i). Both the Mill and Liquid mean squares are significant, though the latter is markedly smaller. Time, tested against Residual (ii), is not significant. However, the method of application interacts strongly with the Mill and Liquid factors. Finally, as was expected, the different wavelengths give very different results. This is obvious from Table 15.50. The interactions *SW, MSW,* and *MW* are also significant. There are other significant interactions (*MT, MLT, MS, LS, MLS, LSW,* and *MLTSW*) but they are not very large. It should be noted that with so many tests being applied, "significance" may well be obtained in a purely random manner.

It is worthwhile noting that, for this example, the Yates' algorithm (with 3 replicates) could have been used. The only additional work would be a proper grouping of the residual terms before calculating the ANOVA.

Example 15.21   In the preparation and testing of laboratory samples of different polyester products the sample is prepared in the following manner. For a single formulation, 2 temperatures ($i = 1,2$) and 2 methods of agitation ($j = 1,2$) are controlled. For each combination of temperature–agitation, 4 batches ($k = 1,2,3,4$) are prepared. Each of these is divided into 3 samples ($l = 1,2,3$). Finally, each sample is again subdivided into 2 subsamples. One subsample is cured for time t_1 and the other for time $t_2(m = 1,2)$. Determine the model equation and the EMS column of the ANOVA table.

This is a mixed design—a hierarchal design with main, sub-, and sub-subgroups, with the sub-subgroups split for the factor Cure Time. The 2×2 combinations of Temperature and Method of Agitation constitute Main Groups in a hierarchal classification, with 4 subgroups (batches) in each of the 4 Main Groups. Each subgroup is divided into 3 sub-subgroups (samples). The $3 \times 4 \times 4 = 48$ sub-subgroups are "Main Plots," each split into subplots according to Cure Time.

The model may be represented by the equation below. The Expected values of Mean Squares and degrees of freedom for each factor are given in Table 15.52.

Note that the Batches within Temperature $\times$ Agitation Method combination mean square (4) should be used as Residual for testing significance of (1), (2), and

TABLE 15.51
ANOVA—PERCENTAGE REFLECTANCE OF PHTHALOCYANINE BLUE

SOURCE	DEGREES OF FREEDOM	SUM OF SQUARES	MEAN SQUARE
Mill (M)	1	48.17	48.17***
Liquid (L)	1	9.38	9.38*
ML	1	0.26	0.26
Whole plot error	8	8.48	1.06
Residual (i)			
Sub-Plots			
Time Aging (T)	1	1.50	1.50
MT	1	2.34	2.34*
LT	1	0.26	0.26
MLT	1	2.04	2.04*
Sub-plot error	8	2.35	0.29
Residual (ii)			
Sub²-Plots			
Spray or Flow Coat (S)	1	0.26	0.26
MS	1	6.00	6.00**
LS	1	5.04	5.04**
MLS	1	5.51	5.51**
TS	1	1.04	1.04
MTS	1	0.01	0.01
LTS	1	0.51	0.51
MLTS	1	0.00	0.00
Sub²-plot error	16	8.50	0.53
Residual (iii)			
Sub³-Plots			
Wave length (W)	1	30566.34	30566.34***
MW	1	3.38	3.38***
LW	1	0.04	0.04
TW	1	0.00	0.00
MLW	1	0.26	0.26
MTW	1	0.51	0.51
LTW	1	0.09	0.09
MLTW	1	0.00	0.00
SW	1	110.51	110.51***
MSW	1	10.67	10.67***
LSW	1	1.50	1.50*
MLSW	1	0.01	0.01
TSW	1	0.67	0.67
MTSW	1	0.51	0.51
LTSW	1	0.09	0.09
MLTSW	1	1.50	1.50*
Sub³-plot error	32	7.68	0.24
Residual (iv)			
Total	95	30805.41	

TABLE 15.52

ANALYSIS OF VARIANCE FOR A HIERARCHAL DESIGN WITH
SPLIT PLOTS

SOURCE	D. F.	MEAN SQUARE	EMS
Temperature (A)	1	(1)	$\sigma^2 + 2\sigma_{ijkl}^2 + 6\sigma_{ijk}^2 + \langle 48\sigma_i^2 \rangle$
Agitation (B)	1	(2)	$\sigma^2 + 2\sigma_{ijkl}^2 + 6\sigma_{ijk}^2 + \langle 48\sigma_j^2 \rangle$
Temperature × Agitation	1	(3)	$\sigma^2 + 2\sigma_{ijkl}^2 + 6\sigma_{ijk}^2 + \langle 24\sigma_{ij}^2 \rangle$
Batch (u) (Within Temperature-Agitation)	12	(4)	$\sigma^2 + 2\sigma_{ijkl}^2 + 6\sigma_{ijk}^2$
Sample (v) (Within Batches)	32	(5)	$\sigma^2 + 2\sigma_{ijkl}^2$ ("main-plot" error)
Cure Time (C)	1	(6)	$\sigma^2 + 3\sigma_{ijkm}^2 + \langle 48\sigma_m^2 \rangle$
Cure Time × Temperature	1	(7)	$\sigma^2 + 3\sigma_{ijkm}^2 + 24\sigma_{im}^2$
Cure Time × Agitation	1	(8)	$\sigma^2 + 3\sigma_{ijkm}^2 + 24\sigma_{jm}^2$
Cure Time × Agitation × Temperature	1	(9)	$\sigma^2 + 3\sigma_{ijkm}^2 + 12\sigma_{ijm}^2$
Cure Time × Batches (Within Temperature × Agitation)	12	(10)	$\sigma^2 + 3\sigma_{ijkm}^2$
$vC(uAB)$ Cure Time × Samples (Within Batches)	32	(11)	σ^2 ("subplot" error)
Total	95		

(3); and the mean square (10) should be used as Residual for testing significance of (6), (7), (8), and (9). These are consequences of representing Batches by random variables in the model.

$$X_{ijklm} = \mu + \alpha_i + \beta_j + (\alpha\beta)_{ij} + U(\alpha\beta)_{k(ij)}$$

$$+ V(U\alpha\beta)_{l(ijk)} + \gamma_m + (\alpha\gamma)_{im} + (\beta\gamma)_{jm}$$

$$+ (\alpha\beta\gamma)_{ijm} + U\gamma(\alpha\beta)_{km(ij)} + V\gamma(U\alpha\beta)_{lm(ijk)}$$

15.8   CONCLUDING REMARKS

The devices of confounding, fractional replication, and split plots which have been discussed in this chapter were all developed in response to practical requirements—reducing residual variation, reducing cost of experimentation, obtaining greater accuracy where it is specially needed, and so on. These devices can be combined in many ways in a single experimental design. Taken together they give us considerable flexibility in constructing designs to meet the particular conditions of each problem as they arise.

Beyond this, there are a number of further elaborations of these kinds of modification of design—partially balanced incomplete blocks, for example, provide a vast number of possible designs. The detailed study of these and similar designs is, however, a matter for specialists and inappropriate to this book. Nevertheless, one must always be prepared to search for a design of unusual nature if the practical conditions seem to preclude the use of a design from a familiar repertoire.

The discussion of basic types of design given in Chapters 13 and 14, and especially in this chapter, should provide a good foundation to assess the kind of design needed in a large proportion of problems, even though it may ultimately be necessary to search through published collections of designs in order to make a final selection.

Distribution-free methods of the kinds described in Chapter 9 can be applied to construct significance tests for most of the hypotheses described in Chapters 13-15. Some care and ingenuity are necessary in choosing methods of assigning rank orders. For example, in a randomized block experiment, deviations from block means might be ordered either separately within each block or taking the experimental data as a whole.

We have not given a detailed development of distribution-free methods applicable to each problem because (*a*) they are not easily adapted to the calculation of estimates, which is an important part of the analysis, and (*b*) one of the most convincing reasons for using distribution-free methods is their robustness with respect to non-normality, but analysis of variance (mean square ratio) tests are also satisfactorily robust in this respect (though not with respect to other departures from standard conditions which, however, also affect distribution-free methods).

REFERENCES

1. Brownlee, K. A., *Statistical Theory and Methodology in Science and Engineering*, 2nd ed., Wiley, New York, 1965.
2. Cochran, W. G. and Gertrude M. Cox, *Experimental Designs*, 2nd ed., Wiley, New York, 1957.
3. Daniel, C., "Use of the Half-Normal Plot in Interpreting Fractional Factorial Two-level Experiments," *Technometrics*, 1 (1959).
4. Davies, O. L., ed., *Statistical Methods in Research and Production*, 2nd ed., Griffin, London; Hafner, New York, 1967.
5. Kempthorne, O., *The Design and Analysis of Experiments*, Wiley, New York, 1952 (reprinted 1973).
6. Neter, J. and W. Wasserman, *Applied Linear Statistical Models*, Irwin, Homewood, Ill., 1974.

7. Peng, K. C., *The Design and Analysis of Scientific Experiments: An Introduction with Some Emphasis on Computation*, Addison-Wesley, Reading, Mass., 1967.

8. Yates, F., *Experimental Design: Selected Papers*, Griffin, London, 1970.

EXERCISES

1. Among the experimental designs appearing in the Exercises of Chapter 14, which are 2^r designs (*i*) without and (*ii*) with replication? Check your calculations for these designs using Yates' algorithm.

2. The data below represent percent yield of a process as a function of agitation speed (*A*), concentration (*B*), and temperature (*C*). The levels of agitation speed are 200 and 400 rpm; of concentration, 2 and 4%; and of temperature, 30 and 50° C. Test whether there are real differences associated with changes in levels of any of the factors. What other conclusions can be drawn from the data? (Use Yates' algorithm.)

PERCENT YIELD

AGITATION SPEED	200 rpm		400 rpm	
CONCENTRATION	2%	4%	2%	4%
TEMPERATURE				
30°C	83.9	87.2	78.4	86.3
50°C	90.1	92.6	88.7	86.2

3. The data below represent the internal dimensions (measured in thousandths of an inch from an origin of 1.634 in.) of a component. The factors are days, operators, and machines. Two replicates are taken for each combination of the 3 factors. By means of Yates' algorithm determine whether any of the factors affect the average internal diameter.

OPERATORS	*A*			*B*			*C*		
MACHINES	1	2	3	1	2	3	1	2	3
Monday	3	1	4	2	1	2	2	2	3
	2	3	3	0	3	3	4	3	5
Wednesday	4	7	8	6	5	7	4	6	3
	6	5	7	4	3	6	6	3	5
Friday	6	3	10	10	9	4	7	8	7
	8	6	12	8	3	11	5	7	6

4. The product yields of an undesirable by-product were measured for 2 different catalysts, each at 2 different pressures. The experiment was carried out at 2 laboratories; 3 replicates were taken at each laboratory. Measurements are expressed in percentages. Use Yates' algorithm, and follow it with an analysis of variance table to test how the factors affect the yield of this by-product.

CATALYST	I		II	
LABORATORY	A	B	A	B
HIGH PRESSURE	53	27	40	45
	43	45	32	12
	45	57	29	69
LOW PRESSURE	42	32	61	54
	95	27	24	60
	60	98	11	26

5. Three factors are being studied to determine their effect on degree of conversion. These are catalyst type (A), catalyst concentration (B), and reaction temperature (C). Use Yates' algorithm to estimate the effect of each factor and interaction. Two replicates are measured for each treatment combination.

	CATALYST 1		CATALYST 2	
CONCENTRATION	0.1%	0.5%	0.1%	0.5%
REACTION TEMPERATURE				
120°C	84	85	61	67
	75	92	60	74
160°C	88	86	63	86
	82	90	59	94

6. A test was conducted on the friction horsepower of engines lubricated with commercial oils. The data below relate to part of the experiment. For the factors Oils (D), SAE Grades (B), temperature (C), and RPM (A) of the engine, determine the values for an ANOVA table. Determine whether the SAE grade effect can be considered linear or quadratic. [From Frazier, Klingel, and Tupa, "Friction and Consumption Characteristics of Motor Oils," *Industrial and Engineering Chemistry* (Oct., 1953).]

RPM (A)		1400			2000			2500	
SAE GRADE (B)	10	20	30	10	20	30	10	20	30

TEMPER-ATURE	OILS									
C_0	D_0	10.6	11.0	11.5	17.3	19.1	18.7	26.4	29.0	29.7
	D_1	10.2	12.1	13.2	18.8	20.4	22.2	26.5	29.4	32.0
	D_2	9.8	11.9	12.1	17.0	20.8	21.3	26.2	28.9	30.4
C_1	D_0	9.0	8.6	9.2	14.5	15.8	16.0	23.4	25.2	23.8
	D_1	8.9	9.4	9.8	15.4	16.9	18.0	23.9	26.1	27.3
	D_2	8.3	9.2	10.0	14.5	17.2	17.1	23.0	25.6	26.5
C_2	D_0	9.8	10.1	11.1	16.2	18.0	18.2	25.2	27.2	27.5
	D_1	9.7	10.4	12.1	17.0	19.3	20.5	25.6	28.2	29.8
	D_2	9.1	10.8	12.4	16.0	18.7	19.6	25.2	28.2	29.6

7. (a) Determine the design of a $(\frac{1}{3})^2$ replicate "in one block" of Exercise 6 with 2 of the defining contrasts AB^2C and BCD^2.

(b) Arrange this design in 3 blocks so that no alias groups containing main effects are confounded.

Analyze the data in each of (a) and (b), testing whether any of the main effects are significant.

8. The data below constitute a $\frac{1}{2}$ replicate of a 2^5 experiment on the insulation properties of a new product called "carpetwall." The 5 factors being investigated are: A, density of the material; B, addition of a specific ingredient; C, moisture content; D, structure of the material; and E, age. Each factor was held at 2 levels (not stated here) for the initial experiment. The data, presented below, represent differential of temperature arising from one fixed application of heat. By means of Yates' algorithm, test whether any of the main effects are significant. The data are in coded units.

(1) = 11	ac = 11	acd = 18	$abce$ = 14
cde = 15	d = 19	ce = 17	ab = 17
ae = 14	abd = 19	bcd = 18	ade = 14
bc = 20	be = 21	$abcde$ = 16	bde = 20

9. Using the defining contrast ABC construct a $\frac{1}{3}$ replicate of the experiment given in Exercise 6. Analyze the resulting data by means of Yates' algorithm.

10. In a study of radioactive decontamination the efficiency of the process was measured by the activity remaining, either alpha or beta. The object was to determine the dependence upon 4 process variables, namely, the amount of barium chloride added (B), the amount of aluminum sulfate added (A), the amount of carbon added, (C) and the final pH (P). Each factor was studied at 2 levels indicated in the table below. Here we consider only the alpha quantity. Carry out an analysis of variance to determine the effect of each factor and its significance. The experiment was conducted in four blocks of eight plots, giving a total sample size of 32. The data are as follows (the factors are in the order $PBAC$). [Data from Barnett and Mead, "A 2^4 Factorial Experiment in Four Blocks of Eight: A Study in Radioactive Decontamination," *Applied Statistics*, **5** (1956)]

BLOCK 1		BLOCK 2		BLOCK 3		BLOCK 4	
TREAT-MENT	YIELD	TREAT-MENT	YIELD	TREAT-MENT	YIELD	TREAT-MENT	YIELD
0011	183	0001	650	1100	273	1000	1193
1111	350	1000	1180	1010	890	0111	156
0101	188	1101	238	0110	370	0010	257
0000	881	0010	191	0000	834	0100	178
0110	225	1110	420	0011	193	1101	254
1100	298	0100	289	1111	389	1011	775
1001	1039	0111	135	0101	163	0001	494
1010	466	1001	781	1001	1146	1110	429

11. In a pilot experiment on heat loss of insulation material, 4 factors (A, B, C, D) were considered, each at 2 levels. Only 4 experiments could be carried out at a single session. Two replicates were desired. The coded data given below are so arranged that the first replicate has as confounding interactions ABC, ACD, and BD, while the second replicate has as confounding interactions BCD, ABD, and AC. Construct an appropriate ANOVA table and indicate which effects and interactions you consider significant.

REPLICATE I

BLOCK	1	2	3	4
	0000 = 6	1000 = 5	0100 = 8	0001 = 6
	0111 = 17	1111 = 15	0011 = 10	0110 = 7
	1010 = 11	0010 = 7	1110 = 17	1011 = 4
	1101 = 12	0101 = 11	1001 = 8	1100 = 7

REPLICATE II

BLOCK	1	2	3	4
	0000 = 3	0100 = 9	0010 = 9	1000 = 6
	0101 = 12	0001 = 6	0111 = 14	1101 = 6
	1011 = 11	1111 = 12	1001 = 7	0011 = 5
	1110 = 17	1010 = 12	1100 = 12	0110 = 13

12. Consider the Latin square of Exercise 6 of Chapter 14. Identify each treatment and determine the defining equation to select these treatments, regarding the design as a fractional replicate. Analyze the data by using a modification of Yates' algorithm.

13. Research Director J. Zilch has selected 3 uncalibrated thermometers. He drew these from a large stock. Three analysts then used them to determine the melting point of a homogeneous sample of hydroquinone following a specified procedure. This entire experiment was performed by the 3 analysts on the 3 thermometers in 3 separate weeks. The results are given in the following table. (a) Use Yates' algorithm to calculate the sums of squares and combine the pseudolinear and quadratic effects for each factor. (b) Determine the EMS. (c) Estimate the variances involved. (d) Test the appropriate hypotheses. (e) Write a brief report.

		THERMOMETER		
ANALYST	REPLICATE	A	B	C
I	1	174.0	173.0	171.5
	2	173.5	173.5	172.5
	3	174.5	173.0	173.0
II	1	173.0	172.0	171.0
	2	173.0	173.0	172.0
	3	173.5	173.5	171.5
III	1	173.5	173.0	173.0
	2	173.0	173.5	173.0
	3	173.0	172.5	172.5

14. Determine all the defining contrasts and confounding interactions for the design exhibited in Table 15.48.

15. Suppose 3 factors (all parametric) are to be studied, each at 2 levels. In carrying out the experiment, it is necessary to run it in 2 blocks of 4. Two replicates are planned. Set up the formulas for the sum of squares and degrees

of freedom for each effect, if the first replicate has blocks confounded with ABC, and the second has blocks confounded with BC. (Use the notation of Section 14.7.)

16. Construct a design for a $\frac{1}{4}$ replicate of a 2^7 experiment in 4 blocks of 8 treatments. Use $ABCDE$ and $CDEFG$ as 2 of the defining contrasts.

17. Determine the elements in the principal block of a 2^{-3} replicate of a 2^7 experiment with $ABCDE$ and $ABFG$ as 2 of the defining contrasts.

18. Set up the defining equations to select a $\frac{1}{5}$ replicate of a 5^4 experiment in 5 blocks of 25 plots each. Determine the treatment combinations in the principal block.

19. The heat treatment of steel casting in an automatic machine consists of heating, quenching, and drawing. The heating could be done by coils on either the upper or lower boss of the casting. An experiment was conducted to determine the effect of heating time (A), quenching time (B), drawing time (C), boss (D), and the position of measurement on the boss (E) on the hardness of the casting. A 2^5 factorial design was used. The levels chosen for each factor were as follows: A, 40 and 63 sec; B, 12 and 30 sec; C, 21 and 55 sec; D, lower or upper boss; and E, lower or upper position of measurement on the boss.

The hardness readings are given below. There are two replicates per cell.

| | | A_0 | | | | A_1 | | | |
| | | C_0 | | C_1 | | C_0 | | C_1 | |
		E_0	E_1	E_0	E_1	E_0	E_1	E_0	E_1
B_0	D_0	71.5	68.0	71.0	69.5	70.5	66.5	70.0	67.0
		70.5	67.5	70.5	68.0	70.0	67.0	69.5	68.0
	D_1	67.5	68.5	66.0	65.5	64.5	63.0	67.0	65.5
		66.0	66.5	66.5	66.0	65.0	65.0	66.5	65.0
B_1	D_0	72.0	70.0	71.5	71.5	71.0	68.0	74.0	70.0
		71.5	69.0	72.5	68.0	69.0	67.5	71.0	69.5
	D_1	69.0	67.5	66.5	64.0	67.0	68.0	69.0	68.0
		67.5	68.0	67.0	65.0	67.0	68.5	69.0	68.0

Conduct an analysis of variance.

(a) Analyze the data by Yates' algorithm.

(b) Determine a 95% confidence interval on the residual variance.

20. For the data of Exercise 19, consider only the first reading in each cell. Use the half-normal plot to analyze the resulting set of data.

21. Analyze the data of Exercise 5 by means of a half-normal plot.
22. An experiment was conducted to discover the important mechanisms by which zinc corrodes. Four factors were considered, namely, solution of liquid in which a zinc specimen was immersed, temperature, length of immersion, and atmosphere above solution. For this experiment there were 4 solutions (A), namely, tap water, 95.4 ppm $ZnCl_2$, 102.7 ppm KCl, and 501 ppm KCl. Temperature was held at 51°C. The atmosphere (B) was air and nitrogen, and time (C) was 24, 72, 120, and 168 hr. The data below represent the sum of the weight loss of 3 specimens in milligrams per square inch. (a) By means of the Yates algorithm, determine the significance of each factor and 2 factor interactions. (b) Is there a linear, quadratic, or cubic effect due to time? (c) Is the effect of tap water different from the other solutions?

WEIGHT LOSS

TIME	TAP WATER		95.4 ppm $ZnCl_2$		102.7 ppm KCl		501 ppm KCl	
	AIR	NITROGEN	AIR	NITROGEN	AIR	NITROGEN	AIR	NITROGEN
24 hr	4.1	1.3	14.2	1.3	15.2	2.8	16.5	2.4
72 hr	8.5	1.7	39.4	3.8	36.1	9.1	47.0	7.7
120 hr	8.0	2.0	9.4	3.3	54.0	10.2	62.2	11.3
168 hr	12.0	2.2	6.6	4.4	86.1	22.8	81.1	15.4

23. The senior analyst, Joseph R. Zilchester, is very suspicious of the data of Exercise 22. He calls upon you to help him find out what is wrong. How would you reexamine the date for consistency?
24. An experiment was designed to determine how the factors Temperature of heating (A), Time of holding at that temperature (T), and Composition of the metal (C) affect the hardenability of cast iron.
    Gray cast iron specimens of 4 different compositions were used. Two specimens of the same composition were heated to 1575°F and 1 of these was held for 5 min and the other for 30 min before quenching. This was repeated for specimens of each of the other 3 compositions. The experiment was repeated using a temperature of 1900°F. To obtain the replicates the entire experiment was repeated. The hardness of the specimens as obtained from measurements is given in the table below.

Replicate	Holding Time (min)	1575°F				1900°F			
		C_1	C_2	C_3	C_4	C_1	C_2	C_3	C_4
I	5	13	8.5	4	6.5	11.5	14	7	8
	30	16.5	19	6	8.5	20	20	5.5	5.5
II	5	9	11.5	5	7.5	16	14	6.5	7.5
	30	16	20	8	9	24	22	6	7

The experiment was designed as a split-plot design with temperature $\times$ replicate as main plot. Perform a complete analysis of variance on these data.

25. Suppose a certain rocket program calls for firing at 3 slant ranges (SR) and with 3 levels of propellant temperature (PT). The yield is the azimuth coordinate of miss distance. Three groups of 3 rounds each were fired for each set of conditions. The data represent this split-plot experiment.
    (a) Construct an appropriate mathematical model.
    (b) Carry out the complete analysis of variance including the calculation of estimates of variance.
    (c) Draw conclusions.

	SR_1			SR_2			SR_3		
	1	2	3	1	2	3	1	2	3
PT_1	−10	−22	−9	−5	−17	−4	11	−10	1
	−13	0	7	−9	6	13	−5	10	20
	14	−5	12	21	0	20	22	6	24
PT_2	−15	−25	−15	−14	−3	14	−9	8	14
	−17	−5	2	15	−1	5	−3	−2	18
	7	−11	5	−11	−20	−10	20	−15	−2
PT_3	−21	−26	−15	−18	−8	0	13	−5	−8
	−23	−8	−5	5	−26	−13	−9	−18	3
	0	−10	0	−10	−10	3	−13	−3	12

26. An experiment was conducted to determine the effect of both hard and soft tubing with or without a particulate core on the wall thickness as the diameter of the tubing is reduced. This reduction is caused by drawing the tubing through conical dies of sizes 0.209, 0.172, 0.142, and 0.123 in., respectively. The measured yield is the ratio of the increased length of tubing to the original length. There were then 3 main factors: tube hardness (H), namely, hard or soft; tube core (F), that is, hollow or filled; and tube reduction (C), that is, ordered pass through the reducing die. Two replications were included. In all there were 8 tube specimens used in the experiment. The table below gives the gauge length ratio. Analyze this split-plot experiment, testing the effect of each factor and estimating the main plot, sub-plot, and sub^2-plot errors.

	H_0 (HARD)				H_1 (SOFT)			
	F_0 (Hollow)		F_1 (Filled)		F_0 (Hollow)		F_1 (Filled)	
Pass	t_0	t_1	t_0	t_1	t_0	t_1	t_0	t_1
1	1.24	1.24	1.29	1.28	1.26	1.26	1.30	1.30
2	1.60	1.59	1.88	1.85	1.63	1.62	1.84	1.84
3	1.96	1.94	2.50	2.45	1.97	1.96	2.46	2.44
4	2.51	2.48	3.57	3.52	2.52	2.52	3.49	3.48

27. Give detailed directions for applying a nonparametric test of significance for equality of factor level effects to the data of a randomized block experiment using the rank orders of the quantities $(X_{ti} - \overline{X}_{t.})$, where X_{ti} represents observed value for the ith factor level in the tth block.

28. Investigate how to apply nonparametric tests in the analysis of a 2^k experiment. Include the analysis of confounded experiments and fractional replicates.

29. In the situation described in Section 15.2.5 obtain the likelihood ratio test criterion for the hypothesis $\sigma_0 = \sigma_1 = \cdots = \sigma_k$ with respect to the alternative that one of $\sigma_1, \sigma_2, \ldots, \sigma_k$ is larger than the remaining $k-1$ σ's, which are all equal.

30. In Exercise 29, assume that $\nu = 2$ and $\sigma_1 = \sigma_2 = \cdots = \sigma_k$. Show that

$$
\Pr\left[\frac{\max\limits_{1 \le j \le k} S_j^2}{\left(\sum\limits_{i=1}^{k} S_i^2 \right)} < K \right] = 1 - \binom{k}{1}(1 - K)^{k-1} + \binom{k}{2}(1 - 2K)^{k-1} - \cdots
$$

the series terminating at $(-1)^l \binom{k}{l}(1 - lK)^{k-k}$, where $lK \le 1 < (l+1)K$.

Use this result to obtain the null hypothesis distribution of the likelihood ratio test criterion of Exercise 29.

CHAPTER 16

Sequential Analysis

16.1 INTRODUCTION

So far we have concerned ourselves, for the most part, with the analysis of data collected according to a predetermined pattern. There have been a few departures from this (for example, in double and multiple sampling) but they have been relatively rare. However, it is evident that information may be obtained, in the course of an investigation, that is of sufficient importance to cause some alteration in plans. For example, an investigation based on the assumption that the population standard deviation of a measured character is about 15 units may need to be replanned if the first 20 or so observations indicate that the standard deviation is quite likely in excess of 40 units. If an estimate of some specified accuracy, for a population parameter, is to be calculated, upward revision of sample size would be imperatively indicated. Many other cases of less blatant need for revision of plans while an investigation is in progress arise in day-to-day statistical work. Although helpful changes in plans can be made on an ad hoc basis, it is very useful to have some more objective rules of procedure, and it is the purpose of this chapter to describe the basic ideas behind these rules.

16.2 DEFINITIONS

We define a *sequential procedure* as any procedure in which the final pattern of the data depends in some way on decisions which are based on the data themselves as they become available.

This is a very broad definition. It includes cases where the total number of observations is known in advance but not the populations from which the individuals to be measured will be chosen. This includes the case described in Chapter 19, where the selection of a stratified sample depends on estimated values of strata standard deviations based on a preliminary sample. It also includes cases where the total number of observations is not fixed in advance but the decision to discontinue an investigation is based (in part, at any rate) on data obtained during the investigation itself. Some sequential procedures lack such a stopping rule and merely indicate the line of inquiry which should next be followed, supposing it is decided to continue.

It can be seen that the addition of such procedures to our available techniques considerably broadens the range of possible methods we can use in planning an investigation. In particular, by intelligent use of these methods, the cost of investigations can be substantially reduced by introducing rules which, in effect, tell us when we have enough evidence to reach a decision and avoid pointless overaccumulation of data.

16.3 SEQUENTIAL TESTS

We first consider a straightforward choice between 2 simple hypotheses, H_0 and H_1 (for example, 2 values p_0, p_1 for a binomial proportion, or 2 values θ_0, θ_1 for the mean of a normal distribution with known variance). Suppose we wish to use a procedure such that

$$\Pr[\text{accept } H_{1-i}|H_i] = \alpha_i \qquad (i = 0, 1) \tag{16.1}$$

Here α_0 and α_1 are probabilities of error. If we regard H_0 as the hypothesis being tested (or "null hypothesis"—see Chapter 7), then α_0 is the probability of "first kind of error" (or the significance level of the test), and α_1 is the probability of "second kind of error" or the "operating characteristic" with respect to H_1. [$(1 - \alpha_1)$ is the *power* with respect to H_1.] Usually, of course, α_0 and α_1 will be rather small (often 0.05 or less) and we can certainly suppose that neither is greater than 0.5.

In earlier chapters we have, from time to time, encountered problems calling for construction of test procedures satisfying conditions like (16.1). Usually, however, there has been an implicit limitation to tests based on sets of data of prespecified size and pattern. If we now admit to consideration other methods of collection of data, we may expect to derive some advantage therefrom.

The form that this advantage takes in the present case is *reduction in the average number of observations needed to satisfy (16.1)*. It will be noted that since the number of observations is no longer necessarily fixed in advance it is necessary to consider the *average* number of observations. Also, this average number can depend on which hypothesis is true—in contrast to a fixed size sample. We shall see that we can construct procedures satisfying (16.1) which need average numbers of observations substantially less than the size of fixed size samples (even if the best possible tests are used in the latter cases), *provided either H_0 or H_1 is valid*. The same will be true when some other hypotheses are valid (especially if these do not differ markedly from H_0 or H_1), but we cannot rely on this being so in *all* possible situations. Sometimes this can be a serious drawback to a sequential test; though its importance should not be overrated, neither should it be forgotten.

16.4   SEQUENTIAL PROBABILITY RATIO TESTS

It will be recalled (Section 7.5) that the best (fixed size sample) test for discriminating between 2 simple hypotheses, H_0 and H_1, was based on the *likelihood ratio* l_{1n}/l_{0n} (where l_{in} represents the probability, or probability density function, of the observed set of values $x_1, x_2, \ldots, x_n$, given that the hypothesis H_i is valid).

It seems natural, therefore, to see whether the ratio l_{1n}/l_{0n} can also be used in the present case. Evidently, small values of the ratio are favorable to H_0, large values to H_1. Consider a procedure defined by the following rules:

(*i*) If $\dfrac{l_{1n}}{l_{0n}} \geqslant B$ accept H_1.

(*ii*) If $\dfrac{l_{1n}}{l_{0n}} \leqslant A$ accept H_0.

(*iii*) If $A < \dfrac{l_{1n}}{l_{0n}} < B$, take one further observation.

This procedure contains rules for deciding whether to continue sampling, and which hypothesis to choose if sampling is not to be continued. What are its properties?

(The following discussion is in a form that applies directly to continuous variables, but it can be modified to apply to discrete and mixed variables in a straightforward fashion.)

We have

$$\Pr\left[\text{accept } H_j | H_i\right] = \sum_n \iint \cdots_{W_{jn}} \int l_{in} \, dx_1 \cdots dx_n \qquad (16.2)$$

where $W_{jn} = $ set of all groups of values $(x_1, x_2, \ldots, x_n)$ which lead to H_j being accepted at final sample size n.

Now from rules (*i*) and (*ii*) we see that

$$\text{in } W_{0n}: \quad l_{1n} \leqslant A l_{0n}$$

and

$$\text{in } W_{1n}: \quad l_{1n} \geqslant Bl_{0n}$$

Inserting these inequalities in (16.2) we obtain:

$$A \cdot \Pr[\text{accept } H_0 | H_0] \geqslant \Pr[\text{accept } H_0 | H_1] \qquad (16.3)$$

$$\Pr[\text{accept } H_1 | H_1] \geqslant B \cdot \Pr[\text{accept } H_1 | H_0] \qquad (16.4)$$

If we now assume that the probability is 1 that the procedure terminates —that is, in effect, we *must* choose either H_0 or H_1—then Pr[accept $H_0 | H_i$] + Pr[accept $H_1 | H_i$] = 1 for any i, and (16.3) and (16.4) become

$$A\big[1 - \Pr[\text{accept } H_1 | H_0]\big] \geqslant \Pr[\text{accept } H_0 | H_1] \qquad (16.5)$$

$$1 - \Pr[\text{accept } H_0 | H_1] \geqslant B \Pr[\text{accept } H_1 | H_0] \qquad (16.6)$$

If further we suppose that the inequalities (*i*) and (*ii*) (in the rule of procedure) can be replaced by equalities without much loss of accuracy, and solve the resulting approximate (16.5) and (16.6) we find

$$A \doteqdot \frac{\Pr[\text{accept } H_0 | H_1]}{1 - \Pr[\text{accept } H_1 | H_0]} \qquad B \doteqdot \frac{1 - \Pr[\text{accept } H_0 | H_1]}{\Pr[\text{accept } H_1 | H_0]}$$

Hence conditions (16.1) would be approximately satisfied by taking

$$A = \frac{\alpha_1}{1 - \alpha_0} \qquad B = \frac{1 - \alpha_1}{\alpha_0} \qquad (16.7)$$

[The assumption that inequalities (16.5) and (16.6) are approximately equalities is likely to be justified if the changes in l_{1n}/l_{0n} as n increases are small. Then the jump from a point between the limits (A, B) will not land far outside the limits.]

Using these values of A and B in rules (*i*) to (*iii*) gives a simple sequential procedure which may be expected to satisfy, approximately, the conditions (16.1). Because it uses the ratio l_{1n}/l_{0n}, it is called the *sequential probability ratio test*.

How effective is it likely to be in reducing the average number of observations? This question is discussed again in Section 16.6, but here we note an interesting argument.

Consider the expected value of l_{1n}/l_{0n} if H_0 is valid *and H_i is accepted*, which is

$$E\left(\frac{l_{1n}}{l_{0n}}\bigg| H_i \text{ accepted, } H_0\right) = \frac{\sum_n \iint \cdots_{W_{in}} \int (l_{1n}/l_{0n})l_{0n}\,dx_1 \cdots dx_n}{\sum_n \iint \cdots_{W_{in}} \int l_{0n}\,dx_1 \cdots dx_n}$$

$$= \frac{\Pr[\text{accept } H_i|H_1]}{\Pr[\text{accept } H_i|H_0]}$$

Similarly,

$$E\left(\frac{l_{0n}}{l_{1n}}\bigg| H_i \text{ accepted, } H_1\right) = \frac{\Pr[\text{accept } H_i|H_0]}{\Pr[\text{accept } H_i|H_1]}$$

Inserting conditions (16.1) we find that *when H_1 is accepted*

$$E\left(\frac{l_{1n}}{l_{0n}}\right) = \frac{1-\alpha_1}{\alpha_0} \qquad \text{if } H_0 \text{ is valid} \tag{16.8}$$

$$E\left(\frac{l_{0n}}{l_{1n}}\right) = \frac{\alpha_0}{1-\alpha_1} \qquad \text{if } H_1 \text{ is valid} \tag{16.9}$$

Similarly, *when H_0 is accepted*

$$E\left(\frac{l_{1n}}{l_{0n}}\right) = \frac{\alpha_1}{1-\alpha_0} \qquad \text{if } H_0 \text{ is valid} \tag{16.10}$$

$$E\left(\frac{l_{0n}}{l_{1n}}\right) = \frac{1-\alpha_0}{\alpha_1} \qquad \text{if } H_1 \text{ is valid} \tag{16.11}$$

Now, ignoring the "overshooting" of boundary values, we stop, and accept H_1, as soon as $l_{1n}/l_{0n} \doteq (1-\alpha_1)/\alpha_0$ or, equivalently,

$$\frac{l_{0n}}{l_{1n}} \doteq \frac{\alpha_0}{(1-\alpha_1)}$$

And we stop, and accept H_0, as soon as $l_{1n}/l_{0n} \doteq \alpha_1/(1-\alpha_0)$ or, equivalently, $l_{0n}/l_{1n} \doteq (1-\alpha_0)/\alpha_1$, if this happens first.

In a heuristic way, then, it can be seen that we are getting approximately the smallest possible samples which satisfy (16.8) to (16.11).

Example 16.1 It is required to discriminate between 2 possible values p_0, p_1 ($p_0 < p_1$) for the probability of an event E. In each of a sequence of independent trials it is observed whether or not E occurs. We want to construct a sequential test procedure for this problem.

$$\text{Let } X_i = 1 \text{ if } E \text{ occurs at the } i\text{th trial}$$

$$X_i = 0 \text{ if } E \text{ does not occur at the } i\text{th trial}$$

Then the likelihood function, after m trials, is

$$\prod_{i=1}^{m} p^{X_i}(1-p)^{1-X_i}$$

where

$$p = \Pr[E]$$

The likelihood ratio is

$$\frac{l_{1m}}{l_{0m}} = \prod_{i=1}^{m} \left[\left(\frac{p_1}{p_0} \right)^{X_i} \left(\frac{1-p_1}{1-p_0} \right)^{1-X_i} \right] = \left(\frac{p_1}{p_0} \right)^{S_m} \left(\frac{1-p_1}{1-p_0} \right)^{m-S_m}$$

where $S_m = \sum_{i=1}^{m} X_i =$ number of times E occurs in the first m trials.

Using the rules (i) to (iii) given at the beginning of this section and (16.7) we have (after taking logarithms) the following procedure:

(i) If $S_m \ln\left[\dfrac{p_1(1-p_0)}{p_0(1-p_1)} \right] \geqslant \ln\left(\dfrac{1-\alpha_1}{\alpha_0} \right) + m \ln\left[\dfrac{1-p_0}{1-p_1} \right]$, accept $p = p_1$.

(ii) If $S_m \ln\left[\dfrac{p_1(1-p_0)}{p_0(1-p_1)} \right] \leqslant \ln\left(\dfrac{\alpha_1}{1-\alpha_0} \right) + m \ln\left[\dfrac{1-p_0}{1-p_1} \right]$, accept $p = p_0$.

(iii) If $\ln\left(\dfrac{\alpha_1}{1-\alpha_0} \right) < S_m \ln\left[\dfrac{p_1(1-p_0)}{p_0(1-p_1)} \right] - m \ln\left[\dfrac{1-p_0}{1-p_1} \right] < \ln\left(\dfrac{1-\alpha_1}{\alpha_0} \right)$

carry out a further trial.

Using this procedure we should have

$$\Pr[\text{accept } p = p_1 | p = p_0] \doteqdot \alpha_0$$

$$\Pr[\text{accept } p = p_0 | p = p_1] \doteqdot \alpha_1$$

Numerical values for α_0 and α_1 will, of course, be chosen in accordance with the conditions of the particular problem under investigation. It is useful to note that the procedure can be represented diagrammatically as in Figure 16.1. The number of trials m is plotted as abscissa, the value of $S_m = \sum_{i=1}^{m} X_i$ as ordinate. So long as

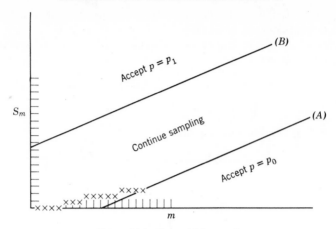

Figure 16.1 Sequential sampling.

the point (m, S_m) lies between the parallel lines,

$$S_m = \frac{\ln\left(\dfrac{\alpha_1}{1-\alpha_0}\right) + m\ln\left(\dfrac{1-p_0}{1-p_1}\right)}{\ln\left[\dfrac{p_1(1-p_0)}{p_0(1-p_1)}\right]} \qquad (A)$$

and

$$S_m = \frac{\ln\left(\dfrac{1-\alpha_1}{\alpha_0}\right) + m\ln\left(\dfrac{1-p_0}{1-p_1}\right)}{\ln\left[\dfrac{p_1(1-p_0)}{p_0(1-p_1)}\right]} \qquad (B)$$

observations are continued. As soon as (m, S_m) falls outside these limits, observation stops and a decision is taken in accordance with the rules of procedure. (With the particular sequence of observations represented in Figure 16.1, the hypothesis $p = p_0$ would be accepted on the sixteenth trial.)

*16.5 OPERATING CHARACTERISTIC OF SEQUENTIAL PROBABILITY RATIO TEST

At this point we introduce some new notation that will help to compress the discussion. The words "sequential probability ratio test" will be abbreviated to s.p.r.t., and if we wish to indicate that it is based on discrimination between simple hypotheses H_0 and H_1 with (approximate) error probabilities α_0 and α_1 respectively, the symbols $S(H_0, \alpha_0; H_1, \alpha_1)$ will be used to denote the s.p.r.t.

The first question we discuss is: what is the probability of "accepting H_1" when a (simple) hypothesis H is true? We know already that this is (approximately) α_0 if

$H = H_0$; $(1 - \alpha_1)$ if $H = H_1$. Now we want to find this probability for a general H.

We do this by showing that $S(H_0, \alpha_0; H_1, \alpha_1)$ is equivalent to another s.p.r.t., $S(H, \alpha; H', \alpha')$, in which "accepting H (or perhaps H')" is the same decision as "accepting H_1" in $S(H_0, \alpha_0; H_1, \alpha_1)$. We already are able to evaluate (approximately) the probability of accepting H (or H') when H is true in $S(H, \alpha; H', \alpha')$—it is $1 - \alpha$ (or α'). This is then also the probability of accepting H_1, using $S(H_0, \alpha_0; H_1, \alpha_1)$ when H is the true hypothesis.

But first we have to find α, H', and α'. To do this we note that if

$$\left[\frac{l(X_1, \ldots, X_m | H_1)}{l(X_1, \ldots, X_m | H_0)} \right]^h = \frac{l(X_1, \ldots, X_m | H')}{l(X_1, \ldots, X_m | H)} \tag{16.12}$$

for some *constant* value of h, then the continuation region (*iii*) of Section 16.4 can be written in the form

$$\left(\frac{\alpha_1}{1 - \alpha_0} \right)^h < \frac{l(X_1, \ldots, X_m | H')}{l(X_1, \ldots, X_m | H)} < \left(\frac{1 - \alpha_1}{\alpha_0} \right)^h \quad \text{(if } h > 0) \tag{16.13}$$

or

$$\left(\frac{1 - \alpha_1}{\alpha_0} \right)^h < \frac{l(X_1, \ldots, X_m | H')}{l(X_1, \ldots, X_m | H)} < \left(\frac{\alpha_1}{1 - \alpha_0} \right)^h \quad \text{(if } h < 0) \tag{16.14}$$

(The case $h = 0$ can be treated as a limit of cases for which $h \neq 0$.)

Supposing for the moment that (16.13) is appropriate, we see that $S(H_0, \alpha_0; H_1, \alpha_1)$ is equivalent to $S(H, \alpha; H', \alpha')$ with

$$\frac{\alpha'}{1 - \alpha} = \left(\frac{\alpha_1}{1 - \alpha_0} \right)^h; \quad \frac{1 - \alpha'}{\alpha} = \left(\frac{1 - \alpha_1}{\alpha_0} \right)^h \tag{16.15}$$

Solving these equations for α, we find

$$\alpha = \frac{1 - \left(\dfrac{\alpha_1}{1 - \alpha_0} \right)^h}{\left(\dfrac{1 - \alpha_1}{\alpha_0} \right)^h - \left(\dfrac{\alpha_1}{1 - \alpha_0} \right)^h} \tag{16.16}$$

This is the (approximate) probability of accepting H' when H is true. Since the events "accept H'" using $S(H, \alpha; H', \alpha')$ and "accept H_1" using $S(H_0, \alpha_0; H_1, \alpha_1)$ are identical, this is also the power of the latter, regarded as a test of the hypothesis H_0, with respect to H. We have

$$\Pr[\text{accept } H_1 | H] \doteqdot \frac{1 - \left(\dfrac{\alpha_1}{1 - \alpha_0} \right)^h}{\left(\dfrac{1 - \alpha_1}{\alpha_0} \right)^h - \left(\dfrac{\alpha_1}{1 - \alpha_0} \right)^h} \tag{16.17}$$

This result has been obtained on the assumption that h is greater than 0. If h is less than 0, inequalities (16.14) apply, but "accept H_1" is now identical with "accept H" using $S(H, \alpha; H', \alpha')$, and we again reach the same formula (16.17). Taking limits as h tends to 0, we obtain the value

$$\frac{\ln\left(\dfrac{1-\alpha_0}{\alpha_1}\right)}{\ln\left(\dfrac{(1-\alpha_0)(1-\alpha_1)}{\alpha_0\alpha_1}\right)}$$

for this case.

The "operating characteristic" (OC) of $S(H_0, \alpha_0; H_1, \alpha_1)$ with respect to H_1 is

$$1 - \Pr[\text{accept } H_1 | H] = \frac{\left(\dfrac{1-\alpha_1}{\alpha_0}\right)^h - 1}{\left(\dfrac{1-\alpha_1}{\alpha_0}\right)^h - \left(\dfrac{\alpha_1}{1-\alpha_0}\right)^h} \tag{16.18}$$

Example 16.2   In the problem discussed in Example 16.1 the likelihood ratio is

$$\left(\frac{p_1}{p_0}\right)^{S_m} \cdot \left(\frac{1-p_1}{1-p_0}\right)^{m - S_m}$$

Since

$$\left[\left(\frac{p_1}{p_0}\right)^{S_m} \left(\frac{1-p_1}{1-p_0}\right)^{m-S_m}\right]^h \cdot p^{S_m}(1-p)^{m-S_m} = p'^{S_m}(1-p')^{m-S_m}$$

provided

$$p' = \left(\frac{p_1}{p_0}\right)^h p$$

and

$$1 - p' = \left(\frac{1-p_1}{1-p_0}\right)^h (1-p)$$

we see that (16.12) is satisfied if h satisfies

$$\left(\frac{p_1}{p_0}\right)^h p + \left(\frac{1-p_1}{1-p_0}\right)^h (1-p) = 1 \tag{16.19}$$

If this is so then H' specifies a probability

$$p' = \left(\frac{p_1}{p_0}\right)^h p$$

Equation (16.19) must usually be solved numerically, but this is not a difficult or very lengthy process.

In the above example, we were able to find an equation for h by fairly direct methods. In general, the following systematic approach can be used.

Equation (16.12) implies that

$$\left[\frac{l(x_1,\ldots,x_m|H_1)}{l(x_1,\ldots,x_m|H_0)} \right]^h l(x_1,\ldots,x_m|H)$$

is a likelihood function. Hence if it is summed, or integrated, over all possible values of x_1, $x_2,\ldots,x_m$ the result should be 1. This can be expressed by the relationship

$$E\left[\left(\frac{l(X_1,\ldots,X_m|H_1)}{l(X_1,\ldots,X_m|H_0)} \right)^h \middle| H \right] = 1 \tag{16.20}$$

which has to be solved for h. This equation is always solved by $h=0$. It can be shown that in general there is just one nonzero value of h satisfying (16.20), and this is the value we use. It can happen that $h=0$ is the only solution; then we must use special formulas, or limits of cases $h \neq 0$, as explained above.

The arguments we have used can only be applied if h is a known constant. In particular h must not depend on m. This will certainly be the case if the X's are mutually independent *and* all have the same distribution. For in this case

$$E\left[\left(\frac{l(X_1,\ldots,X_m|H_1)}{l(X_1,\ldots,X_m|H_0)} \right)^h \middle| H \right] = \prod_{j=1}^{m} E\left[\left(\frac{l(X_j|H_1)}{l(X_j|H_0)} \right)^h \middle| H \right]$$

and (because of the identity of distributions of the X_j's) the same value of h satisfies

$$E\left[\left(\frac{l(X_j|H_1)}{l(X_j|H_0)} \right)^h \middle| H \right] = 1 \text{ for all } j$$

When the X's are not mutually independent and (or) do not have identical distributions, h may very well vary with m. Even in this case useful approximations may be obtained provided h does not vary very much.

16.6 AVERAGE SAMPLE NUMBER OF A S.P.R.T.

The distinguishing feature of a s.p.r.t. procedure is that the final sample size, N, is not determined before starting to take observations. The quantity N, is, in fact, a random variable. We are particularly interested in the expected value of this random variable. This is the average number of observations needed to reach a decision. Since a primary aim of introducing sequential methods is to reduce sample sizes, we are particularly interested in making this expected value small. (This need not, of course, be the sole consideration, as a study of Chapter 11, for example, shows.)

The expected value of N depends, in general, on the hypothesis which is true, H, say, and is therefore denoted by the symbol $E(N|H)$. $E(N|H)$ is called the *average sample number* (often abbreviated to a.s.n.) when H is true. Useful formulas for $E(N|H)$ are available only for the case described toward the end of Section 16.5—where the random variables X_j representing the observations are mutually independent and have a common distribution. It is this case we now consider.

The continuation region for $S(H_0,\alpha_0; H_1,\alpha_1)$ can be written

$$\ln\left(\frac{\alpha_1}{1-\alpha_0}\right) < \sum_{j=1}^{m} Z_j < \ln\left(\frac{1-\alpha_1}{\alpha_0}\right) \tag{16.21}$$

where

$$Z_j = \ln\left[\frac{l(X_j|H_1)}{l(X_j|H_0)}\right]$$

The Z_j's also possess the properties of the X_j's, of being mutually independent and identically distributed.

Now consider the terminating value of $\sum_{j=1}^{N} Z_j$—where N is a random variable representing the sample size determined by some sequential procedure. The expected value of this sum can be shown to be equal to $E(N|H)E(Z|H)$ where Z has the same distribution as each Z_j. (We are assuming here that $E(N|H)$ is not infinite and $E(Z|H)$ is neither infinite nor 0.) This formula appears to be quite a natural one, but needs proving, although we do not give a proof here.

We now obtain an alternative representation of the expected value of $\sum_{j=1}^{N} Z_j$ for a s.p.r.t. by recalling that sampling ceases as soon as this sum falls outside the limits (16.21). If we neglect, as in Section 16.4, the amount by which the boundaries are overshot, it follows that $\sum_{j=1}^{n} Z_j$ has 2 possible values, namely, $\ln[\alpha_1/(1-\alpha_0)]$ and $\ln[(1-\alpha_1)/\alpha_0]$. It takes the first of these values with probability $\Pr[\text{accept } H_0|H]$ and the second with probability $\Pr[\text{accept } H_1|H]$. If we assume that one decision or the other is certain to be taken, then

$$\Pr[\text{accept } H_0|H] = 1 - \Pr[\text{accept } H_1|H]$$

and, equating the 2 formulas for the expected value of $\Sigma_{j=1}^{n} Z_j$ we have

$$E(N|H)E(Z|H) \doteq \{1 - \Pr[\text{accept } H_1|H]\} \ln\left(\frac{\alpha_1}{1-\alpha_0}\right)$$

$$+ \Pr[\text{accept } H_1|H] \ln\left(\frac{1-\alpha_1}{\alpha_0}\right)$$

If $E(Z|H) \neq 0$ we obtain the following approximate formula for average sample number (a.s.n.):

$$E(N|H) \doteq$$

$$\frac{\{1 - \Pr[\text{accept } H_1|H]\} \ln\left(\dfrac{\alpha_1}{1-\alpha_0}\right) + \Pr[\text{accept } H_1|H] \ln\left(\dfrac{1-\alpha_1}{\alpha_0}\right)}{E(Z|H)}$$

$$(16.22)$$

Approximate values for $\Pr[\text{accept } H_1|H]$ can be inserted from (16.17).

Example 16.3 A new container for a certain foodstuff is suspected of insufficient protection against contamination by odors from other foodstuffs. To examine whether this is so, a sample taken from a new container is mixed with $k-1$ similar samples of the same foodstuffs which are known not to be contaminated. These are presented for inspection and the inspector picks out the sample he thinks most likely to be contaminated. The process is repeated; each time a fresh set of k samples (one from a new container) is used. It may be assumed that if an inspector is unable to differentiate between the samples he will select one out of the k samples at random.

It is required to construct a sequential procedure for this purpose, and to choose k so as to minimize the average number of *samples* needed, if there is really no contamination.

If there is no contamination the probability of the sample from the new container being chosen is k^{-1}. If there is contamination, such that a proportion P of inspectors can detect it, the proportion of samples for the new container being chosen on the basis of such detection is P, and a further proportion $(1-P)$ will be chosen by guesswork, giving a total probability

$$P + (1-P)k^{-1}$$

Hence we can use the s.p.r.t. constructed in Example 16.1, with $p_0 = k^{-1}$, $p_1 = P + (1-P)k^{-1}$. From the results of (16.22), the average number of *inspectors* needed *when there is really no contamination* is approximately

$$\frac{(1-\alpha_0)\ln\left(\dfrac{\alpha_1}{1-\alpha_0}\right)+\alpha_0\ln\left(\dfrac{1-\alpha_1}{\alpha_0}\right)}{k^{-1}\ln(p_1/p_0)+(1-k^{-1})\ln[(1-p_1)/(1-p_0)]}$$

The average number of samples needed will be k times this quantity. To minimize this we must *maximize* the value of

$$g_0(k)=k^{-2}\left\{\ln\left(\frac{p_1}{p_0}\right)+(k-1)\ln\left[\frac{(1-p_1)}{(1-p_0)}\right]\right\}$$

$$=k^{-2}[\ln(kP+1-P)+(k-1)\ln(1-P)] \qquad (16.23)$$

Some numerical values of $g_0(k)\times0.4343$ (using logarithms to base 10), for a few values of P are shown in Table 16.1.

TABLE 16.1
VALUES OF $0.4343g_0(k)$ FROM (16.23)

$k =$	$P = 0.5$	$P = 0.75$	$P = 0.9$	$P = 0.95$
2	0.0312	0.0898	0.1803	0.2527
3	0.0334	0.0896	0.1725	0.2377
4	0.0316	0.0809	0.1520	0.2074
5	0.0291	0.0722	0.1335	0.1809

For larger values of P, $k=2$ is best; below $P=0.75$ (approximately) $k=3$ should give smaller average number of samples.

If there is contamination and the proportion of inspectors able to detect contamination is in fact P the average number of samples is (approximately) minimized if

$$g_1(k)=k^{-2}[(kP+1-P)\ln(kP+1-P)+(k-1)(1-P)\ln(1-P)] \quad (16.24)$$

is maximized.

Some numerical values of $0.434g_1(k)$ are shown in Table 16.2 (again using logarithms to base 10).

In this case $k=3$ is best for larger values of P, while for P less than about 0.7, $k=4$ is slightly better. Taking an overall view, $k=3$ seems to be the safest choice.

It is worthwhile to note that in each case the values of $g_0(k)$ and $g_1(k)$ increase with P. This indicates that the average number of samples decreases as P increases, and is to be expected since it is less difficult to detect a stronger degree of contamination.

TABLE 16.2

VALUES OF $0.4343g_1(k)$ FROM (16.24)

$k =$	$P = 0.5$	$P = 0.75$	$P = 0.9$	$P = 0.95$
2	0.0284	0.0688	0.1074	0.1251
3	0.0334	0.0771	0.1169	0.1345
4	0.0340	0.0758	0.1126	0.1287
5	0.0332	0.0722	0.1059	0.1204

We now come to consider the case where the *only* solution of (16.20) is $h = 0$. We have already seen, from considerations of continuity, that the probability of accepting H_1 in this case is approximately

$$w = \frac{\ln\left(\dfrac{1-\alpha_0}{\alpha_1}\right)}{\ln\left[\dfrac{(1-\alpha_0)(1-\alpha_1)}{\alpha_0\alpha_1}\right]} \tag{16.25}$$

We cannot use (16.22) in this case because the numerator and denominator are each 0. Instead we equate 2 formulas for the expected value of the *square* of the logarithm of the likelihood ratio, giving

$$E(N|h=0) \doteq \frac{(1-w)\left[\ln\left(\dfrac{\alpha_1}{1-\alpha_0}\right)\right]^2 + w\left[\ln\left(\dfrac{1-\alpha_1}{\alpha_0}\right)\right]^2}{E(Z^2|h=0)} \tag{16.26}$$

where w is given by (16.25).

Example 16.4 To illustrate we discuss the problem stated in Example 16.3 for the case $h = 0$. When $h = 0$, the expected value of the logarithm of the likelihood ratio is 0. In the notation used in Example 16.1 and 16.2 this means that $h = 0$ corresponds to the value of p satisfying the equation

$$p\ln\frac{p_1}{p_0} + (1-p)\ln\frac{1-p_1}{1-p_0} = 0$$

Using the notation of Example 16.3, this becomes

$$p\ln(kP+1-P) + (1-p)\ln(1-P) = 0 \tag{16.27}$$

and so

$$p = \frac{-\ln(1-P)}{\ln\left[1 + \dfrac{kP}{1-P}\right]}$$

The numerator of (16.26) depends only on α_0 and α_1. Hence the average number of *inspectors* is inversely proportional to

$$p[\ln(kP+1-P)]^2 + (1-p)[\ln(1-P)]^2$$

and the average number of *samples* is inversely proportional to

$$g(k) = k^{-1}\left[p[\ln(kP+1-P)]^2 + (1-p)[\ln(1-p)]^2 \right]$$

$$= -k^{-1}[\ln(kP+1-P)][\ln(1-P)] \tag{16.28}$$

Some numerical values of $(0.4343)^2 g(k)$ are shown in Table 16.3.

These figures confirm our impression that $k = 3$ is a good choice.

We have not found it necessary, in this investigation, to calculate explicitly the values of p for which $h = 0$. If this is necessary, it can be done by a simple application of (16.27).

TABLE 16.3

VALUES OF $(0.4343)^2 \, g(k)$ FROM (16.28)

$k =$	$P = 0.50$	$P = 0.75$	$P = 0.9$	$P = 0.95$
2	0.0265	0.0732	0.1394	0.1887
3	0.0302	0.0799	0.1491	0.2005
4	0.0299	0.0770	0.1421	0.1904
5	0.0287	0.0725	0.1326	0.1773

*16.7 APPROXIMATIONS IN THE THEORY OF S.P.R.T.'S

The point at which approximation enters the theory of sequential tests as developed in this chapter is the assumption that the limits $[\alpha_1(1-\alpha_0)^{-1}$ or $\alpha_0^{-1}(1-\alpha_1)]$ are attained exactly by the likelihood ratio. This assumption is not only used in obtaining (16.22) [and (16.25)] for the average sample number, but is also involved in the probabilities of reaching a given decision (for example, to accept H_1). In fact we have only *inequalities* for these probabilities. If the *actual* values of Pr[accept $H_1|H_0$] and Pr[accept $H_0|H_1$] are α_0', α_1', respectively, then (16.5) and (16.6) give

$$\frac{\alpha_1}{1-\alpha_0}(1-\alpha_0') \geqslant \alpha_1' \quad \text{or} \quad \alpha_1 \geqslant \alpha_1' + \alpha_0'\alpha_1 - \alpha_0\alpha_1' \tag{16.29}$$

and

$$1 - \alpha_1' \geqslant \frac{1 - \alpha_1}{\alpha_0} \alpha_0' \quad \text{or} \quad \alpha_0 \geqslant \alpha_0' - \alpha_0' \alpha_1 + \alpha_0 \alpha_1' \tag{16.30}$$

Adding (16.29) and (16.30) together we find

$$\alpha_0' + \alpha_1' \leqslant \alpha_0 + \alpha_1 \tag{16.31}$$

This is a reassuring result. It shows that the sum of the 2 *nominal* probabilities of error used in constructing the s.p.r.t. exceeds the sum of the *actual* values of these probabilities of error. Hence since α_0 and α_1 are usually not very large, neither α_0' nor α_1' can be much in excess of the corresponding nominal values; and if one probability is underestimated the other must be less than the nominal value.

In fact, the actual error probabilities are usually substantially less than the nominal values by something like 30 percent. On the other hand, actual a.s.n.'s (when H_0 or H_1 is valid) are usually higher, by some 20 to 30 percent, than the values given by (16.22). The errors tend to balance out. The a.s.n. given by (16.22) is not as bad an approximation to the average sample size needed to give actual error probabilities of the specified amounts [i.e., if we use nominal error probabilities equal to $1.45\alpha_0$, $1.45\alpha_1$, say, then actual values should be about α_0, α_1 and actual a.s.n. about the value given by (16.22) as written].

In deriving (16.16) we have used a s.p.r.t. $S(H', \alpha'; H, \alpha)$ in which α' and α may quite possibly not be small quantities. There is the possibility of more substantial inaccuracy here, but (16.31) is still valid.

Coming to (16.22) [also (16.25)] we see that there are 2 places in which approximation can appear, though both stem from the same source of neglecting overshoot of the boundary values by the likelihood ratio. First, the values of Pr[accept $H_1|H$] will be inexact for the reasons already described. Second, even if this probability were known exactly, the expected value of the logarithm of the likelihood ratio would not be

$$\left\{ 1 - \Pr[\text{accept } H_1|H] \right\} \ln \left(\frac{\alpha_1}{1 - \alpha_0} \right) + \left\{ \Pr[\text{accept } H_1|H] \right\} \ln \left(\frac{1 - \alpha_1}{\alpha_0} \right)$$

In this formula $\ln \left(\dfrac{\alpha_1}{1 - \alpha_0} \right)$ should be replaced by

$$E \left[\ln \left(\frac{l_{1N}}{l_{0N}} \right) \middle| H \text{ valid, } H_0 \text{ accepted at } N \text{th observation} \right]$$

and $\ln \left(\dfrac{1 - \alpha_1}{\alpha_0} \right)$ should be replaced by

$$E \left[\ln \left(\frac{l_{1N}}{l_{0N}} \right) \middle| H \text{ valid, } H_1 \text{ accepted at } N \text{th observation} \right]$$

In the case where the X's are mutually independent and identically distributed, it is possible to find limits for the effect of this last kind of approximation. For instance, in this case,

$$\ln\left(\frac{l_{1N}}{l_{0N}}\right) = Z_1 + Z_2 + \cdots + Z_n = T_n$$

Consider the last step from T_{n-1} to T_n and suppose that H_1 is accepted on the nth observation. Then $T_{n-1} < \ln[(1-\alpha_1)/\alpha_0] \leqslant T_n$, and we have the situation represented in Figure 16.2. If we let $\ln[(1-\alpha_1)/\alpha_0] - T_{n-1} = Z'_n$, then the expected value of the overshoot $T_n - \ln[(1-\alpha_1)/\alpha_0]$ (given that there *is* an overshoot), is

$$E(Z - Z'_n \mid Z > Z'_n)$$

where Z is a random variable distributed as the logarithm of the likelihood ratio of a single X. Hence the average amount of overshoot is never greater than

$$\max_{\zeta > 0} E(Z - \zeta \mid Z > \zeta, H)$$

Similarly, considering the case when H_0 is accepted, we find that the average overshoot is between 0 and $\min_{\zeta > 0} E(Z + \zeta \mid Z < -\zeta, H)$. The numerator of the right side of (16.22), therefore, lies between

$$\left\{1 - \Pr[\text{accept } H_1 \mid H]\right\}\left[\ln\left(\frac{\alpha_1}{1-\alpha_0}\right) + \min_{\zeta > 0} E(Z + \zeta \mid Z < -\zeta, H)\right]$$

$$+ \left\{\Pr[\text{accept } H_1 \mid H]\right\} \cdot \ln\left(\frac{1-\alpha_1}{\alpha_1}\right) \qquad (16.32)$$

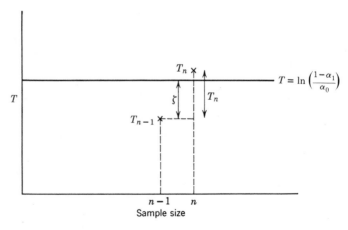

Figure 16.2 A step in the sequential process.

and

$$\{1 - \Pr[\text{accept } H_1 | H]\} \ln\left(\frac{\alpha_1}{1 - \alpha_0}\right) + \{\Pr[\text{accept } H_1 | H]\}$$

$$\times \left[\ln\left(\frac{1 - \alpha_1}{\alpha_0}\right) + \max_{\zeta > 0} E(Z - \zeta | Z > \zeta, H)\right] \quad (16.33)$$

*Example 16.5 We consider the calculation of the bounds (16.32) and (16.33) for the s.p.r.t. comparing the hypotheses $H_0(\theta = \theta_0)$ and $H_1(\theta = \theta_1)$, with $0 < \theta_0 < \theta_1$, on the expected value θ of independent variables X_i, each following the exponential distribution

$$p_X(x) = \theta^{-1} e^{-x/\theta} \quad (x > 0)$$

The s.p.r.t. continuation region is defined by

$$\frac{\alpha_1}{1 - \alpha_0} < \left(\frac{\theta_0}{\theta_1}\right)^n \exp\left[-(\theta_1^{-1} - \theta_0^{-1}) \sum_{i=1}^{n} X_i\right] < \frac{1 - \alpha_1}{\alpha_0}$$

or

$$\frac{\ln\left(\frac{\alpha_1}{1 - \alpha_0}\right) + n \ln\left(\frac{\theta_1}{\theta_0}\right)}{\theta_0^{-1} - \theta_1^{-1}} < \sum_{i=1}^{n} X_i < \frac{\ln\left(\frac{1 - \alpha_1}{\alpha_0}\right) + n \ln\left(\frac{\theta_1}{\theta_0}\right)}{\theta_0^{-1} - \theta_1^{-1}}$$

The natural logarithm of the likelihood ratio for a single X is

$$Z = (\theta_0^{-1} - \theta_1^{-1})X - \ln\left(\frac{\theta_1}{\theta_0}\right)$$

Hence the event $Z > \zeta$ is equivalent to the event $X > x(\zeta)$ where

$$x(\zeta) = \left[\zeta + \ln\left(\frac{\theta_1}{\theta_0}\right)\right]\left[\theta_0^{-1} - \theta_1^{-1}\right]^{-1} \quad (16.34)$$

Hence

$$E(Z - \zeta | Z > \zeta, \theta) = \theta_0^{-1} - \theta_1^{-1} E(X | X > x(\zeta), \theta) - \ln\left(\frac{\theta_1}{\theta_0}\right) - \zeta$$

$$= \theta_0^{-1} - \theta_1^{-1}(x(\zeta) + \theta) - \ln\left(\frac{\theta_1}{\theta_0}\right) - \zeta$$

Inserting the value of $x(\zeta)$ from (16.34), we find

$$E(Z - \zeta | Z > \zeta, \theta) = \theta(\theta_0^{-1} - \theta_1^{-1})$$

which is *independent* of ζ. A similar result is obtained for $E(Z+\zeta|Z<-\zeta,\theta)$.
This is a very special case, but it means that we can allow exactly for the overshoot in this case, insofar as it affects the numerator of the right side of (16.22).

16.8 COMPARISON WITH TESTS USING SAMPLES OF FIXED SIZE

Since a primary aim of sequential procedures is to reduce (on the average) the size of sample necessary, it is of some importance to know the relative sizes of sample needed (that is, average sample number), using sequential (in particular, s.p.r.t.) procedures and fixed size sample procedures. In order to make the procedures comparable we compare s.p.r.t.'s $S(H_0,\alpha_0; H_1,\alpha_1)$ with fixed size sample procedures with the error probabilities

$$\Pr[\text{accept } H_{1-i}|H_i]=\alpha_i \qquad (i=0,1)$$

It should be noted that we are not comparing procedures with the *same* error probabilities, since α_0,α_1 are only the *nominal* values of these probabilities for $S(H_0,\alpha_0; H_1,\alpha_1)$. As we have seen in the last section (inequality (16.7)] the actual error probabilities $\alpha_0',\ \alpha_1'$ are usually rather smaller than α_0,α_1, so the comparison will be rather unfair to the sequential procedure.

We concentrate our attention on the problem of discrimination between the values θ_0,θ_1 for the mean of a normal population, the standard deviation, σ_0, being supposed known. This special case is of interest in itself, but also, as we show below, the results of comparison of sample sizes for this case may be expected to apply, at any rate in broad outline, to many other cases.

We suppose that the sequence of observations can be represented by random variables X_1,X_2,X_3, and so on, which are mutually independent, and that the probability density function of X_i is

$$p_{X_i}(x_i|\theta)=\left(\sqrt{2\pi}\ \sigma_0\right)^{-1}\exp\left[-\frac{1}{2}\left(\frac{x_i-\theta}{\sigma_0}\right)^2\right]$$

for all x_i. The likelihood function for the first m observations is

$$p(x_1,\ldots,x_m|\theta)=\prod_{i=1}^{m} p(x_i|\theta)$$

and the likelihood ratio used in the s.p.r.t. is

$$\frac{p(x_1,\ldots,x_m|\theta_1)}{p(x_1,\ldots,x_m|\theta_0)}=\exp\left\{\frac{1}{2\sigma_0^2}\sum_{i=1}^{m}\left[(x_i-\theta_0)^2-(x_i-\theta_1)^2\right]\right\}$$

$$=\exp\left\{\frac{\theta_1-\theta_0}{\sigma_0^2}\left[\sum_{i=1}^{m} x_i-\frac{m}{2}(\theta_0+\theta_1)\right]\right\} \qquad (16.35)$$

If $\theta_1 > \theta_0$, the continuation region is

$$\left(\frac{\sigma_0^2}{\theta_1-\theta_0}\right)\ln\left(\frac{\alpha_1}{1-\alpha_0}\right) + \frac{m}{2}(\theta_0+\theta_1) < \sum_{i=1}^{m} X_i$$

$$< \left(\frac{\sigma_0^2}{\theta_1-\theta_0}\right)\ln\left(\frac{1-\alpha_1}{\alpha_0}\right) + \frac{m}{2}(\theta_0+\theta_1) \quad (16.36)$$

This can be represented by a diagram similar to Figure 16.1, in which $\sum_{i=1}^{m} X_i$ is plotted against m. The slope of the lines is $\frac{1}{2}(\theta_0+\theta_1)$.

As a matter of practical convenience we may plot $\sum_{i=1}^{m} Y_i$ against m, where $Y_i = X_i - \frac{1}{2}(\theta_0+\theta_1)$. Then the limiting lines become horizontal with

$$\sum_{i=1}^{m} Y_i = \begin{cases} -(\theta_1-\theta_0)^{-1}\sigma_0^2 \ln\left(\dfrac{1-\alpha_0}{\alpha_1}\right) & \text{(lower limit)} \\[2em] (\theta_1-\theta_0)^{-1}\sigma_0^2 \ln\left(\dfrac{1-\alpha_1}{\alpha_0}\right) & \text{(upper limit)} \end{cases}$$

To evaluate h [solving (16.20)] we need to solve

$$\frac{1}{\sqrt{(2\pi)}\,\sigma_0} \int_{-\infty}^{\infty} \exp\left[\frac{h}{2\sigma_0^2}\left[(x-\theta_0)^2-(x-\theta_1)^2\right]-\frac{1}{2\sigma_0^2}(x-\theta)^2\right]dx = 1$$

$$(16.37)$$

The left side of this equation can be arranged to give

$$\frac{1}{\sqrt{(2\pi)}\,\sigma_0} \int_{-\infty}^{\infty} \exp\left[-\frac{1}{2\sigma_0^2}\left[x^2 - 2x\left[h(\theta_1-\theta_0)+\theta\right] + h(\theta_1^2-\theta_0^2)+\theta^2\right]\right]dx$$

$$= \exp\left[\frac{1}{2\sigma_0^2}\left\{\left[h(\theta_1-\theta_0)+\theta\right]^2 - h(\theta_1^2-\theta_0^2)-\theta^2\right\}\right]$$

$$= \exp\left[\frac{1}{2\sigma_0^2}h(\theta_1-\theta_0)\left[h(\theta_1-\theta_0)-\theta_1-\theta_0+2\theta\right]\right]$$

This is equal to 1 if $h=0$ (of course), and if

$$h = \frac{\theta_1+\theta_0-2\theta}{\theta_1-\theta_0} \quad (16.38)$$

Hence from (16.16),

$$\Pr[\text{accept } \theta = \theta_1 | \theta]$$

$$\doteq \frac{1-\left(\dfrac{\alpha_1}{1-\alpha_0}\right)^{(\theta_1+\theta_0-2\theta)/(\theta_1-\theta_0)}}{\left(\dfrac{1-\alpha_1}{\alpha_0}\right)^{(\theta_1+\theta_0-2\theta)/(\theta_1-\theta_0)} - \left(\dfrac{\alpha_1}{1-\alpha_0}\right)^{(\theta_1+\theta_0-2\theta)/(\theta_1-\theta_0)}} \quad (16.39)$$

In order to evaluate the approximate formula (16.22) for the average sample number, we also need to evaluate the expected value of

$$\ln\left[\frac{p_X(X|\theta_1)}{p_X(X|\theta_0)}\right] = \left(\frac{\theta_1 - \theta_0}{\sigma_0^2}\right)\left[X - \tfrac{1}{2}(\theta_1 + \theta_0)\right]$$

This is evidently equal to

$$\left(\frac{\theta_1 - \theta_0}{\sigma_0^2}\right)\cdot\left[\theta - \tfrac{1}{2}(\theta_1 + \theta_0)\right]$$

or (in this case)

$$= \frac{h\left[(\theta_1 - \theta_0)/\sigma_0\right]^2}{2}$$

So the average sample number (a.s.n.) is approximately

$$\frac{\left[\left(\frac{1-\alpha_1}{\theta_0}\right)^h - 1\right]\ln\left(\frac{\alpha_1}{1-\alpha_0}\right) + \left[1 - \left(\frac{\alpha_1}{1-\alpha_0}\right)^h\right]\ln\left(\frac{1-\alpha_1}{\alpha_0}\right)}{-\tfrac{1}{2}h\left[(\theta_1 - \theta_0)/\sigma_0\right]^2}$$

$$\times \frac{1}{\left(\frac{1-\alpha_1}{\alpha_0}\right)^h - \left(\frac{\alpha_1}{1-\alpha_0}\right)^h}$$

where $h = (\theta_1 + \theta_0 - 2\theta)/(\theta_1 - \theta_0)$.

In Table 16.4 some typical values of the a.s.n. are given, (for $\theta = \theta_1$ and $\theta = \theta_0$) and compared with the least sample sizes (fixed) for which $\Pr[\text{accept } H_{1-i}|H_i] = \alpha_i$, $(i = 0, 1)$, can be satisfied.

TABLE 16.4

AVERAGE SAMPLE NUMBERS (A.S.N.) FOR S.P.R.T.
AND FIXED SAMPLE SIZES

		$\alpha_0 = \alpha_1 = 0.05$				$\alpha_0 = \alpha_1 = 0.01$			
		$\dfrac{\theta - \theta_0}{\theta_1 - \theta_0}$					$\dfrac{\theta - \theta_0}{\theta_1 - \theta_0}$				
		0	0.5	1.0	1.5	Fixed Size	0	0.5	1.0	1.5	Fixed Size
		$(\theta = \theta_0)$		$(\theta = \theta_1)$			$(\theta = \theta_0)$		$(\theta = \theta_1)$		
$\dfrac{\theta_1 - \theta_0}{\sigma_0}$	0.5	21.2	34.7	21.2	11.7	43.3	36.0	84.5	36.0	18.4	86.6
	1.0	5.3	8.7	5.3	(2.9)	10.8	9.0	21.1	9.0	4.6	21.6
	2.0	(1.3)	(2.2)	(1.3)	(0.7)	2.7	(2.3)	5.3	(2.3)	(1.1)	5.4

The sample sizes in parentheses are so small that little reliance can be placed on the accuracy of the approximate formulas. The values are given for purposes of formal comparison.

It will be seen that a saving of about $\frac{1}{3}$ to $\frac{2}{3}$ is effected by the sequential procedure, when $\theta = \theta_0$ or $\theta = \theta_1$.

For values of θ between θ_0 and θ_1 the position is not quite so clear-cut. Indeed it can happen that the a.s.n. for the sequential procedure *exceeds* the fixed sample size.

If $\alpha_0 = \alpha_1 = \alpha$, for instance, the maximum average sample number corresponds to $h = 0$; that is, $\theta = \frac{1}{2}(\theta_0 + \theta_1)$, and is approximately

$$\frac{\left[\ln\left(\frac{1-\alpha}{\alpha} \right) \right]^2}{\left(\frac{\theta_1 - \theta_0}{\sigma_0} \right)^2} \tag{16.40}$$

For a fixed size sample to satisfy the conditions $\Pr[\text{reject } H_i | H_i] \leq \alpha_i$ $(i = 0, 1)$ the sample size n must satisfy the condition

$$\frac{\theta_1 - \theta_0}{\left(\sigma_0 / \sqrt{n} \right)} > 2\lambda_\alpha \tag{16.41}$$

where $\Phi(\lambda_\alpha) = 1 - \alpha$. The least possible value of n is thus approximately $4\lambda_\alpha^2 [(\theta_1 - \theta_0)/\sigma_0]^{-2}$.

Comparing this with (16.40), we see that the average sample number of the s.p.r.t. will exceed the corresponding fixed sample size if

$$\ln\left(\frac{1-\alpha}{\alpha} \right) > 2\lambda_\alpha \tag{16.42}$$

Table 16.5 shows that for $\alpha < 0.008$ this is, in fact, the case.

TABLE 16.5

$\ln \dfrac{1-\alpha}{\alpha}$ VERSUS $2\lambda_\alpha$

α	$\ln\left(\dfrac{1-\alpha}{\alpha} \right)$	$2\lambda_\alpha$
0.2	1.386	1.683
0.15	1.735	2.073
0.1	2.197	2.563
0.05	2.944	3.290
0.01	4.595	4.652
0.009	4.701	4.731
0.008	4.820	4.818
0.007	4.955	4.915

So although the s.p.r.t. gives savings of $\frac{1}{3}$ to $\frac{2}{3}$ when H_0 or H_1 is true, it can lead to an *average* sample size exceeding the corresponding fixed sample size under certain conditions.

Even when H_0 or H_1 is true it is always possible that a particular sequence may need considerably more observations than the fixed sample size before a conclusion is reached. The occasional occurrence of very long sequences of observations before a decision is reached can be very inconvenient. To reduce this inconvenience *truncated* sequential procedures have been introduced. These are described in Section 16.10.

The preceding comparisons need to be interpreted with some care, taking into account the fact that all the formulas used are approximate (see Section 16.7). However, it may be noted that the inaccuracies in error probabilities and average sample number tend to cancel each other.

Example 16.6   Suppose that

$$p_{X_j}(x_j|H_1) = \theta_1^{-1} e^{-x_j/\theta_1} \qquad (0 < x_j)$$

$$p_{X_j}(x_j|H_0) = \theta_0^{-1} e^{-x_j/\theta_0} \qquad (0 < x_j)$$

where $\theta_1 > \theta_0$.
The fixed size sample procedure will then be

Reject H_0 if $\prod_{j=1}^{m} \left[(\theta_0/\theta_1) e^{-X_j(\theta_1^{-1} - \theta_0^{-1})} \right] > K_0$ where K_0 is a constant.

The inequality can be rearranged to read

$$\sum_{j=1}^{m} X_j > K_1$$

Since $\sum_{j=1}^{m} X_j$ is distributed as $\frac{1}{2}\theta_0 \cdot (\chi^2$ with $2m$ degrees of freedom) when H_0 is true, K_1 is chosen to be $\frac{1}{2}\theta_0 \cdot$ (upper $100\alpha_0$ percent point of χ^2 distribution with $2m$ degrees of freedom) in order to make the significance level of the test equal to α_0.

In order to make the probability of rejecting H_0 to be at least $100(1 - \alpha_1)$ percent when H_1 is true, we require

$$\frac{\theta_1}{2} \chi^2_{2m, \alpha_1} > \frac{\theta_0}{2} \chi^2_{2m, 1-\alpha_0}$$

that is,

$$\frac{\chi^2_{2m, 1-\alpha_0}}{\chi^2_{2m, \alpha_1}} < \frac{\theta_1}{\theta_0} \tag{16.43}$$

If we take $\alpha_0 = \alpha_1 = 0.05$; $\theta_1/\theta_0 = 2$, we find that the least possible value of m is 23.

Now consider the corresponding s.p.r.t. $S(\theta_0, 0.05; 2\theta_0, 0.05)$. The test procedure is

Accept $H_0(\theta = \theta_0)$ if

$$\sum_{j=1}^{m} X_j < 2\theta_0 \left[\ln\left(\frac{0.05}{0.95} \right) + m \ln 2 \right]$$

Accept $H_1(\theta = 2\theta_0)$ if

$$\sum_{j=1}^{m} X_j > 2\theta_0 \left[\ln\left(\frac{0.95}{0.05} \right) + m \ln 2 \right]$$

Otherwise, take another observation.
The logarithm of the likelihood ratio $[p_X(X|H_1)/p_X(X|H_0)]$ is

$$Z = \frac{X}{2\theta_0} - \ln 2$$

Hence

$$E(Z|\theta_0) = \tfrac{1}{2} - \ln 2; \quad E(Z|2\theta_0) = 1 - \ln 2$$

Therefore, the average sample number is approximately

$$\frac{0.9\,(\ln 19)}{\ln 2 - \tfrac{1}{2}} = 13.7 \quad \text{when } \theta = \theta_0$$

$$\frac{0.9\,(\ln 19)}{1 - \ln 2} = 8.6 \quad \text{when } \theta = 2\theta_0$$

The s.p.r.t. thus gives a saving of about 40 percent when $\theta = \theta_0$, $62\tfrac{1}{2}$ percent when $\theta = 2\theta_0$.

To complete this example we consider the case $h = 0$. If $\alpha_0 = \alpha_1 = \alpha$, the probability of "accepting $\theta = \theta_0$" is $\tfrac{1}{2}$ (approx.) when $h = 0$. Hence the expected value of the square of the logarithm of the likelihood ratio is $(1/2) \times 2[\ln((1 - \alpha)/\alpha)]^2 = [\ln((1 - \alpha)/\alpha)]^2$. The value of θ for which $h = 0$ will make the expected value of Z equal to 0; hence it is $2\theta_0 \ln 2$. The expected value of Z^2 for this value of θ is $(\theta_0^{-1}/2)^2(2\theta_0 \ln 2)^2 = (\ln 2)^2$. Hence the a.s.n. is approximately

$$\left[\frac{\ln\left(\dfrac{1 - \alpha}{\alpha} \right)}{\ln 2} \right]^2$$

Putting $\alpha = 0.05$, we get the value 18.0.

16.9 A GENERAL LIMITING CASE

Now consider the s.p.r.t. discriminating between 2 simple hypotheses H_0, H_1 about the distribution of each of a sequence of independent, identically distributed variables $X_1, X_2, \ldots$. The likelihood ratio is then

$$\prod_{j=1}^{m} \left[\frac{l(X_j|H_1)}{l(X_j|H_0)} \right]$$

and its logarithm is $\Sigma_{j=1}^{m} Z_j$, where

$$Z_j = \ln \left[l(X_j|H_1)/l(X_j|H_0) \right]$$

The Z_j's are independent and identically distributed. If we suppose that, when a hypothesis H is true, the expected value $E(Z|H)$ and variance $\text{var}(Z|H)$ of each Z_j are finite, then, except when m is small, we would expect $\Sigma_{j=1}^{m} Z_j$ to be approximately normally distributed. Since the fixed size sample procedure is also based on the likelihood ratio—or, equivalently, on the sum $\Sigma_{j=1}^{m} Z_j$—we would expect the same kind of comparison between the average sample number of s.p.r.t.'s with the fixed size sample as was obtained in the preceding section. It will be remembered that, in the case considered in that section, each Z_j was (exactly) normally distributed.

It follows that the relative advantage of the s.p.r.t. with respect to average size of sample, which was indicated by the results of Section 16.8, might be expected in very many cases, provided that

   (*i*) the successive observations can be represented by random variables which are independent and identically distributed,
   (*ii*) the average size of sample is not too small, and
   (*iii*) the expected values and variances of the Z's are related in approximately the same way as in Section 16.8.

These conditions are, of course, not always satisfied. In Example 16.6, for instance, as we have already seen,

$$Z = \tfrac{1}{2}\theta_0^{-1}X - \ln 2$$

so that if

$$p_X(x) = \theta^{-1}e^{-x/\theta} \qquad (0 < x; \, 0 < \theta)$$

then

$$\text{var}(Z) = \tfrac{1}{4}\theta_0^{-2}\text{var}(X) = \tfrac{1}{4}\left(\frac{\theta}{\theta_0} \right)^2$$

which depends on θ, contradicting part of condition (*iii*).

16.10   MODIFIED PROCEDURES

We have already referred, in Section 16.8, to the possibility of modifying s.p.r.t. procedures by *truncation*. The distribution of sample size for a s.p.r.t. is very skew—a typical case is shown in [16.3]—and it is possible to get very large samples occasionally. To avoid the inconvenience of such large samples a modified s.p.r.t., which concludes at a preassigned sample size, N, for instance (whether or not the formal s.p.r.t. would indicate further sampling) is used. Decisions are made in accordance with standard rules until the limiting sample size, N, is reached. At this stage some more or less arbitrary rule is used to make a decision; commonly that hypothesis is accepted which has the larger likelihood. Provided N is chosen so that this arbitrary rule is used only in a fairly small (for example, 5 percent) proportion of cases it can reasonable be expected that the modified procedure will have properties very similar to those of the strict s.p.r.t. test. Since these are, in any case, known only approximately, the further approximation introduced by a good truncation procedure should not be very serious.

In order to select a "good" truncation procedure, a useful working rule is to take N between 3 and 4 times the average sample number when H_0 or H_1 is true. For safety the larger of the 2 average sample numbers should be used.

Another practically useful form of modification is *grouping*. The standard s.p.r.t. procedure necessitates stopping after every single observation and deciding whether it is necessary to continue sampling. This is often a considerable waste of time. To reduce this, observations may be taken in groups. Usually it is easiest to take groups of constant size—of say, 3, 4, or 5 observations each. In some cases sampling may be commenced by taking a group of observations, thereafter continuing with smaller groups, or single observations. If the average sample number is substantially larger than the size of the first group of observations, this modification will alter the properties of the standard s.p.r.t. very little.

Formally, indeed, the use of groups of observations makes *no* difference. In the formula for average sample number, for example, the denominator, $E(Z|H)$ is m times as big (if groups of m observations are used) while the numerator is unchanged. So the average number of *groups* of *size m* is $(1/m)$ times the average number of single observations—in other words, the average number of observations is the same. This simply emphasizes the fact that these formulas are *approximate*, for it is obvious that average sample number will, in reality, increase if groups of observations are used —because decisions that would have been reached at the earlier observations of a group will be delayed at least until the end of the group. But it does appear that, *provided* the expected number of *groups* is not too small, grouping has little effect on the properties of a sequential procedure.

Example 16.7 Suppose $X_1, X_2, X_3, X_4, \ldots$ is a sequence of independent random variables taken cyclically from 3 normal populations with common mean and standard deviations $\sigma_1, \sigma_2, \sigma_3$, respectively. (The values of σ_1, σ_2, and σ_3 are assumed known.) This model might apply, for example, to a situation where samples are taken in turn from the output of 3 machines. Construct a sequential probability ratio test discriminating between stated values θ_0 and $\theta_1 (> \theta_0)$ for θ, and obtain approximations to the operating characteristic of this test, and to its average sample number.

We have

$$
p_{X_{3+j}}(x_{3r+j}|\theta) = \left(\sqrt{2\pi}\,\sigma_j\right)^{-1} \exp\left[-\tfrac{1}{2}\sigma_j^{-2}(x_{3r+j} - \theta)^2 \right] \quad (j = 1, 2, 3)
$$

and

$$
p_{X_{3r+1}, X_{3r+2}, X_{3r+3}}(x_{3r+1}, x_{3r+2}, x_{3r+3})
$$

$$
= (2\pi)^{-3/2} \left(\prod_{j=1}^{3} \sigma_j \right)^{-1} \exp\left[-\tfrac{1}{2} \sum_{j=1}^{3} \sigma_j^{-2}(x_{3r+j} - \theta)^2 \right]
$$

Hence the likelihood ratio, when the sample size is $m = 3R + i$, is

$$
\frac{\exp\left[-\tfrac{1}{2} \sum_{r=0}^{R-1} \sum_{j=1}^{3} \sigma_j^{-2}(X_{3r+j} - \theta_1)^2 - \tfrac{1}{2} \sum_{j=1}^{i} \sigma_j^{-2}(X_{3R+j} - \theta_1)^2 \right]}{\exp\left[-\tfrac{1}{2} \sum_{r=0}^{R-1} \sum_{j=1}^{3} \sigma_j^{-2}(X_{3r+j} - \theta_0)^2 - \tfrac{1}{2} \sum_{j=1}^{i} \sigma_j^{-2}(X_{3R+j} - \theta_0)^2 \right]}
$$

Taking logarithms we obtain the continuation inequalities for the s.p.r.t. in the form

$$
(\theta_1 - \theta_0)^{-1} \log\left(\frac{\alpha_1}{1 - \alpha_0} \right) + \tfrac{1}{2}(\theta_1 + \theta_0)\left[R \sum_{j=1}^{3} \sigma_j^{-2} + \sum_{j=1}^{i} \sigma_j^{-2} \right]
$$

$$
< \sum_{j=1}^{3} \sigma_j^{-2} \sum_{r=0}^{R-1} X_{3r+j} + \sum_{j=1}^{i} \sigma_j^{-2} X_{3R+j}
$$

$$
< (\theta_1 - \theta_0)^{-1} \log\left(\frac{1 - \alpha_1}{\alpha_0} \right) + \tfrac{1}{2}(\theta_1 + \theta_0)\left[R \sum_{j=1}^{3} \sigma_j^{-2} + \sum_{j=1}^{i} \sigma_j^{-2} \right]
$$

To find approximations to the operating characteristic and the average sample number we treat each set of 3 successive variables $(X_{3r+1}, X_{3r+2}, X_{3r+3})$ as a single "observation," which we may denote by $[y_{r+1}]$. The logarithm of the likelihood ratio for such an "observation" is

$$Z_{r+1} = (\theta_1 - \theta_0)\left[\sum_{j=1}^{3} \sigma_j^{-2} X_{3r+j} - \tfrac{1}{2}(\theta_1 + \theta_0)\sum_{j=1}^{3}\sigma_j^{-2}\right]$$

and so

$$E(Z_{r+1}|\theta) = (\theta_1 - \theta_0)\left[\theta - \tfrac{1}{2}(\theta_1 + \theta_0)\right]\sum_{j=1}^{3}\sigma_j^{-2}$$

From (16.38) we see that

$$h = (\theta_1 + \theta_0 - 2\theta)/(\theta_1 - \theta_0)$$

makes

$$E\left[\left(\frac{p_X(X|\theta_1)}{p_X(X|\theta_0)}\right)^h\Big|\theta\right] = 1$$

for *any* value of σ. So this value of h also satisfies

$$E\left[\left(\frac{p_{Y_{r+1}}([Y_{r+1}]|\theta_1)}{p_{Y_{r+1}}([Y_{r+1}]|\theta_0)}\right)^h\Big|\theta\right] = 1$$

Hence the approximate operating characteristic is again that implied by (16.39). For the average number of groups of 3 in the sample, (16.40) must be modified by replacing $-h[(\theta_1 - \theta_0)/\sigma_0]^2$ in the denominator by

$$-h\left[(\theta_1 - \theta_0)^2\sum_{j=1}^{3}\sigma_j^{-2}\right]$$

To obtain the average sample number, the resulting formula must, of course, be multiplied by 3.

The general remarks on formulas for grouped sequential samples apply to these results. That is, they should give good approximation if the expected number of *groups* (in this case, of size 3) is not too small.

S.p.r.t.'s can be used in the same way as ordinary significance tests with H_0 as the null hypothesis, and α_0 as the (nominal) significance level. In order to construct the test it is necessary to choose (more or less arbitrarily) some particular alternative H_1 and corresponding (nominal) power $(1 - \alpha_1)$. However, the test $S(H_0, \alpha_0; H_1, \alpha_1)$ can be regarded as a test of H_0 with respect to a set of alternatives including, but not limited to, H_1.

Such a test is usually of one-sided type (using this term as in Chapter 8). For example, the test [which we denote by $S(\theta_0, \alpha_0; \theta_1, \alpha_1,)$] with continuation region defined by (16.36) (with $\theta_1 > \theta_0$) gives a useful test of the hypothesis $\theta = \theta_0$ with respect to alternatives $\theta > \theta_0$, but not with respect to $\theta < \theta_0$.

In order to get a two-sided sequential test we consider two alternatives to H_0, namely,

$$H_1 : \theta = \theta_1 = \theta_0 + \delta_1 \quad \text{and} \quad H_{-1} : \theta = \theta_{-1} = \theta_0 - \delta_{-1}$$

with $\delta_1, \delta_{-1} > 0$. Very often we take $\delta_1 = \delta_{-1}$, though this is not essential.

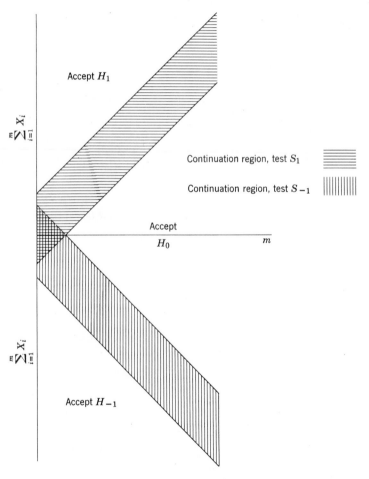

Figure 16.3 Choice among 3 hypotheses.

We then construct two s.p.r.t.'s

$$S_1 \equiv S(\theta_0, \alpha'_0; \theta_1, \alpha_1)$$

and

$$S_{-1} \equiv S(\theta_0, \alpha''_0; \theta_{-1}, \alpha_{-1})$$

and apply them concurrently to the sequence of observed values of $X_1, X_2, \ldots$. We accept H_0 if and only if H_0 is accepted by both S_1 and S_{-1}. Usually it is possible to ignore the possibility that S_1 will lead to acceptance of H_1 and S_{-1} to acceptance of H_{-1} for the same sequence of data. So the additional rule to accept H_j if S_j accepts H_j ($j = -1, 1$) gives a complete decision scheme. This is represented graphically in Figure 16.3. This figure is symmetrical about $\Sigma X_i = \theta_0$, corresponding to the case $\delta_1 = \delta_{-1}, \alpha'_0 = \alpha''_0, \alpha_1 = \alpha_{-1}$ (which is most frequently used). Note that H_0 is accepted if *both* broken lines are crossed, as well as if the "Accept H''_0" region is entered.

Power of the test procedure with respect to H_j is approximately $(1 - \alpha_j)$ ($j = -1, 1$), and the significance level (probability of rejecting H_0 when H_0 is valid) is approximately $(\alpha'_0 + \alpha''_0)$. So in order to get a (nominal) significance level, α_0, we have to choose α'_0, α''_0 to make $\alpha'_0 + \alpha''_0 = \alpha_0$.

The method can be extended to discrimination among more than 3 hypotheses, but is rarely so applied.

*16.11 DISCRIMINATION BETWEEN COMPOSITE HYPOTHESES

The construction of a s.p.r.t. is based on the likelihood ratio

$$\frac{l(x_1, \ldots, x_m | H_1)}{l(x_1, \ldots, x_m | H_0)}$$

If H_0 and H_1 are not simple hypotheses, it will, in general, not be possible to evaluate this likelihood ratio because $l(x_1, \ldots, x_m | H)$ will not be determined exactly by the hypothesis H_i. In a number of cases, however, it is possible to circumvent this difficulty by replacing the original problem by a slightly different one, even though this sometimes involves the loss of a certain amount of information.

As a first example, consider the problem of discriminating between values σ_0 and $\sigma_1 (> \sigma_0)$ for the standard deviation of a normal population with unknown mean. Because the mean (θ_1, say) is not known, the hypotheses $H_0(\sigma = \sigma_0)$ and $H_1(\sigma = \sigma_1)$ are not simple, but composite. The likelihood

$$l(x_1, x_2, \ldots, x_m | H_j) = (2\pi)^{-m/2} \sigma_i^{-m} \exp\left[-\tfrac{1}{2}\sigma_i^{-2} \sum_{j=1}^{m} (x_j - \theta)^2 \right]$$

depends on the unknown parameter θ. However, if we consider, instead of the sequence $X_1, X_2, \ldots$, the sequence $Y_1, Y_2, \ldots$, where

$$Y_1 = \frac{X_2 - X_1}{\sqrt{2}}$$

$$Y_2 = \frac{2X_3 - X_1 - X_2}{\sqrt{2 \cdot 3}}$$

$$Y_3 = \frac{3X_4 - X_1 - X_2 - X_3}{\sqrt{3 \cdot 4}}$$

and so on, then

$$l(y_1, y_2, \ldots, y_{m-1} | H_i) = (2\pi)^{-(m-1)/2} \sigma_i^{-(m-1)} \exp\left[-\tfrac{1}{2}\sigma_i^{-2} \sum_{j=1}^{m-1} y_j^2 \right]$$

and, for the Y's, H_0 and H_1 are simple hypotheses. Therefore, an s.p.r.t. comparing H_0 and H_1 can be constructed in the usual way.

Next, suppose we do not know the value (say, σ_1) of the standard deviation, and wish to compare hypotheses H'_0, H'_1 specifying that the value of the mean is $\theta = \theta_0, \theta = \theta_1$, respectively. Again, the likelihood cannot be evaluated explicitly. This time we have to lose some information to reduce H'_0 to a simple hypothesis, by replacing $X_1, X_2, \ldots$ by $Y'_1, Y'_2, \ldots$, where $Y'_j = 1$ if $X_j \geqslant \theta_0$, $Y'_j = 0$ if $X_j < \theta_0$. Then if H'_0 is true, $Y'_1, Y'_2, \ldots$ are binomial variables with $p = \tfrac{1}{2}$. There is the further difficulty that when H'_1 is true

$$p = 1 - \Phi\left(\frac{\theta_1 - \theta_0}{\sigma} \right)$$

and this depends on σ, which is unknown. So we must "guess" a value $p = p_1$ for our alternative hypothesis H'_1. A s.p.r.t. comparing H'_0 and H'_1 will have (approximately) the specified "significance level" ($\alpha_0 = $ probability of rejecting H_0 when it is true). Its power with respect to H'_1 will depend on H''_1 and the value chosen for α_1.

An alternative approach to this problem, which appears to lose less information, is to base the test on the sequence $T_2, T_3, \ldots$, where

$$T_2 = \frac{\tfrac{1}{2}(X_1 + X_2) - \theta_0}{S_2/\sqrt{2}} \; ; \quad T_3 = \frac{\tfrac{1}{3}(X_1 + X_2 + X_3) - \theta_0}{S_3/\sqrt{3}}$$

and generally

$$T_m = \frac{m^{-1}(X_1 + X_2 + \cdots + X_m) - \theta_0}{S_m/\sqrt{m}}$$

where

$$S_m^2 = (m-1)^{-1} \sum_{j=1}^{m} \left[X_j - m^{-1}(X_2 + \cdots + X_m) \right]^2$$

It can be shown (see [16.2]) that

$$\frac{l(T_2, T_3, \ldots, T_m | H_1'')}{l(T_2, T_3, \ldots, T_m | H_0')} = \frac{l(T_m | H_1'')}{l(T_m | H_0')}$$

and so the likelihood ratio can be evaluated as the ratio of a noncentral t_{m-1} distribution with noncentrality parameter $\sqrt{m} \, (\theta_1 - \theta_0)/\sigma$ to a central t_{m-1} distribution. The continuation region can be expressed in the form of an interval of values for T_m. Owing to the complexity of the expression for the likelihood ratio it is necessary to use tables to determine the limits of the continuation interval.

Suitable tables are given by K. J. Arnold, ed., in *Tables to Facilitate Sequential t-Tests*, (National Bureau of Standards, Applied Mathematics Series No. 7), U.S. Department of Commerce, Washington, D.C., 1951; and also in [15.5].

It will be noted that this method also required that the hypothesis H_1' be replaced by another, H_1'', defining the value of the ratio $(\theta_1 - \theta_0)/\sigma$.

Similar methods are available for constructing sequential tests discriminating between 2 specified values of the correlation coefficient, and for applying sequential tests in the analysis of variance. In the first case a good approximate test is obtained by using Fisher's z' transformation (see Section 12.3.4). In the second case the position may be loosely summarized as follows:

At each stage in the experiment carry out the standard analysis of variance test. Calculate the appropriate likelihood ratio—this will usually be the ratio of a noncentral F probability density function to a central or noncentral F. Use the standard limits $[\alpha_1/(1 - \alpha_0)$ and $(1 - \alpha_1)/\alpha_0]$ for the likelihood ratio to construct the continuation region.

Suitable tables are given by W. D. Ray in "Sequential Analysis Applied to Certain Experimental Designs in the Analysis of Variance," *Biometrika*, **43** (1956).

It should be noted that this method can be extended to apply to a series of different experimental designs, provided the appropriate noncentrality parameters can be evaluated in each case. For the likelihood ratio of the combined experiments is the product of the likelihood ratios for each successive experiment (assuming them independent) and the s.p.r.t. procedure is simply based on the continuation region

$$\frac{\alpha_1}{1 - \alpha_0} < \text{likelihood ratio} < \frac{1 - \alpha_1}{\alpha_0}$$

however the likelihood ratio is obtained.

Therefore, the results of series of randomized block, Latin square, and other forms of experiments can be regarded as a sequence of "observations." Although the observed mean square ratios do not in general have the same likelihood ratio function, this does not affect the applicability of the theory of the construction of s.p.r.t.'s based on such series of observations.

*16.12 OTHER SEQUENTIAL METHODS

It must not be supposed that sequential probability ratio tests exhaust the available sequential procedures. As we have already stated, sequential methods include *any* procedure in which the final pattern of data is *not predetermined* before starting the investigation. This is quite a wide definition. It includes double (or multiple) sampling inspection procedures such as those discussed in Section 10.6. In this section we briefly mention some of the methods included in its scope. For a fuller appreciation of the details, references given at the end of the chapter should be studied.

*16.12.1   Two-Stage Procedures

The standard confidence interval for the mean of a normal population with known standard deviation, σ, based on a random sample of n observations, $X_1, X_2, \ldots, X_n$, has limits

$$\bar{X} \pm \frac{\lambda_{\alpha/2}\sigma}{\sqrt{n}}$$

where

$$\Phi(\lambda_{\alpha/2}) = 1 - \frac{\alpha}{2}$$

and $(1 - \alpha)$ is the confidence coefficient. The length of this interval is $2\lambda_{\alpha/2}\sigma/\sqrt{n}$. If we desire the length to be less than a given value, L, say, we simply choose n so that

$$\frac{\lambda_{\alpha/2}\sigma}{\sqrt{n}} < L$$

that is,

$$n > \left(\frac{\lambda_{\alpha/2}\sigma}{L}\right)^2$$

However, if σ is not known, we have to use the confidence interval

$$\bar{X} \pm \frac{t_{n-1,1-\alpha/2}S}{\sqrt{n}}$$

where $S^2 = (n-1)^{-1}\sum_{i=1}^{n}(X_i - \bar{X})^2$ and $t_{n-1,1-\alpha/2}$ is the upper 50α percent point of the t distribution with $(n-1)$ degrees of freedom. The length of this interval is

$$\frac{2t_{n-1,1-\alpha/2}S}{\sqrt{n}}$$

and it cannot be predicted in advance, since S is not known until the observations have been obtained. How, then, can we construct a confidence interval with preassigned confidence coefficient $(1 - \alpha)$, which is of length not greater than $2L$?

This cannot be done with a single sample, but it can be done if 2 samples are taken. The standard deviation (σ) is estimated from the first sample, and this estimate is used in deciding *how big* the second sample should be. This is done in a natural, straightforward manner.

Denote the mean and variance of the first sample (of size n_1) by $\bar{X}_1, S_1^2$, respectively. Assuming the population distribution to be normal, $\bar{X}_1$ and S_1^2 are mutually independent. If a second random sample of size N_2 is taken, where N_2 depends on S_1^2, but not on $\bar{X}_1$, then the mean of the second sample $\bar{X}_2$ is independent of $\bar{X}_1$. Formally it is natural to construct a confidence interval [with confidence coefficient $(1-\alpha)$] with limits

$$\frac{n_1\bar{X}_1 + N_2\bar{X}_2}{n_1 + N_2} \pm \frac{t_{n_1-1, 1-\alpha/2}S_1}{\sqrt{n_1 + N_2}}$$

The length of this interval is $2t_{n_1-1, 1-\alpha/2}S_1/\sqrt{n_1 + N_2}$. To make this length less than or equal to $2L$ we must choose n_2 so that

$$t_{n_1-1, 1-\alpha/2}S_1/\sqrt{n_1 + N_2} \leqslant L$$

that is,

$$n_2 \geqslant \left(\frac{t_{n_1-1, 1-\alpha/2}}{L}\right)^2 S_1^2 - n_1 \qquad (16.44)$$

If n_1 is already greater than $(t_{n_1-1, 1-\alpha/2}/L)^2 \cdot S_1^2$, no further sample need be chosen.

It should be clearly understood that we have not *proved* that our procedure does indeed give us a confidence interval for the mean with confidence coefficient $(1-\alpha)$. The procedure is natural, and would certainly give the required confidence interval if $\bar{X}_2$ were independent of S_1^2. However, the distribution of $\bar{X}_2$ depends on N_2, which itself is a random variable depending on S_1^2. Nevertheless, it can be shown that the procedure described above is a good practical working rule.

It is interesting to consider the distribution of N_2, the size of the second sample. N_2 is, of course, a discrete random variable, taking values $0, 1, 2, \ldots$ and so on.

$$\Pr[N_2 = 0] = \Pr\left[S_1^2 \leqslant \left(\frac{L}{t_{n_1-1, 1-\alpha/2}}\right)^2 n_1\right]$$

$$= \Pr\left[\chi^2_{n_1-1} < n_1(n_1-1)\phi^2\right] \qquad (16.45)$$

where $\phi = L/(\sigma t_{n_1-1, 1-\alpha/2})$.
Generally, for $r > 0$

$$\Pr[N_2 = r] = \Pr\left[r - 1 < \phi^{-2}S_1^2 - n_1 \leqslant r\right]$$

This can be written

$$\Pr[N_2 = r] = \Pr\left[(n_1-1)(r+n_1-1)\phi^2 < \chi^2_{n_1-1} \leqslant (n_1-1)(r+n_1)\phi^2\right]$$

$$(16.46)$$

Therefore, the probability distribution of N_2 is given by integrals, over successive ranges of values, of the probability density function of χ^2 with $n_1 - 1$ degrees of freedom.

Approximately N_2 can be considered as a *continuous* variable distributed as $(n_1 - 1)^{-1}\phi^{-2}\chi^2_{n_1 - 1} - n_1$ for $\chi^2_{n_1 - 1} \geqslant n_1(n_1 - 1)\phi^2$, combined with a *discrete* probability [given by (16.45)] of taking the value 0. The expected value of N_2, using this approximation is

$$\left[(n_1 - 1)^{-1}\phi^{-2}E\left(\chi^2_{n_1 - 1} | \chi^2_{n_1 - 1} > n_1(n_1 - 1)\phi^2 \right) - n_1 \right]$$

$$\times \Pr\left[\chi^2_{n_1 - 1} > n_1(n_1 - 1)\phi^2 \right] \quad (16.47)$$

Since

$$E\left(\chi^2_{n_1 - 1} | \chi^2_{n_1 - 1} > x \right)\Pr\left[\chi^2_{n_1 - 1} > x \right]$$

$$= \frac{1}{2^{(n_1 - 1)/2}\Gamma\left[(n_1 - 1)/2 \right]} \int_x^\infty (x)x^{\frac{1}{2}(n_1 - 1) - 1}e^{-x/2}dx$$

$$= \frac{n_1 - 1}{2^{(n_1 + 1)/2}\Gamma\left[(n_1 + 1)/2 \right]} \int_x^\infty x^{\frac{1}{2}(n_1 + 1) - 1}e^{-x/2}dx$$

$$= (n_1 - 1)\Pr\left[\chi^2_{n_1 + 1} > x \right]$$

expression (16.47) can be written in the form

$$\phi^{-2}\Pr\left[\chi^2_{n_1 + 1} > n_1(n_1 - 1)\phi^2 \right] - n_1 \Pr\left[\chi^2_{n_1 - 1} > n_1(n_1 - 1)\phi^2 \right]$$

A rather better approximation is

$$\phi^{-2}\Pr\left[\chi^2_{n_1 + 1} > n_1(n_1 - 1)\phi^2 \right] - (n_1 - \tfrac{1}{2})\Pr\left[\chi^2_{n_1 - 1} > n_1(n_1 - 1)\phi^2 \right] \quad (16.48)$$

[Remember that $\phi = L/(\sigma t_{n_1 - 1, 1 - a/2})$]

Some approximate values of $E(N_2|n_1, \phi)$, calculated from (16.48), are given in Table 16.6. It will be noted that the expected value of the total sample size, $n_1 + E(N_2|n_1, \phi)$, considered as a function of n_1, has a very flat minimum. Hence, although it is not possible in practice to know the optimum value of n_1, since this depends on ϕ and so on σ (which is unknown), we can hope sometimes to guess a value that will give a total sample size (on average) not greatly in excess of the minimum possible. Evidently it is a reasonable precaution to take n_1 rather on the small side since $n_1 + E(N_2|n_1, \phi)$ increases with

$$|n_1 - \text{optimum } n_1|$$

more rapidly for n_1 greater than the optimum value.

TABLE 16.6

L/σ	n_1	ϕ	$E(N_2|n_1,\phi)$	$n_1 + E(N_2|\,n_1,\phi)$
0.5	8	0.211417	14.99	22.99
	15	0.233101	4.94	19.94
	23	0.241080	0.52	23.52
	30	0.244499	0.02	30.02
1.0	2	0.078703	160.00	162.00
	4	0.314268	6.93	10.93
	5	0.360231	3.78	8.78
	6	0.388954	2.12	8.12
1.5	2	0.118055	70.36	72.36
	3	0.348594	6.06	9.06
	4	0.471402	1.8	5.8

16.12.2   Stochastic Approximation

Suppose that we know that Y is a random variable with conditional expected value $E(Y|x)=g(x)$ where x can be measured exactly. Suppose further that the variance of Y (given x) is constant (and finite)—equal to σ^2, say.

The reader should recognize $g(x)$ as the *regression function* of Y on x (Section 12.2) and the condition in the second sentence as a condition of *homoscedasticity*.

We suppose that we can observe a succession of independent pairs of observations (x_i, Y_i) $(i=1,2,\ldots)$ on x and Y. The formula representing the model described above is then

$$Y_i = g(x_i) + Z_i \tag{16.49}$$

where the Z_i's are mutually independent random variables, with $E(Z_i)=0$; var(Z_i) $=\sigma^2$ for each i.

In Chapter 12 we have described methods of fitting the regression function $g(x)$. The application of these methods depended on a knowledge of the form of the function $g(x)$—a polynomial, or a trigonometrical series, and so on. Essentially we reduced the problem to that of estimating certain unknown parameters in the expression (of supposedly known form) for $g(x)$. Now we make less detailed assumptions about $g(x)$—simply that it is increasing, or continuous, for example—and we are not interested so much in estimating a complete expression for $g(x)$ as in estimating certain special values of x determined by $g(x)$, for example, the value of x for which $g(x)$ is a maximum, or for which $g(x)$ is equal to zero.

*This method was introduced by Robbins and Monro in "A Stochastic Approximation Method," *Annals of Mathematical Statistics*, **22** (1951).

Let us suppose, then, that $g(x)$ is a continuous increasing function of x, and we wish to find the value $x(\theta)$ satisfying the equation

$$g[x(\theta)] = \theta$$

where θ is a known constant.

Such a problem arises, for example, if $g(x)$ is the probability of a component failing when a stress x is applied to it and we wish to estimate the stress $x(\frac{1}{20})$ at which there will be 5 percent failures.

The method of *stochastic approximation* is a rule for choosing the values of X sequentially so that X_n approaches $x(\theta)$ as n increases. The rule is to calculate X_{i+1} from the formula

$$X_{i+1} = X_i + a_i(\theta - Y_i) \tag{16.50}$$

where $a_1, a_2, \ldots$ are positive numbers chosen so that $\Sigma_i a_i$ diverges and $\Sigma_i a_i^2$ converges.

If $g(x)$ satisfies certain rather natural conditions (in addition to being monotonic) then it can be shown that the mean square error after m stages, $E[\{X_m - x(\theta)\}^2 | X_0]$, tends to 0 as $m \to \infty$.

The first condition on the a_i's (Σa_i diverges) may be regarded as ensuring that it will be *possible* to reach the value $x = x(\theta)$ from *any* initial value x_1, even though $E(\theta - Y_i)$ is bounded. The second condition (convergence of $\Sigma_i a_i^2$) may similarly be regarded as helping to ensure that x_m does "settle down" to some value as m increases.

The simplest suitable sequence of values for a_i is obtained by taking a_i proportional to i^{-1}. However, it is not necessary to adhere strictly to the rule $a_i = c/i$, where c is a constant. Values of c varying between fixed limits can also be used, if convenient.

If a constant c is used the choice of the actual value should be made, bearing in mind the following considerations:

(a) the larger c is, the slower the convergence of the sequence $x_1, X_2, \ldots$;

(b) the smaller c is, the longer it will take to reach the neighborhood of $x(\theta)$ from a given initial value x_1;

and, of course, conversely. If there is reason to suppose that x_1 is a pretty good guess at $x(\theta)$ a rather small value of c would be appropriate. If there is some uncertainty on this point it may be decided to start with rather large values of c, decreasing to a fixed value later in the series of experiments.

The properties of the estimators $\{X_m\}$ depend on the form of the function $g(\cdot)$. If we can suppose that $a_j = cj^{-1}(c > 0)$, and $g(x)$ is a linear function of x, so that

$$Y = \alpha + \beta(X - \theta) + Z \quad \text{(with } \beta > 0)$$

where $E(Z) = 0$; $\text{var}(Z) = \sigma^2$, then it can be shown that, when the initial value of X is x_1,

$$E(X_m - \theta) = (x_1 - \theta) \prod_{j=1}^{m-1} (1 - \beta c j^{-1}) \tag{16.51}$$

(Note that x_1 is not here regarded as a random variable.) Also

$$\operatorname{var}(X_m) = \sigma^2 c^2 \sum_{j=1}^{m-1} \left[j^{-1} \prod_{k=j+1}^{m-1} (1 - \beta c k^{-1}) \right]^2 \tag{16.52}$$

(with the convention $\prod_{j=m}^{m-1}(\cdot) = 1$).

As $m \to \infty$, $E(X_m - \theta) \to 0$ and $m \operatorname{var}(X_m) \to (2\beta c - 1)^{-1} c^2 \sigma^2$ provided $\beta c > \frac{1}{2}$. These asymptotic results also apply under more general conditions with β replaced by $dg/dx|_{x=x(\theta_0)}$.

The theory given above applies only if the values of the a_i's are fixed *before* the series of experiments is started. However, in practice, no harm is done if the values of c are varied in the light of the experimental results in the initial stages of the investigation, provided the requirements of ultimate divergence of $\sum_i a_i$ and convergence of $\sum_i a_i^2$ are kept well in mind.

If a maximum (or minimum) value of $g(x)$ is sought by this method, the problem is posed in terms of finding a value of x such that $g'(x) = 0$. Two (or more) values of x are used to obtain an estimate of $g'(x)$ which is then used in the same way as Y was for the above analysis.

These methods can be extended to functions $g(X_1, \ldots, X_r)$ of several random variables. When used to find values $x_1, \ldots, x_r$ maximizing (or minimizing) $g(x_1, \ldots, x_r)$ they form the basis for applications in which optimal operating conditions are sought. Here $g(\cdot)$ might be a function representing "profit" or "yield" and the x's are variables representing various operating conditions. This is a very important field of application, but one that is beyond the scope of this book.

REFERENCES

1. Armitage, P., *Sequential Medical Trials*, Blackwell, Oxford, 1960.

2. Cox, D. R., "Sequential Tests for Composite Hypotheses," *Proceedings of the Cambridge Philosophical Society*, **48**, 1952.

3. Ghosh, B. K., *Sequential Tests of Statistical Hypotheses*, Addison-Wesley, Reading, Mass., 1970.

4. Johnson, N. L., "Sequential Analysis—A Survey," *Journal of the Royal Statistical Society, Series A*, **124** (1961).

5. Wald, A., *Sequential Analysis*, Wiley, New York, 1947.

6. Wasan, M. T., *Stochastic Approximation*, Cambridge University Press, 1969.

7. Wetherill, G. B., *Sequential Methods in Statistics*, Wiley, New York, 1966.

EXERCISES

1. Construct a sequential sampling plan for attribute sampling for $p_0 = 0.01$, $p_1 = 0.02$, $\alpha_0 = 0.01$, and $\alpha_1 = 0.02$.

2. Sketch the average sample number curve for the plan of Exercise 1.

3. Construct a sequential sampling plan to test the hypothesis H_0: Population mean equals 1.36 with $\alpha_0 = 0.05$ against an alternative hypothesis H_1: $\theta = 1.43$ with $\alpha_1 = 0.01$. Assume that the underlying distribution is normal with variance 0.50.

4. For Exercise 3 determine the OC curve and the a.s.n. curve.

5. Construct a sequential sampling plan to test the hypothesis H_0: $\theta = 4.1$ with $\alpha = 0.01$ against an alternative hypothesis H_1: $\theta = 3.6$ with $\alpha_1 = 0.01$. Assume that the underlying distribution is normal with mean θ and (known) standard deviation θ.

6. A population is known to contain N individuals. Construct a sequential sampling plan to discriminate between the hypotheses that the number of individuals in the population possessing a certain defect is equal to D_0, or $D_1 (> D_0)$.

7. Taking $N = 5, D_0 = 2, D_1 = 4$ as an example, discuss practical modifications of the plan for Exercise 6.

8. It is required that the average pH (θ) be less than 8.25. Assuming pH to be normally distributed with standard deviation σ, construct a sequential test for H_0: $\theta \leqslant 8.25$, H_1: $\theta > 8.31$, $\alpha_0 = 0.05$, $\alpha_1 = 0.05$, and $\sigma = 0.04$. Carry out the test using sample points as follows:

$$8.28, 8.25, 8.24, 8.29, 8.23, 8.25, 8.24, 8.26, 8.27, 8.24,$$

$$8.26, 8.28, 8.27, 8.24, 8.25, 8.26, 8.27, 8.24, 8.23, 8.25.$$

(Use as many as necessary.)

9. In the situation of Exercise 8, examine the effects of supposing the distribution of pH to be
   (a) a mixture in equal proportions of two normal distributions, with means $\theta - 0.01, \theta + 0.01$, and with equal variances such that the overall variance is 0.04, as in Exercise 8;
   (b) a double exponential, with density function proportional to $\exp(-|x - \theta|/\phi)$, and ϕ chosen so that the variance is 0.04.

10. Determine inequalities for acceptance and rejection for a sequential sampling plan if one wishes to reject lots with 2% defective at most only 1% of the time and accept lots with 3% defective at most 5% of the time. Sketch the lines represented by these equations, indicating the acceptance and rejection region.

11. In the sampling plan of Exercise 10, how large a sample is necessary to accept the lot if no defectives are found? Sketch the average sample number curve. Sketch the OC curve.

12. (a) Using the data of Exercise 6 of Chapter 10, apply a sequential probability ratio test of the hypothesis that the mean is 0.5000, with respect to two-sided alternatives.
   Assume normal variation, take $\sigma = 0.0002$, $\alpha = \beta = 0.01$, and alternative hypotheses H_1: $\mu = 0.5003$; H_{-1}: $\mu = 0.4997$
   (b) Answer (a), but using only the signs of the differences between the observed values and 0.5000.

13. Joe Zilch is a veteran inspector of electrical components. For a certain product it is known that his measurements are unbiased and have a standard deviation

of 0.02 ohms. He carries out routine tests of sets of 5 items each and presents each of the 5 measurements. This is repeated 3 times each day, at 9:00 A.M., 12 NOON, and 3:00 P.M. It is suspected that once each day (at the same time) his son Joe Zilch, Jr. is, in fact, making the measurements. It is believed that these measurements are still unbiased, but that the standard deviation of the son's measurements is greater than of the father's measurements.

Suggest a sequential procedure for investigating whether this is so. If the average number of days needed to reach a conclusion, on the assumption that Joe Zilch, Jr., is not, in fact, taking measurements, is to be not more than 10, and the probability of an incorrect conclusion not more than 5% in this case, find the least standard deviation of Joe Zilch, Jr.'s measurements for these to be a 5% chance of detecting these. (Results of Exercise 20 are relevant to this.)

14. Obtain the standard s.p.r.t. procedure discriminating between 2 hypotheses about the value of a binomial proportion p, satisfying the conditions

$$\Pr[\text{accept } p = p_1 | p = p_0] \doteq \alpha_0$$

$$\Pr[\text{accept } p = p_0 | p = p_1] \doteq \alpha_1$$

Consider the special case when p_0 is less than $p_1, \alpha_0 = p_0$, and $\alpha_1 = 1 - p_1$. Show that in this case the s.p.r.t. is in fact a fixed size sample procedure.

Discuss the accuracy of the standard approximate formulas for a.s.n. and power in this case.

15. Compare the approximate a.s.n. functions for the tests in (a) and (b) of Exercise 12.

16. $X_1, X_2, \ldots, X_n, \ldots$ are mutually independent random variables each distributed normally with expected value ξ and standard deviation σ; and

$$U_j = \left\{ \sum_{i=1}^{j} X_i - j X_{j+1} \right\} [j(j+1)]^{-1/2} \qquad (j = 1, 2, \ldots)$$

(a) Show that $U_1, U_2, \ldots$ are mutually independent, normal, and each has expected value 0 and standard deviation σ.

(b) Using the result (a), construct the s.p.r.t. $S(\sigma = \sigma_0, \alpha_0; \ \sigma = \sigma_1, \alpha_1)$ in the notation of Section 16.5.

(c) Show that the probability of accepting the hypothesis $\sigma = \sigma_1$ is given approximately by (16.17) with h the nonzero root (if there is one) of the equation

$$\left(\frac{\sigma_0}{\sigma_1} \right)^{2h} = 1 + h \left[\left(\frac{\sigma}{\sigma_1} \right)^2 - \left(\frac{\sigma}{\sigma_0} \right)^2 \right]$$

(d) Obtain an approximate formula for the average sample number of this s.p.r.t.

17. Construct s.p.r.t.'s discriminating between
(a) values of θ_0, θ_1 of the expected value (θ) of a Poisson distribution
(b) values of θ_0, θ_1 of the expected value (θ) of an exponential distribution.

18. Show that the probability of accepting the hypothesis $\theta = \theta_1$, in each of Exercises 16, 17(a), and 17(b) is given approximately by (16.17) with h the nonzero root (if there is one) of an equation of the form

$$y^h = 1 + hx(y - 1)$$

with x and y chosen appropriately.

SOLUTIONS OF THE EQUATION $y^h = 1 + hx(y - 1)$

y	0.1	0.2	0.3	0.4	x 0.5	0.6	0.8	1.2	1.5	2.0
1.2	-49.99	-24.72	-15.72	-10.73	-7.41	-4.96	-1.46	2.89	5.07	7.73
1.4	-24.99	-12.30	-7.71	-5.14	-3.42	-2.14	-0.300	2.00	3.16	4.57
1.6	-16.66	-8.15	-5.03	-3.27	-2.08	-1.19	0.089	1.70	2.52	3.52
1.8	-12.49	-6.07	-3.69	-2.33	-1.41	-0.714	0.286	1.55	2.20	2.98
2.0	-9.99	-4.82	-2.88	-1.76	-1.00	-0.427	0.404	1.46	2.00	2.66
2.2	-8.32	-3.99	-2.34	-1.38	-0.728	-0.234	0.484	1.40	1.87	2.44
2.4	-7.13	-3.39	-1.95	-1.11	-0.532	-0.095	0.541	1.36	1.77	2.29
2.6	-6.23	-2.94	-1.65	-0.903	-0.384	0.0098	0.585	1.32	1.70	2.18
2.8	-5.54	-2.58	-1.42	-0.742	-0.267	0.092	0.619	1.30	1.65	2.08
3.0	-4.98	-2.30	-1.24	-0.612	-0.174	0.158	0.646	1.28	1.60	2.00

19. Using the table of solutions of the equation $y^h = 1 + hx(y - 1)$ obtain approximate values for the a.s.n. of the tests in Exercise 17(a) with

$$\frac{\theta_1}{\theta_0} = 0.5; \qquad \alpha = \beta = 0.025$$

at $\theta = \theta_0$; $\theta = \theta_1$; $\theta = 0.6\theta_0$; and $\theta = 0.8\theta_0$.
Interpret your results in terms of the test in Exercise 17(b).

20. As in Exercise 19, but with

$$a = 0.05, \qquad \beta = 0.01.$$

Also interpret your result in terms of the test in Exercise 16.

Exercises 21 and 22 are based on "A Simple Sequential Procedure to Test Whether Average Conditions Achieve a Certain Standard," by R. C. Tomlinson, *Applied Statistics*, **6** (1957).

21. It is possible to observe a succession of independent random variables $X_1, X_2, \ldots$, each normally distributed with (unknown) mean θ and (known) standard deviation σ. The following test procedure has been suggested to discriminate between the hypothesis $\theta = \theta'_0 = \theta - 1.96\sigma$ and $\theta = \theta'' = \theta_0 + 1.96\sigma$.
   Observe the sequence of independent random variables $X_1, X_2, \ldots$. As soon as there is an observed value x_i less than θ'_0 (greater than θ''_0) or 2 successive observed values x_i, x_{i+1} between θ_0 and $\theta'_0(\theta''_0)$, accept the hypothesis $\theta = \theta'_0(\theta = \theta''_0)$. The procedure terminates as soon as a decision is reached.

Show that if the population mean is equal to θ_0'' the probability of a correct decision is 0.999 (very nearly) and the average sample number is 1.53.

22. Generalize the procedure described in Exercise 21 to cover the cases when the common distributions of each of X_1, X_2, etc. can be non-normal continuous distributions, and $\theta_0' < \theta_0 < \theta_0''$ need not satisfy the condition $2\theta_0 = \theta_0' + \theta_0''$.

    Show that the a.s.n. of this test procedure is never greater than 3. Point out any outstanding disadvantages of test procedures of this kind.

23. Suppose the prior probabilities of the hypotheses H_0, H_1 are known to be $\omega_0, \omega_1(\omega_0 + \omega_1 = 1)$. Obtain equations for determining α_0, α_1 so that a sequential test $S(H_0, \alpha_0; H_1, \alpha_1)$ shall minimize the average sample number, subject to the condition that the expected proportion of errors has a specified value α—that is, $\omega_0\alpha_0 + \omega_1\alpha_1 = \alpha$.

24. Show that if H_0, H_1 specify 2 values for the mean of a normal population with known variance then α_0 and α_1 in Exercise 23 should satisfy

$$\omega_0^2\alpha_0(1 - \alpha_0) = \omega_1^2\alpha_1(1 - \alpha_1)$$

Investigate whether there are other distributions for which this is so.

25. Extend the results of Exercises 23 and 24 to the case when the total expected cost

$$\omega_0 c_0 \alpha_0 + \omega_1 c_1 \alpha_1 + c \times (\text{average sample number})$$

is to be minimized.

26. Each number of a sequence of independent random variables $X_1, X_2, X_3, \ldots$ has the same expected value ξ and standard deviation σ. The distribution of $Y_i = (X_i - \xi)/\sigma$ does not depend on ξ or σ, and it is known that

$$E\left[(Y_i - Y_j)^{-2}\right] = \lambda \quad \text{for} \quad i \neq j$$

First, X_1 and X_2 are observed. Then a further N random variables $X_3, X_4, \ldots, X_{N+2}$ are observed where N is the least integer exceeding $M(X_1 - X_2)^2$, M being a positive number. The statistic

$$\bar{X} = N^{-1} \sum_{j=1}^{n} X_{2+j}$$

is then calculated.

(a) Show that

$$\Pr\left[|\bar{X} - \xi| < D\right] \geqslant 1 - \frac{\lambda}{MD^2}$$

for any positive number D.

(b) Explain the relevance of this result to 2-stage sampling procedures of the kind discussed in Section 16.12.1. How would you try to extend this result to widen its field of application?

27. $Z_1, Z_2, \ldots Z_n, \ldots$ are mutually independent random variables with a common distribution. The Z's are observed in sequence until

$$(a) \sum_{j=1}^{N} Z_j \leqslant a \qquad \text{or} \qquad (b) \sum_{j=1}^{N} Z_j \geqslant b \quad (b > a)$$

$P(a, b)$ denotes the probability that (a) is true when observation ceases. Show that

$$P(a, b) = E[P(a - Z', b - Z')] \Pr[a < Z' < b] + \Pr[Z \leqslant a]$$

where Z is a random variable having the same distribution as each Z_j, while Z' has the same distribution, but truncated to $a < Z' < b$. If $P(a, b) = (e^{bh} - 1)/(e^{bh} - e^{ah})$, show that h must satisfy the equation

$$E[\exp(Z''h)] = 1$$

where Z'' has the same distribution as Z for $a < Z'' < b$ and

$$\Pr[Z'' = a] = \Pr[Z \leqslant a]; \quad \Pr[Z'' = b] = \Pr[Z \geqslant b]$$

Explain the relevance of these results to the theory of sequential probability ratio tests.

CHAPTER 17

Multivariate Observations

17.1 INTRODUCTION

We have already, in Chapters 12 and 13, described some methods of analyzing data in which more than one character is measured on each individual. The methods there developed, though useful, were restricted to the *multiple linear regression* technique and certain specializations thereof. Further applications are discussed in Chapter 18.

Now we describe some further techniques, rather broader in scope, which are specially designed to be applicable to such *multivariate* data. This field of statistics is called *multivariate analysis*—it includes multiple linear regression, but also many other techniques. Indeed, we describe only a small part of the available methods. A specialist treatment of the subject is not suitable here; interested readers should consult references to obtain further detailed information. Nevertheless, it is hoped that the discussion that follows will suffice to indicate the types of method available for a large proportion of problems arising in the course of analysis of multivariate data.

Many of the procedures described are practicable only with the help of computers. Appropriate programs are, fortunately, widely available. However, for bivariate data such computational assistance is often not essential.

17.2 DISCRIMINANT ANALYSIS

An essential feature of multivariate data is that more information is available, in respect of each individual subjected to measurement, than is the case when only one character is measured. It is natural to consider in what situations we can make special use of this extra abundance of information. A type of situation where this is evidently possible arises when we are trying to assign an individual to one of two or more classes.

The more characters measured for each individual, the more information is available, and we might hope to utilize this information so as to increase, perhaps substantially, the probability of correct classification.

If we have *sufficient* information about the problem a simple and effective procedure can be constructed which should minimize the proportion of times an incorrect classification is made. The information needed is as follows:

(a) For each of the k possible classes we need the joint distribution of the q observed variables.

(b) We also need the proportions of individuals to be expected from each of the k classes (the "prior probabilities" of these classes).

It will be realized that exact information on these matters will be available but rarely. However, in very many cases there will be some relevant information. This may suffice to give a clear enough picture for procedures to be constructed which, though not optimal, are "good."

We will first suppose that requirement (a) has been formulated for continuous random variables. The joint probability density function of the variables $X_1, X_2, \ldots, X_q$ in the jth class (which is supposed known) will be denoted by $f_j(x_1, x_2, \ldots, x_q)$, for any integer j between 1 and k inclusive.

The information required by (b) can be summarized in the prior probabilities $P_1, P_2, \ldots, P_k$ (with $\sum_{j=1}^{k} P_j = 1$); P_j is the prior probability for the jth class.

17.3 OPTIMAL DISCRIMINATION

Any rule for assigning an individual to one of the k classes is equivalent to dividing the set of all possible values of $(X_1, X_2, \ldots, X_q)$ into k mutually exclusive subsets $R_1, R_2, \ldots, R_k$, and assigning the individual to the jth class if $(X_1, X_2, \ldots, X_q)$ is in the subset R_j. (Geometrically minded readers will find it helpful to replace "set" by "space" in the last sentence.) The probability of *correct* classification is then

$$\sum_{j=1}^{k} \Pr[\text{individual is from } j\text{th class}]$$

$$\times \Pr[\text{individual is assigned to } j\text{th class} \,|\, \text{individual is from } j\text{th class}]$$

$$= \sum_{j=1}^{k} P_j \int \int_{R_j} \cdots \int f_j(x_1, x_2, \ldots, x_q) \, dx_1 \, dx_2 \cdots dx_q \qquad (17.1)$$

This can be interpreted as the integral, *over all values of* $(x_1, x_2, \ldots, x_q)$, of a function of $x_1, x_2, \ldots, x_q$, which is equal to

$$P_1 f_1 (x_1, x_2, \ldots, x_q) \text{ in } R_1$$

$$P_2 f_2 (x_1, x_2, \ldots, x_q) \text{ in } R_2$$

and generally

$$P_j f_j (x_1, x_2, \ldots, x_q) \text{ in } R_j \quad (j = 1, 2, \ldots, k)$$

Our problem is to choose $R_1, R_2, \ldots, R_k$ so as to maximize this integral (which is equal to the probability of correct classification). For any combination of values of $(x_1, x_2, \ldots, x_q)$ we have a choice of the k values $P_j f_j(x_1, x_2, \ldots, x_q)$ for our integrand. Evidently the integral will be maximized if we always choose the largest of these k values.

This means that R_j is defined as the set in which $P_j f_j(x_1, x_2, \ldots, x_q)$ is not exceeded by any of the k quantities:

$$P_1 f_1 (x_1, x_2, \ldots, x_q), \ldots, P_k f_k (x_1, x_2, \ldots, x_q)$$

Sometimes 2 or more of these quantities may be equal to each other and greater than any of the remainder. In such cases assignment to one of the possible R's may be arbitrary. In the case of continuous random variables (with which we are dealing at present) this will not have any effect on the overall probabilistic properties of the procedure.

The merit of a discriminant procedure is assessed by the probability of correct classification. When making such an assessment one should remember that it is always possible to achieve a probability of correct classification, equal to the greatest of the P_j's, simply by assigning every individual to the class corresponding to this greatest value.

17.4   WEIGHTED LOSSES

Sometimes it is possible to differentiate between decisions according to their relative importance (see some remarks on this in Chapter 11). Generally we can assign a "loss" W_{ij} to the decision to allot an individual to the ith class when it really belongs to the jth class. Then we will seek to minimize

$$\sum_j P_j \sum_i W_{ij} \Pr[\text{assign to } i\text{th class} \mid j\text{th class}]$$

(The procedure outlined in Section 17.3 is equivalent to taking

$$W_{jj} = 0, \qquad W_{ij} = 1 \text{ if } i \neq j)$$

This sum is equal to

$$\sum_i \int\int_{R_i} \cdots \int \left\{ \sum_j P_j W_{ij} f_j (x_1, x_2, \ldots, x_q) \right\} dx_1 dx_2 \cdots dx_q$$

and is minimized by assigning to R_i those points where

$$S_i = \sum_j P_j W_{ij} f_j (x_1, x_2, \ldots, x_q)$$

is the minimum of the k quantities $S_1, S_2, \ldots, S_k$.

The methods described above apply in a similar fashion to discrete random variables, as is shown in the following example.

Example 17.1 Large batches of assortments of certain prepacked foodstuffs are known to contain products A, B, C, and D in proportions as shown in the table below

TYPE OF BATCH	PRODUCT A	B	C	D
I	0.1	0.1	0.3	0.5
II	0.1	0.2	0.2	0.5
III	0.1	0.2	0.1	0.6

The value of A is $4.00 per unit
        B is $3.00 per unit
        C is $2.00 per unit
        D is $0.50 per unit

It is known that one-quarter of all batches are of type I, one-quarter of type II, and one-half of type III.

Construct rules for assignment to one of the 3 types of batches, based on the observed numbers of A, B, C, and D units in a random sample of 20 units from a batch, such that

(i) the probability of correct assignment is maximized, or

(ii) the expected (absolute) difference between actual value (per unit) and assessed value (according to type of batch) is minimized.

(i) *Maximizing Probability of Correct Assignment.* Since the batches are "large," we may assume that the probability of observing N_A units of A, N_B of B, N_C of C, and N_D of D, in a random sample of 20 units is very nearly equal to

$$\frac{20!}{N_A! N_B! N_C! N_D!} p_A^{N_A} p_B^{N_B} p_C^{N_C} p_D^{N_D}$$

where $N_A + N_B + N_C + N_D = 20$ and p_A, p_B, p_C, p_D are the proportions of A, B, C, D, respectively, in the batch. This corresponds to taking $q = 4$ in the general theory and replacing X_1, X_2, X_3, X_4 by N_A, N_B, N_C, N_D, respectively. (Since $N_A + N_B + N_C + N_D = 20$, we really need use only 3 variables, but the formulas are more symmetrical in appearance if we keep all 4.)

We assign the batch to type I if

$$\tfrac{1}{4}(0.1)^{N_A}(0.1)^{N_B}(0.3)^{N_C}(0.5)^{N_D}$$

is greater than either

$$\tfrac{1}{4}(0.1)^{N_A}(0.2)^{N_B}(0.2)^{N_C}(0.5)^{N_D} \quad \text{or} \quad \tfrac{1}{2}(0.1)^{N_A}(0.2)^{N_B}(0.1)^{N_C}(0.6)^{N_D}$$

Taking logarithms, and simplifying by the omission of the term involving N_A (which is the same in all 3 expressions), we find that the batch is assigned to types I, II, or III according as the greatest of

$$N_B \log 0.1 + N_C \log 0.3 + N_D \log 0.5 = -N_B - 0.523 N_C - 0.301 N_D$$

$$N_B \log 0.2 + N_C \log 0.2 + N_D \log 0.5 = -0.699(N_B + N_C) - 0.301 N_D$$

and

$$N_B \log 0.2 + N_C \log 0.1 + N_D \log 0.6 + \log 2 = -0.699 N_B - N_C - 0.222 N_D + 0.301$$

is the first, second, or third of these quantities, respectively. [For practical application we might add $N_B + N_C + N_D$ to each quantity, giving us

$$0.477 N_C + 0.699 N_D, 0.301(N_B + N_C) + 0.699 N_D, 0.301 N_B + 0.778 N_D + 0.301$$

respectively.]

(*ii*) *Minimizing Expected Difference between Assessed and Actual Value.* In this case we have losses [W_{ij} in the notation of (17.4)] when a type j batch is assigned to type i, as shown below

$j^{\backslash i}$	I	II	III
I	0	0.10	0.05
II	0.10	0	0.15
III	0.05	0.15	0

(This table is formed by noting that the value per unit of type I is $\$(4 \times 0.1 + 3 \times 0.1 + 2 \times 0.3 + 0.5 \times 0.5) = \$1.55$; the value per unit of type II is \$1.65, and of type III, \$1.50.)

So now we have to find the *minimum* of

(I) $\qquad \tfrac{1}{4}(0.10)(0.2)^{N_B}(0.2)^{N_C}(0.5)^{N_D} + \tfrac{1}{2}(0.05)(0.2)^{N_B}(0.1)^{N_C}(0.6)^{N_D}$

(II) $\qquad \tfrac{1}{4}(0.10)(0.1)^{N_B}(0.3)^{N_C}(0.5)^{N_D} + \tfrac{1}{2}(0.15)(0.2)^{N_B}(0.1)^{N_C}(0.6)^{N_D}$

and

(III) $\qquad \tfrac{1}{4}(0.05)(0.1)^{N_B}(0.3)^{N_C}(0.5)^{N_D} + \tfrac{1}{4}(0.15)(0.2)^{N_B}(0.2)^{N_C}(0.5)^{N_D}$

(The common factor $(0.1)^{N_A}$ has been omitted for convenience.)

If (I) is the minimum, the batch is assigned to type I; similarly, if (II) or (III) is minimum, the batch is assigned to type II or type III, respectively.

Suppose now $N_A = 2, N_B = 2, N_C = 5, N_D = 11$; we would choose type I by the first criterion (maximizing probability of correct assignment), since

$$0.477 N_C + 0.699 N_D = 10.074$$

$$0.301 (N_B + N_C) + 0.699 N_D = 9.796$$

$$0.301 N_B + 0.778 N_D + 0.301 = 9.461$$

By the second criterion (minimizing expected difference between assessed and actual value), we would again choose type I, since the differences (with 0.1^{N_A} omitted) are 19.25×10^{-11}, 40.55×10^{-11}, and 38.27×10^{-11} for I, II, and III, respectively.

17.5   LACK OF INFORMATION

Quite often we may know the distributions in each of the groups with sufficient accuracy, but have little idea of the values of the P's. If we are classifying a series of individuals, the resultant data can be used to give improved estimates of the P's.

First, consider the case where there are just 2 groups (so that $k = 2$). If the probability of correct assignment for the first group is p_1; and for the second group p_2, then the expected proportion of assignments to the first group is

$$P_1 p_1 + (1 - P_1)(1 - p_2) = P_1 (p_1 + p_2 - 1) + 1 - p_2$$

Hence we may use

$$\frac{(\text{observed proportion assigned to first group}) - (1 - p_2)}{p_1 + p_2 - 1}$$

as an improved estimate of P_1. [The numerator may also be written "$p_2 - $(observed proportion assigned to second group)."]

If there are k groups, with observed proportions $\hat{P}_1, \hat{P}_2, \ldots, \hat{P}_k$, then we use the fact that the expected value of $\hat{P}_j$, the proportion assigned to the jth class, is

$$P_j p_j + \sum_{i \neq j} P_i \int \int_{R_j} \cdots \int f_i(x_1, x_2, \ldots, x_q)\, dx_1 dx_2 \cdots dx_q$$

$$= P_j p_j + \sum_{i \neq j} P_i a_{ij}, \quad \text{say} \quad \left(\text{Note that } p_j + \sum_{i \neq j} a_{ij} = 1. \right)$$

Then we solve the k equations

$$\hat{P}_j = P_j p_j + \sum_{i \neq j} P_i a_{ij} \tag{17.2}$$

for $P_1, P_2, \ldots, P_k$. Similar equations are obtained when the variables are discrete.

It is usually a straightforward matter (though sometimes tedious) to evaluate the quantities p_j and a_{ij}. For instance, in Example 17.1, these quantities can be calculated by summing the probabilities

$$\frac{20! \, p_A^{N_A} p_B^{N_B} p_C^{N_C} p_D^{N_D}}{N_A! N_B! N_C! N_D!}$$

over the appropriate sets of values of N_A, N_B, N_C, and N_D.

The improved estimates of the P's can then be used to modify the rules for assigning individuals to groups in accordance with the methods described in Sections 17.2 and 17.3.

The position is rather more difficult if sufficient information is not available about the distributions of the variables in each of the groups. Then it is necessary to estimate these distributions in some way. This may be done using regular random samples from each of the populations $\Pi_1, \ldots, \Pi_k$. If the form of distribution in each population is reasonably well-established, values of parameters have to be established from the relevant random sample.

Discriminant functions obtained using estimated population distribution functions will not, in general, be as effective as those which would be obtained using correct formulas for the distribution functions. Consequently, calculated values for probability of misclassification, or expected loss, are inaccurate, also, and usually they are overoptimistic. Such values may be roughly corrected in certain specific cases for which formulas are available. (See [17.4] for some important cases.) Even these corrections only allow (usually approximately) for bias in the estimated values; variability still remains.

17.6 DISCRIMINATION USING NORMAL VARIABLES

In many cases it will be sufficiently accurate to use a multinormal distribution for the variables $X_1, X_2, \ldots, X_q$. Thus we have (See Chapter 5) in the jth group,

$$f_j(X_1, X_2, \ldots, X_q) = \frac{|\mathbf{C}_j|^{1/2}}{(2\pi)^{q/2}} \exp\left[-\tfrac{1}{2} \sum_r^q \sum_s^q c_{rsj}(x_r - \xi_{rj})(x_s - \xi_{sj}) \right]$$

ξ_{rj} is the expected value of X_r in the jth group; the matrix $\mathbf{C}_j = (c_{rsj})$ is the inverse of the matrix of variances and covariances of the X's in the jth group.

We again take the case of 2 groups ($j = 1, 2$) first. If we wish to maximize the probability of correct assignment, we assign to the first group if $P_1 f_1 > P_2 f_2$, that is,

$$\ln\left(\frac{f_1}{f_2}\right) > \ln\left(\frac{P_2}{P_1}\right)$$

Now

$$\ln\frac{f_1(X_1, X_2, \ldots, X_q)}{f_2(X_1, X_2, \ldots, X_q)} = \tfrac{1}{2}\ln\frac{|C_1|}{|C_2|} - \tfrac{1}{2}\left[\sum_r^q \sum_s^q c_{rs1}(X_r - \xi_1)(X_s - \xi_{s1})\right.$$

$$\left. - \sum_r^q \sum_s^q c_{rs2}(X_r - \xi_{r2})(X_s - \xi_{s2})\right]$$

Therefore, we assign to the first group if

$$\sum_r^q \sum_s^q \left[c_{rs2}(X_r - \xi_{r2})(X_s - \xi_{s2}) - c_{rs1}(X_r - \xi_{r1})(X_s - \xi_{s1})\right]$$

$$> 2\ln\left(\frac{P_2|C_2|^{1/2}}{P_1|C_1|^{1/2}}\right) \quad (17.3)$$

On the left side of (17.3) there is a *quadratic* function of the observed values of $X_1, X_2, \ldots, X_q$. Assignment to the first or second group depends on the value of this function. It is called a *quadratic discriminant function*, or, more briefly, a *quadratic discriminator*.

If the variance–covariance matrices in the 2 groups are identical, so that $c_{rs1} = c_{rs2} (= c_{rs})$ for all values of r and all values of s, then $|C_1| = |C_2|$ and (17.3) becomes

$$\sum_r^q \sum_s^q c_{rs}\left[(X_r - \xi_{r2})(X_s - \xi_{s2}) - (X_r - \xi_{r1})(X_s - \xi_{s1})\right] > 2\ln\left(\frac{P_2}{P_1}\right)$$

The left side of this inequality can be rearranged to read

$$\sum_r^q \sum_s^q c_{rs}\left[X_r(\xi_{s1} - \xi_{s2}) + X_s(\xi_{r1} - \xi_{r2}) + \xi_{r2}\xi_{s2} - \xi_{r1}\xi_{s1}\right]$$

and the whole inequality can be put in the form

$$\sum_{r=1}^q \lambda_r X_r > K \quad (17.4)$$

where K is some constant (not depending on the X's) and

$$\lambda_r = \sum_{s=1}^{q} c_{rs}(\xi_{s1} - \xi_{s2})$$ (17.5)

We now have a *linear discriminant function* or *linear discriminator*, with coefficients given by (17.5). Although K is a rather complicated function of the ξ's, c's, and P's, we can obtain a suitable value for K from direct consideration of the statistic $D = \sum_{r=1}^{q} \lambda_r X_r$. In the jth group ($j = 1, 2$) the expected value of this statistic is $\sum_{r=1}^{q} \lambda_r \xi_{rj}$. In either group (on our assumption of a common variance–covariance matrix) the variance of the statistic is

$$\sigma_D^2 = \sum_{r=1}^{q} \lambda_r^2 [\text{variance of } X_r] + 2 \sum_{r<s}^{q-1} \sum^{q} \lambda_r \lambda_s [\text{covariance of } X_r \text{ and } X_s]$$ (17.6)

Hence if we take as the critical value of D the midpoint between its expected values in the 2 groups, which is

$$\tfrac{1}{2} \sum_{r=1}^{q} \lambda_r(\xi_{r1} + \xi_{r2}) = \delta_0, \quad \text{say,}$$

then the probability of misclassification is the same, whether the individual comes from the first or the second group.
    If

$$\sum_{r=1}^{q} \lambda_r \xi_{r1} < \delta_0$$

we naturally assign the individual to the first group if D is less than the critical value δ_0. The probability of misclassification is then

$$1 - \Phi(\Delta) = \Phi(-\Delta)$$

where

$$\Delta = \frac{\delta_0 - \sum_{r=1}^{q} \lambda_r \xi_{r1}}{\sigma_D} = -\frac{\delta_0 - \sum_{r=1}^{q} \lambda_r \xi_{r2}}{\sigma_D}$$

and σ_D is given by (17.6).
    Hence, whatever be the values of P_1 and P_2, the probability of correct classification is

$$\Phi(\Delta)$$

This means that even if we have no knowledge whatever of P_1 and P_2 we can still keep the probability of misclassification at a known level. Knowledge of P_1 and P_2 would, of course, usually make it possible to reduce the probability of misclassification. However, it is very useful to be able to control probability of error in those cases, which are all too frequent, where our knowledge is not sufficiently extensive to permit of the construction of optimal procedures.

This method of controlling error probabilities can be applied generally to the case $k = 2$ (2 "groups" or populations), provided the true distributions are known, and is not restricted to the case of multinormal distributions.

Example 17.2   It is known that lots of an electrical component come from one of 2 sources. It is also established that the probability that a component selected at random from a lot from source (j) lasts for a time t, or more, is represented fairly accurately by the formula $e^{-\lambda_j t}$ ($j = 1$ or 2). The numerical values of λ_1 and λ_2 are known. In order to assign a lot to one of the 2 sources, each component in a random sample of n components is put into use simultaneously and the lifetimes of the first k components to fail are observed. These k observed values of the random variables $T_1', T_2', \ldots, T_k' (T_1' < T_2' < \cdots < T_k')$ are to be used to assign the lot to source (1) or source (2).

The joint probability density function of $T_1', T_2', \ldots, T_k'$, given that the lot comes from source (j), is

$$p_{T_1', T_2', \ldots, T_k'}(t_1', t_2', \ldots, t_k' | \lambda_j) = \lambda_j^k \exp\left[-\lambda_j \left\{ \sum_{i=1}^{k-1} t_i' + (n-k+1)t_k' \right\} \right]$$

$$(0 < t_1' < t_2' < \cdots < t_k')$$

Hence we assign the batch to source (1) if

$$\frac{p(t_1', t_2', \ldots, t_k' | \lambda_1)}{p(t_1', t_2', \ldots, t_k' | \lambda_2)} = \left(\frac{\lambda_1}{\lambda_2} \right)^k \exp\left[-(\lambda_1 - \lambda_2) \left\{ \sum_{i=1}^{k-1} T_i' + (n-k+1)T_k' \right\} \right] > K$$

where K is a suitably chosen constant.

If $\lambda_1 > \lambda_2$, this inequality may be written in the form

$$\sum_{i=1}^{k-1} T_i' + (n-k+1)T_k' < K'$$

where K' is a constant. [Actually $K' = [k \ln(\lambda_1/\lambda_2) - \ln K](\lambda_1 - \lambda_2)^{-1}$.]

If values of the prior probabilities of the lot coming from source (1) or source (2) can be used (for example, if there is a good estimate of the relative numbers of lots supplied from the 2 sources) the value of K will be determined from the formula

$$K = \frac{\text{prior probability of source (2)}}{\text{prior probability of source (1)}}$$

Even in the absence of this knowledge, it is still possible to attain a known error probability, by choosing K' so that

$$\Pr\left[\sum_{i=1}^{k-1} T_i' + (n-k+1)T_k' < K'|\lambda_2\right] = \Pr\left[\sum_{i=1}^{k-1} T_i' + (n-k+1)T_k' > K'|\lambda_1\right]$$

The common value of these 2 probabilities is then the probability of error. From Example 6.2, it is known that if the lot is from source (j), then

$$\sum_{i=1}^{k-1} T_i' + (n-k+1)T_k'$$

is approximately distributed as $(2\lambda_j)^{-1}$. (χ^2 with $2k$ degrees of freedom). Hence, to satisfy the above condition, K' must be chosen so that

$$\Pr\left[\chi_{2k}^2 < 2\lambda_2 K'\right] = \Pr\left[\chi_{2k}^2 > 2\lambda_1 K'\right]$$

The following calculations, based on Table E of the Appendix, illustrates some possibilities. The ratio of the upper to the lower 100α percent points of the appropriate χ^2 distribution give the ratio of λ_1 to λ_2 for which the error probability is α. Taking $\alpha = 0.10$ we find the values shown in Table 17.1.

TABLE 17.1

CRITICAL VALUES OF RATIOS OF FAILURE RATES
(Probability of Detection = 0.90)

k	λ_1/λ_2	k	λ_1/λ_2
1	21.85	10	2.28
2	7.31	12	2.12
3	4.83	15	1.95
4	3.83	20	1.78
5	3.29	25	1.68
6	2.94	30	1.60
8	2.53		

By increasing k the discriminatory power is increased.

It is interesting to note that the sample size (n) itself does not enter into these calculations (though it does enter into the calculation of the criterion $\sum_{i=1}^{k-1} T_i' + (n-k+1)T_k'$). Since the expected value of T_k' decreases as n increases, it is possible to reduce the average time to complete the observations, by increasing the sample size, without affecting the discriminatory power of the procedure.

Example 17.3 It is known that 4 variates X_1, X_2, X_3, X_4 have a joint multinormal distribution in each of 2 populations π_1 and π_2. It is further known that their variance–covariance matrix is the same in each of the 2 populations. The values of the variances and covariances are known to a good degree of accuracy. The linear discriminator

$$F = 0.1X_1 + 0.7X_2 + 0.5X_3 + 0.3X_4$$

has expected value 1.75 in π_1 and 2.55 in π_2. Its variance in either population is 0.36. Two hundred individuals chosen at random from a large aggregate believed to be a mixture of individuals from π_1 and π_2 gave values of F from which the following statistics were calculated:

Number of values of F less than 2.15: 60
Arithmetic mean of the 200 values of F: 2.40

(*i*) Are these data consistent with the hypothesis that the aggregate is a mixture of individuals from π_1 and π_2?

(*ii*) Suggest a discriminating procedure which should minimize the probability of incorrect assignment, assuming that each individual does belong to either π_1 or π_2. Estimate the minimum probability of incorrect assignment. (University of London, B.Sc. (Special) Examination, 1962.*)

(*i*) If individuals come from either π_1 or π_2, the probability of incorrect assignment, using the critical value 2.15 $[= \frac{1}{2}(1.75 + 2.55)]$, is

$$1 - \Phi\left(\frac{0.4}{0.6}\right) = 0.2525$$

If the aggregate is a mixture of individuals from π_1 and π_2 in the ratio $p : (1 - p)$, the expected proportion of observed F values below 2.15 would be

$$0.7475p + 0.2525(1 - p) = 0.2525 + 0.4950p$$

and the expected value of the arithmetic mean of the F values would be

$$1.75p + 2.55(1 - p) = 2.55 - 0.80p$$

Therefore, 2 unbiased estimators $\hat{p}_1$, $\hat{p}_2$ of p can be obtained by solving

$$0.2525 + 0.4950\hat{p}_1 = \frac{60}{200}$$

and

$$2.55 - 0.80\hat{p}_2 = 2.40$$

*The University of London does not in any way commit itself to approval of solutions to this or other problems taken (with permission) from its examinations.

respectively. Solving these equation, we find $\hat{p}_1 = 0.0960$, $\hat{p}_2 = 0.1875$. These values appear to differ considerably. To help assess the significance of this observed difference, the variance of $(\hat{p}_1 - \hat{p}_2)$ might be estimated.

The variance of the number of values of F less than 2.15 is

$$200(0.2525 + 0.4950p)(0.7475 - 0.4950p)$$

Therefore,

$$\text{var}(\hat{p}_1) = \frac{(0.2525 + 0.4950p)(0.7475 - 0.4950p)}{200 \times (0.4950)^2}$$

$$= 0.00385 + 0.005p(1-p)$$

The variance of F is $0.36 + (2.55 - 1.75)^2 p(1-p) = 0.36 + 0.64p(1-p)$ (see Section 5.10.3). The variance of the arithmetic mean of 200 values of F is $\text{var}(F)/200$, and so $\text{var}(\hat{p}_2) = \frac{1}{200} \times \text{var}(F)/(0.80)^2 = 0.00281 + 0.05p(1-p)$. To estimate the variance of $(\hat{p}_1 - \hat{p}_2)$ it is also necessary to estimate the covariance between $\hat{p}_1$ and $\hat{p}_2$, because

$$\text{var}(\hat{p}_1 - \hat{p}_2) = \text{var}(\hat{p}_1) - 2\text{cov}(\hat{p}_1, \hat{p}_2) + \text{var}(\hat{p}_2)$$

It can be seen that

$$\text{cov}(\hat{p}_1, \hat{p}_2) = -\frac{\text{cov}\left(N, \sum_{i=1}^{200} F_i\right)}{200^2(0.4950)(0.80)}$$

where N represents the number of observed values of F less than 2.15, and $\sum_{i=1}^{200} F_i$ represents the sum of the observed values of F. Introducing a characteristic variable Z, taking the values 1 if F is less than 2.15, and 0 otherwise, we have

$$\text{cov}\left(N, \sum_{i=1}^{200} F_i\right) = 200\,\text{cov}(Z, F)$$

Now

$$\text{cov}(Z, F) = E[ZF] - E[Z]E[F]$$

$$E[Z] = 0.2525 + 0.4950p$$

$$E[F] = 2.55 - 0.80p$$

and

$$E[ZF] = E[F|F < 2.15]\Pr[F < 2.15]$$

$$= (0.6\sqrt{2\pi})^{-1} \int_{-\infty}^{2.15} F \left\{ p \exp\left[-\frac{(F-1.75)^2}{0.72} \right] \right.$$

$$\left. + (1-p)\exp\left[-\frac{(F-2.55)^2}{0.72} \right] \right\} dF$$

$$= p\left[(1.75)\Phi\left(\frac{0.4}{0.6}\right) du - (0.6)\frac{1}{\sqrt{2\pi}} e^{\frac{1}{2}\left(\frac{0.4}{0.6}\right)^2} \right]$$

$$+ (1-p)\left[(2.55)\Phi\left(-\frac{0.4}{0.6}\right) - (0.6)\frac{1}{\sqrt{2\pi}} e^{-\frac{1}{2}\left(\frac{0.4}{0.6}\right)^2} \right]$$

$$= 2.55(1-p) - (2.55 - 4.30p)\Phi\left(\frac{0.4}{0.6}\right) - (0.6)\frac{1}{\sqrt{2\pi}} e^{-\frac{1}{2}\left(\frac{0.4}{0.6}\right)^2}$$

whence $E[ZF] = 2.55(1-p) - (2.55 - 4.30p)(0.7475) - (0.6)(0.3194)$
$$= 0.4522 + 0.6642p$$

Hence

$$\text{cov}(Z, F) = 0.4522 + 0.6642p - (0.2525 + 0.4950p)(2.55 - 0.80p)$$

$$= -0.1894 - 0.3960p + 0.3960p^2$$

Therefore

$$\text{cov}(\hat{p}_1, \hat{p}_2) = \frac{0.3960p(1-p) + 0.1894}{(200)(0.4950)(0.80)} = 0.005p(1-p) + 0.0024$$

and

$$\text{var}(\hat{p}_1 - \hat{p}_2) = 0.00385 + 0.005p(1-p)$$

$$- 0.01p(1-p) - 0.00478$$

$$+ 0.00281 + 0.005p(1-p)$$

$$= 0.00188$$

The standard deviation of $(\hat{p}_1 - \hat{p}_2)$ is 0.043, whatever be the value of p. The observed value of $\hat{p}_1 - \hat{p}_2$ is $0.1875 - 0.0960 = 0.0915$, which is more than twice the standard deviation.

This result might well be taken to indicate that the aggregate contains individuals from populations other than π_1 or π_2, though in default of more precise theory on the distribution of $\hat{p}_1 - \hat{p}_2$ it is not possible to be very definite on this point.

(*ii*) Supposing, nevertheless, that the aggregate is a mixture of π_1 and π_2 the optimal discrimination will be attained by assigning an individual to π_1 if

$$\frac{p}{0.6\sqrt{2\pi}}\exp\left[-\frac{(F-1.75)^2}{0.72}\right] > \frac{1-p}{0.6\sqrt{2\pi}}\exp\left[-\frac{(F-2.55)^2}{0.72}\right]$$

and otherwise assigning the individual to π_2. This means that an individual is assigned to π_1 if

$$\frac{1}{0.72}\left[(F-1.75)^2-(F-2.55)^2\right] < \ln\left(\frac{p}{1-p}\right)$$

that is,

$$F < 2.15 + 0.45\ln\left(\frac{p}{1-p}\right)$$

To obtain a numerical value for the limit an estimate of p must be used. If, as suggested in Section 17.5, $\hat{p}_2(=0.1875)$ is used we get the value 1.49; if $p=0.1$ (close to $\hat{p}_1$) the value 1.16 is obtained.

The table below shows the probabilities of misclassification for 6 possible combinations of p and the limiting value of F.

	LIMIT OF F		
p	1.16	1.49	2.15
0.1	0.093	0.102	0.252
0.1875	0.166	0.157	0.252

It can be seen that any value between 1.15 and 1.50, say, could be used as limit without affecting the probability of error to a great extent.

(Note that the solution of this problem does not use the coefficients of X_1, X_2, X_3, and X_4 in F explicitly.)

17.7   TEST PROCEDURES BASED ON THE MULTINORMAL DISTRIBUTION

The multinormal distribution has already been discussed in Section 5.11.2. Here we briefly discuss a few test procedures based on this distribution. Although multivariate data are frequently encountered, it is very often preferred to use univariate tests, supplemented by the regression techniques we have already discussed, for their analysis. Many tests appropriate to multivariate data have been elaborated, but they are mostly of the socalled "portmanteau" type—covering so many different possible forms of departure from the hypothesis tested that there is some suspicion that they may not be very sensitive to any one particular kind of departure, relative to some more specific test. Furthermore, we are usually interested in finding out (at any rate, qualitatively) the actual form of departure from the hypothesis tested. This is essentially a problem of estimation rather than of hypothesis testing; nevertheless, specific tests usually do give a more detailed picture than a "portmanteau" test. The procedures we will describe below are the most generally used of these "portmanteau" tests.

First we suppose that we have a random sample of n individuals, and that we know the value of every element in the variance–covariance matrix $\mathbf{V} = (\sigma_{ij})$. The joint probability density function of random variables $X_{1l}, X_{2l}, \ldots, X_{ql}$ representing measurements of characters $X_1, X_2, \ldots, X_q$, respectively, on the lth member of the sample is

$$p_{X_{1l}, X_{2l}, \ldots, X_{ql}} (x_{1l}, x_{2l}, \ldots, x_{ql})$$

$$= (2\pi)^{-q/2} |\mathbf{V}|^{-1/2} \exp\left[-\tfrac{1}{2} \sum_{i=1}^{q} \sum_{j=1}^{q} \sigma^{ij} (x_{il} - \xi_i)(x_{jl} - \xi_j) \right] \quad (17.7)$$

where $\boldsymbol{\xi} = (\xi_1, \xi_2, \ldots, \xi_q)$ is the vector of (unknown) expected values of X_1, $X_2, \ldots, X_q$, respectively, and σ^{ij} is the element in the ith row and jth column of $\mathbf{V}^{-1}$. If $\mathbf{V}$ is known, then, of course, the values of the quantities σ^{ij} are also known.

The joint probability density function of the qn random variables representing all the measurements on the individuals in the sample is

$$\prod_{l=1}^{n} p_{X_{1l}, X_{2l}, \ldots, X_{ql}} (x_{1l}, x_{2l}, \ldots, x_{ql})$$

where $p_{X_{1l}, X_{2l}, \ldots, X_{ql}}(x_{1l}, x_{2l}, \ldots, x_{ql})$ is given by (17.7).

If we wish to test the hypothesis that the expected values have certain specified values, so that

$$\boldsymbol{\xi} = \boldsymbol{\xi}_0$$

where $\boldsymbol{\xi}_0$ is a completely specified *vector* we can use the likelihood ratio method to construct a test. (Note that in this case the hypothesis tested—H_0: $\boldsymbol{\xi} = \boldsymbol{\xi}_0$—is simple, but the alternative $\boldsymbol{\xi} \neq \boldsymbol{\xi}_0$ is not.) The critical region for this test is defined by an inequality of the form

$$n \sum_{i=1}^{q} \sum_{j=1}^{q} \sigma^{ij} \left(\overline{X}_i - \xi_{i0} \right)\left(\overline{X}_j - \xi_{j0} \right) > K \quad (17.8)$$

where $\bar{X}_i = n^{-1}\sum_{l=1}^{n} X_{il}$ and K is a suitable chosen constant. The level of significance can be varied by varying K.

If the hypothesis H_0 is true the joint probability density function of the sample means $\bar{X}_1, \bar{X}_2, \ldots, \bar{X}_q$ is

$$p_{\bar{X}_1, \bar{X}_2, \ldots, \bar{X}_q}\left(\bar{x}_1, \bar{x}_2, \ldots, \bar{x}_q\right)$$

$$= (2\pi)^{-q/2} n^{q/2} |V|^{-1/2} \exp\left[-\frac{n}{2}\sum_{i=1}^{q}\sum_{j=1}^{q}\sigma^{ij}\left(\bar{x}_i - \xi_{i0}\right)\left(\bar{x}_j - \xi_{j0}\right)\right]$$

By transforming to new variables $Y_1, Y_2, \ldots, Y_q$ which are linear functions of the deviations $(\bar{X}_i - \xi_{i0})$ in such a way that

$$n\sum_{i=1}^{q}\sum_{j=1}^{q}\sigma^{ij}\left(\bar{X}_i - \xi_{i0}\right)\left(\bar{X}_j - \xi_{j0}\right) = \sum_{i=1}^{q} Y_i^2$$

it can be shown that, if H_0 is true then this quantity is distributed as χ^2 with q degrees of freedom. More generally, it is distributed as noncentral χ^2 with q degrees of freedom and noncentrality parameter

$$n\sum_{i=1}^{q}\sum_{j=1}^{q}\sigma^{ij}(\xi_i - \xi_{i0})(\xi_j - \xi_{j0})$$

Hence to obtain a level of significance α, we take $K = \chi^2_{q,1-\alpha}$. The inequality (17.8) is then $T^2 > K$. The statistic

$$T^2 = n\sum_{i=1}^{q}\sum_{j=1}^{q}\sigma^{ij}\left(\bar{X}_i - \xi_{i0}\right)\left(\bar{X}_j - \xi_{j0}\right) \tag{17.9}$$

is called Hotelling's T^2 after Harold Hotelling. If $q = 1$, we have

$$T^2 = \frac{n\left(\bar{X}_1 - \xi_{10}\right)^2}{\sigma^2}$$

where $\sigma^2 = \sigma_{11}$.

Example 17.4 The inner and outer diameters (ID and OD, respectively), of a tube are known to be normally distributed with correlation 0.6. Each dimension has standard deviation 0.04 units. It is required to test whether the mean ID (x_1) is equal to 3 units and the mean OD (x_2) is equal to 4 units.

Evidently 2 separate tests could be applied, one to measurements of ID and one to measurements of OD. The tests would not be independent, so there would be some difficulty in deciding precisely how to combine the results, but they would suffice for most practical purposes. Also, they would immediately indicate which dimension has the more serious departure from the specified mean values 3 (for ID) and 4 (for OD).

However, assuming the joint distribution to be bivariate normal, it is possible to use Hotelling's T^2 test, which does require only a single criterion to be calculated. To calculate this statistic we must first invert the variance–covariance matrix

$$\begin{pmatrix} \sigma_1^2 & \sigma_{12} \\ \sigma_{12} & \sigma_2^2 \end{pmatrix} = \begin{pmatrix} 0.00160 & 0.00096 \\ 0.00096 & 0.00160 \end{pmatrix}$$

The inverse matrix is

$$\frac{1}{(0.00160)(0.64)} \begin{pmatrix} 1 & -0.6 \\ -0.6 & 1 \end{pmatrix}$$

If we have a random sample of size n giving sample mean ID equal to $\overline{X}_1$, and sample mean OD equal to $\overline{X}_2$, then we can use the criterion (17.9)

$$T^2 = \frac{n}{0.001024} \left[\left(\overline{X}_1 - 3\right)^2 - 1.2\left(\overline{X}_1 - 3\right)\left(\overline{X}_2 - 4\right) + \left(\overline{X}_2 - 4\right)^2 \right]$$

Upper significance limits for this statistic are $\chi^2_{2,0.95} = 5.991$ and $\chi^2_{2,0.99} = 9.210$.

If the true population means of ID and OD are ξ_1, ξ_2, respectively, then T^2 will be distributed as noncentral χ^2 with 2 degrees of freedom and noncentrality parameter

$$\lambda = \frac{n}{0.001024} \left[(\xi_1 - 3)^2 - 1.2(\xi_1 - 3)(\xi_2 - 4) + (\xi_2 - 4)^2 \right]$$

Pairs of values (ξ_1, ξ_2) giving the same value for λ have equal chances of being detected by the T^2 test. The chance of detection increases with λ. Note that $(\xi_1 - \delta, \xi_2 - \delta)$ has less chance of being detected than $(\xi_1 - \delta, \xi_2 + \delta)$ or $(\xi_1 + \delta, \xi_2 - \delta)$. The latter 2 cases have equal chances of detection.

For any given (nonzero) pair of values (ξ_1, ξ_2), λ increases with n. By taking a large enough sample, the chance of detection can be made as large as desired.

Note that the region defined by $T^2 < \chi^2_{2, 1-\alpha}$ is a $100(1 - \alpha)$ percent confidence region for (ξ_1, ξ_2). A straightforward modification of the T^2 test, as described above, provides a test of the hypothesis that 2 multinormal populations with variance–covariance matrices $V_1 \equiv (\sigma_{ij}^{(1)})$ and $V_2 \equiv (\sigma_{ij}^{(2)})$ have identical expected value vectors. This hypothesis means that each of the measured characters has the same expected value in the 2 populations.

We simply consider the differences between the sample means in the 2 populations for each character. Let X_{itl} represent the value of the character X_i for the lth individual in the random sample (containing n_t individuals in all) from the tth population. Then, in place of the mean $\overline{X}_i$ we consider the difference $D_i = \overline{X}_{i1.} - \overline{X}_{i2.}$ where $\overline{X}_{it.} = n_t^{-1} \Sigma_{l=1}^{n_t} X_{itl}$. The statistics

$D_1, D_2, \ldots, D_q$ have the joint probability density function

$$p_{D_1, D_2, \ldots, D_q}(d_1, d_2, \ldots, d_q)$$

$$= (2\pi)^{-q/2} |\overline{V}|^{-1/2} \exp\left[-\tfrac{1}{2} \sum_{i=1}^{q} \sum_{j=1}^{q} \bar{\sigma}^{ij}(d_i - \delta_i)(d_j - \delta_j) \right]$$

where

$$\delta_i = (\text{expected value of } X_i \text{ in first population})$$

$$- (\text{expected value of } X_i \text{ in second population})$$

$$\overline{V} = (\bar{\sigma}_{ij}), \qquad \overline{V}^{-1} = (\bar{\sigma}^{ij})$$

$$\bar{\sigma}_{ij} = \text{cov}(D_i, D_j) = n_1^{-1} \sigma_{ij}^{(1)} + n_2^{-1} \sigma_{ij}^{(2)}$$

The hypothesis of identity of expected value vectors can then be stated in the simple form $\delta_1 = \delta_2 = \cdots = \delta_q = 0$. We are thus led to use the criterion

$$T^2 = \sum_{i=1}^{q} \sum_{j=1}^{q} \bar{\sigma}^{ij} D_i D_j \qquad (17.10)$$

T^2 is distributed as a noncentral χ^2 with q degrees of freedom and noncentrality parameter $\lambda = \sum_{i=1}^{q} \sum_{j=1}^{q} \bar{\sigma}^{ij} \delta_i \delta_j$.

If the 2 populations have identical variance-covariance matrices, then

$$\sigma_{ij}^{(1)} = \sigma_{ij}^{(2)} = \sigma_{ij}, \quad \text{for example,}$$

$$\bar{\sigma}_{ij} = \frac{n_1 + n_2}{n_1 n_2} \sigma_{ij}$$

and

$$\bar{\sigma}^{ij} = \frac{n_1 n_2}{n_1 + n_2} \sigma^{ij}$$

Hence, in this case

$$T^2 = \frac{n_1 n_2}{n_1 + n_2} \sum_{i=1}^{q} \sum_{j=1}^{q} \sigma^{ij} D_i D_j \qquad (17.11)$$

and the noncentrality parameter is

$$\lambda = \frac{n_1 n_2}{n_1 + n_2} \sum_{i=1}^{q} \sum_{j=1}^{q} \sigma^{ji} \delta_i \delta_j \qquad (17.12)$$

*17.7.1   More Than Two Populations

Now suppose we have k (>2) populations with known variance–covariance matrices. For simplicity we restrict ourselves to the case when all k of these matrices are exactly the same as each other. (The more general case can be treated in a similar fashion, but the algebraic expressions used are sometimes rather heavy.)

Suppose we have random samples of sizes $n_1, n_2, \ldots, n_k$, respectively. If we want to test the hypothesis that the q expected values are, in the first population, $\xi_{11}, \xi_{21}, \ldots, \xi_{q1}$, in the second population, $\xi_{12}, \xi_{22}, \ldots, \xi_{q2}$, and generally in the tth population $\xi_{1t}, \xi_{2t}, \ldots, \xi_{qt}$ (where t can take any integer value from 1 to k inclusive) we can combine together the criteria

$$T_t^2 = n_t \sum_{i=1}^{q} \sum_{j=1}^{q} \sigma^{ij} \left(\overline{X}_{it.} - \xi_{it} \right) \left(\overline{X}_{jt.} - \xi_{jt} \right)$$

(where $\overline{X}_{it.}$ denotes the sample mean of the ith variable in the ith population) into the single criterion $\sum_{t=1}^{k} T_t^2$. If the hypothesis tested is true, the statistics T_t^2 are a mutually independent set, and each is distributed as χ^2 with q degrees of freedom. Hence $\sum_{t=1}^{k} T_t^2$ will be distributed as χ^2 with kq degrees of freedom if the mean value vectors are $(\xi_{1t}, \xi_{2t}, \ldots, \xi_{qt})$ as specified by the hypothesis.

This hypothesis is not encountered very often. However, with slight modification the more common hypothesis

$$\xi_{i1} = \xi_{i2} = \cdots = \xi_{ik} \qquad \text{for all } i = 1, 2, \ldots, q$$

can be tested by this kind of statistic. By an appropriate transformation we can show that this is equivalent to a hypothesis stating that each of $(k-1)$ sets of multinormal variables has a *zero* expected value vector—just as the hypothesis "$\xi_{i1} = \xi_{i2}$ for all i" was reduced to the hypothesis "$\delta_i = 0$ for all i." The hypothesis (that all k populations have the same expected values vector) is tested by criterion

$$\sum_{t=1}^{k} n_t \sum_{i=1}^{q} \sum_{j=1}^{q} \sigma^{ij} \left(\overline{X}_{it.} - \overline{X}_{i..} \right) \left(\overline{X}_{jt.} - \overline{X}_{j..} \right) \qquad (17.13)$$

where

$$\overline{X}_{i..} = \frac{\sum_{t=1}^{k} n_t \overline{X}_{it.}}{\sum_{t=1}^{k} n_t}$$

This criterion is distributed as noncentral χ^2 with $(k-1)q$ degrees of freedom and noncentrality parameter $\sum_{t=1}^{k} n_t \sum_{i=1}^{q} \sum_{j=1}^{q} \sigma^{ij} (\xi_{it} - \xi_i)(\xi_{jt} - \xi_j)$ where $\xi_i = (\sum_{t=1}^{k} n_t \xi_{it})/(\sum_{t=1}^{k} n_t)$. If the hypothesis is true, the distribution is that of a central χ^2, and the upper percentage points of χ^2 with $(k-1)q$ degrees of freedom can be used as significance limits for the criterion.

*17.7.2   Procedures when the Variance–Covariance Matrix is not Known

Even if the variances and covariances are not known with complete accuracy, it will very often be possible to use reasonably approximate values. Some univariate tests (such as the t tests) may be used. If, however, it is really desired to apply "portmanteau" tests in such cases, then the criterion of the preceding section cannot be used. For the case of a single T^2 the modification is simple. The σ_{ij}'s are replaced by unbiased estimators S_{ij}. For the single-sample case

$$S_{ij} = (n-1)^{-1} \sum_{l=1}^{n} \left(X_{il} - \bar{X}_{i.} \right)\left(X_{jl} - \bar{X}_{j.} \right)$$

For the 2-sample case

$$S_{ij} = (n_1 + n_2 - 2)^{-1} \sum_{t=1}^{2} \sum_{l=1}^{n_t} \left(X_{itl} - \bar{X}_{it.} \right)\left(X_{jtl} - \bar{X}_{jt.} \right)$$

Then

$$T'^2 = \frac{n}{q} \sum\sum S^{ij} \left(\bar{X}_i - \xi_{i0} \right)\left(\bar{X}_j - \xi_{j0} \right) \text{ for the single sample case} \qquad (17.14)$$

or

$$T'^2 = \frac{n_1 n_2}{(n_1 + n_2)q} \sum\sum S^{ij} (D_i - \delta_i)(D_j - \delta_j) \text{ for the 2-sample case} \qquad (17.15)$$

are used.

These are known as generalized Hotelling's T^2 statistics.

Note that if $q = 1$, using the notation of Chapters 7 and 8,

$$T'^2 = n\left(\bar{X}_1 - \xi_{10} \right)^2 / S^2 \text{ for the single sample}$$

$$T'^2 = \left[\bar{X}_1 - \bar{X}_2 - \delta \right]^2 \left[\frac{(n_1 - 1)S_1^2 + (n_2 - 1)S_2^2}{n_1 + n_2 - 2} \left(\frac{1}{n_1} + \frac{1}{n_2} \right) \right]^{-1} \text{ for two samples}$$

If the hypothesis tested is true, then T'^2 is distributed as F with q, $(n - q)$ or q, $(n_1 + n_2 - q - 1)$ degrees of freedom, respectively.

It would be possible similarly to use a criterion of the form

$$\sum_{t=1}^{k} \frac{n_t}{q} \sum_{i=1}^{q} \sum_{j=1}^{q} S^{ij} \left(\bar{X}_{it.} - \bar{X}_{i..} \right)\left(\bar{X}_{jt.} - \bar{X}_{j..} \right) \qquad (17.16)$$

to test the hypothesis $\xi_{i1} = \xi_{i2} = \cdots = \xi_{ik}$ for all i. If the hypothesis is true this hypothesis is distributed as F with kq, $(\sum_{t=1}^{k} n_t - k - q + 1)$ degrees of freedom. (It is assumed that each n_t is bigger than q.)

However, it should be noted that applying the likelihood method, as modified for composite hypotheses, leads to the criterion

$$\frac{|S_{ij(a)}|}{|S_{ij(r)}|} \tag{17.17}$$

where $|S_{ij(a)}|$ is a determinant with element in the ith row and jth column

$$S_{ij(a)} = \left(\sum_{t=1}^{k} n_t\right)^{-1} \sum_{t=1}^{k} \sum_{l=1}^{n_t} \left(X_{itl} - \bar{X}_{it.}\right)\left(X_{jtl} - \bar{X}_{jt.}\right)$$

and similarly $|S_{ij(r)}|$ is a determinant with elements

$$S_{ij(r)} = \left(\sum_{t=1}^{k} n_t\right)^{-1} \sum_{t=1}^{k} \sum_{l=1}^{n_t} \left(X_{itl} - \bar{X}_{i..}\right)\left(X_{jtl} - \bar{X}_{j..}\right)$$

We can write

$$S_{ij(r)} = S_{ij(a)} + S_{ij(b)}$$

where

$$S_{ij(b)} = \left(\sum_{t=1}^{k} n_t\right)^{-1} \sum_{t=1}^{k} n_t\left(\bar{X}_{it.} - \bar{X}_{i..}\right)\left(\bar{X}_{jt.} - \bar{X}_{j..}\right)$$

In the univariate case we would have the criterion

$$S_{11(a)}/(S_{11(a)} + S_{11(b)}) \tag{17.18}$$

and $S_{11(b)}/S_{11(a)}$ is equivalent to this criterion. However, in the multivariate case, $|S_{ij(b)}|/|S_{ij(a)}|$ is not an equivalent criterion to $|S_{ij(a)}|/|S_{ij(a)} + S_{ij(b)}|$.

Example 17.5 Suppose that there are available pairs of measurements of characters X_1, X_2 on each of n randomly chosen individuals, and, furthermore, that these measurements can be represented by random variables X_{1i}, X_{2i} ($i = 1, 2, \ldots, n$) having the same joint bivariate normal distribution for all i. ((X_{1i}, X_{2i}) and ($X_{1i'}, X_{2i'}$) are supposed to be mutually independent if $i \neq i'$.) It is desired to test the hypothesis H_0 which specifies that

$$\text{var}(X_{1i}) = \text{var}(X_{2i})$$

and

$$E(X_{1i}) = E(X_{2i})$$

It is rather simpler to work in terms of the new (but equivalent) variables

$$Y_{1i} = \frac{X_{1i} - X_{2i}}{\sqrt{2}}$$

$$Y_{2i} = \frac{X_{1i} + X_{2i}}{\sqrt{2}}$$

The hypothesis H_0 can be regarded as being defined by the conditions:

$$Y_{1i} \quad \text{and} \quad Y_{2i} \quad \text{are uncorrelated}$$

and

$$E(Y_{1i}) = 0$$

If the likelihood ratio (see Section 7.9) is calculated it is found to be

$$\left\{ (1 - R_Y^2) \left[1 + \left(\frac{n\overline{Y}_1^2}{S_{Y_1}^2} \right) \right]^{-1} \right\}^{n/2}$$

where

$$R_Y^2 = \left[\sum_{i=1}^{n} (Y_{1i} - \overline{Y}_1)(Y_{2i} - \overline{Y}_2) \right]^2 \left[\sum_{i=1}^{n} (Y_{1i} - \overline{Y}_1)^2 \right]^{-1} \left[\sum_{i=1}^{n} (Y_{2i} - \overline{Y}_2)^2 \right]^{-1}$$

$$\overline{Y}_1 = n^{-1} \sum_{i=1}^{n} Y_{1i}; \quad \overline{Y}_2 = n^{-1} \sum_{i=1}^{n} Y_{2i}; \quad S_{Y_1}^2 = \sum_{i=1}^{n} (Y_{1i} - \overline{Y}_1)^2$$

Hence

$$L_0 = (1 - R_Y^2) \left[1 + \left(\frac{n\overline{Y}_1^2}{S_{Y_1}^2} \right) \right]^{-1}$$

can be used as a criterion.

Writing

$$L_1 = 1 - R_Y^2; \quad L_2 = \left\{ 1 + \frac{n\overline{Y}_1^2}{S_{Y_1}^2} \right\}^{-1}$$

it can be shown, using (13.88) and (5.42) that

$$p_{L_1}(l_1 | H_0) = \frac{1}{B\left[\frac{1}{2}(n-2), \frac{1}{2}\right]} l_1^{(n-4)/2} (1 - l_1)^{-1/2} \quad (0 < l_1 < 1)$$

$$p_{L_2}(l_2 | H_0) = \frac{1}{B\left[\frac{1}{2}(n-1), \frac{1}{2}\right]} l_2^{(n-3)/2} (1 - l_2)^{-1/2} \quad (0 < l_2 < 1)$$

Furthermore, L_1 and L_2 are mutually independent when H_0 is true. Making the transformation $l_0 = l_1 l_2$; $l' = l_2$ which has the inverse $l_1 = l_0/l'$, $l_2 = l'$ and the Jacobian $\partial(l_1, l_2)/\partial(l_0, l') = l'^{-1}$, we find

$$p_{L_0, L'}(l_0, l' | H_0) = \frac{l_0^{(n-4)/2}(l' - l_0)^{-1/2}(1 - l')^{-1/2}}{B\left[\frac{1}{2}(n-2), \frac{1}{2}\right] B\left[\frac{1}{2}(n-1), \frac{1}{2}\right]} \quad (0 < l_0 < l' < 1)$$

Integrating out l'

$$p_{L_0}(l_0|H_0) = \frac{l_0^{(n-4)/2}}{B\left[\frac{1}{2}(n-2),\frac{1}{2}\right]B\left[\frac{1}{2}(n-1),\frac{1}{2}\right]} \int_{l_0}^1 (l'-l_0)^{-1/2}(1-l')^{-1/2}dl'$$

$$= \frac{1}{2}(n-2)l_0^{(n-4)/2} \quad (0 < l_0 < 1)$$

This particularly simple result means that the 100α percent significance limit $L_{0,\alpha}$ of L_0 satisfies the formula

$$\frac{1}{2}(n-2)\int_0^{L_{0,\alpha}} l^{(n-4)/2}dl = \alpha$$

and so can be calculated directly from the formula

$$L_{0,\alpha} = \alpha^{2/(n-2)}$$

17.7.3   Test of Equality of Variance–Covariance Matrices

In much of the above analysis it has been assumed that 2 or more multinormal populations have identical variance–covariance matrices. Sometimes it may be necessary to test whether this is a reasonable assumption, just as we may test equality of variances to check whether some of the conditions for applying analysis of variance techniques are reasonable (see Chapter 13).

A test of this hypothesis can be obtained by applying the likelihood ratio method as modified for composite hypotheses. If the results of measurements on each individual in random samples of size $n_1, n_2, \ldots, n_k$ respectively, are available, the likelihood ratio is found to be equal to

$$\frac{\prod_{t=1}^k |S'_{ijt}|^{n_t/2}}{|S'_{ij}|^{\Sigma n_t/2}}$$

where $|S'_{ijt}|$ denotes the determinant with the element in the ith row and jth column equal to $S'_{ijt} = n_t^{-1}\Sigma_{l=1}^{n_t}(X_{itl} - \bar{X}_{it.})(X_{jtl} - \bar{X}_{jt.})$, and $|S'_{ij}|$ denotes the determinant with element in the ith row and jth column equal to

$$S'_{ij} = \left(\sum_{t=1}^k n_t\right)^{-1} \sum_{t=1}^k \sum_{l=1}^{n_t} (X_{itl} - \bar{X}_{it.})(X_{jtl} - \bar{X}_{jt.})$$

Analogously to the univariate case we use as criterion the natural modification

$$L = \frac{\prod_{t=1}^k |S_{ijt}|^{(n_t-1)/2}}{|S_{ij}|^{\frac{1}{2}\sum_{t=1}^k (n_t-1)}}$$

where

$$S_{ijt} = (n_t - 1)^{-1} \sum_{l=1}^{n_t} \left(X_{itl} - \overline{X}_{it.} \right) \left(X_{jtl} - \overline{X}_{jt.} \right)$$

and

$$S_{ij} = \left[\sum_{t=1}^{k} (n_t - 1) \right]^{-1} \sum_{t=1}^{k} \sum_{l=1}^{n_t} \left(X_{itl} - \overline{X}_{it.} \right) \left(X_{jtl} - \overline{X}_{jt.} \right)$$

If the variance–covariance matrices of the k populations *are* identical then (provided $q \geqslant 2$)

$$-2 \left[1 - \frac{(2q^2 + 3q - 1)}{6(q+1)(k-1)} \left[\sum_{t=1}^{k} \frac{1}{n_t - 1} - \frac{1}{\sum_{t=1}^{k} (n_t - 1)} \right] \right] \ln L \quad (17.19)$$

is approximately distributed as χ^2 with $\frac{1}{2}(q+1)(k-1)$ degrees of freedom.

Just as in the univariate case, this test of homogeneity of variance–covariance matrices is not very robust. If the population joint distributions are not multinormal, the nominal significance levels may be quite inaccurate.

Example 17.6 In the same circumstances as those described in Example 17.5 suppose that it is desired to test the hypothesis H_0' which specifies

$$\text{var}(X_{1i}) = \text{var}(X_{2i})$$

and the correlation between X_{1i} and X_{2i} has a specified value ρ_0.

Introducing the random variables

$$Y_{1i} = \frac{X_{1i} - X_{2i}}{\sqrt{2}}; \qquad Y_{2i} = \frac{X_{1i} + X_{2i}}{\sqrt{2}}$$

as in Example 17.5, H_0' can be seen to be equivalent to specifying that

$$Y_{1i} \quad \text{and} \quad Y_{2i} \quad \text{are uncorrelated}$$

$$\text{var}(Y_{1i}) = \left(\frac{1 - \rho_0}{1 + \rho_0} \right) \text{var}(Y_{2i})$$

The likelihood ratio for testing H_0' is found to be

$$\left[(1-R_Y^2) \left[\frac{4(S_{Y_2}^2/S_{Y_1}^2)}{\left(\dfrac{1+\rho_0}{1-\rho_0}\right)\left(1+\dfrac{1-\rho_0}{1+\rho_0}\cdot\dfrac{S_{Y_2}^2}{S_{Y_1}^2}\right)^2} \right] \right]^{n/2}$$

Therefore,

$$L_0' = (1-R_Y^2) \times \frac{4(S_{Y_2}^2/S_{Y_1}^2)}{\left(\dfrac{1+\rho_0}{1-\rho_0}\right)\left(1+\dfrac{1-\rho_0}{1+\rho_0}\cdot\dfrac{S_{Y_2}^2}{S_{Y_1}^2}\right)^2}$$

may be used as a criterion.
The distribution of

$$L_3 = \frac{4(S_{Y_2}^2/S_{Y_1}^2)}{\left(\dfrac{1+\rho_0}{1-\rho_0}\right)\left(1+\dfrac{1-\rho_0}{1+\rho_0}\cdot\dfrac{S_{Y_2}^2}{S_{Y_1}^2}\right)^2}$$

when H_0' is true can be shown to be the same as that of L_2 when H_0 is true. Further, R_Y^2 and $S_{Y_2}^2/S_{Y_1}^2$ are mutually independent. Therefore, following exactly the same analysis as in Example 17.5 the conclusion is reached that

$$p_{L_0}(l_0'|H_0') = \tfrac{1}{2}(n-2){l_0'}^{(n-4)/2} \qquad (0 < l_0' < 1)$$

and the significance limits for L_0' can be found from the same simple formula as those for L_0.

17.8   MULTIVARIATE ANALYSIS OF VARIANCE (MANOVA)

All the univariate analysis of variance (ANOVA) methods described in Chapters 13 to 15 can be extended in a straightforward way to cases when more than one variate is measured on each individual. The resultant techniques (multivariate analysis of variance—often abbreviated as MANOVA) are appropriate to testing hypotheses on parameters in the models relevant to each of the variables involved. It is a restriction of the method that the hypotheses specify the same relations among the parameters for each variable. For example, in the analysis of a 2^3 factorial with factors A, B, and C the extension of the univariate hypothesis that the $A \times B$ interaction is negligible (for some particular variable X_1, say), is that the $A \times B$ interaction is negligible for *each* variable $X_1, \ldots, X_q$, assuming a linear model for each variable and a structure of expected values the same for each variable.

In order to construct the likelihood test of such a multivariate hypothesis we construct a variance–covariance matrix for each line in the (univariate) analysis of variance table. Taking, for example, a 2^3 factorial with r replications, the model is

$$X_{uijkl} = \xi_u + \rho_{ul} + \alpha_{ui} + \beta_{uj} + \gamma_{uk} + (\beta\gamma)_{ujk}$$
$$+ (\alpha\gamma)_{uik} + (\alpha\beta)_{uij} + (\alpha\beta\gamma)_{uijk} + Z_{uijkl} \qquad (17.20)$$

where $\quad u = 1, \ldots, q$ (u denotes variable) $\quad i, j, k = 1, 2 \quad l = 1, \ldots, r$

The hypothesis of zero $A \times B$ interaction specifies $(\alpha\beta)_{uij} = 0$ for all u, all i, and all j.

The $A \times B$ matrix, $\mathbf{S}_{AB}$, has for its u, vth element

$$\frac{1}{8r} \left(\overline{X}_{u22..} + \overline{X}_{u11..} - \overline{X}_{u12..} - \overline{X}_{u21..} \right) \left(\overline{X}_{v22..} + \overline{X}_{v11..} - \overline{X}_{v12..} - \overline{X}_{v21..} \right)$$

The Total matrix $\mathbf{S}_T$ has, for its u, vth element

$$\sum_{i=1}^{2} \sum_{j=1}^{2} \sum_{k=1}^{2} \sum_{l=1}^{r} \left(X_{uijkl} - \overline{X}_{u\ldots} \right) \left(X_{vijkl} - \overline{X}_{v\ldots} \right)$$

The Residual matrix is

$$\mathbf{S}_R = \mathbf{S}_T - \mathbf{S}_A - \mathbf{S}_B - \mathbf{S}_C - \mathbf{S}_{BC} - \mathbf{S}_{BA} - \mathbf{S}_{ABC} - \mathbf{S}_{Rep}$$

where $\mathbf{S}_A$, $\mathbf{S}_B$, etc. are calculated in a manner similar to $\mathbf{S}_{AB}$. $\mathbf{S}_{Rep}$—the Between Replication matrix—has u, vth element

$$\frac{1}{8r} \sum_{l=1}^{r} \left(\overline{X}_{u\ldots l} - \overline{X}_{u\ldots} \right) \left(\overline{X}_{v\ldots l} - \overline{X}_{v\ldots} \right)$$

The test criterion is the ratio of the determinant $|\mathbf{S}_R|$ to the determinant $|\mathbf{S}_R + \mathbf{S}_{AB}|$. Low values of the ratio $|\mathbf{S}_R|/|\mathbf{S}_R + \mathbf{S}_{AB}|$ are regarded as evidence for nonzero values of some, at least of the $(\alpha\beta)_{uij}$'s.

The null hypothesis distribution depends on the numbers of degrees of freedom, ν_2, say, and also on the number of variables, q. There are tables (see [17.7] and [17.8]) of lower percentage points of the ratio (which is often denoted by U_{q, ν_2, ν_R}). For practical purposes a useful approximate formula is

$$-\left\{\nu_R - \tfrac{1}{2}(q - \nu_2 + 1)\right\} \log U_{q,\nu_2,\nu_R,\alpha} \doteq \chi^2_{q\nu_2,\,1-\alpha} \qquad (17.21)$$

This approximation can be improved by using multipliers interpolated from values given in a table constructed by M. Schatzoff [17.7]. An extract from this table is given as Table J of the Appendix.

Note that this table is entered with α, q, ν_2 and a quantity $M = \nu_R - q + 1$. For large M the multiplier is very nearly (and slightly greater than) 1. For M fairly large ($M \geqslant 10$, say) the multiplier for $\alpha = 0.05$ and $\alpha = 0.01$ do not differ greatly, and for $\nu_2 \geqslant 6$ and $M \geqslant 5$ variation with q is not great (for $q \leqslant 5$).

For the bivariate case, when $q = 2$, we can use the result that

$$\frac{1 - \sqrt{U_{2,\nu_2,\nu_R}}}{\sqrt{U_{2,\nu_2,\nu_R}}} \cdot \frac{\nu_R - 1}{\nu_2} \qquad (17.22)$$

is distributed as F with $2\nu_2$, $2(\nu_R - 1)$ degrees of freedom.

This result also gives the distribution of $U_{q,2,\nu_R}$ for general q because $U_{\nu_2,q,\nu_2 + \nu_R - 2}$ has the same distribution as U_{q,ν_2,ν_R}. In particular, $U_{q,2,\nu_R}$ has the same distribution as $U_{2,q,\nu_2 + \nu_R - 2}$ and (17.22) can be used to evaluate this latter distribution. This relationship also extends the effective range of Table J of the Appendix.

The method applies quite generally. If the main A effect is to be tested (i.e., to test the hypothesis that the main A effect is 0 for each of the q variables) the test criterion $|\mathbf{S}_R|/|\mathbf{S}_R + \mathbf{S}_A|$ is used. (There is another example in Section 17.7.2, just before Example 17.5.)

An alternative criterion for testing a general linear hypothesis H is the maximum root, $\hat{\lambda}_1$, say, of the determinantal equation

$$|\mathbf{S}_{(H)} - \lambda(\mathbf{S}_R + \mathbf{S}_{(H)})| = 0 \qquad (17.23)$$

where $\mathbf{S}_{(H)}$ is the matrix of sums of squares and products corresponding to "departure from H."

Very often $\hat{\theta}_1 = \hat{\lambda}_1(1 - \hat{\lambda}_1)^{-1}$, which is the maximum root of the equation

$$|\mathbf{S}_{(H)} - \theta \mathbf{S}_R| = 0 \qquad (17.24)$$

is used as the criterion. Since $\hat{\theta}_1$ and $\hat{\lambda}_1$ are monotonically related, tests based on them are equivalent to each other.

D. F. Morrison [17.5] gives a number of examples of tests based on this criterion, which was originally suggested by S. N. Roy [17.6].

Further possible criteria have been suggested. They are all functions of roots of the equation (17.23) [or (17.24)].

It is natural to wonder which test criterion (the likelihood ratio, maximum root, or some other) to use in a particular situation. Ideally the answer to this would follow from a comparison of powers of the tests with respect to a number of likely alternative hypotheses. With the large number of available parameters in multivariate problems it is often difficult to specify likely alternatives with much confidence. Investigations have shown that the likelihood ratio test provides fairly good all-round power; the maximum root test provides better power against "one-dimensional" alternatives—for example, when just one population mean vector differs from the others in a group—but is weaker against other alternatives.

Example 17.7 We will use Table J to evaluate approximately $U_{10,6,16,0.05}$

$$U_{10,6,16,0.05} = U_{6,10,12,0.05}$$

In evaluating $U_{6,10,12,0.05}$, $M = 12 - 6 + 1 = 7$. From Table J, interpolating with respect to $12M^{-1}$, and keeping $M = 7$, $v_2 = 10$, $\alpha = 0.05$, we find

$$K \doteq 1.082 \quad \text{for } q = 3$$

$$K \doteq 1.080 \quad \text{for } q = 4$$

$$K \doteq 1.080 \quad \text{for } q = 5$$

Extrapolating to $q = 6$ we take $K \doteq 1.079$ and find

$$- \{ 16 - \tfrac{1}{2}(10 - 6 + 1) \} \ln U_{10,6,16,0.05} \doteq 1.079 \times \chi^2_{60,0.95}$$

$$U_{10,6,16,0.05} \doteq \exp\left(\frac{-1.079 \times 79.08}{13.5} \right) = \exp(-6.3205) = 0.001800$$

From [17.7] we find the exact value 0.001865.

It is a defect of most multivariate tests that they test simultaneously a considerable number of relationships among parameters. In the general linear hypothesis case they test separate and similar relationships among parameters relating to each of the q measured variates and give no direct detailed information on each specific relationship. There are, as pointed out in Section 17.7, univariate tests involving many parameters (for example, tests of equality among a large number of population means), which also suffer from this defect, but the situation is much aggravated in the multivariate case.

Among practical remedies suggested for this state of affairs we only mention a system put forward by K. R. Gabriel [17.2]. He proposes that the test criterion used be evaluated not only for each possible subset of the q variates, but for each possible subset of the experimental design. For example, if equality of population mean vectors among k populations is being tested, test criteria would be evaluated for all possible

$$\left[1+\binom{q}{q-1}+\binom{q}{q-2}+\cdots+\binom{q}{1}\right]\left[1+\binom{k}{k-1}+\cdots+\binom{k}{2}\right]$$

$$=(2^q-1)(2^k-k-1)$$

combinations of subsets of the q variates and the k populations (with, of course, at least 2 populations). Conclusions are then drawn from a study of the resultant criteria values. A method of assessment for the case when the criterion is the maximum root is given in [17.2]. Example 17.8, based on part of the data used in [17.2] illustrates the procedure.

Example 17.8   The values of the maximum root criterion ($\hat{\theta}_1$) for all subsets of 3 characters X, Y, Z measured on anteaters' skulls in a few localities (1, Santa Maria, Colombia; 2, Minas Gerais, Brazil; 3, Matto Grosso, Brazil; 6, Mexico) and all subsets of at least 2 of the localities are set out below. (They are taken from more extensive data, for 6 localities, in [17.2].)

Locations	XYZ	XY	XZ	YZ	X	Y	Z
1 2 3 6	0.679	0.600	0.677	0.620	0.559	0.481	0.224
1 2 3 ·	0.675	0.570	0.673	0.614	0.501	0.397	0.003
1 2 · 6	0.593	0.533	0.592	0.533	0.505	0.439	0.221
1 · 3 6	0.591	0.529	0.589	0.526	0.498	0.430	0.223
· 2 3 6	0.275	0.088	0.275	0.226	0.018	0.035	0.184
1 2 · ·	0.565	0.452	0.563	0.501	0.388	0.294	0.002
1 · 3 ·	0.568	0.457	0.565	0.500	0.387	0.291	0.002
1 · · 6	0.378	0.377	0.377	0.347	0.375	0.344	0.218
· 2 3 ·	0.019	0.011	0.018	0.015	0.009	0.007	0.000
· 2 · 6	0.242	0.070	0.241	0.221	0.001	0.010	0.138
· · 3 6	0.225	0.077	0.225	0.211	0.015	0.035	0.164

The number of skulls measured were:

Location	1	2	3	4
Number of skulls	21	6	9	5

In this case there are $(2^3-1)(2^4-4-1)=77$ values according to Gabriel's scheme of analysis. Each of these values is compared with the critical value (upper percentage point) appropriate when all the data are used. In the present case, using [17.3], the appropriate 5 percent critical value (for $q=3$, $k=4$ and residual degrees of freedom 37) is 0.303 approximately (the 1 percent value is approximately 0.376). Calculated values which are greater than 0.303 are underlined.

We do not give here further details of Gabriel's method, but the analysis so far given brings out quite clearly (*a*) the greater selectivity of measures X and Y, (*b*) the fact that location 1 is a major contributor to differences among locations.

REFERENCES

1. Anderson, T. W., *An Introduction to Multivariate Statistical Analysis*, 2nd ed., Wiley, New York, 1958

2. Gabriel, K. R., "Simultaneous Test Procedures in Multivariate Analysis," *Biometrika*, **55**, 489–504 (1968).

3. Heck, D. L., "Charts of Some Upper Percentage Points of the Distribution of the Largest Characteristic Root," *Annals of Mathematical Statistics*, **31**, 625–642 (1960).

4. Lachenbruch, P. A., *Discriminant Analysis*, Hafner, New York, 1975, (Chapter 2).

5. Morrison, D. F., *Multivariate Statistical Methods*, Wiley, New York, 1967.

6. Roy, S. N., *Some Aspects of Multivariate Analysis*, Wiley, New York, Indian Statistical Institute, Calcutta, 1957.

7. Schatzoff, M., "Exact Distribution of Wilks's Likelihood Ratio Criterion," *Biometrika*, **53**, 347–358 (1966).

8. Wall, F. J., *The Generalized Variance Ratio or U-Statistic*, Dikewood Corporation, Albuquerque, New Mexico, 1967.

9. Bishop, Y. M. M., S. E. Fienberg and P. W. Holland, *Discrete Multivariate Analysis: Theory and Practice*, MIT Press, Cambridge, Mass., 1975.

10. Gnanadesikan, R., *Methods for Statistical Data Analysis of Multivariate Observations*, New York, Wiley, 1977.

11. Kendall, Sir Maurice G., *Multivariate Analysis*, Griffin, London, 1975.

EXERCISES

1. The means and standard deviations of 3 measurements (X_1, refractive index; X_2, density; X_3, proportion of silicon) in samples of glass from 2 factories are

	FACTORY 1		FACTORY 2	
	MEAN	S.D.	MEAN	S.D.
X_1	1.37	0.03	1.41	0.03
X_2	1.72	0.07	1.70	0.07
X_3	0.65	0.06	0.66	0.06

Correlations (the same for each factory) are: between X_1 and X_2, 0.75; X_1 and X_3, 0.65; X_2 and X_3, 0.90.

A set of samples is known to come from one of the factories. How many of the samples should be measured to be 99% certain of a correct decision as to which factory produced them? It may be assumed that the joint distribution of the 3 characters for product from the same factory is multinormal.

2. How would your answer to Exercise 1 be affected by the following additional items of information?

(a) Factory 1 produces twice as much as Factory 2, and the chances that the set of samples come from these factories are in the same ratio.

(b) There will be time to take only 2 of the 3 measurements (X_1, X_2, and X_3).

3. Construct a sequential procedure appropriate to the conditions of Exercise 1.

4. Suppose that in Exercise 1, the cost of measuring X_1 is 2 units, of measuring X_2 is 1 unit, and X_3 is 4 units. Let the cost of incorrect classification be C units. How big should C be to make it worthwhile using all 3 variables, X_1, X_2, and X_3, in discriminating the source of
   (i) a single sample;
   (ii) a set of n samples (it being known that they are all from the same source)?

5. Answer Exercises 1 and 2, with Factory 2 standard deviations half as big as those shown in Exercise 1, but Factory 1 standard deviations unchanged.

6. Under the conditions stated in Exercise 1, we have a set of samples of glass known to contain products of Factories 1 and 2 mixed in unknown proportions
   (a) Obtain the best discriminator.
   (b) Out of 100 samples, 80 are allotted to Factory 1. Why is $80/100$ not necessarily an unbiased estimate of the proportion of samples from Factory 1?
   (c) Using the data described in (b) suggest an improvement in the discriminatory procedure.

7. Answer Exercise 6 under the conditions of Exercise 5.

8. Under the conditions described in Exercise 1 a single sample of glass gives $X_1 = 1.39$, $X_2 = 1.75$, $X_3 = 0.61$. Test the hypothesis that the expected values of X_1, X_2, and X_3 are those given for Factory 1.

9. Suppose that the values given in Exercise 8 are in fact arithmetic means of values of X_1, X_2, and X_3 observed on a random sample of n pieces. How big should n be for the hypothesis (of Exercise 8) to be rejected at the 1% level? Why would it not necessarily follow that the samples should be allotted to Factory 2 in this case?

10. Show how the T^2 statistic [as given in (17.9)] can be used to construct a confidence region for the expected values of a set of multinormal variables, given measurements on a random sample of size n.

11. Recall the random (component-of-variance) model (13.12).
    (a) Show that the joint distribution of $X_{t1}, X_{t2}, \ldots, X_{tn_t}$ is multinormal.
    (b) Express the hypothesis of no between groups variation in terms of a hypothesis about the variance–covariance matrix of the X_{ti}'s.
    (c) Obtain the likelihood ratio criterion for testing this hypothesis (with respect to alternatives specifying nonzero between group variation).

12. What is a linear discriminant function? F is a linear discriminant function used for differentiating between two populations Π_1, Π_2. The expected value of F in Π_1 is 0.71; in Π_2 it is 13.27. The standard deviation of F (in either population) is 1.10. (These values can be taken as, effectively, exact.) An individual is assigned to Π_1 if $F < 11.99$, to Π_2 if $F > 11.99$. Out of 500 individuals chosen at random from a large population consisting of a mixture of Π_1 and Π_2, 380 individuals are assigned to Π_1 and 120 to Π_2.

   Estimate how many individuals have been misclassified, and suggest an improved critical value for F, to be used in the classification of further individuals drawn from the same mixed population.

   You may assume F to be approximately normally distributed in each of Π_1 and Π_2. [Universtiy of London B.Sc. (Special) Examination 1960.]

13. The Zilch Company wishes to supply glass in competition with Factories 1 and 2 of Exercise 1. They find difficulty in meeting specification, but claim that their product is "indistinguishable" from that of Factories 1 and 2. To test this, equal (and large) numbers of samples from Factories 1 and 2 and from Zilch (that is, one-third of each kind), are mixed. Assuming that variances and covariances for Zilch are the same as for Factories 1 and 2, and that the average values of X_2 and X_3 are 1.65 and 0.70, respectively, find limits between which the average value of X_1 should fall for less than 40% of the Zilch product to be so assigned

  (a) if optimum assignment to all 3 sources is planned,

  (b) if Factories 1 and 2 are treated as a single "non-Zilch" source.

14. In order to help reach a decision as to which of 2 authors, A and B, wrote a certain manuscript, a number of works of undisputed authorship are studied. It is found that the number of occurrences of each of a number (k) of words in 2000 words of text can be represented by Poisson distributions for each author. The means of the distributions are not the same for each author; their values can be regarded as having been sufficiently well established by the study. Furthermore, calculations can be based on the assumption that the numbers of occurrences of the different words can be represented by mutually independent random variables (though a more accurate representation would quite possibly allow for dependence).

  The number of occurrences of each of the k words in 2000θ words of manuscripts is observed.

  Obtain the form of discriminant function to choose between the 2 authors.

15. In the preceding exercise the means for 8 words were

WORD	A	B	WORD	A	B
1	4.26	6.83	5	7.42	8.06
2	13.18	14.30	6	21.68	13.94
3	10.52	6.28	7	2.30	1.07
4	8.22	5.62	8	14.04	15.20

How big should θ be for it to be possible to ensure a minimum probability of 0.95 of correct choice of author, in the absence of information about prior probabilities of authorship? (A normal approximation may be used to simplify the calculations.)

CHAPTER 18

Response Surfaces

18.1 INTRODUCTION

In Chapters 13 to 15 we considered the analysis of the variation within an experiment either to estimate functions of means (and sometimes of variances) or to test hypotheses concerning means and variances, or possibly for both purposes. We described techniques—of blocking and confounding and also fractional experimentation—that are used to obtain information and to develop inferences eliminating some kinds of extraneous variation. In Chapter 17 we considered multivariate observations. Now we return again to polynomial regression (studied in Chapter 12) and apply some of the methods of the earlier chapters with the aim of solving problems of determination of optimum conditions.

In the operation of a plant, for example, we are frequently concerned with the expected value of an important variable (such as overall yield, profit rate, or proportion of unsatisfactory product) as a function of a number of variables defining conditions of operation of the plant (such as type of material, period of reaction, temperature, or pressure). These latter can be, and often are, controlled to produce desired values of the former. Very often these desired values are maxima (for example, yield and profit rate) or minima (for example, proportion of unsatisfactory product). Here, however, we first consider the problem of determining the relationship generally without special reference to the determination of maxima, minima or other special sets of values of the controlled variable.

If $x_1, x_2, \ldots, x_q$ represent the values of q controlled variables, then the expected value of a variable Y, given $x_1, x_2, \ldots, x_q$, can be written as a function $\eta(x_1, x_2, \ldots, x_q)$ of the values of the controlled variables. This is recognized as the multiple regression of Y on $x_1, x_2, \ldots, x_{q-1}$ and x_q. The surface represented by the equation

$$y = \eta(x_1, x_2, \ldots, x_q) \tag{18.1}$$

is the *regression surface* of Y on $x_1, x_2, \ldots, x_{q-1}$, and x_q. It is, however, the custom in this connection to call it a *response surface*. The relation between random variables $y_t (t = 1, 2, \ldots, n)$ representing n observed values of Y corresponding to actual sets of values $x_{1t}, x_{2t}, \ldots, x_{qt}$ of the controlled variables $x_1, x_2, \ldots, x_q$, then satisfy the condition

$$Y_t = \eta(x_{1t}, x_{2t}, \ldots, x_{qt}) + Z_t \tag{18.2}$$

where the Z_t's are random variables each with expected value 0.

We further impose the conditions that

(a) each Z_t has the same variance, σ^2,

(b) the Z_t's are a mutually independent set of random variables, and

(c) the Z's are each normally distributed.

These can be recognized as the conditions (stated in Chapter 13) under which standard analysis of variance techniques can be applied. As will be seen below the techniques of multiple *linear* regression alone enable us to get a good idea of the shape of the response surface, at any rate over part of the range of variation of the controlled variables. In fact, multiple linear regression techniques can be applied whenever $\eta(x_1, x_2, \ldots, x_q)$ can be written (at any rate approximately) in the form

$$\eta(x_1, x_2, \ldots, x_q) = \sum \beta_j f_j (x_1, x_2, \ldots, x_q) \tag{18.3}$$

where the f_j's are explicit functions of $x_1, x_2, \ldots, x_q$.

18.2 ILLUSTRATIONS OF RESPONSE SURFACES

Suppose the experimenter has selected a region of experimentation in which he feels confident of finding an optimum. He first estimates a linear (first-order) equation for the surface, which is then tested for adequacy. If it is adequate he examines the coefficients of the input variables $(x_1, x_2, \ldots, x_q)$ and uses these to see in which direction he should move to go toward the optimum. If the first-order equation is not adequate he turns to a second-order equation, of the form

$$\eta(x_1, x_2, \ldots, x_q) = \beta_0 + \sum_j \beta_j x_j + \sum_{j < j'} \sum \beta_{jj'} x_j x_{j'} \tag{18.4}$$

The fitted response function would be

$$y = B_0 + \sum_j B_j x_j + \sum_{j < j'} \sum B_{jj'} x_j x_{j'} \tag{18.5}$$

This fitted surface may attain (1) a maximum, (2) a minimum, or (3) a saddle point in the region of interest, or (4) it may do none of these things. [It may even attain more than one of (1) to (3) and/or attain some or all of them more than once, but this does not often happen.]

Contours around a typical maximum, minimum, and saddle point are shown in Figures 18.1, 18.2, and 18.3, respectively. At an optimal (minimum or maximum) point the rates of change, $\partial\eta/\partial x_j$ (for all j) are equal to 0. For this reason the optimum is sometimes called a "stationary point." However, these conditions are not *sufficient* to *ensure* that we have an optimal (maximum or minimum) point. In Figure 18.1, as one moves away from the stationary point, the function (or response) decreases. Likewise, for Figure 18.2 the response increases as one moves away. But in Figure 18.3, the stationary point is a "saddle point," and leaving it does decrease the response if x_1 and x_2 both increase or both decrease but not otherwise. This illustrates the fact that we may not reach an optimum by solution of the equations $\partial\eta/\partial x_j = 0$. We need to take a closer look at the contour lines around the point(s) corresponding to the solutions.

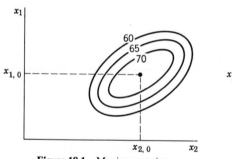

Figure 18.1 Maximum point.

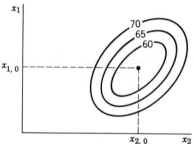

Figure 18.2 Minimum point.

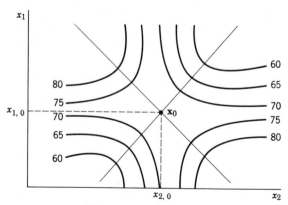

Figure 18.3 Saddle point.

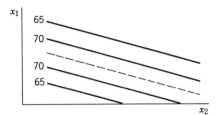

Figure 18.4 Stationary ridge system.

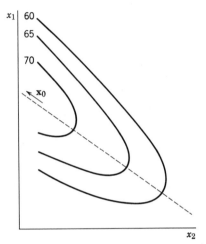

Figure 18.5 Rising ridge system.

Two other possibilities are a stationary ridge (Figure 18.4) and a rising ridge (Figure 18.5). In the first of these there is an infinite set of values (along the broken line) each of which will satisfy the requirement of optimum. In the second the path to reach the optimum is along a particular line, or ridge; in fact this situation arises with a surface like that in Figure 18.1, when the field of experimentation does not extend to the actual maximum at $(x_{1,0}, x_{2,0})$.

18.3 RECAPITULATION OF SOME MULTIPLE LINEAR REGRESSION TECHNIQUES

We now recall some of the standard techniques employed in connection with the analysis of multiple linear regression. The model used is

$$Y_t = \beta_0 + \sum_{j=1}^{q} \beta_j x_{jt} + Z_t \qquad (t = 1, 2, \ldots, k) \qquad (18.6)$$

Here $(x_{1t}, x_{2t}, \ldots, x_{qt}, Y_t)$ represent the values of q "independent variables" $x_1, x_2, \ldots, x_q$ and 1 "dependent" variable, Y, for the tth individual in a random sample of size k. The Z_t's are mutually independent random variables; each is normally distributed with expected value 0 and variance σ^2. The β's are unknown constants. Our first task is to estimate the value of each of the β's. The method of least squares leads to the set of equations

$$\hat{\beta}_0 = \bar{Y} - \sum_{j=1}^{q} \hat{\beta}_j \bar{x}_j$$

$$\sum_{j=1}^{q} \hat{\beta}_j A_{lj} = A_{yl} \qquad (l = 1, 2, \ldots, q)$$

for the estimators $\hat{\beta}_0, \hat{\beta}_1, \hat{\beta}_2, \ldots, \hat{\beta}_q$ of $\beta_0, \beta_1, \beta_2, \ldots, \beta_q$, respectively, where

$$A_{jl} = A_{lj} = \sum_{t=1}^{k} (x_{jt} - \bar{x}_j)(x_{lt} - \bar{x}_l); \qquad A_{yl} = \sum_{t=1}^{k} (x_{lt} - \bar{x}_l)(\bar{Y}_t - \bar{Y})$$

The variance–covariance matrix of the $\hat{\beta}$'s is $\sigma^2 \mathbf{A}^{-1}$ where $\mathbf{A}$ is the $q \times q$ matrix of elements A_{lj}.

The analysis of variance for multiple linear regression, with the model described above, contains the sums of squares

(i) $\sum_{t=1}^{k} [\hat{\beta}_1(x_{1t} - \bar{x}_1) + \hat{\beta}_2(x_{2t} - \bar{x}_2) + \cdots + \hat{\beta}_q(x_{qt} - \bar{x}_q)]^2$, the sum of squares "Due to Multiple Linear Regression on $x_1, x_2, \ldots, x_q$," which has q degrees of freedom, and

(ii) $\sum_{t=1}^{k} [Y_t - \bar{Y} - \hat{\beta}_1(x_{1t} - \bar{x}_1) - \cdots - \hat{\beta}_q(x_{qt} - \bar{x}_q)]^2$, the sum of squares "About Multiple Linear Regression," which has $k - q - 1$ degrees of freedom.

The latter sum of squares gives a mean square with expected value equal to σ^2 if the model (18.6) is appropriate, and so it is used as a Residual sum of squares. If, however, the model (18.6) is modified by the insertion of constants δ_t representing "lack of fit" of the proposed model, and giving the model

$$Y_t = \beta_0 + \sum_{j=1}^{q} \beta_j x_{jt} + \delta_t + Z_t \qquad (t = 1, 2, \ldots, k) \qquad (18.7)$$

then the expected value of the mean square will be greater than σ^2 by an amount depending on the squares of the δ_t's. It is not possible to test whether the δ_t's are 0 (that is, whether the linear model is adequate) unless a separate, genuinely unbiased estimate of σ^2 alone can be obtained. This can be done if, for at least some sets of values $x_{1t}, x_{2t}, \ldots, x_{qt}$, more than 1 value of Y is available. Then we have the model

$$Y_{ti} = \beta_0 + \sum_{j=1}^{q} \beta_j x_{jt} + \delta_t + Z_{ti} \qquad (t=1,\ldots,k; \quad i=1,\ldots,n_t) \qquad (18.8)$$

with at least one n_t greater than 1.

Then

$$\bar{Y}_{t.} = \beta_0 + \sum_{j=1}^{q} \beta_j x_{jt} + \delta_t + \bar{Z}_{t.}$$

and so $Y_{ti} - \bar{Y}_{t.} = Z_{ti} - \bar{Z}_{t.}$ and the "Within Groups" (or, as it is more usually termed, "Within Arrays") mean square $(\Sigma_{t=1}^{k}[n_t - 1])^{-1}$ $\times \Sigma_{t=1}^{k} \Sigma_{i=1}^{n_t} (Y_{ti} - \bar{Y}_{t.})^2$ gives an unbiased estimator of σ^2, and has $\Sigma_{t=1}^{k}[n_t - 1] = N - k$ degrees of freedom. (See Section 13.5.)

The remaining ("Between Arrays") sum of squares $\Sigma_{t=1}^{k} n_t (\bar{Y}_{t.} - \bar{Y}_{..})^2$ can be analyzed into

$$\sum_{t=1}^{k} n_t \left[\hat{\beta}_1 (x_{1t} - \bar{x}_1) + \cdots + \hat{\beta}_q (x_{qt} - \bar{x}_q) \right]^2$$

$$+ \sum_{t=1}^{k} n_t \left[\bar{Y}_{t.} - \hat{\beta}_1 (x_{1t} - \bar{x}_1) - \cdots - \hat{\beta}_q (x_{qt} - \bar{x}_q) \right]^2$$

"Due to Multiple Linear Regression" + "About Multiple Linear Regression" or "Departure from Linearity"

with q and $(k - q - 1)$ degrees of freedom respectively, just as before. Now, however, we can test whether the δ_t's are all 0, by comparing the ratio

$$\frac{\text{About Multiple Linear Regression mean square}}{\text{Within Arrays mean square}}$$

with the F distribution with $(k - q - 1)$, $(N - k)$ degrees of freedom.

Repetition of some sets of values of the x's thus makes it possible to calculate an unbiased estimate of the residual variance σ^2, and also to test the adequacy of the formula for the expected value of Y, which is used in the model.

The analysis of variance is shown in Table 18.1.

TABLE 18.1
ANOVA FOR MULTIPLE LINEAR REGRESSION

SOURCE	DEGREES OF FREEDOM	SUM OF SQUARES	EXPECTED VALUE OF MEAN SQUARE
Multiple Linear Regression on $x_1, x_2, \ldots, x_q$	q	$\displaystyle\sum_{t=1}^{k} n_t \left[\sum_{j=1}^{q} \hat{\beta}_j (x_{jt} - \bar{x}_j) \right]^2$	$\displaystyle\sigma^2 + q^{-1} \sum_{t=1}^{k} n_t \left[\sum_{j=1}^{q} \beta_j (x_{jt} - \bar{x}_j) \right]^2$
Departure from Linearity	$k - q - 1$	$\displaystyle\sum_{t=1}^{k} n_t \left[\bar{Y}_t - \bar{Y} - \sum_{j=1}^{q} \hat{\beta}_j (x_{jt} - \bar{x}_t) \right]^2$	$\displaystyle\sigma^2 + (k-q-1)^{-1} \sum_{t=1}^{k} n_t D_t^2$
Residual	$N - k$	$\displaystyle\sum_{t=1}^{k} \sum_{i=1}^{n_t} (Y_{ti} - \bar{Y}_t)^2$	σ^2
Total	$N - 1$	$\displaystyle\sum_{t=1}^{k} \sum_{i=1}^{n_t} (Y_{ti} - \bar{Y}_{..})^2$	

Note that the Multiple Linear Regression sum of squares can be divided into sums of squares for multiple linear regression on x_1, on x_2 given x_1, on x_3 given x_1 and x_2, and so on. In particular, one could have a sum of squares (with r degrees of freedom) for regression on $x_1, x_2, \ldots, x_r$ and a sum of squares [with $(q-r)$ degrees of freedom] for regression on $x_{r+1}, x_{r+2}, \ldots, x_q$, given $x_1, x_2, \ldots, x_r$.

18.4 POLYNOMIAL RESPONSE SURFACES

We restrict ourselves to the case when η is a polynomial in the x's. Even with the further restriction of not proceeding beyond terms of second order, this enables us to give an adequate treatment of most cases, provided we remember that we may have to restrict the range of variation of some of all of our controlled variables to make the approximation adequate.

Including all terms of second order we have the model

$$Y_t = \beta_0 + \beta_1 x_{1t} + \beta_2 x_{2t} + \cdots + \beta_q x_{qt}$$

$$+ \beta_{11} x_{1t}^2 + \beta_{22} x_{2t}^2 + \cdots + \beta_{qq} x_{qt}^2$$

$$+ \beta_{12} x_{1t} x_{2t} + \beta_{13} x_{1t} x_{3t} + \cdots + \beta_{q-1,q} x_{q-1,t} x_{qt} + Z_t \qquad (18.9)$$

(The notation for the subscripts of the β's is self-explanatory, and is very convenient.)

Our problem is essentially the simple one of fitting a multiple linear regression, that is, of estimating the values of the β's. This can be done by using the technique of multiple linear regression. The analysis of variance resulting from this application of multiple linear regression technique to model (18.9) will contain a Residual sum of squares with $n - [1 + q + q + \frac{1}{2}q(q-1)] = n - \frac{1}{2}(q+1)(q+2)$ degrees of freedom. This will not allow us, however, to test whether the model is an adequate representation of the situation. The "estimate of σ^2" provided by the Residual mean square can be, in fact, inflated by departures of reality from the model assumed. In order to be able to obtain an estimate of σ^2 unaffected by such departures we must have 2 or more independent observations of Y for at least one fixed set of values of $x_1, x_2, \ldots, x_q$. The "Within Fixed Sets of Values of $(x_1, x_2, \ldots, x_q)$" mean square for such repeated observations of Y provides an unbiased estimate of σ^2. This is useful in itself, and also the weighted sum of squares of mean values of Y for fixed sets of $(x_1, x_2, \ldots, x_q)$ about the fitted regression can be compared with it to test whether the model used is adequate.

Thus suppose we decide to use the second-order model

$$Y_{tl} = \beta_0 + \beta_1 x_{1t} + \cdots + \beta_q x_{qt} + \beta_{11} x_{1t}^2 + \cdots$$

$$+ \beta_{qq} x_{qt}^2 \beta_{12} x_{1t} x_{2t} + \cdots + \beta_{q-1,q} x_{q-1,t} x_{qt} + Z_{tl}$$

with $t = 1, \ldots, k$; $l = 1, \ldots, n_t$ $(n_t > 1$ for at least one t) and the fitted values of β_i, β_{ij} are B_i, B_{ij}, respectively. Then

$$\frac{\sum_{t=1}^{k} \sum_{l=1}^{n_t} \left(Y_{tl} - \bar{Y}_{t.} \right)^2}{N - k}$$

is an estimator of σ^2 (where $N = \sum_{t=1}^{k} n_t$) and the ratio of

$$\frac{\sum_{t=1}^{k} n_t \left(\bar{Y}_{t.} - B_0 - B_1 x_{1t} - \cdots - B_{q-1,q} x_{q-1,t} x_{qt} \right)^2}{k - (q+1)(q+2)/2}$$

to this estimator of σ^2 can be compared with the F distribution with $[k - (q+1)(q+2)/2]$, $(N-k)$ degrees of freedom to test the adequacy of the model. (Note that if $k < (q+1)(q+2)/2$ we cannot estimate the β's by the method of least squares.) If desired the Regression sum of squares can be split up into Linear Regression (with q degrees of freedom) and Quadratic, given Linear, Regression [with $q(q+1)/2$ degrees of freedom] sums of squares.

18.5 FIRST AND SECOND ORDER DESIGNS

In Chapters 14 and 15 we have paid considerable attention to the use of factorial designs. In cases when each factor can be measured numerically, these designs can be represented by points in a space with as many dimensions as there are factors. For example, a complete 2^2 experiment with factors

Temperature—10 and 15°C
Time—1 and 2 min

can be represented by the 4 points in Figure 18.6.

By appropriate (linear) scaling of variables any 2^2 experiment can be represented in a standard form with experimental points at $(1,1)$, $(1,-1)$, $(-1,1)$ and $(-1,-1)$, as in Figure 18.7. [Other standard forms, for example, the 4 points $(0,0)$, $(0,1)$, $(1,0)$, $(1,1)$, are also used.]

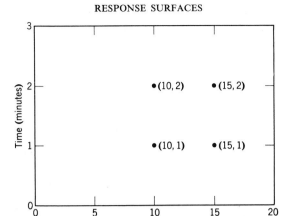

Figure 18.6   Time and temperature in a 2^2 design.

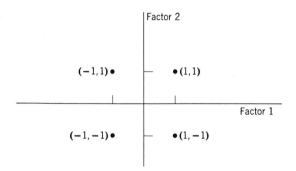

Figure 18.7   A 2^2 design centered at $(0,0)$.

If 1 factor (or more) has more than 2 levels, we cannot always reduce the design to a single standard from by linear scaling of variables. This can be done, however, if the levels of such factors are *equally spaced*. One possible standard form for a 2×3 experiment is shown in Figure 18.8.

If there are more factors, direct geometrical (or diagrammatic) representation is not so simple, but the corresponding concept is useful. As an example, a representation of a complete 2^3 factorial is shown in Figure 18.9.

Other factorial experiments can be represented in a similar way. For example, a confounded 2^3 experiment, confounding the triple interaction, is obtained if the 4 circled points in Figure 18.9 are assigned to one block

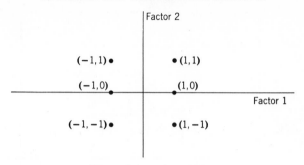

Figure 18.8   A 2×3 design.

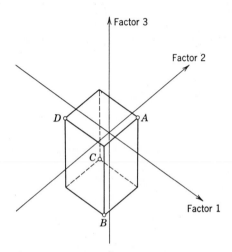

Figure 18.9   A 2^3 experimental design.

and the remaining 4 points are assigned to the other block (the principal block). The 4 circled points alone, of course, represent a half-replicate fractional replication with the triple interaction as defining interaction. (Point A corresponds to the upper level for all 3 factors; points B, C, and D correspond to the upper level of the first, second, and third factors alone, respectively.)

In the following discussions we work in terms of standard forms. This makes the exposition clearer, and in any specific case, actual levels to be used can be obtained by applying an appropriate linear transformation.

Experimental designs from which the coefficients $\beta_1, \beta_2, \ldots, \beta_q$ of the first-order terms can be estimated are called *first-order* designs. For such a design it must be possible to obtain proper estimates of the β's in the model

$$Y_l = \beta_0 + \beta_1 x_{1l} + \cdots + \beta_q x_{ql} + Z_l \tag{18.10}$$

The 2^q complete factorial is such a design.

For the system of linear scaling shown in Figure 18.7 and in model (18.10), the estimators, B_j of the coefficients and the variances of these estimators simplify to

$$B_j = \frac{\sum\limits_l Y_l x_{jl}}{\sum\limits_l x_{jl}^2} \qquad \operatorname{var}(B_j) = \frac{\sigma^2}{\sum\limits_l x_{jl}^2} \tag{18.11}$$

where σ^2 is the error variance.

If some of the coefficients β_{ij} in model (18.9) are not 0, the model is called a *second-order* model. Evidently a second-order *design* is necessarily also a first-order design. It is, however, sometimes the custom to consider as first-order designs only those designs that will provide estimates of $\beta_0, \beta_1, \ldots, \beta_q$, but not of any 2 factor or higher-order coefficients.

In model (18.9) the terms $\beta_i x_i$ and $\beta_{ii} x_i^2$ correspond to linear and quadratic main effects of the ith factor. The term $\beta_{ij} x_i x_j$ corresponds to interaction between the ith and jth factors. It is a special kind of interaction because it is supposed to be proportional to $x_i x_j$, but it does represent a departure from additivity of the effects of changing the levels of the ith and the jth factors. We might thus expect to obtain some estimate of β_{12} from a 2^2 design, although we cannot estimate the quadratic coefficients β_{11}, β_{22} from such a design. However, this can be done with a 3^2 design. The 2^2 design is thus called a first-order design, and the 3^2 design a second-order design. It is easy to see that it is not possible to estimate the complete set of coefficients in a third-order model from a 3^2 complete factorial. This could be done with a 4^2 design, but it should be clear that we need not restrict ourselves to designs which are simple complete factorials. In fact, we need not restrict ourselves to equally spaced values for the levels of the controlled variables, or, if we do admit this limitation, we need not have identical replications at each level. For example, a simple extension of the 2^2 factorial design shown in Figure 18.7 is to add 2 or more points at the origin. With such a design it is possible to use the replicated results for $x_1 = x_2 = 0$ to provide an unbiased estimate of σ^2, and also to test the adequacy of the first-order model. This is shown in Example 18.1 in which the experiment contains a 2^3 design with 2 center points.

Example 18.1 In 1957–1958, a research laboratory of the The Goodyear Tire and Rubber Company undertook a project to improve a product named Triform. This type of rubber was being used for paddings on the dashboards of automobiles, as well as in many other situations. Different specifications of several automobile manufacturers had to be met. These specifications included conditions on such responses as tensile strength, elongation, hardness, "dimple recovery" (recovery to initial form after impact), stiffness, and several other properties. In order to determine the nature of the response surfaces using as few experiments as possible, a factorial design with additional center points was utilized. First a linear fit was attempted and tested. Further experimentation gave data for additional points from which was obtained a quadratic fit.

We consider here only the property of tear strength (lb/in) and 3 controlled variables. These were: (1) percentage of component A in rubber, (2) percentage of component A in resin, and (3) percentage modifier. In this example we investigate only the linear fit. Ten data points were selected as shown in Table 18.2. In this table are given the original data as well as the transformed coordinates, x_1, x_2, x_3. Note that the points comprise the 8 treatment combinations of a 2^3 factorial design plus 2 center points (corresponding to 10 percent of component A in rubber, 20 percent in resin, and 0.2 percent modifier). The latter 2 are used to obtain an estimate of the error variance.

TABLE 18.2
INITIAL EXPERIMENT ON TRIFORM

SAMPLE NUMBER	COMPOUND A IN RUBBER	COMPOUND A IN RESIN	MODIFIER	x_1	x_2	x_3	Y
1	0	10	0.1	-1	-1	-1	407
2	20	10	0.1	1	-1	-1	230
3	0	30	0.1	-1	1	-1	322
4	20	30	0.1	1	1	-1	250
5	0	10	0.3	-1	-1	1	421
6	20	10	0.3	1	-1	1	243
7	0	30	0.3	-1	1	1	371
8	20	30	0.3	1	1	1	259
9	10	20	0.2	0	0	0	421
10	10	20	0.2	0	0	0	435

Note that with the proper origin the design is such that the sums Σx_i and $\Sigma x_i x_{i'} (i \neq i')$ are 0. The advantages of this property of *orthogonality*, will become apparent. The other sums are $\Sigma x_i^2 = 8$ for $i = 1, 2, 3$, $\Sigma y = 3359$, $\Sigma x_1 y = -539$, $\Sigma x_2 y = -99$, $\Sigma x_3 y = 85$, and $N = 10$. Solving directly, one obtains the linear equation

$$y_x = 335.9 - 67.375 x_1 - 12.375 x_2 + 10.625 x_3$$

We now test for goodness of fit of this linear equation. The analysis of variance in Table 18.3 contains the quantities needed to apply this test. Note that in this

TABLE 18.3

ANOVA OF LINEAR FIT OF TRIFORM DATA

SOURCE	SUM OF SQUARES	DEGREES OF FREEDOM	MEAN SQUARE	MEAN SQUARE RATIO
Regression on				
x_1		36,315	1	
x_2		1,225	1	
x_3		903	1	
Lack of Fit	25,382	5	5,076	51.8
Error	98	1	98	
Residual		25,480	6	
Total		63,923	9	

table, the Residual sum of squares is the sum of the Error sum of squares (from the 2 center points) and the Lack of Fit sum of squares. Also the sums of squares for regression on x_1, x_2, and x_3 are mutually orthogonal—this follows from the orthogonality of the design, which we have already noted. One first tests the Lack of Fit mean square against the Error mean square. This variance ratio (51.8) indicates quite strongly that the linear equation does not adequately describe the surface. To obtain a better fit it was decided to try fitting a quadratic regression. To do this it was necessary to carry out some supplementary experiments, which will be described in the next example.

In assessing experimental designs when they are to be used for fitting response surfaces, we are usually less interested in the accuracy of estimation of individual coefficients (β's) than in the accuracy of estimation of $\eta(x_1, x_2, \ldots, x_q)$, the expected value of the dependent variable Y. A natural measure to use is the variance of

$$\hat{\eta}(x_1, x_2, \ldots, x_q) = \hat{\beta}_0 + \hat{\beta}_1 x_1 + \cdots + \hat{\beta}_q x_q + \hat{\beta}_{11} x_1^2 + \cdots + \hat{\beta}_{q-1,q} x_{q-1} x_q$$

This is a function, $V(x_1, x_2, \ldots, x_q)$, say, of $x_1, x_2, \ldots, x_q$. The form of the function depends on the experimental design.

Two (or more) designs can be compared by comparing the functions $V(x_1, x_2, \ldots, x_q)$ for each design. If a single design is replicated n times the corresponding V function is divided by n, and so the variance of $\hat{\eta}(x_1, x_2, \ldots, x_q)$ is reduced by the factor n^{-1} for every set of values $x_1, x_2, \ldots, x_q$. For other cases, in particular when we are trying to decide among several designs with the same total number of sets of observations, the comparison is not so simple. $V(x_1, x_2, \ldots, x_q)$ is smaller (for a given design) for some sets of values of $x_1, x_2, \ldots, x_q$, and larger for other sets of values. This greater complexity is largely inescapable. If a considerable amount of effort is to be expended on the investigation it is worthwhile spending some time on a careful assessment of the properties of various designs, as reflected in their V functions.

Example 18.2   In Example 18.1 there was strong evidence that the linear fit was not adequate. Seven additional points were taken. These points did not produce a completely orthogonal design. (Note that one does not have all zeros off the diagonal in Table 18.6.) However, this example is presented in order to show that although orthogonality is desirable it is not essential.

The additional points in standard form are given in Table 18.4.

TABLE 18.4

SEVEN ADDITIONAL EXPERIMENTAL POINTS FOR TRIFORM DATA

SAMPLE NUMBER	x_1	x_2	x_3	Y
11	-0.5	0.7	0	359
12	0.5	0.7	0	360
13	-0.5	1.3	0	321
14	0.5	1.3	0	314
15	0	1.0	0	380
16	0	1.0	0	376
17	0	1.0	0	354

(The same transformation is used as in Example 17.7, so that the first experiment used 5 per cent component A in rubber, 27 per cent in resin, 0.2 per cent modifier, and so on.)

The estimated regression equations based on the 17 observations (10 from the original series and 7 additional points) have coefficients given by terms in the matrix of Table 18.6.

The numerical values for Table 18.5 are given in Table 18.6. Some of the symmetry of Example 17.7 is still retained, in that we have a number of zeros in the table. However, we do not have complete orthogonality. (A further discussion of orthogonal designs is given later.)

Inverting the matrix of Table 18.6 and solving for the estimates of the parameters, one obtains the regression equation

$$y_x = 422.5 - 63.0x_1 - 13.2x_2 + 10.6x_3 - 121.9x_1^2$$
$$- 36.4x_2^2 + 48.7x_3^2 + 25.2x_1x_2 - 5.1x_1x_3 + 3.9x_2x_3$$

One must now determine whether the quadratic fit is adequate and, finally, the significance of the regression coefficients. Table 18.7 is the ANOVA table for this fitted regression. The Error sum of squares is determined from the 2 points at $(0,0,0)$ and the 3 points at $(0,1,0)$. That is, each set provides an estimate of error which is then combined into a pooled estimate with 3 degrees of freedom.

TABLE 18.5

SUMS, SUMS OF SQUARES AND SUMS OF CROSSPRODUCTS FOR A SECOND ORDER DESIGN

N	Σx_1	Σx_2	Σx_3	Σx_1^2	Σx_2^2	Σx_3^2	$\Sigma x_1 x_2$	$\Sigma x_1 x_3$	$\Sigma x_2 x_3$	Σy
N	Σx_1	Σx_2	Σx_3	Σx_1^2	Σx_2^2	Σx_3^2	$\Sigma x_1 x_2$	$\Sigma x_1 x_3$	$\Sigma x_2 x_3$	Σy
	Σx_1^2	$\Sigma x_1 x_2$	$\Sigma x_1 x_3$	Σx_1^3	$\Sigma x_1 x_2^2$	$\Sigma x_1 x_3^2$	$\Sigma x_1^2 x_2$	$\Sigma x_1^2 x_3$	$\Sigma x_1 x_2 x_3$	$\Sigma x_1 y$
		Σx_2^2	$\Sigma x_2 x_3$	$\Sigma x_1^2 x_2$	Σx_2^3	$\Sigma x_2 x_3^2$	$\Sigma x_1 x_2^2$	$\Sigma x_1 x_2 x_3$	$\Sigma x_2^2 x_3$	$\Sigma x_2 y$
			Σx_3^2	$\Sigma x_1^2 x_3$	$\Sigma x_2^2 x_3$	Σx_3^3	$\Sigma x_1 x_2 x_3$	$\Sigma x_1 x_3^2$	$\Sigma x_2 x_3^2$	$\Sigma x_3 y$
				Σx_1^4	$\Sigma x_1^2 x_2^2$	$\Sigma x_1^2 x_3^2$	$\Sigma x_1^3 x_2$	$\Sigma x_1^3 x_3$	$\Sigma x_1^2 x_2 x_3$	$\Sigma x_1^2 y$
					Σx_2^4	$\Sigma x_2^2 x_3^2$	$\Sigma x_1 x_2^3$	$\Sigma x_1 x_2^2 x_3$	$\Sigma x_2^3 x_3$	$\Sigma x_2^2 y$
						Σx_3^4	$\Sigma x_1 x_2 x_3^2$	$\Sigma x_1 x_3^3$	$\Sigma x_2 x_3^3$	$\Sigma x_3^2 y$
							$\Sigma x_1^2 x_2^2$	$\Sigma x_1^2 x_2 x_3$	$\Sigma x_1 x_2^2 x_3$	$\Sigma x_1 x_2 y$
								$\Sigma x_1^2 x_3^2$	$\Sigma x_1 x_2 x_3^2$	$\Sigma x_1 x_3 y$
									$\Sigma x_2^2 x_3^2$	$\Sigma x_2 x_3 y$

TABLE 18.6

SAMPLE SUMS, SUMS OF SQUARES, AND SUMS OF CROSSPRODUCTS

17	0	7	0	9	15.36	8	0	0	0	5823
	9	0	0	0	0	0	1	0	0	−542
		15.36	0	1	8.08	0	0	0	0	2339
			8	0	0	0	0	0	0	85
				8.25	9.09	8	0	0	0	2842
					17.1924	8	0	0	0	5038
						8	0	0	0	2503
							9.09	0	0	166
								8	0	−41
									8	31

From Table 18.7 it can be seen that the ratio of the Lack of Fit mean square to the Error mean square is not significant. The critical F ratio (at 5 percent significance) is $F_{4,3,0.95} = 9.12$. Hence the quadratic fit appears to be satisfactory. One then might combine the Lack of Fit and Error sums of squares into a pooled Residual sum of squares; the 1 percent significance limit for individual mean square ratios would then be $F_{1,7,0.99} = 12.75$. Of the "interaction" terms, only $x_1 x_2$ shows up strongly. The only linear term which is significant is x_1.

Note that none of the terms $(x_3, x_3^2, x_1 x_3, x_2 x_3)$ involving x_3 gives a significantly large mean square. This might be interpreted to mean that we need not include x_3 (percent modifier) in the regression equation. If x_3 were not included in the regression equation the new equation, estimated from these data, would be

$$y_x = 421.3 - 61.0 x_1 - 13.7 x_2 - 67.5 x_1^2 - 41.4 x_2^2 + 25.0 x_1 x_2$$

The objectives of the experiment were adequately met by the 17 experimental units. The proportions of composition A in the rubber and in the resin and the

TABLE 18.7
ANALYSIS OF VARIANCE OF THE REGRESSION EQUATION

SOURCE	SUM OF SQUARES	DEGREES OF FREEDOM	MEAN SQUARE	RATIO OF MEAN SQUARES
Regression on:				
x_1	32,639.24	1	32,639.24	$87.4^{**}\left(=\dfrac{32,639.24}{373.4}\right)$
x_2	278.06	1	278.06	
x_3	903.55	1	903.55	
x_1^2	22,169.36	1		
x_2^2	3,895.88	1		
x_3^2	339.51	1		
Total Square Terms	26,404.75	3	8,801.6	$23.6^{**}(=8801.6/373.4)$
$x_1 x_2$	5,698.48	1		$15.3^{**}\left(=\dfrac{5698.48}{373.4}\right)$
$x_1 x_3$	210.33	1		
$x_2 x_3$	120.38	1		
Lack of Fit	2,124.12	4	531.0	$3.25\left(=\dfrac{531.0}{163.3}\right)$
Error	490.00	3	163.3	
Residual	2,614.12	7	373.4	
Total	63,868.81	16		

concentration of modifier as indicated by the equations of the response surfaces enabled The Goodyear Tire and Rubber Company to satisfy the numerous specifications imposed by the consumers. In arriving at an optimum to satisfy the different specifications, a number of response surfaces were constructed from the equations. Once the surfaces have been represented graphically some judgment is required (especially if the optimum properties conflict) to combine the desired properties.* Figures 18.10, 18.11, and 18.12 are graphical representations of the response surfaces for 4 different dependent variables. In each of the figures the rectangular area designated by broken lines is the region of experimentation. For all 3 surfaces, the modifier was fixed at 0.2 percent. The "yield" variables are tear strength (based on the data in this example), tensile strength, elongation, and "dimple recovery," which is measured by the proportion (measured as percent) to which a test piece recovers its original form after suffering a standard impact.

* Under some circumstances it is possible to combine properties into a so-called "figure of merit." This has been done with varying degrees of success, but is not discussed here.

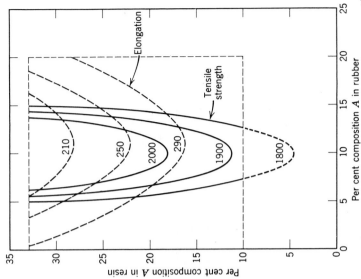

Figure 18.11 Response surface representation of tensile strength (psi) and elongation (percent)—modifier at 0.2 percent.

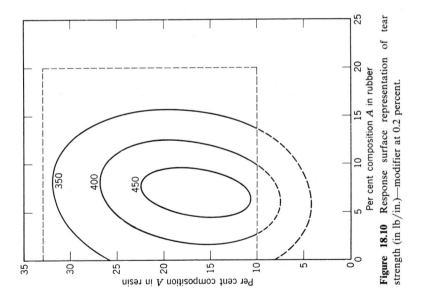

Figure 18.10 Response surface representation of tear strength (in lb/in.)—modifier at 0.2 percent.

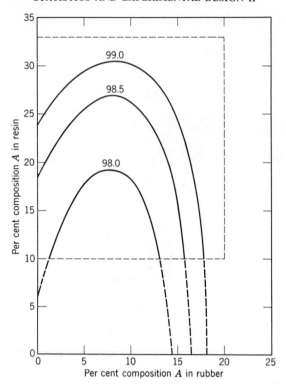

Figure 18.12 Response surface representation of dimple recovery (in percent)—modifier at 0.2 percent.

If experimentation is costly then it is worth considering a *simplex* design. This is a design with $(q+1)$ experimental points—just sufficient to provide estimators for the $(q+1)$ parameters $\beta_0, \beta_1, \ldots, \beta_q$ in a first-order model with q controlled variables. The experimental points are at the vertices of a simplex in q dimensions. If $q=2$ this is an equilateral triangle; if $q=3$ it is a tetrahedron. The design may be rotated about the origin, so that a variety of combinations of factor levels is possible. The design is orthogonal, so the estimators of the β's are mutually uncorrelated.

It is a defect of this design that there is no residual sum of squares, so that it is not possible to estimate residual variance, nor to test for lack of fit. An estimate of residual variance can be obtained by adding 2 or more experimental points at the origin. This also provides 1 degree of freedom for lack of fit. Although useful, this will, of necessity, provide a test for only one specific kind of lack of fit.

18.6 ROTATABLE DESIGNS OF FIRST ORDER

One way in which the comparison is sometimes simplified is by the use of *rotatable designs*. These are designs which are such that $V(x_1, x_2, \ldots, x_q)$ is a function of $\sum_{i=1}^{q} x_i^2$ only. This means that the variance of $\hat{\eta}(x_1, x_2, \ldots, x_q)$ depends only on the "distance" of the point $(x_1, x_2, \ldots, x_q)$ from the point $(0, 0, \ldots, 0)$. (The latter point is the center of the experimental system and will, of course, correspond to nonzero values of the actual variables in most cases.) The fact that the variance depends only on the distance $(\sum_{i=1}^{q} x_i^2)^{1/2}$ means that it is unaltered if the design is rotated through an arbitrary angle about the point $(0, 0, \ldots, 0)$. This is the reason the designs are called rotatable. It can be useful to have such designs available and a brief description of some such designs is given in the next section. These designs can be useful when the units of the controlled variables are given naturally. A good example is when the spatial variation of a variable (for instance, temperature) is being investigated. It is natural to measure each dimension in the same units of length. Rotatable designs will give estimates of the average value of the dependent variable which have constant variance at a constant distance from the center of the design. The latter can be sited at a convenient point in the space (for example, a furnace) being studied.

However, it should be realized that, in general, the "distance" depends on the units in which the controlled variables are measured, and that changes in the relative magnitude of these units change the locus of "points" at constant "distance" from the origin. Therefore, although there are cases in which rotatable designs are useful technical devices, in very many cases the choice of units is subject to more or less arbitrary human choice, and the "rotatability" of a design has no absolute meaning.

The estimated value of Y using the fitted multiple regression can be written

$$\hat{y}_x = \overline{Y} + B_1 x_1 + B_2 x_2 + \cdots + B_q x_q$$

where the B's are calculated according to the least squares formulas, (remembering that we are taking the design mean value of each controllable variable $x_1, x_2, \ldots, x_q$ to be 0). Then the variance of $\hat{y}_x$ is

$$\mathrm{var}(\hat{y}_x) = \sigma^2 \left[\frac{1}{n} + \sum_h \sum_l x_h x_l \, \mathrm{cov}(b_h, b_l) \right] = \sigma^2 \left[\frac{1}{n} + \mathbf{X} \mathbf{C}^{-1} \mathbf{X}' \right] \quad (18.12)$$

where $\mathbf{C}$ is the $q \times q$ matrix with (h, l)th element $\sum_{i=1}^{n} x_{hi} x_{li}$, n is the size of the design, and $\mathbf{X}$ is the vector $(x_1, x_2, \ldots, x_q)$.

For the design to be rotatable this variance must depend only on $\sum_{l=1}^{q} x_l^2$. This means that $\mathbf{C}^{-1}$ must be a multiple of the identity matrix; in other words, $\mathbf{C}^{-1}$ is proportional to

$$\begin{bmatrix} 1 & 0 & \cdots & 0 \\ 0 & 1 & \cdots & 0 \\ \cdot & \cdot & \cdots & \cdot \\ 0 & 0 & \cdots & 1 \end{bmatrix}$$

This in turn implies that $\mathbf{C}$ is also proportional to the identity matrix. This means that the design must be such that

$$\sum_{i=1}^{n} x_{li}^2 \quad \text{does not depend on } l(=1,2,\ldots,q)$$

and

$$\sum_{i=1}^{n} x_{li}x_{hi}=0 \quad \text{if } l \neq h$$

(It is also necessary, of course, that $\sum_{i=1}^{n} x_{li}=0$ for all l, so that the centroid of the design is at the origin.)

For first-order designs, therefore, the conditions for rotatability are equivalent to the conditions that each partial regression coefficient shall be estimated with the same error variance, and that the estimators of different regression coefficients shall be uncorrelated. Complete balanced fractional designs do satisfy these conditions; but there are many others which also satisfy them. Of particular interest among these designs are those which form part of rotatable designs of *second* order. If such a rotatable design of first order has been planned, it is a straightforward matter to supplement it to produce a rotatable design of second order.

Example 18.3   For 2 factors a rotatable design of first order can be constructed by taking the vertices of a regular polygon. For a pentagon and hexagon one has for the (transformed) data the experimental points as shown in Figure 18.13. The

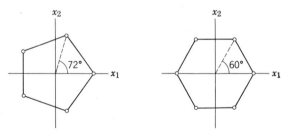

Figure 18.13   (*a*) Pentagonal design; (*b*) hexagonal design.

TABLE 18.8

Designs using Coordinates of the Vertices
of a Pentagon and a Hexagon

| PENTAGON | | HEXAGON | |
x_1	x_2	x_1	x_2
1.000	0	1.000	0
0.309	0.951	0.500	0.866
−0.809	0.588	−0.500	0.866
−0.809	−0.588	−1.000	0
0.309	−0.951	−0.500	−0.866
		0.500	−0.866

coordinates of the designs are given in Table 18.8. Additional points may be added at the center $(0,0)$.

18.7   ROTATABLE DESIGNS OF SECOND ORDER

Rotatable designs of first order can be formed by taking combinations of values of the 2 controlled variables corresponding to points at the vertices of regular polygons. For second-order designs, these can be supplemented by observations at the origin (the center of the design). It is often recommended that there should be more than 1 observation at the origin because these can be used to provide an estimate of residual variance. It is worth noting that such an estimate can also be obtained from repeated observations at the vertices of the regular polygon. However, to retain rotatability there must be equal numbers of observations at each vertex, so that the minimum number of additional observations is equal to the number of vertices of the regular polygon, while there need be only 2 observations at the center. On the other hand, replication at each vertex does provide some information on possible heteroscedasticity of the residual variation. There can be situations where such information is of particular importance.

Details of methods of construction of other rotatable designs of second order are not given here. References [18.1], [18.2], and [18.4] provide a useful introduction to the subject. A few examples of rotatable designs of second order are given, however, to indicate the types of design that meet the stipulated conditions.

Apart from designs with observations at the vertices of regular polygons, centered at the origin, together with 2 or more observations at the origin, there are a number of designs with values of the controlled variables corresponding to equally spaced points on each of 2 circles centered at the

TABLE 18.9

ORTHOGONAL ROTATABLE DESIGNS OF ORDER TWO

Number of points on outer circle	6	7	8	7	8	8
Number of points on inner circle	5	5	5	6	6	7
Ratio of radii	0.204	0.267	0.304	0.189	0.250	0.176

origin. (The designs already discussed can be regarded as special cases with the smaller circle of 0 radius.) For the designs also to be orthogonal (in the sense of giving uncorrelated estimators of the partial regression coefficients) the radii of the circles should be in definite ratios depending on the numbers of points on each circle. Some values (taken from [18.2]) are given in Table 18.9.

Example 18.4 Construct an orthogonal rotatable design of order 2 with 12 points (excluding center points). For a concentric circle design we choose (from Table 18.9) 7 equidistant points on the outer circle and 5 equidistant points on the inner circle. From Table 18.9 the ratio of radii is 0.267. Figure 18.14 presents these values in graphical form. The coordinates of the points are given in Table 18.10.

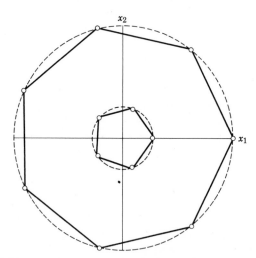

Figure 18.14 An orthogonal rotatable design of order 2, with 12 points.

TABLE 18.10

COORDINATES OF EQUIDISTANT POINTS ON CONCENTRIC
CIRCLES OF ORTHOGONAL ROTATABLE DESIGN OF ORDER TWO

OUTER CIRCLE

x_1	1.000	0.622	-0.220	-0.903	-0.903	-0.220	0.622
x_2	0	0.783	0.973	0.431	-0.431	-0.973	-0.783

INNER CIRCLE

x_1	0.267	0.083	-0.216	-0.216	0.083
x_2	0	0.254	0.157	-0.157	-0.254

18.8   CENTRAL COMPOSITE DESIGNS

A completely balanced rotatable design of second order for q controllable variables can be constructed using the following combinations of values of the q controllable variables. [It is well to keep in mind that each variable is supposed measured on a suitably chosen scale so that the center of the design is at the origin—the point $(0, 0, \ldots, 0)$.]

We first take all $2q$ points with coordinates

$$(\pm \alpha, 0, 0, \ldots, 0), (0, \pm \alpha, 0, \ldots, 0), \ldots, (0, 0, 0, \ldots, \pm \alpha)$$

and add to these the 2^q points $(\pm \beta, \pm \beta, \pm \beta, \ldots, \pm \beta)$. The constants α and β are chosen so that $\alpha / \beta = 2^{q/4}$. These designs are sometimes called *central composite designs*.

It will be noted that the second set of 2^q points corresponds to a complete 2^q factorial design. The number of observations called for may be reduced by retaining only those points corresponding to a 2^{-p} fractional replicate of the 2^q experiment in which no main effects or 2-factor interactions appear among the defining contrasts. We then take

$$\frac{\alpha}{\beta} = 2^{(q-p)/4}$$

These designs do not provide an estimate of residual variance. This can be obtained if 2 or more observations are added, corresponding to values of the controllable variables at the origin $(0, 0, 0, \ldots, 0)$. By choosing a suitable number of replications at the origin it is also possible to produce

TABLE 18.11

FURTHER ORTHOGONAL ROTATABLE DESIGNS

Dimensions (q)	2	3	4	5	5	6	6	7	7	8	8	8
Replicate of 2^q	Full	Full	Full	Full	Half	Full	Half	Full	Half	Full	Half	Qtr.
α/β	1.414	1.682	2.000	2.378	2.000	2.828	2.378	3.364	2.828	4.000	3.364	2.828
Number of observations at center	8	9	12	17	10	24	15	35	22	52	33	20

designs which are orthogonal as well as rotatable. Some examples (from [18.2]) are shown in Table 18.11.

The number of central observations needed to produce orthogonality becomes large quite rapidly as q increases. However, it should be remembered that rotatability is attained with *any* number of observations at $(0,0,0,\ldots,0)$. Orthogonality (in the present context) is just an additional property which is of some convenience, but may be foregone if it is deemed too expensive to attain.

Chapter 8A of [18.4] contains a useful discussion of the problems treated in the last few sections. [It should be noted that the composite designs introduced in this reference are not intended to produce orthogonal (uncorrelated) estimators of the partial regression coefficients, but to make the variance of $\hat{\eta}$ the same for $\sum x_j^2 = 1$ as for $x_1 = x_2 = \cdots = x_q = 0$.]

A somewhat fuller discussion of these results can be found in Chapters 5 to 7 of [18.5].

Example 18.5 Construct a central composite design for $q = 3$. From Table 18.11 we have the points $(\pm\alpha, 0, 0)$, $(0, \pm\alpha, 0)$, $(0, 0, \pm\alpha)$; $(\pm\beta, \pm\beta, \pm\beta)$; and 9 center points. $(2+2+2+8+9=23$ points in all). Further, $\alpha/\beta = 1.682$. If we let $\beta = 1$, then $\alpha = 1.682$. The 23 points are given in Table 18.12. Figure 18.15 shows the experimental points.

TABLE 18.12

COORDINATES OF EXPERIMENTAL POINTS
IN A CENTRAL COMPOSITE DESIGN WITH $q = 3$

x_1	x_2	x_3	x_1	x_2	x_3	x_1	x_2	x_3
-1.682	0	0	-1	-1	-1	0	0	0
1.682	0	0	1	-1	-1	0	0	0
0	-1.682	0	-1	1	-1	0	0	0
0	1.682	0	1	1	-1	0	0	0
0	0	-1.682	-1	-1	1	0	0	0
0	0	1.682	1	-1	1	0	0	0
			-1	1	1	0	0	0
			1	1	1	0	0	0
						0	0	0

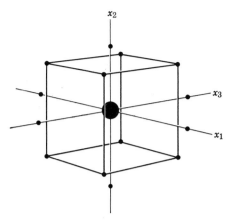

Figure 18.15 Experimental points of a central composite design with $q = 3$.

Example 18.6 The results of a preliminary investiagtion of an industrial fermentation process have been described briefly in Example 15.3. Two culture medium ingredients (x_1 and x_2) and 2 environmental variables (x_4 and x_5) were controlled.

Further experiments demonstrated that the yield passes through a maximum if x_1 is varied over a sufficiently wide range. This maximum, however, depends on x_2 as well. Similarly, a conditional optimum for x_2 appears to exist which depends on the value of x_1.

Previous experience suggested that a similar situation holds for the variables x_4 and x_5.

It was decided, after examining the results of the experiment described in Example 15.3, to try to estimate equations of response surfaces for Y (yield per "design unit") in terms of x_1, x_2, x_4, x_5, and a new factor x_3, representing another culture medium ingredient, expected to have a generally beneficial effect.

*Practical considerations** suggested the most satisfactory proceudre to be an examination of the 5 factors in 2 groups of 3, with x_2 common to both since its role seemed to be especially complex. These experiments would then serve as a guide to the design of a 4- or 5-factor experiment. [This course had the advantages that the smaller experiments are more easily analyzed and interpreted, and can more easily be represented graphically. The main disadvantage is that some of the 2-factor interactions ($x_1 x_4, x_1 x_5, x_3 x_4, x_3 x_5$) are ignored.]

A 3-factor orthogonal composite design was used. This consisted of a 2^3 factorial experiment, supplemented with a center point. There were duplicate observations for each combination of levels, as tabulated in Tables 18.13 and 18.15. In the first experiment factors x_1, x_2, and x_3 were varied in the manner shown in Table 18.13,

*If possible, it would have been preferable to examine all 5 factors simultaneously.

TABLE 18.13

Optimization of Medium Composition at Fixed Levels of x_4 and x_5†

TREAT- MENT NUMBER	x_1	x_2	x_3	DUPLICATE YIELDS	TREAT- MENT NUMBER	x_1	x_2	x_3	DUPLICATE YIELDS
1	−1	−1	−1	62.4	9	−1.215	0	0	102.4
				65.6					103.2
2	+1	−1	−1	131.2	10	+1.215	0	0	176.8
				134.4					175.2
3	−1	+1	−1	95.2	11	0	−1.215	0	112.8
				100.8					113.6
4	+1	+1	−1	136.8	12	0	+1.215	0	158.4
				156.0					159.2
5	−1	−1	+1	110.4	13	0	0	−1.215	111.2
				109.6					116.8
6	+1	−1	+1	160.0	14	0	0	+1.215	161.6
				163.2					163.2
7	−1	+1	+1	147.2	15	0	0	0	147.2
				132.0					145.6
8	+1	+1	+1	168.8					
				172.0					

† ($x_4 = 0$ Design Units; $x_5 = 0.75$ Design Units for all preparations.)

while factors x_4 and x_5 were held constant at the levels shown. The actual ingredient concentrations which correspond to the design levels of treatments 1 through 8 were so chosen that effects of comparable magnitude would be found for each of the three variables. The same units and center point were used as in Example 15.3.

The fitted equation was

$$\hat{y}_{x(1)} = 147.11 + 26.35x_1 + 12.91x_2 + 18.19x_3$$

$$- 5.15x_1x_2 - 4.35x_1x_3 - 1.15x_2x_3$$

$$- 5.36x_1^2 - 7.66x_2^2 - 6.17x_3^2$$

where $\hat{y}_{x(1)}$ denotes estimated product yield, in "yield units (Y.U.)" and x_1, x_2, and x_3 are expressed in "design units."

The equation describes the experimental system extremely well; regression accounts for 97 percent of the Total sum of squares of the data (Table 18.17). The small but statistically significant discrepancy between Treatments and Regression sums of squares indicates that one or more higher-order terms exists. Each of the terms of the equation except that for the interaction between x_2 and x_3 are statistically significant. The negative quadratic term for each of the variables indicates that each has an optimum level. The significant interactions of x_1 with the

TABLE 18.14

ANOVA OF REGRESSION ON THREE CULTURE MEDIUM INGREDIENTS

SOURCE	SUM OF SQUARES	DEGREES OF FREEDOM	MEAN SQUARE	MEAN SQUARE RATIO
Regression on:				
x_1	15,203.46	1	15,203.46	639.1***
x_2	3,651.40	1	3,651.40	153.5***
x_3	7,246.72	1	7,246.72	304.6***
$x_1 x_2$	424.36	1	424.36	17.8***
$x_1 x_3$	302.76	1	302.76	12.7**
$x_2 x_3$	21.16	1	21.16	0.9
x_1^2	250.66	1	250.66	10.5**
x_2^2	512.16	1	512.16	21.5***
x_3^2	332.40	1	332.40	14.0**
Total regressions	27,945.08	9	3,105.01	
Lack of fit	411.70	5	82.34	3.46*
Treatments	28,356.78	14	2,025.48	
Error (residual)	356.80	15	23.787	
Total	28,713.58	29		

Significance limits for F: $F_{5,15,0.95} = 2.90$, $F_{1,15,0.99} = 8.68$, $F_{1,15,0.999} = 16.59$

other 2 variables is a reflection of the fact that x_1 supplies some of the nutritional factors which are also contained in x_2 and x_3.

Equating to 0 the partial derivatives of the fitted equation with respect to the variables gives consistent equations with a unique solution. These solutions give the following estimates for the composition of the optimum medium: $x_1 = +2.13$, $x_2 = +0.0719$; $x_3 = +0.715$. When these values are substituted in the equation, a yield of 180 Y.U. is predicted as the maximum that can be attained with these ingredients at the levels of x_4 and x_5 chosen. Since the point lies somewhat outside the experimental region, the estimates may be somewhat in error.

The relationship of yield to the experimental variables is represented diagrammatically in Figure 18.16. Contours of equal yield are ellipsoids which enclose the maximum point like the layers of an onion. If the concentrations of 2 of the medium ingredients are chosen arbitrarily, some level of the third ingredient can be found which is superior to any other; but the estimated yield will be less than 180 Y.U.

Factors x_2, x_4, and x_5 were explored in an experiment of the design shown in Table 18.15. Levels of x_1 and x_3 were held constant. The experiment was done simultaneously with the one just described, so that other conditions are identical for the 2 experiments. Further, results of the first design were not available in time

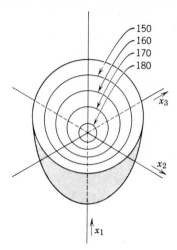

Figure 18.16 Yield of industrial fermentation as a function of 3 culture medium ingredients.

TABLE 18.15

OPTIMIZATION OF ENVIRONMENTAL FACTORS x_4 AND x_5

TREAT-MENT NUMBER	x_2	x_4	x_5	DUPLICATE YIELDS	TREAT-MENT NUMBER	x_2	x_4	x_5	DUPLICATE YIELDS
16	−1	−1	−1	132.0 133.0	24	−1.215	0	0	145.6 148.0
17	+1	−1	−1	146.0 157.0	25	+1.215	0	0	158.4 154.4
18	−1	+1	−1	112.7 114.7	26	0	−1.215	0	151.5 142.9
19	+1	+1	−1	135.3 129.3	27	0	+1.215	0	137.6 126.7
20	−1	−1	+1	136.0 120.0	28	0	0	−1.215	143.2 135.2
21	+1	−1	+1	165.0 165.0	29	0	0	+1.215	149.6 151.2
22	−1	+1	+1	123.3 116.7	30	0	0	0	141.6 136.8
23	+1	+1	+1	158.0 160.0					

to plan the second. The equation fitted to the data was

$$\hat{y}_{x(2)} = 146.88 + 11.29x_2 - 6.36x_4 + 5.09x_5 + 0.20x_2x_4$$

$$+ 4.80x_2x_5 + 3.00x_4x_5 + 1.39x_2^2 - 6.24x_4^2 - 3.29x_5^2$$

The equation describes the data moderately well. The sum of squares for regression accounts for 84 percent of the total. The sizable "lack of fit" sum of squares indicates that one or more higher-order terms exists (Table 18.16). The lack of significance for the quadratic term in x_2 appears to indicate that no optimum level exists near the area of the experiment, although one might have been expected from the evidence of the previous experiment. This apparent paradox was clarified by examining the partial derivatives with respect to x_4 and x_5:

$$\frac{\partial \hat{y}_{x(2)}}{\partial x_4} = -6.36 + 0.20x_2 - 12.48x_4 + 3.00x_5$$

$$\frac{\partial \hat{y}_{x(2)}}{\partial x_5} = +5.09 + 4.80x_2 + 3.00x_4 - 6.58x_5$$

TABLE 18.16

ANOVA OF REGRESSION ON THREE ENVIRONMENTAL VARIABLES

SOURCE	SUM OF SQUARES	DEGREES OF FREEDOM	MEAN SQUARE	MEAN SQUARE RATIO
Regression on:				
x_2	2792.0	1	2792.0	108.8***
x_4	887.4	1	887.4	34.6***
x_5	566.6	1	566.6	22.1***
x_2x_4	0.4	1	0.4	0.0
x_2x_5	201.0	1	201.0	7.8*
x_4x_5	78.5	1	78.5	3.1
x_2^2	31.1	1	31.1	1.2
x_4^2	622.6	1	622.6	24.3***
x_5^2	173.0	1	173.0	6.7*
Total regressions	5352.6	9	594.7	
Lack of fit	638.8	5	127.8	4.98**
Treatments	5991.4	14	428.0	
Duplicates (residual)	384.8	15	25.653	
Total	6376.2	29		

Significance limits for F: $F_{5,15,0.99} = 4.56$, $F_{1,15,0.95} = 4.54$
$F_{1,15,0.99} = 8.68$, $F_{1,15,0.999} = 16.59$

The negative terms demonstrate that both x_4 and x_5 can have optimum levels. For x_5, this value depends upon the levels of x_2 and x_4; the optimum levels of x_4 depends almost entirely upon that of x_5. Equating the partial derivatives to zero and solving for x_4 and x_5 the resulting equations are

$$x_4 = 0.215x_2 - 0.364$$

$$x_5 = 0.827x_2 + 0.608$$

These equations define a "ridge" of conditions for x_2, x_4, and x_5, as illustrated in Figure 18.17. By appropriate adjustment of the variables, one might "climb" the ridge, if there were not the complications which must arise from the changing responses of x_1 and x_3 as the other factors are changed.

Factor x_2 appears to have a central role in the system, for its level influences the response of the system to all of the other factors. It is a surprising consequence that an equation of the form

$$\hat{y} = B_0 + B_2x_2 + B_{22}x_2^2$$

is obtained when the last 2 equations are substituted in the fitted equation. Thus the response along the "ridge" is predicted to be a function primarily of the *square*

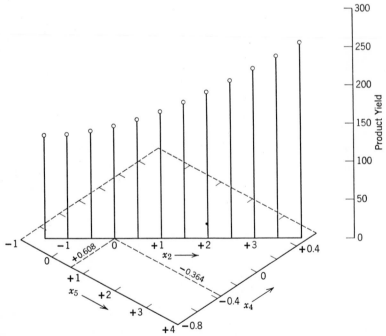

Figure 18.17 Yield of industrial fermentation as a function of 1 culture medium ingredient (x_2) and 2 environmental variables (x_4 and x_5).

of x_2. It is at this point that we become aware of the shortcomings of the model, for it is obvious that the last equation cannot be unreservedly true, as the model predicts. Undoubtedly a mechanism exists by which x_2 reaches an optimum value, but this point is not very close to the experimental area. In short, the yield depends upon the relative magnitudes of B_2 and B_{22}.

On the basis of these experiments, an additional experiment was made in which all 5 variables were examined. The medium composition for the central point of the design was essentially that predicted by the experiment in which x_1, x_2, and x_3 were studied. Factors x_4 and x_5 were adjusted to span wider ranges, and were moved to somewhat higher center-point values. Yields of 200 yield units occurred at the center point; yields as high as 225 units were found in some of the other treatment combinations.

18.9 ROTATABLE DESIGNS OF HIGHER ORDER

The argument used in Section 18.7 also applies when higher-order designs are used. We simply identify each power (or product of powers), of the controlled variable with a new "controlled variable." Thus if we have a second-order formula

$$y_x = \overline{Y} + \beta_1 x_1 + \cdots + \beta_q x_q + \beta_{11} x_1^2 + \cdots + \beta_{qq} x_q^2$$
$$+ \beta_{12} x_1 x_2 + \cdots + \beta_{q-1,q} x_{q-1} x_q$$

we can regard x_1^2 as a new variable x_{q+1} (say)

$$x_2^2 \quad \text{as a new variable} \quad x_{q+2} \text{ (say)}$$

and so on. (And we can identify β_{11} with β_{q+1}, β_{22} with β_{q+2}, and so on, if we wish.)

The elements of the matrix C [see (18.12)] will now be various sums of products of powers of the quantities x_{li}. The notation

$$[a_1 a_2 a_3 \cdots a_q] = n^{-1} \sum_{i=1}^{n} x_{1i}^{a_1} x_{2i}^{a_2} \ldots x_{qi}^{a_q} \tag{18.13}$$

(for the "design" product moments) is conventionally used in this connection. Conditions for rotatability can be expressed in terms of these quantities. We note that we still suppose that we have arranged to have

$$[100\ldots00] = [010\ldots00]\ldots = [000\ldots01] = 0$$

It can be shown that a mth-order design is rotatable if and only if

$$[a_1 a_2 a_3 \cdots a_q] = 0 \quad \text{if any } a_j \text{ is odd} \tag{18.14}$$

and

$$[a_1 a_2 a_3 \cdots a_q] = g_a \prod_{j=1}^{q} \left[\frac{a_j!}{\left(\frac{1}{2}a_j\right)!} \right] \tag{18.15}$$

where all a_j's are even for $a = a_1 + a_2 + \cdots + a_q$ between 1 and $2m$ inclusive. [Here g_a is a number depending only on a.]

In particular, for second-order designs ($m = 2$), (18.15) reduces to

$$n^{-1} \sum_{i=1}^{n} x_{ri}^4 = 12 g_a \qquad (r = 1, 2, \ldots, q)$$

$$n^{-1} \sum_{i=1}^{n} x_{ri}^2 x_{si}^2 = 4 g_a \qquad (r \neq s; \; r, s = 1, 2, \ldots, q)$$

from which we see that

$$\sum_{i=1}^{n} x_{ri}^4 = 3 \sum_{i=1}^{n} x_{ri}^2 x_{si}^2$$

Of course (18.14) must also be satisfied so that

$$\sum_{i=1}^{n} x_{1i}^{a_1} x_{2i}^{a_2} \ldots x_{qi}^{a_q} = 0$$

when any a_j is odd, for $1 \leqslant a_1 + a_2 + \cdots + a_q \leqslant 4$, for the design to be rotatable.

We can always scale the x's so that, in addition to satisfying $\Sigma x_{ri} = 0$ we also have

$$n^{-1} \sum_{i=1}^{n} x_{ri}^2 = 1 \qquad \text{for all } r \; \left(\text{i.e., } g_2 = \tfrac{1}{2} \right)$$

If this is done, and the conditions (18.14) and (18.15) are met, then it can be shown that

$$\operatorname{var}(\hat{y}_x) = n^{-1} \sigma^2 \left\{ 32(q+2) g_4^2 + 8(q+2) g_4 (4g_4 - 1) \sum_{j=1}^{q} x_j^2 \right.$$

$$\left. + \left[4(q+1) g_4 - (q-1) \right] \left(\sum_{j=1}^{q} x_j^2 \right)^2 \right\}$$

$$\times \left[8 g_4 \{ 4(q+2) g_4 - q \} \right]^{-1} \tag{18.16}$$

Note that

(*i*) this variance does indeed depend only on $\sum_{j=1}^{q} x_j^2$ and

(*ii*) if $g_4 = \frac{1}{4} q(q+2)$ the variance is infinite.

[In case (ii) it is not possible to estimate the β's from the least square equations.]

The symbol λ_a is often used for $2^{(1/2)a} g_a$.

REFERENCES

1. Box, G. E. P., "The Exploration and Exploitation of Response Surfaces: Some General Considerations and Examples," *Biometrics*, **10**, 16–60 (1954).

2. Box, G. E. P. and J. S. Hunter, "Multi-factor Experimental Designs for Exploring Response Surfaces," *Annals of Mathematical Statistics*, **28**, 195–241 (1957).

3. Box, M. J. and N. R. Draper, "On Minimum-Point Second Order Designs," *Technometrics*, **16**, 613–616 (1975).

4. Cochran, W. G. and G. M. Cox, *Experimental Designs*, 2nd ed., Wiley, New York, 1957.

5. Myers, R. H., *Response Surface Methodology*, Allyn and Bacon, Boston, 1971.

EXERCISES

1. A first-order model is proposed for the yield of a process dependent upon temperature and time. The experimental levels are 260 and 300°C and 2 and 6 hr. The data are as follows. Two observations are taken at each of the 4 combinations of levels.

Temperature	Time	Yield	Temperature	Time	Yield
260	2	30	280	2	15
260	2	35	280	2	17
260	6	26	280	6	37
260	6	23	280	6	30

Estimate the linear regression equation. By means of an analysis of variance table test for lack of fit and significance of regression coefficients.

2. As a preliminary test to determine the adequacy of a linear fit, 2 controlled variables were used at each of 2 levels—forming a 2^2 experiment. To these were added 3 "center points," giving a total sample of size 7. The data are as follows:

x_1	−1	1	−1	1	0	0	0
x_2	−1	−1	1	1	0	0	0
YIELD	56	80	64	94	76	72	70

Estimate the linear regression equation. Test the fit by the analysis of variance.

3. The data below represent a $\frac{1}{2}$ replicate of a 2^4 factorial design with 4 center points. Estimate the linear regression equation and test by the analysis of variance.

y	x_1	x_2	x_3	x_4	y	x_1	x_2	x_3	x_4
10	-1	-1	-1	-1	30	-1	-1	1	1
21	1	1	-1	-1	39	1	1	1	1
26	1	-1	1	-1	22	0	0	0	0
31	-1	1	1	-1	24	0	0	0	0
21	1	-1	-1	1	26	0	0	0	0
25	-1	1	-1	1	25	0	0	0	0

4. An applied statistician is requested to help design an experiment to determine the optimum conditions of yield dependent upon four controlled variables. He is allowed a budget sueficient for 25 experiments. He would like to obtain a quadratic equation for regression of yield on the controlled variables together with a good estimate of the error. Suggest how he might allocate these 25 experiments. What questions must be answered before he can reasonably begin the actual experimentation?

5. Three elements, copper (x_1), molybdenum (x_2), and iron (x_3) are being studied with respect to their effect on growth (y in arbitrary units) of lettuce in water culture. A central composite rotatable design for $k = 3$ variables (but with only 6 center points) was utilized (with transformed controlled variables) as follows:

x_1	x_2	x_3	y	x_1	x_2	x_3	y
-1	-1	-1	16.44	0	-1.68	0	25.39
1	-1	-1	12.50	0	1.68	0	18.16
-1	1	-1	16.10	0	0	-1.68	7.37
1	1	-1	6.92	0	0	1.68	11.99
-1	-1	1	14.90	0	0	0	22.22
1	-1	1	7.83	0	0	0	19.49
-1	1	1	19.90	0	0	0	22.76
1	1	1	4.68	0	0	0	24.27
-1.68	0	0	17.65	0	0	0	27.88
1.68	0	0	0.20	0	0	0	27.53

Original references are: R. J. Hader, et al., "An Investigation of Some of the Relationships of Copper, Iron, and Molybdenum in the Growth and Nutrition of Lettuce, I," *Proc. Soil Sci. Soc. Amer.*, **21** (1957). D. P. Moore, et al., "An Investigation of Some of the Relationships of Copper, Iron, and Molybdenum in the Growth and Nutrition of Lettuce, II," *Proc. Soil Sci. Soc. Amer.*, **21** (1957). These data are discussed in [18.4].

Determine the second-order regression equation and test by analysis of variance (*i*) the lack of fit, (*ii*) the second order terms and (*iii*) the first order terms.

6. Construct a central composite orthogonal rotatable design for $q=2$ controlled variables.

7. Give a table from which values of the controlled variables can be found for a central composite orthogonal rotatable design for $q=5$ controlled variables, which includes a $\frac{1}{2}$ replicate of a 2^5 design. (Use *ABCDE* as the defining contrast.)

8. Construct an orthogonal rotatable design of order 2 for two controlled variables made up of points on 2 concentric circles with a total of 14 points on the circles. Discuss points arising in the analysis of data obtained with this experimental design.

9. An experiment was conducted in order to find a method of improving a process (see [18.1]). Two solids were fused at high temperature to form a third substance. The 3 controlled variables were (1) fusion temperature, (2) fusion time, and (3) molar ratio of the 2 solids. The response variable was the cost of the process (in arbitrary units). A *non-central* composite design was evolved as given in the following table. Estimate the second-order regression from the data.

x_1	x_2	x_3	y	x_1	x_2	x_3	y
1	1	1	37	−3	−1	−1	90
1	1	−1	70	−1	−3	−1	52
1	−1	1	70	−1	−1	−3	16
1	−1	−1	39	0	0	0	38
−1	1	1	64	−3	−3	−3	48
−1	1	−1	74	−1	−1	0	33
−1	−1	1	48				
−1	−1	−1	18				

10. An octagonal design with 4 center points is given below. Estimate the second-order regression equation and test each term for significance by analysis of variance. Draw conclusions.

EXPERIMENT NUMBER	x_1	x_2	y	EXPERIMENT NUMBER	x_1	x_2	y
1	−1	−1	80.8	7	$\sqrt{2}$	0	81.7
2	1	−1	85.1	8	$-\sqrt{2}$	0	82.9
3	−1	1	82.9	9	0	$\sqrt{2}$	57.7
4	1	1	71.9	10	0	$-\sqrt{2}$	84.7
5	0	0	82.9	11	0	0	83.8
6	0	0	81.1	12	0	0	80.9

11. Construct a simplex design for 4 control variates, with the center at $(100, 50, 0, -10)$ and such that the variance of the estimator of β_1 is 4 times the variance of each of the estimators of β_2, β_3, and β_4.

12. Obtain the $\mathbf{C}$ matrix for the q-dimensional composite design with
    (a) 2^q experimental points $(\pm \beta, \pm \beta, \ldots, \pm \beta)$;
    (b) $2q$ experimental points $(\pm \alpha, 0, \ldots, 0)(0, \pm \alpha, \ldots, 0) \ldots (0, 0, \ldots, \pm \alpha)$;
    (c) m experimental points at the origin.
    (Assume a second-order model.)

13. Joe Zilch has constructed an experimental design for 2 control variables with 26 experimental points with coordinates $(\pm 2, \pm 1)$, $(\pm 1, \pm 4)$, $(\pm 1, \pm 2)$ $(\pm 1, \pm 1)(\pm 1, -3)(0, 4)(0, \pm 3)$, $(0, 2)$, $(0, -1)$, $(-2, 0)$. $((\pm a, \pm b)$ means the 4 points (a, b), $(-a, b)$, $(a, -b)$, $(-a, -b))$.

    His colleague comments that Zilch has clearly suggested this design out of personal vanity and points out that it is not symmetrical. However, the design can be made symmetrical by moving just 3 points.

    Give reasons for the colleague's comments, and obtain the variance function for a symmetrical design that can be obtained in the stated manner.

CHAPTER 19

Sample Structures

19.1 INTRODUCTION

In statistical work we frequently need to estimate the average value (or, equivalently, the total) of one or more characters in a population. We have already seen that random sampling can be used to provide an estimator (the sample arithmetic mean) which is unbiased, the accuracy of which can be estimated using the sample variance. This method of estimation is very widely used, but it is often possible to obtain an estimator (of the population mean) of greater accuracy with the same sample size or, more generally, with the same total expenditure. To do this, we must choose the sample in a special way, making use of such information as we have about the structure of the population under investigation.

Details of techniques of this kind are developed later in this chapter. Here we outline, in general terms, the kind of information it is useful to have in order to decide on a suitable form of sample. The most important thing is to be able to divide the population into a number of parts, which are conventionally called *strata*. It is desirable that the values of the characters we are measuring should vary as little as possible from individual to individual within the same stratum. This is equivalent to saying that we want the mean values to vary as much as possible from stratum to stratum. This equivalence can be appreciated from the following equation [compare (13.8)]:

$$\sum_{i=1}^{k} \sum_{j=1}^{M_i} (x_{ij} - \xi)^2 = \sum_{i=1}^{k} \sum_{j=1}^{M_i} (x_{ij} - \xi_i)^2 + \sum_{i=1}^{k} M_i (\xi_i - \xi)^2 \qquad (19.1)$$

where k = number of strata

M_i = number of individuals in ith stratum

x_{ij} = value of character X for jth individual in ith stratum

$$\xi_i = M_i^{-1} \sum_{j=1}^{M_i} x_{ij} = \text{mean value of } X \text{ in } i\text{th stratum},$$

$$\xi = M_0^{-1} \sum_{i=1}^{k} M_i \xi_i = \text{mean value of } X \text{ in the population},$$

$$M_0 = \sum_{i=1}^{k} M_i = \text{number of individuals in the population}.$$

The left side of (19.1) is fixed, for a given population. If we decrease the first term (representing within strata variation) we must increase the second term (representing between strata variation).

By random sampling *within* strata we can estimate each stratum mean ξ_i with accuracy depending on the *within strata* variation. These estimators can then be combined by the formula:

$$\text{Estimator of population mean} = M_0^{-1} \sum_{i=1}^{k} M_i (\text{estimator of } \xi_i) \quad (19.2)$$

Assuming the estimators of the ξ_i's to be mutually independent

$$\text{var}(\text{estimator of population mean}) = M_0^{-2} \sum_{i=1}^{k} M_i^2 \text{var}(\text{estimator of } \xi_i)$$

$$(19.3)$$

If we have succeeded in choosing strata so that var(estimator of ξ_i) is small for each i, then the estimator (19.2) will also have a small variance.

In practice there are limits to our ability to reduce the within strata variation. One reason for this is that we often need to estimate population means for more than one variable. It is very unlikely that the same system of strata will minimize within strata variation for *each* of a number of different characters. Apart from this there is nearly always the practical limitation that the strata must be clearly determined and it must be reasonably simple to take a random sample from each stratum. Very often natural subdivisions are used. For example, a geographical area might be subdivided by counties; a population of farms by size of farm; industrial output by batches or by date of production. Usually there is some choice of strata, but is is a limited choice, and we must make the best of the possibilities open to us, rather than attempt to exhaust every possibility of minimizing within strata variability even at the expense of excessive increase in complexity of sampling.

19.2 PROPORTIONAL SAMPLING

We suppose that the population in which we are interested has been divided into k strata $\Pi_1, \Pi_2, \ldots, \Pi_k$, and that it is possible to obtain random samples from all strata though not necessarily at equal cost per individual.

Suppose we take random samples of size $n_1, n_2, \ldots, n_k$ from $\Pi_1, \Pi_2, \ldots, \Pi_k$, respectively, and measure the value of X for each individual in these samples. We denote the total sample size by n_0, so that

$$\sum_{i=1}^{k} n_i = n_0$$

and the random variable representing the value of X for the rth individual in the sample from the ith stratum (Π_i) will be denoted by X_{ir}. Then the sample arithmetic mean

$$\overline{X}_i = n_i^{-1} \sum_{r=1}^{n_i} X_{ir}$$

is an unbiased estimator of the stratum mean ξ_i and

$$\overline{X} = M_0^{-1} \sum_{i=1}^{k} M_i \overline{X}_i \qquad (19.4)$$

is an unbiased estimator of the population mean ξ.

The variance of $\overline{X}_i$ is then

$$\text{var}(\overline{X}_i) = (n_i^{-1} - M_i^{-1})\sigma_i^2$$

where

$$\sigma_i^2 = (M_i - 1)^{-1} \sum_{j=1}^{M_i} (x_{ij} - \xi_i)^2$$

and the variance of $\overline{X}$ is

$$\text{var}(\overline{X}) = M_0^{-2} \sum_{i=1}^{k} M_i^2 \text{var}(\overline{X}_i) = M_0^{-2} \sum_{i=1}^{k} M_i^2 (n_i^{-1} - M_i^{-1})\sigma_i^2$$

$$= M_0^{-2} \sum_{i=1}^{k} M_i^2 n_i^{-1} \sigma_i^2 - M_0^{-2} \sum_{i=1}^{k} M_i \sigma_i^2 \qquad (19.5)$$

If the sample sizes n_i are *proportional* to the sizes M_i of the strata, so that $n_i = n_0 M_i / M_0$, then (19.5) becomes

$$\mathrm{var}(\overline{X}) = \left(n_0^{-1} - M_0^{-1}\right)\left(M_0^{-1}\sum_{i=1}^{k} M_i\sigma_i^2\right) \qquad (19.6)$$

If a random sample of size n_0 had been chosen from the whole population, disregarding stratification, the variance of the arithmetic mean of the sample values would be

$$\left(n_0^{-1} - M_0^{-1}\right)\sigma_0^2$$

where

$$\sigma_0^2 = (M_0 - 1)^{-1}\sum_{i=1}^{k}\sum_{j=1}^{M_i}(x_{ij} - \xi)^2$$

From (19.1) and the definition of σ_i^2

$$\sigma_0^2 = (M_0 - 1)^{-1}\left[\sum_{i=1}^{k}(M_i - 1)\sigma_i^2 + \sum_{i=1}^{k}M_i(\xi_i - \xi)^2\right] \qquad (19.7)$$

If (as is the case very often) the stratum sizes M_i are fairly large, then

$$(M_0 - 1)^{-1}\sum_{i=1}^{k}(M_i - 1)\sigma_i^2 \doteqdot M_0^{-1}\sum_{i=1}^{k}M_i\sigma_i^2$$

and so

$$\left(n_0^{-1} - M_0^{-1}\right)\sigma_0^2 \geqslant \mathrm{var}(\overline{X}) \qquad \text{as given by (19.6)}$$

Thus we can usually expect a *proportional sample* to give an estimator of ξ with smaller variance than a simple random sample of the same size. (It is *just* possible that the reverse could be the case, but very unlikely; this could only happen if the strata were badly chosen so that $\xi_i \doteqdot \xi$.)

It is important to notice that a proportional sample can be constructed with no prior knowledge of the way in which X varies from stratum to stratum, or within strata. The same proportional sample can therefore be repeated to produce an increase in accuracy (as compared with an unrestricted random sample) for any one of a whole series of variables.

Note that for a proportional sample (19.4) becomes

$$\overline{X} = n_0^{-1}\sum_{i=1}^{k}n_i\overline{X}_i, \quad \text{and so } \overline{X} \text{ is the sample arithmetic mean}$$

As we see later, it is usually possible to sample in such a way as to get greater accuracy than that provided by a proportional sample for any given character, but the structure of such a sample is likely to vary from character to character. On the other hand, a proportional sample can be expected to produce *some* improvement for *each one* of a number of characters.

It is often necessary to introduce considerations of cost in constructing a stratified sample. We consider here only the simple case where the cost of sampling is c_i per individual in the ith stratum. It is quite possible that c_i will vary with i. For example, a stratum concentrated in a town is usually less expensive to sample (per individual) than a stratum spread out over a relatively inaccessible and sparsely settled region.

If the total cost is fixed at C then we have to satisfy the condition

$$\sum_{i=1}^{k} n_i c_i = C$$

For proportional sampling we would have

$$n_i = \frac{CM_i}{\sum_{i=1}^{k} c_i M_i}$$

The variance of $\overline{X}$ would then be

$$\left(M_0^{-1} C^{-1} \sum_{i=1}^{k} c_i M_i - M_0^{-1} \right) \left(M_0^{-1} \sum_{i=1}^{k} M_i \sigma_i^2 \right) \tag{19.8}$$

Example 19.1 A machine is producing cigarettes at the rate of one thousand per minute. At intervals of 1 min a cigarette is weighed. The average weights of the 60 cigarettes measured in successive hours are shown in Table 19.1.

(*i*) Estimate the average weight over the whole 7 hr.

(*ii*) Estimate the standard error of this estimate.

(*iii*) Estimate the standard error of an estimate of average weight based on a sample of 420 cigarettes taken at random intervals during the 7 hr.

Here we can treat the hours as "strata." (If data were given for smaller intervals, for example, half-hours, these could be treated as strata.) In 1 hr, $60 \times 1000 = 60,000$ cigarettes are produced, and so we take $M_1 = M_2 = \cdots = M_7 = 60,000$ and $M_0 = 7(60,000) = 420,000$. The sampling is evidently proportional, with $n_1 = n_2 = \cdots = n_7 = 60$ and $n_0 = 7(60) = 420$.

An unbiased estimate of average weight is $\overline{X} = \frac{1}{7}(0.031 + 0.033 + \cdots + 0.030) = 0.03114$ oz. The variance of $\overline{X}$ is, according to (19.6),

$$\left(\frac{1}{420} - \frac{1}{420000} \right) \left(\frac{1}{420000} \right) (60000) \sum_{i=1}^{7} \sigma_i^2 \doteq \frac{1}{(420)(7)} \sum_{i=1}^{7} \sigma_i^2$$

TABLE 19.1
AVERAGE AND VARIANCE OF WEIGHT
OF CIGARETTE SAMPLES

HOUR	AVERAGE WEIGHT (OZ)	VARIANCE (OZ^2) $\times 10^6$
1	0.031	2.1
2	0.033	4.2
3	0.034	3.7
4	0.029	5.6
5	0.031	2.4
6	0.030	3.0
7	0.030	4.6

σ_i^2 is estimated by the sample variance for the ith hour. Hence we estimate the variance of $\bar{X}$ as

$$\frac{10^{-6}}{(420)(7)}[2.1+4.2+\cdots+4.6]=10^{-6}(0.008707) \text{ oz}^2$$

Taking the square root, the standard error is estimated as

$$10^{-3}(0.093) \text{ oz}$$

If a random sample of 420 cigarettes is taken from the whole 7 hr production, the variance of the arithmetic mean is

$$\left(\frac{1}{420}-\frac{1}{420000}\right)\sigma_0^2 \doteqdot \frac{\sigma_0^2}{420}$$

(This formula implies that there are no variations, of period about 1 min, which would materially affect the results.) Very nearly,

$$\sigma_0^2 \doteqdot \frac{1}{7}\sum_{i=1}^{7}\sigma_i^2+\frac{1}{7}\sum_{i=1}^{7}(\xi_i-\xi)^2$$

We have already seen how σ_i^2 can be estimated. Similarly, ξ_i can be estimated by the observed average weight in the ith hour (and ξ by the overall average 0.03114 in the present case).

Inserting these values, we obtain as an estimate of σ_0^2

$$\frac{10^{-6}}{7}[2.1+4.2+\cdots+4.6]+\frac{1}{7}\left[(0.031-0.03114)^2+\cdots+(0.030-0.03114)^2\right]$$

$$=3.657(10^{-6})+2.694(10^{-6})=6.351(10^{-6})$$

Hence we estimate the variance of the arithmetic mean as

$$\left(\frac{1}{420}\right)(6.351)(10^{-6}) = 0.01512 \times 10^{-6} \; oz^2$$

Taking the square root, we obtain an estimated standard error equal to $(0.123)10^{-3}$ oz.

This is about 30 percent greater than the estimated standard error for proportional sampling.

19.3 OPTIMUM SAMPLING

We now return to (19.5) for the variance of $\overline{X}(= M_0^{-1}\Sigma_{i=1}^k M_i \overline{X}_i)$. The only part of this formula that depends on the sample sizes n_i is $\Sigma_{i=1}^k M_i^2 n_i^{-1}\sigma_i^2$. If the n_i's are chosen to minimize this quantity they will also minimize $\mathrm{var}(\overline{X})$.

Suppose that the total sample size n_0 is given, and that we wish to minimize $\mathrm{var}(\overline{X})$. Then we have to minimize $\Sigma_{i=1}^k M_i^2 n_i^{-1}\sigma_i^2$, subject to the conditions $n_i \geqslant 0$, $\Sigma_{i=1}^k n_i = n_0$. This can be changed to a problem of finding an unconditioned minimum by writing $n_k = n_0 - n_1 - n_2 - \cdots - n_{k-1}$ and seeking the minimum of

$$f(n_1, n_2, \ldots, n_{k-1}) = \sum_{i=1}^{k-1} M_i^2 n_i^{-1}\sigma_i^2 + M_k^2 \left(n_0 - \sum_{i=1}^{k-1} n_i\right)^{-1} \sigma_k^2$$

Differentiating with respect to n_j and equating to 0, we find

$$\frac{M_j^2 \sigma_j^2}{n_j^2} = \frac{M_k^2 \sigma_k^2}{n_k^2} \qquad \text{for } j = 1, 2, \ldots, k-1$$

This means that n_j should be proportional to $M_j \sigma_j (j = 1, 2, \ldots, k)$. By further differentiation it can be confirmed that these values of n_j do in fact minimize $\Sigma_{i=1}^k M_i^2 n_i^{-1}\sigma_i^2$ subject to $\Sigma_{j=1}^k n_i = n_0$

The appropriate values of n_j,

$$n_j = n_0 \frac{M_j \sigma_j}{\sum_{i=1}^k M_i \sigma_i} \qquad (19.9)$$

are said to give *optimum sampling* for the variable concerned. In practice it is not usually possible to attain these values exactly because formula (19.9) gives fractional values. So as far as possible the nearest integer values are used.

It will be noted that the constitution of the optimum sample depends on the values of $\sigma_1, \sigma_2, \ldots, \sigma_k$. It is therefore generally not the same for different variables. As was pointed out in Section 19.2, proportional sampling does not suffer from this drawback. In particular cases, however, where the ratios $\sigma_1 : \sigma_2 : \sigma_3 : \ldots : \sigma_k$ do not vary greatly from variable to variable, it is possible to use a "compromise" system—based on an "average" set of ratios of σ's—which will still give more accuracy than proportional sampling (though less than "optimum" sampling) for each variable separately.

If the cost of sampling is brought into account, then (in the notation of Section 19.2) we may wish to minimize $\sum_{i=1}^{k} M_i^2 n_i^{-1} \sigma_i^2$ subject to a fixed total cost $C = \sum_{i=1}^{k} n_i C_i$. We find that n_j should then be proportional to $M_j \sigma_j / \sqrt{c_j}$ and in place of formula (19.9) we have

$$
n_j = n_0 \frac{M_j \sigma_j / \sqrt{c_j}}{\sum_{i=1}^{k} \left[M_i \sigma_i / \sqrt{c_i} \right]}
\tag{19.10}
$$

The remarks about the interpretation and use of formula (19.9) apply where appropriate to (19.10).

Of course, it is often the case that the values of the σ_i's are not known at the beginning of an investigation. In some cases an *approximate* knowledge of the ratios of the σ_i's can be used to obtain a sampling system which is *approximately* optimum. If this prior knowledge is felt to be insufficiently precise, recourse may be had to a "pilot sample." This is a preliminary sample, the results of which are used in planning the main inquiry. Estimates of the σ_i^2's are provided by the within strata sample variances (as, in fact, were used in Example 19.1). If there are about 30 or more individuals from each stratum, then theoretical investigations have shown that the sample estimates of σ_i^2 can be used in formulas (19.9) or (19.10) with a high probability of considerable improvement in accuracy of estimation of population mean, compared with that obtained by simple random, or by proportional, sampling.

Example 19.2 A machine processing a foodstuff is adjusted so that it automatically sorts its output into 4 grades. In Grade I the proportion of underweight product is between 0 and 2 percent; for the other grades the proportions underweight are

GRADE	PER CENT UNDERWEIGHT
II	2–5
III	5–10
IV	10–20

The product in each grade is thoroughly mixed and sold in boxes of 144 items. Over a certain period of time the proportions of Grades I, II, III, and IV are 15, 45, 30, and 10 percent, respectively.

In order to estimate the average proportion of underweight items it is decided to completely examine 100 boxes and count the number of underweight items in each. It is necessary to decide how many boxes of each grade should be examined.

In this case we do not know the exact numbers (of boxes) in each Grade, but we do know their ratios (15, 45, 30, and 10 percent). Neither do we know the standard deviation of the number of underweight items in a box, but we know that in Grade I the standard deviation is between 0 and $\sqrt{(144)(0.98)(0.02)} = 1.68$.

Similarly, in Grade II the limits for standard deviation are

$$\sqrt{(144)(0.98)(0.02)} = 1.68$$

and

$$\sqrt{(144)(0.95)(0.05)} = 2.62$$

Similar calculations give limits 2.62—3.60 in Grade III and 3.60—4.80 in Grade IV.

Optimum proportions for sampling will therefore be in ratios between the limits shown below:

Grade I: $0.15 \times (0, 1.68) = (0, 0.252)$
Grade II: $0.45 \times (1.68, 2.62) = (0.756, 1.179)$
Grade III: $0.30 \times (2.62, 3.60) = (0.786, 1.080)$
Grade IV: $0.10 \times (3.60, 4.80) = (0.360, 0.480)$

If we take the mid-point of each range as a rough guide we obtain the ratios

$$0.126, 0.967, 0.933, 0.420$$

Therefore, we suggest taking

$$100 \times \frac{0.126}{0.126 + 0.967 + 0.933 + 0.420} \doteqdot 5 \text{ boxes from Grade I}$$

Similar calculations (rounding off to suitable whole numbers) lead to 40 boxes from Grade II, 38 from Grade III, and 17 from Grade IV. The overall proportion underweight would be estimated as

$$\frac{1}{144} \left[\frac{\text{(Grade I)}}{5} \cdot 0.15 + \frac{\text{(Grade II)}}{40} \cdot 0.45 + \frac{\text{(Grade III)}}{38} \cdot 0.30 + \frac{\text{(Grade IV)}}{17} \cdot 0.10 \right]$$

where (Grade j) denotes the total number of underweight items found in Grade j boxes.

Example 19.3 The following remarks are taken from an article by John C. Tanner ["The Sampling of Road Traffic," *Applied Statistics*, **6** (1957)], discussing methods of estimating the number of road junctions in England and Wales.

"The National Grid used by the Ordnance Survey superposes on the map of Great Britain a rectangular grid of 100-km squares, each square of which is subdivided into one hundred 10-km squares. England and Wales are completely covered by 28 of the 100-km squares, but a number of them consist mainly of sea, and by superposing a number of these, the effective number is reduced to 20.

A 5 percent sample was drawn with five 10-km squares chosen at random from each of the 100-km squares. There were thus 100 sample squares...." The frequency distribution of the number of junctions in the 100 sample squares was as shown in Table 19.2.... "Although the number of junctions varies widely from one sample to another the proportions of various types vary much less.... More precise results would have been obtained by using a larger number of smaller sample squares, say 10,000 1-km squares instead of 100 40-km ones."

TABLE 19.2

FREQUENCY DISTRIBUTION OF NUMBER OF JUNCTIONS

NUMBER OF JUNCTIONS PER 10-KM SQUARE	FREQUENCY	NUMBER OF JUNCTIONS PER 10-KM SQUARE	FREQUENCY
0	24	8	2
1	17	9	2
2	10	10–19	5
3	6	20–29	3
4	5	30–39	1
5	8	40–49	2
6	7	149	1
7	7		

We comment on this passage from the point of view developed in this chapter.

We can regard the 100-km squares as strata, each containing 100 sampling units (10-km squares). Then the sampling method used can be regarded as proportional sampling with 20 strata and a sampling fraction of 1 in 20. The estimated total number of road junctions in England and Wales is

$$[\text{Total number of junctions in sample}] \times 20$$

The standard deviation of this estimate is

$$\left\{ \frac{5(100-5)}{100-1} \sum_{i=1}^{20} (\text{variance between 10-km squares within } i\text{th 100-km square}) \right\}^{1/2}$$

The estimate of the number of junctions by the second proposed method would also be

$$[\text{total number of junctions in sample}] \times 20$$

since the total area examined will be the same (10000 km^2) for each method. The standard deviation of this estimate would be

$$\left\{ \frac{500(10000-500)}{10000-1} \sum_{i=1}^{20} \begin{pmatrix} \text{variance between 1-km squares} \\ \text{within } i\text{th 100-km square} \end{pmatrix} \right\}^{1/2}$$

if 500 1-km squares were chosen from each 100-km square. This is smaller than the standard deviation for the method actually used if

$$\frac{\text{average variance between 1-km squares in the same 100-km squares}}{\text{average variance between 10-km squares in the same 100-km squares}}$$

is less than $(5 \times 95/99)/(500 \times 9500/9999) = 0.0101$.

This is on the assumption that proportional sampling be retained. However, the author suggests "...the sample might profitably have been weighted by varying the number of sample squares from one 100-km square to another, with more of them in the 100-km squares containing greater mileages."

In view of formula (19.9) this implies an assumption (quite probably correct) that the standard deviation of number of road junctions increases with the average number of junctions.

*19.3.1 Stratification Subsequent to Sampling

It can sometimes happen that the possibility of stratification arises after a sample has been collected. This may happen because some set of records becomes available classifying the population according to some other character—for example, as a result of the completion of a census, or some other investigation. It may even happen that the observations themselves may suggest a form of stratification previously available, but thought to be unprofitable. This will be the case if differences between observed values for individuals in different putative strata are large compared with differences for individuals in the same putative stratum. [See (19.7).] If this circumstance had been foreseen, of course, stratified sampling could have been employed *ab initio*, and the number of individuals chosen from each stratum could have been controlled. Even if this is not the case, however, we can still use formula (19.2) to give an estimate of the population mean. This will not usually be as accurate as the estimate that would have been obtained by using optimum, or even proportional, sampling, but it is quite likely to be more accurate than the simple arithmetic mean of all the sample values.

It is easy to see that the estimator so obtained is unbiased, provided at least one individual is available from each stratum. The variance of the estimator is [applying formula (19.3) or (19.5)]

$$M_0^{-2} \sum_{i=1}^{k} M_i^2 \left[E\left(N_i^{-1}\right) - M_i^{-1} \right] \sigma_i^2 \tag{19.11}$$

where $E(N_i^{-1})$ is the expected value of N_i subject to the condition that none of the N_i's are equal to 0. (It is interesting to note that, assuming the total sample size, n, fixed, the stratum means $\overline{X}_i$ are uncorrelated; but they are not independent, since the N_i's are linked by the condition $\sum_{i=1}^{k} N_i = n$.)

However, for an *actually realized* sample, (19.5) applies, and should be used in assessing attained accuracy. Formula (19.11) gives the overall average variance, and should be used only to provide information on average accuracy to be expected, if a decision is to be made to use some method of sampling involving (though not necessarily limited to) random selections unrelated to strata.

It may be felt that the procedure described above is inconsistent with the principle, which we have emphasized from time to time, that hypotheses suggested by the data cannot properly be made the subject of formal statistical tests. However, at present we are concerned with obtaining as good an *estimate* as possible from the available data, not with *testing* a hypothesis. In this connection it should further be noted (as has already been pointed out) that data *should* be examined to see if they *suggest* fresh hypotheses. Statistical tests of these fresh hypotheses, however, should be made on the basis of data collected from further experiments.

19.4   USE OF CONCOMITANT VARIABLES

It is evident that some characters are easier to measure than others. For example tensile strength of a material is usually measured much more easily at 15°C than at 600°C. Very often this relative ease of measurement is reflected in lower cost per measurement. If there is a fairly close relationship between the 2 (or more) characters, it may be more economical to expend a substantial part of our effort on measuring the characters which it is cheaper to measure, and use the information so obtained to estimate a character which is more expensive to measure. The additional characters used in this way are called *concomitant characters* or concomitant variables.

If there is some evidence of relationship between the character (in which we are primarily interested) which is more expensive to measure, and the other character(s) then some part of our effort must be directed to estimating this relationship. On the other hand, if this relationship is well established, we can decide, in advance, how to apply our sampling effort, provided we have good estimates of the costs of measurement of each character. It may happen that we will not measure the character in which we are primarily interested at all; if we do measure this character we may not measure the others unless we also wish to obtain some direct information on these characters.

We first discuss a situation in which we are interested in estimating the total of a character X in a population divided into strata $\Pi_1, \Pi_2, \ldots, \Pi_k$. We suppose that there is a single possible concomitant variable Y and that it is known that the regression of X on Y is linear, such that in Π_i the distribution of X, given Y, has mean $\alpha + \beta Y$ and standard deviation

$\sigma_{iX}\sqrt{1-\rho^2}$ where ρ, the correlation coefficient between X and Y, is known, as also are α and β.

Suppose we take a sample of size n_0 with n_i individuals from $\Pi_i (i = 1, 2, \ldots, k)$ and measure the value of Y for each individual. Let Y_{ij} denote the value of Y for the jth measured individual from the ith stratum. Then we estimate the total X in the population by the formula

$$M_0\alpha + \beta \text{ (estimate of total } Y) = M_0\alpha + \beta \sum_{i=1}^{k} M_i \overline{Y}_i \qquad (19.12)$$

where

$$\overline{Y}_i = n_i^{-1} \sum_{j=1}^{n_i} Y_{ij}$$

The variance of this quantity is

$$\beta^2 \sum_{i=1}^{k} M_i^2 \left(n_i^{-1} - M_i^{-1}\right)\sigma_{iY}^2 \qquad (19.13)$$

where σ_{iY}^2 is the variance of Y in the ith stratum.

If we had measured X in each case we would have estimated total X by

$$\sum_{i=1}^{k} M_i \overline{X}_i$$

and the variance of this estimator would be

$$\sum_{i=1}^{k} M_i^2 \left(n_i^{-1} - M_i^{-1}\right)\sigma_{iX}^2 \qquad (19.14)$$

There is a relationship between σ_{iX}^2 and σ_{iY}^2 which is implied by the assumptions

$$E(X|Y) = \alpha + \beta Y$$

$$\text{Var}(X|Y) = \sigma_{iX}^2 (1-\rho^2)$$

$$\sigma_{iX}^2 = \beta^2\sigma_{iY}^2 + \sigma_{iX}^2 (1-\rho^2)$$

that is,

$$\sigma_{iY}^2 = \frac{\rho^2\sigma_{iX}^2}{\beta^2}$$

Hence (19.13) can be written

$$\rho^2 \sum_{i=1}^{k} M_i^2 \left(n_i^{-1} - M_i^{-1} \right) \sigma_{iX}^2 \qquad (19.15)$$

Since $|\rho| \leqslant 1$ this can never be greater than the value given by (19.14). Hence we have the rather surprising result that *the use of the concomitant variable Y will never give a less accurate estimator of total X in the population than would direct estimates of X taken on a sample of the same constitution.* If (as we have supposed) Y is cheaper to measure than X, it seems clear that we should not measure X at all, but only Y. A further remarkable result is that measurement of Y gives a more accurate estimate of total X, the smaller the absolute magnitude of the correlation coefficient between X and Y. This is opposite to what might be expected; it is natural to think that the *closer* the linear relation between X and Y, the *more* accurate would be the estimate given by (19.12).

Another way of looking at the position, which confirms the conclusions we have reached, is as follows. There are 2 natural statistics which we might use as estimators of ξ_i, the mean of X in the ith stratum. These are

$$\overline{X}_i \quad \text{and} \quad \alpha + \beta \overline{Y}_i$$

The correlation between these 2 estimators is ρ; their variances are

$$\sigma_{iX}^2 \left(n_i^{-1} - M_i^{-1} \right) \quad \text{and} \quad \rho^2 \sigma_{iX}^2 \left(n_i^{-1} - M_i^{-1} \right)$$

respectively. Consider now the construction of linear combinations of $\overline{X}_i$ and $\alpha + \beta \overline{Y}_i$ to be unbiased estimators of ξ_i. These will be of the form

$$K\overline{X}_i + (1 - K)\left(\alpha + \beta \overline{Y}_i \right)$$

The variance of this statistic is

$$\sigma_{iX}^2 \left(n_i^{-1} - M_i^{-1} \right) \left[K^2 + (1 - K)^2 \rho^2 + 2K(1 - K)\rho^2 \right]$$

$$= \sigma_{iX}^2 \left(n_i^{-1} - M_i^{-1} \right) \left[K^2 + (1 - K^2)\rho^2 \right]$$

This variance is minimized by taking $K = 0$; that is, by using $\alpha + \beta \overline{Y}_i$ and ignoring $\overline{X}_i$, the sample mean of values of X from the ith stratum.

The reasons for these seemingly paradoxical results lie in the assumptions we have made. We have supposed that we know *exactly* the values of α and β in the equation

$$E(X|Y) = \alpha + \beta Y$$

If the correlation coefficient ρ were equal to 0, β would be 0, and we would in fact *know*

$$E(X) = \alpha$$

and so

$$\text{Total } X = M_0\alpha$$

exactly—that is, with zero variance, as given by (19.15).

For a more realistic model, therefore, we must use less sweeping assumptions, not presupposing such precise knowledge about the relationship between X and Y. (At this point it is well to note that it is the assumed *precision* that has led to our rather paradoxical results, not the supposed *simplicity* of the relationship. Similar results would be obtained if a polynomial regression with completely *known* coefficients were assumed, or a set of separate *known* regressions, a different one for each stratum.)

19.4.1 Adjustment by Use of Linear Regression

It is sometimes not at all possible to be satisfied that the sample (for some, or all, of the strata) is randomly chosen. We cannot feel very confident, in such cases, that $\overline{X}_i$ is an unbiased estimator of ξ_i. If, however, we have observed the values of Y, as well as X for each individual, we can remove the bias provided
(1) there is linear regression of X on Y in the stratum, and the slope, β, of this regression is known, and
(2) the mean value, η_i, of Y in the stratum is known.
Then

$$\overline{X}_i + \beta\left(\eta_i - \overline{Y}_i\right)$$

is an unbiased estimator of ξ_i.

If the linear regression

$$E(X|y) = \alpha + \beta y \tag{19.16}$$

were *completely* known then, of course, there would be no need to estimate ξ_i. It could be calculated as $\alpha + \beta\eta_i$.

The comments of this section apply, with straightforward modifications, when each stratum has a linear regression with different slopes, or even more generally, when each stratum has a separate regression which need not be linear.

Example 19.4 Suppose that it is known that for a certain alloy the distribution of Y (tensile strength at 600°C) given X (tensile strength at 15°C) is normal with expected value $(0.87X + 0.14)$ units and a standard deviation 0.13 units, over the range of values of X usually encountered. By taking n sample pieces of alloy and measuring X it is desired to estimate the mean tensile strength at 600°C and to obtain an estimate of the standard deviation of this estimate. The measurement of tensile strength at 15°C is subject to a random unbiased error of standard deviation 0.02 units.

Denoting the observed values of X by $X'_1, X'_2, \ldots, X'_n$ and their arithmetic mean by $\bar{X}' = n^{-1}\sum_{i=1}^{n} X'_i$, the natural estimator of mean tensile strength at 600°C is

$$0.87\bar{X}' + 0.14$$

Assuming that the errors of measurement can be represented by independent random variables, the standard deviation of this estimator is

$$\frac{(0.87)(0.02)}{\sqrt{n}} = \frac{0.0174}{\sqrt{n}}$$

Note that the standard deviation of Y, given X, (equal to 0.13) does not enter into this calculation. If it is desired to estimate the population *variance* of Y, then an unbiased estimator of this quantity is

$$(0.87)^2\left[(n-1)^{-1}\sum_{i=1}^{n} \left(X'_i - \bar{X}'\right)^2 - (0.02)^2 \right] + (0.13)^2$$

$$= \frac{0.7569}{n-1}\sum_{i=1}^{n} \left(X'_i - \bar{X}'\right)^2 + 0.0166$$

Example 19.5 Results of measurements of characters X and Y on each of 30 individuals chosen (not necessarily at random) from a well-defined population are summarized below:

$$S_X^2 = 1.10 \qquad s_y^2 = 1.36$$

Fitted linear regression $\quad E(X|y) = 1.07 + 0.54y$

The cost of measuring X is c_X per individual, and the cost of measuring Y is c_Y per individual. It is desired to select a new *random* sample from which an estimate of the mean value of X in the population can be obtained which has a variance V approximately. Investigate whether it is likely to be better to measure X directly, or to measure X and calculate a regression estimate of X.

It will make the argument clearer to express it in terms of symbols, and then insert appropriate numerical values. From the first sample (of n individuals) the regression line

$$E[X|y] = \bar{X} + B(y - \bar{y})$$

is fitted. From Section 13.17

$$\text{var}(\overline{X}) = \sigma_X^2 (1-\rho^2) n^{-1}$$

$$\text{var}(B) = \sigma_X^2 (1-\rho^2) \left[\sum_{i=1}^{n} (y_i - \bar{y})^2 \right]^{-1}$$

and

$$\text{cov}(B, \overline{X}) = 0$$

where σ_X = population standard deviation of X

ρ = linear correlation coefficient between X and y

$$\bar{y} = n^{-1} \sum_{i=1}^{n} y_i$$

and $y_1, y_2, \ldots, y_n$ are the observed values of Y in the first sample. (Note that this formula treats these values as fixed; the sample is not supposed to be a random sample from the population, but it is assumed that the selection of individuals does not depend on the value of Y for the chosen individuals.)

If X is measured in a second *random* sample of size n' the arithmetic mean of the n' observed values will be used as estimator of the mean population value of X. The variance of this estimator is σ_X^2 / n'. To make the variance less than V, n' must be at least σ_X^2 / V.

If Y is measured, instead of X, in the second sample, then the regression estimator

$$\overline{X} + B(\overline{Y}' - \bar{y})$$

is used, where $\overline{Y}'$ is the arithmetic mean of the n' observed values of Y in the second sample.

Note that $\overline{Y}'$ is regarded as a random variable, whereas $\bar{y}$ is regarded as fixed, in this discussion.

The variance of this estimator is

$$\text{var}(\overline{X}) + \text{var}\left[B(\overline{Y}' - \bar{y}) \right]$$

$$= \frac{\sigma_X^2 (1-\rho^2)}{n} + E(B^2) E\left[(\overline{Y}' - \bar{y})^2 \right] - [E(B)]^2 \left[E(\overline{Y}' - \bar{y}) \right]^2$$

$$= \frac{\sigma_X^2 (1-\rho^2)}{n} + \left[\beta^2 + \frac{\sigma_X^2 (1-\rho^2)}{\sum_{i=1}^{n} (y_i - \bar{y})^2} \right] \left[\frac{\sigma_Y^2}{n'} + (\eta - \bar{y})^2 \right] - \beta^2 (\eta - \bar{y})^2$$

where $\beta = \rho\sigma_X/\sigma_Y$, and η, σ_Y are the population mean and standard deviation, respectively, of Y. This formula simplifies to

$$\sigma_X^2 \left[\frac{1-\rho^2}{n} + \frac{\rho^2}{n'} + \frac{(1-\rho^2)\{(\eta-\bar{y})^2 + \sigma_Y^2/n'\}}{\sum\limits_{i=1}^{n}(y_i-\bar{y})^2} \right]$$

Unless estimates of η and σ_Y are available, it is not possible to estimate the multiplier of σ_X^2. However, it can be seen that the variance can never be less than

$$\sigma_X^2 \left[\frac{1-\rho^2}{n} + \frac{\rho^2}{n'} \right]$$

so that n' must be at least $\rho^2[(V/\sigma_X^2) - (1-\rho^2)/n]^{-1}$ for the variance to be less than V. It will then certainly not be advantageous to measure Y, instead of X, if

$$\left(\frac{\sigma_X^2}{V} \right) \cdot c_X < \rho^2 \left[\frac{V}{\sigma_X^2} - \frac{1-\rho^2}{n} \right]^{-1} c_Y$$

or if

$$\frac{c_Y}{c_X} > \rho^{-2} \left[1 - \left(\frac{\sigma_X^2}{V} \right) \cdot \frac{1-\rho^2}{n} \right]$$

It can be seen from this inequality that the more accurate an estimate is required (that is, the smaller V is) the smaller must c_Y be to make it advantageous to measure Y. For the data given, for example, ρ may be estimated by the sample correlation coefficient

$$R = \frac{BS_y}{S_X} = 0.60$$

and $n = 30$, so the inequality becomes

$$c_Y/c_X > 2.778 - 0.059 \left(\frac{\sigma_X^2}{V} \right)$$

and for $V < 0.021\sigma_X^2$ it is certainly not advantageous to measure Y instead of X.

(This discussion ignores the effects of the essentially integer-valued nature of n. Further, it is necessary to have some idea of the value of σ_X^2 to apply the results.)

19.4.2   Ratio Estimators

Another common way in which observations of a concomitant variable Y are used to obtain an estimate of the population mean (or total) of the variable X is by the construction of *ratio estimators*.

As in the case of *regression estimators*, which have just been described, certain special conditions need to be satisfied before ratio estimates can be used. The essential condition is that the average value of X for a given value of Y is a constant multiple of the value of Y. In symbols

$$E(X|y) = \beta y \tag{19.17}$$

It will be appreciated that this is a more restrictive condition than the condition of linear regression. However, there are cases where it appears to be reasonable to assume that (19.17) is satisfied.

If (19.17) is satisfied, and we know the mean (or total) of Y in the population, then we estimate β from our data and estimate the mean X in the population as

(estimate of β) $\times$ (mean Y in population)

If it is assumed that the variance of X for Y fixed is constant (does not depend on y) then the least squares estimator of β is $(\Sigma Xy)/(\Sigma y^2)$, the summation signs referring to all values in the sample. Quite often, however, the simpler formula

$$\frac{\Sigma X}{\Sigma y}$$

is used. This may be regarded as the least square estimator which is appropriate if the variance of X, given y, is proportional to y.

Both regression and ratio estimators depend very heavily on the assumptions underlying their use. If these assumptions are incorrect, the estimators can be seriously biased. The assumptions are of similar nature, that is,

$$E(X|y) = \alpha + \beta y \qquad \text{for regression estimators}$$

and

$$E(X|y) = \beta y \qquad \text{for ratio estimators}$$

Some inaccuracy in these assumptions can be consistent with usefulness of the estimators, but substantial variation from these models can render the estimators very misleading. There is considerable scope for judgment on the applicability of the estimators.

Example 19.6 Suppose you are given the same information as in Example 19.5, but are asked to investigate the possibility of using a ratio estimator.

A preliminary investigation can be made to test the hypothesis that α in the regression equation

$$E(X|y) = \alpha + \beta y$$

is equal to 0. The value of the likelihood ratio criterion is

$$\frac{28(1.07)^2}{1.10(1-0.36)(29)} x \left(\frac{1}{30} + \frac{\bar{Y}^2}{29 \times 1.36} \right)^{-1} = \frac{47.11}{1+63.97\,\bar{Y}^2}$$

which is to be compared with the F distribution with $1,28$ degrees of freedom.

Since $F_{1,28,0.95} = 4.20$ and $F_{1,28,0.99} = 7.64$, the test gives a significant result (indicating $\alpha \neq 0$) at 5 percent level if

$$|\bar{Y}| < \sqrt{\frac{(47.106/4.20)-1}{63.97}} = 0.40$$

and at the 1 percent level if

$$|\bar{Y}| < \sqrt{\frac{(47.106/7.64)-1}{63.97}} = 0.28$$

19.4.3 Use of Sample in Utilization of Ratio and Regression Estimators

If values of the characters X and Y are measured on some (or all) of the individuals in the sample, we can use the information provided by these measurements to help decide whether to use regression or ratio estimators. Further, if we do decide to use such estimators, we can use the sample values to estimate the values of α and β (for regression estimators) or for β (for ratio estimators). It must be remembered, of course, that we also need to know the total (or average) value of Y in the whole population (or stratum) to use such estimators.

The basic assumption, for both estimators, is that the regression of X on Y is linear. For the use of ratio estimators we further require that in the regression equation

$$E(X|y) = \alpha + \beta y$$

the parameter α is equal to 0. Statistical tests of the appropriate hypotheses can be used to help decide whether these assumptions are reasonable. Beyond the mere question of whether or not there is "significant departure" from the hypotheses, it is important to be able to assess how accurate the estimated regression or ratio estimates are likely to be. This depends on the variability about the regression lines. The smaller this variability (that is, the higher, numerically, the linear correlation coefficient between X and Y), the more accurate should be the estimate.

It is important to realize that the calculation of a regression or ratio estimate may imply extrapolation beyond the range of values of Y available in the sample. It is possible, for example, that sample values of Y may range from 5 to 10 units, while population values may range from 3 to 30 units. The formulas used in regression or ratio estimators assume that the *same* linear relation applies over the range of *Y in the population*. Insofar as this assumption is not justified, bias can be introduced by the use of such estimators. This is a particularly important point to remember, for it is often the case that regression or ratio estimators are used to "correct" a value of mean *X* obtained from a nonrandomly selected sample.

We once more reiterate that the above considerations apply whether each stratum is treated separately or the whole population is assumed to have a common regression. The latter procedure is less troublesome, but the former is safer and, on the whole, to be recommended as insurance against unsuspected variation in regression from stratum to stratum.

19.5 MULTISTAGE AND CLUSTER SAMPLING

We now return to consider cases in which only 1 character is measured on each individual. We concentrate our attention on variations in methods of selecting individuals to be measured—or, as they are often termed, methods of sampling. In this connection we are particularly concerned with practical circumstances and with the accuracy of estimates of mean value (or equivalently, total) of the measured character in the population. We have already discussed some important aspects of the construction of samples in Sections 19.3 and 19.4. Here we are concerned with similar problems in rather more complicated situations.

So far we have considered populations split up into a number of strata. Although this is apparently a rather simple structure, it covers a remarkably broad range of cases encountered in practice. It is important to remember that complexity of the system of classification dividing the population into strata need not correspond to complexity in the sampling. If we have a population divided into strata we can apply the general theory for simple stratified populations, even though several factors (for example, geographical location, population density, average income, etc.) may have been used in arriving at the stratification. It is well to remember, of course, that stratification according to a complex scheme is likely to be more costly (because it is more troublesome to effect) than a simpler kind of stratification. Unless it seems very likely that substantially more accurate estimators will be obtained, complex schemes should be eschewed.

A really new factor is introduced, however, when we come to consider how the members of the sample are to be selected. So far we have described samples "of individuals" with the implicit assumption that the sample is chosen one individual at a time, each individual being chosen at random from the total number available in the stratum. In many cases, however, individuals are naturally grouped in *clusters*. For example, human beings are grouped in families, or, if the family is to be regarded as the individual for sampling purposes, in blocks in residential areas. Even whole cities may be considered as clusters in this sense. Apart from such human situations, individual articles such as engineering components appear as a matter of routine in "clusters" such as lots, batches, or boxes.

It is often practically convenient to select a sample in two stages. First, a number of clusters are chosen (at random, if possible). The random samples may be taken of individuals belonging to each of the chosen clusters. [If each cluster contains subclusters—as a township contains blocks, for example—then 3 (or more) stages may be used.] Such systems of sampling are called *multistage sampling*. As a particular case we may measure all individuals belonging to each chosen cluster. In this case there is, in fact, only one stage of sampling. To distinguish this system from simple random sampling it is called *cluster sampling*.

There is evidently considerable scope for choice in the details of construction of a multistage sample. A sample of specified total size can be made up in many different ways, by using few clusters with large samples from each cluster, or more clusters with smaller samples from each cluster. Also, it is necessary to consider whether the number of individuals chosen shall vary with the number of individuals in (or size of) each cluster in the first-stage cluster. If 3 (or more) stages of sampling are to be used, with clusters, subclusters, etc., the variety of possibilities becomes very great.

Here we consider only the simpler cases of clusters of equal size, indicating general principles. References [19.2], [19.3], [19.6], [19.7], [19.8], and [19.9] contain more detailed treatment of these problems.

*19.5.1 Clusters of Equal Size

Clusters are often not of equal sizes. This is particularly true of sampling human populations. However, manufactured articles are commonly grouped in lots or batches of equal (or very nearly equal) size so there is some justification for making a study of this special case.

The following discussion ignores stratification. If the population is stratified, the discussion should be regarded as relating to one particular stratum. Each stratum will be subjected to similar study.

Let x_{jl} denote the value of the measured character X for the lth individual in the jth cluster. The subscript j runs over the values $1, 2, \ldots, N$, where N is the number of clusters; the subscript l runs over the values $1, 2, \ldots, K$, where K is the size of each cluster. The total number of individuals in the population (or stratum) is $NK = M$. The population mean of X is

$$\xi = M^{-1} \sum_{j=1}^{N} \sum_{l=1}^{K} x_{jl} = N^{-1} \sum_{j=1}^{N} \bar{x}_{j.}$$

where

$$\bar{x}_{j.} = K^{-1} \sum_{l=1}^{K} x_{jl}$$

is the mean value of X in the jth cluster.

Suppose that n clusters are chosen at random from the N clusters, and that k individuals are chosen at random from the K individuals in each of the selected clusters. The symbol $X_{j'l'}$ is used to denote the value of X for the l'th individual in the sample from the j'th cluster in the sample of clusters. Then the arithmetic mean of $nk = m$ sample values of X is

$$\bar{X} = m^{-1} \sum_{j'=1}^{n} \sum_{l'=1}^{k} X_{j'l'}$$

It is easy to see that $\bar{X}$ (as given above) is an unbiased estimator of the population mean, ξ. The variance of $\bar{X}$ can be shown to be

$$\operatorname{var}(\bar{X}) = m^{-1} \left[\left(1 - \frac{k}{K}\right) \operatorname{var}(X_W) + \left(1 - \frac{n}{N}\right) k \operatorname{var}(\bar{X}_S) \right] \qquad (19.18)$$

where

$$\operatorname{var}(X_W) = [N(K-1)]^{-1} \sum_{j=1}^{N} \sum_{l=1}^{K} \left(x_{jl} - \bar{x}_{j.}\right)^2$$

$$= \text{average variance within clusters} \qquad (19.19)$$

and

$$\operatorname{var}(\bar{X}_S) = (N-1)^{-1} \sum_{j=1}^{N} \left(\bar{x}_{j.} - \xi\right)^2$$

$$= \text{variance between clusters} \qquad (19.20)$$

The variance of the mean of a sample chosen by simple random sampling (without replacement) is

$$m^{-1}\left(1 - \frac{m}{M}\right)\left[(M-1)^{-1} \sum_{j=1}^{N} \sum_{l=1}^{K} (x_{jl} - \xi)^2\right] = m^{-1}\left(1 - \frac{m}{M}\right) \operatorname{var}(X)$$

Hence cluster sampling gives a smaller variance for the arithmetic mean if

$$\left(1 - \frac{m}{M}\right) \operatorname{var}(X) > \left(1 - \frac{k}{K}\right) \operatorname{var}(X\,W) + \left(1 - \frac{n}{N}\right) k \operatorname{var}(\bar{X}_S) \qquad (19.21)$$

Note that if the "clusters" are simply random samples from the population, then

$$\operatorname{var}(\bar{X}_S) = \frac{\operatorname{var}(X)}{K}$$

and

$$\mathrm{var}(X_W) = \mathrm{var}(X)$$

The right side of equation (19.18) is then

$$m^{-1}\mathrm{var}(X)\left[1 - \frac{k}{K} + \left(1 - \frac{n}{N}\right)\cdot\frac{k}{K}\right] = m^{-1}\mathrm{var}(X)\left[1 - \frac{nk}{NK}\right]$$

$$= m^{-1}\mathrm{var}(X)\cdot\left(1 - \frac{m}{M}\right)$$

and, as would be expected, the variances of estimators based on multistage and random sampling are equal.

This shows that cluster sampling reduces the variance of the sample mean (for specified total sample size $m = nk$) if

$$\mathrm{var}(\overline{X}_S) < \frac{\mathrm{var}(X)}{K} \tag{19.22}$$

The smaller the variation between cluster means, the greater the advantage of multistage sampling. An alternative way of stating this condition is that the correlation between individuals within a cluster should be negative and as small (that is near to -1) as possible. Unfortunately, it often happens that there is positive correlation between individuals within the same cluster. In such cases the variance of the sample arithmetic mean is *increased* by multistage sampling. The justification for cluster sampling must then be sought in its simplicity and economy.

There is one special, but important, case where a reduction in variance of sample arithmetic mean can be expected from multistage sampling. This is where the clusters are families. For many characters (especially age, sex, income, and similar characters), variation between family means is considerably less than would be expected from the variation over the whole population. Other cases are encountered, from time to time, where conditions are especially favorable for multistage sampling, but mostly considerations of economy are needed to justify its use.

If it is decided to use a multistage sample of fixed total size $m = nk$, we have to decide on values of n and k. Naturally, we would like to choose n and k to minimize

$$\left(1 - \frac{k}{K}\right)\mathrm{var}(X_W) + \left(1 - \frac{n}{N}\right)k\,\mathrm{var}(\overline{X}_S)$$

subject to the condition $nk = m$. Putting $n = m/k$, we see that this expression is equal to

$$\mathrm{var}(X_W) - \frac{m}{N}\mathrm{var}(\overline{X}_S) - k\left[\frac{\mathrm{var}(X_W)}{K} - \mathrm{var}(\overline{X}_S)\right]$$

So, if multistage sampling does reduce the variance of the sample mean (for which the condition $\mathrm{var}(X_W)/K > \mathrm{var}(\overline{X}_S)$ must be satisfied) it is best to take k as large as possible. In other words, we use $k = K$ (if possible) and apply cluster sampling.

On the other hand, if $\mathrm{var}(X_W)/K < \mathrm{var}(\overline{X}_S)$, k should be taken as small as possible. If there are at least m clusters (that is, $N \geqslant m$), this means that we take $k = 1$—that is, only 1 individual per cluster.

In the construction of multistage sampling schemes, cost considerations are of special importance because, as has already been pointed out, it is often only on account of economy that multistage sampling is justified. A simple approximation to sampling cost is

$$C = nc_1 + nkc_2 \tag{19.23}$$

where

$$c_1 = \text{cost of including a cluster in the sample}$$

$$c_2 = \text{cost per individual sampled in a cluster}$$

Typically, c_1 is considerably greater than c_2. If C is kept fixed, we seek to minimize

$$(nk)^{-1}\left[\left(1 - \frac{k}{K}\right)\mathrm{var}(X_W) + \left(1 - \frac{n}{N}\right)k\,\mathrm{var}(\overline{X}_S)\right]$$

with respect to k, subject to the condition

$$C = nc_1 + nkc_2$$

If $\mathrm{var}(X_W)/K > \mathrm{var}(\overline{X}_S)$, then the additional cost of including extra clusters in the samples reinforces the arguments leading to the conclusion that k should be taken as large as possible—that is, cluster sampling should be used.

However, if $\mathrm{var}(X_W)/K < \mathrm{var}(\overline{X}_S)$, there is a value of k which minimizes the variance. Replacing n by $C(c_1 + c_2 k)^{-1}$, we obtain the following expression for the variance:

$$C^{-1}\left(c_2 - c_1 K^{-1} + c_1 k^{-1} - c_2 k K^{-1}\right)\mathrm{var}(X_W) + \left[C^{-1}(c_1 + c_2 k) - N^{-1}\right]\mathrm{var}(\overline{X}_S)$$

Equating to 0 the differential coefficient of this quantity with respect to k, we obtain

$$C^{-1}\left(-c_1 k^{-2} - c_2 K^{-1}\right)\mathrm{var}(X_W) + C^{-1}c_2\,\mathrm{var}(\overline{X}_S) = 0$$

or

$$k^2 = \frac{c_1}{c_2}\left[\frac{\mathrm{var}(\overline{X}_S)}{\mathrm{var}(X_W)/K} - 1\right]^{-1} \tag{19.24}$$

The second derivative with respect to k is $2C^{-1}c_1 k^{-3}$, and this is positive if the positive square root of the right side of (19.24) is taken for the value of k. This therefore corresponds to a minimum value of the variance. (In practice, of course, k must take an integer value, and so (19.24) is not satisfied exactly.) It is worthwhile noting that as the ratio

$$\frac{\text{var}(\bar{X}_S)}{\text{var}(X_W)\cdot K^{-1}}$$

tends to 1, the value of k^2 tends to infinity, as we would expect, since cluster sampling is indicated when this ratio is less than 1. In fact, cluster sampling may be indicated even when the ratio is greater than 1. If the value of k calculated from (19.24) is greater than K (the number of individuals per cluster), then the best we can do is to take $k = K$, that is, apply cluster sampling.

Example 19.7   A component is manufactured in small lots of 100 items each. To estimate the average quality of production it has been the practice to observe the value of a critical X on each of 10 items chosen at random from each lot. This sampling system was calculated to be the optimum, assuming that an equal number of individuals are to be taken from each lot. You are asked to advise whether the sampling system could be advantageously modified by reducing the number of lots inspected if
   (*i*) the cost of sampling (per lot) is to remain the same
   (*ii*) the cost of sampling (per lot) is to be halved
The cost function should be assumed to be of form (19.23).

The way to answer this problem becomes plain if we note that the right side of (19.24) does not depend on the total cost C. It follows that $k = 10$ items per lot will be optimum, *whatever the proportion of lots inspected*. Hence if C remains fixed [(*i*) above] the present system of taking 10 items from every lot is the best that can be done. If C is to be halved [(*ii*) above] then the number, n, of lots to be sampled is related to the total number (N) of lots by the relation

$$\tfrac{1}{2}(Nc_1 + 10Nc_2) = nc_1 + 10nc_2$$

so a proportion $\dfrac{n}{N} = \dfrac{1}{2}$ of all lots should be sampled, 10 items being taken from each.

*19.5.2   Clusters of Unequal Sizes

For *cluster sampling* (as opposed to multistage sampling) we can simply regard each cluster as a single sampling unit, record the total of observed values in each cluster included in the sample, and estimate the total in the population as

$$(\text{Total in sample}) \times \frac{\text{Number of clusters in population}}{\text{Number of clusters in sample}}$$

This formula applies even though there is variation in the number of individuals per cluster from one cluster to another. It should be noted, however that the "population variance" is now the variance between totals of clusters. If there is considerable variation in size from cluster to cluster, it is likely that there will be considerable variation in total from cluster to cluster.

(The above discussion is based on the implicit assumption that clusters are chosen at random, each cluster having an equal chance of being chosen. If clusters are of unequal sizes then the precise number of *individuals* in the sample will not be known beforehand.)

In order to reduce the variability, 2-stage sampling can be used, with the number of individuals chosen from each of the clusters in the sample proportional to the size of the cluster. Exact proportionality is unlikely to be attainable, unless all cluster sizes are in simple ratios. However, if the clusters are fairly large, a close approach to exact proportionality is possible.

In planning a 2-stage sample of this kind we have to choose 2 sampling fractions

(*i*) the proportion of clusters to choose (1 in m)

(*ii*) the proportion of individuals to be chosen from each selected cluster (1 in m')

All schemes with a given value of the product mm' will have the same average number of individuals per sample, that is, $(1/mm') \times$ (number of individuals in the population). Generally speaking, it is better to increase the number of clusters and reduce the number of individuals per cluster (that is, decrease m and increase m'). However, m' should not be increased so that it is greater than the size of any cluster, if possible.

Sometimes it is possible to control the size of clusters. In sampling a city, for example, clusters can be single blocks, or 2, 3, or more neighboring blocks. By increasing cluster size, it becomes possible to use larger values of m'. However, it should be remembered that increase in cluster size is likely to be associated with increase in magnitude of variation from cluster to cluster.

19.6   MISCELLANEOUS TECHNIQUES

We conclude this chapter by giving a brief account of some special techniques that are used from time to time in the organization of sampling investigations. These techniques have developed in response to particular practical requirements, just as the basic modifications of experimental design originated from the necessity to meet certain practical conditions (see Section 15.3, for example). The practical circumstances were usually those of sampling specified human population as, for example, in a census, or a market research investigation, or in special investigations to assess the condition of a population in some important respect (such as health, economical status, or opinion). The techniques so developed are not primarily intended for use in the more common types of statistical investigation in research in engineering or the physical sciences. However, a knowledge of the kinds of methods available can prove useful on many occasions. Lack of sufficiently detailed knowledge may be corrected, when necessary, by study of works (such as references [19.2], [19.3], [19.6], [19.7], [19.8], and [19.9]) more especially concerned with "sample surveys," etc.

19.6.1 Systematic Sampling

If the individuals in a population (or in a stratum) are already arranged in a list (as, for example, in a telephone directory, or a list of numbers of a society, or an inventory), or can easily be so arranged, it is very simple to draw a sample of 1 in m of the population by choosing every mth name in the list. The only "random" step in such a procedure is the selection of the first individual. This is chosen at random from the first m individuals on the list. (If the number of individuals in the population is not an exact multiple of m, we cannot draw an exactly 1 in m sample. We will suppose that the population size is an exact multiple of m.)

It is clear that this is not a random sample. It is, in fact, a very special type of *cluster* sample. The population is divided into m clusters each of size $(1/m) \times$ (population size). The first cluster contains the first, $(m+1)$th, $(2m+1)$th, etc., individuals in the list, the second contains the 2nd, $(m+2)$th, $(2m+2)$th, etc., individuals, and so on. The sample then consists of just 1 cluster, and all individuals in the cluster are measured.

The advantages and disadvantages of systematic sampling can be assessed on the same grounds as for cluster sampling. That is, systematic sampling is most useful when there is as much variation as possible within each "cluster" (that is, set of individuals spaced m apart on the list) and as little variation as possible between the means of different "clusters." Since the sample actually obtained contains measurements on individuals in only one cluster, it is not possible to estimate variability between clusters from the sample. In other words, a systematic sample does not provide data from which unbiased estimates of variances of estimators based on the sample can be calculated. (It is possible to calculate formal estimates based on certain special assumptions—such as assuming that random sampling formulas can be applied—but the results of such calculations must be regarded as mere guides, which may be untrustworthy.)

Example 19.8 A mechanical sampler can be adjusted to withdraw for inspection every mth item of routine production for values of m between 50 and 200 inclusive. It is at present set $m = 100$. Figures obtained by measuring resistance per unit length (ohms/cm) on each of the last 30 items withdrawn for inspection are (values given were coded by subtracting 1.5 from observed values and then multiplying by 100):

$$-5, \quad 1, \quad -3, \quad 5, \quad -5, \quad 5, \quad 0, \quad 9, \quad -1, \quad 11, \quad 5, \quad 15, \quad 4, \quad 9, \quad 6,$$
$$11, \quad 7, \quad 7, \quad 8, \quad 13, \quad 9, \quad 12, \quad 10, \quad 14, \quad 13, \quad 10, \quad 15, \quad 10, \quad 8, \quad 8$$

It has been proposed to increase m to 200. Comment on this proposal.

In the initial stages a periodicity of about 200 items combined with a slowly increasing trend in average value is indicated by the data. Both the periodicity and the trend become less marked in the second half of the data. However, we cannot be confident that the periodicity will not reappear. Since the period appears to be about 200 items we should certainly try to avoid taking a systematic sample with $m = 200$.

19.6.2 Multiphase Sampling

This is a technique applied when information is required in respect of each of a number of characters, but it is not possible to obtain the full range of information from each individual in the sample. In such cases, information on certain characters is obtained only from some (a "subsample") of the individuals in the sample. The subsampling is usually effected by a more or less systematic proportional sampling scheme—selecting every mth individual in the sample, for example.

Multiphase sampling has found special application in connection with very large-scale investigations, such as censuses. In such cases even the subsamples are large enough for considerable confidence to be felt in estimates calculated from them. Also, the values of related characters, observed over the *whole* sample (or in a complete census, the whole population) can be used to check and perhaps improve the accuracy of estimates based on the subsample.

In the simplest form of multiphase sampling, a subsample of individuals is subjected to extra measurements (for example, by being asked to answer extra questions on a supplementary census form). This method has been used in United Kindgom censuses. In some United State censuses, on the other hand, every individual answers some special questions, so that there are a number (usually 5) of different subsamples. The latter system has the advantage that more information can be obtained without each person answering many additional questions. It may, however be rather troublesome to administer in certain circumstances.

19.6.3 Interpenetrating Samples

This method of control on accuracy of sampling has been especially developed by Indian statisticians, notably by P. C. Mahalanobis [19.5]. It is intended primarily as a check on the observers' accuracy in making and recording measurements. Essentially it consists of arranging that each part of the investigation is performed in duplicate (occasionally more than twice) by independent observers. Preferably, the observers should be ignorant of the checking procedure, but it is evident that this will often be difficult to achieve in practice.

From each of the 2 (or more) parallel samples, estimates of population parameters can be computed. It is also possible to estimate the standard deviations of these estimates. Comparison of differences between estimators of the same parameter with the estimated standard deviation of the difference between them can indicate the existence of gross inconsistencies. This can be used as a check on the accuracy with which the sample has been selected and measured. Furthermore, if there is evidence of inconsistency, this evidence may also point to the general nature of this inconsistency. More detailed investigation can then be set up (if felt to be justifiable) to identify the causes more precisely, and perhaps to eliminate them. Removal of causes (such as inadequate training, poor supervision, laziness, etc.) is facilitated by their identification.

This checking procedure can be applied separately to each stratum within a population. Occasionally, some economy may be effected by restricting the second (checking) sample to measure fewer characters, or even by making it a smaller sample. In the latter case it is of course necessary to make appropriate allowance for the sample sizes in assessing what is a "reasonable" difference between the estimates obtained from the 2 samples.

19.6.4 Inadequate Responses

In most inquiries in physical and engineering science we are fairly certain that each planned measurement will in fact be obtained, provided care is taken to keep requirements within the capabilities of existing staff and equipment. There can be cases, however, where unforeseen emergencies arise and it is not possible to obtain all the planned measurements. In Section 13.16.3 we discussed one way of dealing with such situations, when our main interest is in hypothesis testing. We showed how to modify our analysis to make it applicable to incomplete data. However, especially in sampling human populations, measurement may be lacking because of direct refusal to reply to the inquiries, or because the chosen individual is not available ("not-at-home") when it is desirable to make contact. There is a strong suspicion in such cases that lack of response may be related to the values of some of the characters measured and so it is not justifiable simply to disregard the lacking observations and complete the analysis using the available data. Such a procedure might well produce substantially biased estimators.

If the proportion of nonresponse is small—less than 5 percent for example—it may be possible to ignore it in practice, provided the possible effect of extraordinary (and unsuspected) variation in the omitted 5 percent is recognized.

However, in investigations of urban human populations, for example, the proportion of nonresponse to a simple sample inquiry is quite often as large as 25 percent or more. It would be unrealistic to ignore completely the possible effect of inclusion of the missing values in such cases. Considerable ingenuity has been shown in devising methods of dealing with this problem, commonly called the "problem of call-backs." This name indicates the essential requirement in most such cases—that of making a second (and perhaps a third or more) approach to the individuals from whom no response was obtained in the original inquiry. Occasionally call-backs may be avoided by using observations of concomitant variable to estimate what the missing values would be. Suitable concomitant variable observations are, however, often not available.

Successive "calls," or equivalent endeavors to obtain a response, usually produce more responses, but there usually remain some nonresponses that cannot be eliminated. No really satisfactory estimates of the response for these "hard-core" cases are available, but consideration of the trend of average values with increasing number of calls necessary to elucidate a response sometimes suggests a rough estimate that can be used with some confidence. It should be remembered that errors which would be serious in a final estimate are not so important when occurring in an estimate applied to a small part of the population only. In the latter case the effect of the error on final estimates for the whole population is much reduced by the process of combination with the more accurate results for the rest of the population.

Bearing this in mind we can seek to make rough estimates of the responses for individuals from whom no response has been obtained, even after several call-backs.

Also, broad limits on final results can be obtained by ascribing extreme responses to the hardcore not-at-home's. (If the latter are relatively few, even extreme responses on their part often have little effect on overall averages.)

19.7 CONCLUDING REMARKS

It is in discussing topics such as those covered in this chapter that the inadequacies of the written word as a medium of instruction are particularly noticeable. It is true of all statistical techniques that they must be applied to be understood, but this is especially so with sampling scheme construction. Theoretical work can go a good deal further in fields of application where the nature of the model, even if not definitely established, is at any rate known in general outline.

In sample construction, however, theory can do little more than seize on certain possibilities and work out their likely consequences. We are then left to try to relate our real problem to one or more of these more or less standard cases.

Fortunately, few inquiries in physical or engineering science have to meet the difficulties presented by population heterogeneities, of more or less unexpected types, which are of relatively common occurrence in research in economic, political, and social science. While appreciating our good fortune, we should be constantly alert lest we overlook some properties peculiar to a particular problem in unjustified confidence that they cannot affect methods that have proved valuable on so many occasions.

The contents of this chapter will, it is hoped, provide a guide to some of the points to be borne in mind. The references provide useful additional reading, but in this field nothing can approach practical experience for instructional value.

REFERENCES

1. Cochran, W. G., *Sampling Techniques*, Wiley, New York, 1953.
2. Hansen, M. H., W. N. Hurwitz, and W. G. Madow, *Sample Surveys*, Vols. 1 and 2, *Methods and Theory*, Wiley, New York, 1956.
3. Kish, L., *Survey Sampling*, Wiley, New York, 1965.
4. Konijn, H. S., *Statistical Theory of Sample Survey Design and Practice*, American Elsevier, New York, 1973.
5. Mahalanobis, P. C., "On Large-Scale Sample Surveys," *Philosophical Transactions of the Royal Society, London*, **231B**, (1944).
6. Murthy, M., *Sampling Theory and Methods*, Statistical Publishing Co., Calcutta, 1967.
7. Som, R. K., *A Manual of Sampling Techniques*, Crane-Russak, New York, 1973.
8. Sukhatme, P. V., *Sampling Theory of Surveys with Practical Applications*, Iowa State College Press, 1960.
9. Yates, F., *Sampling Methods for Censuses and Surveys*, 3rd ed., Griffin, London, 1960.

EXERCISES

The first 7 exercises refer to a population with 6 strata; the strata sizes and (approximate) standard deviations for 2 characters, X and Y, are shown in the following table:

STRATUM	SIZE	X STANDARD DEVIATION	Y STANDARD DEVIATION
I	40	50	85
II	100	20	50
III	150	25	25
IV	450	10	70
V	60	10	25
VI	300	45	100

1. A sample of size 100 is to be chosen. Obtain the optimum structures of such a sample if it is desired to minimize the variance of the estimator of (*a*) total X, (*b*) total Y, in the population.

2. You are asked to suggest a constitution of a single sample of 100 individuals that will be used to obtain estimates of both total X and total Y in the population. It is suggested that a simple (nearly) proportional sample should be taken. Can you suggest an improvement on this?
3. Give answers to Exercises 1 and 2, with sample size 600.
4. Costs of measurement (per individual) in the 6 strata are shown in the table below.

	COSTS OF MEASURING	
STRATUM	X	Y
I	1	4
II	4	4
III	4	4
IV	1	$2\frac{1}{4}$
V	$2\frac{1}{4}$	$2\frac{1}{4}$
VI	$6\frac{1}{4}$	9

A sum of 350 is available for taking a sample. Obtain the optimum sample structures if it is desired to minimize the variance of estimators of
(a) average value of X;
(b) average value of Y, in the population.
5. In each stratum, cost of measuring both X and Y is equal to 75% of the sum of costs of measuring X and Y separately in that stratum. It is desired to select a sample so as to minimize the greater of the variances of the estimators of average X and average Y in the population. Obtain the structure of the sample, if the total sum available for sampling is 500 and both X and Y will be measured on each individual in the sample.
6. If, in Exercise 5, the possibility of measuring only X, or only Y, on some individuals in the sample is admitted, discuss, in general terms, its effect on the solution.
7. It has been suggested that better results might be obtained by splitting up one of the 6 strata into 2 strata. Which of the 6 strata is the most promising to consider in this connection? Discuss how to assess the improvement likely to result from splitting up this stratum.
8. The probability density function of a continuous random variable X is $f(x)$. It is desired to divide a population into k strata defined by values of X according to the scheme

Stratum (1)	$X \leqslant x_1$
Stratum (2)	$x_1 < X \leqslant x_2$
$\vdots$	
Stratum $(k-1)$	$x_{k-2} < X \leqslant x_{k-1}$
Stratum (k)	$X > x_{k-1}$

The method of division should be such that if a proportional sample is taken (number in stratum (t) proportional to

$$\int_{x_{t-1}}^{x_t} f(x)\,dx$$

putting x_0, x_k equal to lower and upper bound of X, respectively) the variance of the estimate of $E(X)$,

$$\sum_{t=1}^{k} \left[\int_{x_{t-1}}^{x_t} f(x)\,dx \right] \times [\text{sample mean in stratum } (t)]$$

is as small as possible, for given total sample size. Show that the points of division $x_1, x_2, \ldots, x_{k-1}$ should be chosen so that

$x_i = $ unweighted arithmetic mean of population means in

strata (i) and $(i+1)$

[This result is not directly useful in practice since it requires a knowledge of $f(x)$; and if $f(x)$ is known, $E(X)$ can be calculated without recourse to sampling. However, the result can be used in constructing stratification schemes based on variate values, indicating, at any rate roughly, the kind of stratification likely to be of most use.]

9. A population contains N individuals. The values of characters X and Y on the ith individual are x_i, y_i, respectively. A sample of n different individuals is chosen at random from the population and the values (X_j, Y_j) $(j = 1, 2, \ldots, n)$ of x and y for these individuals are recorded.

Show that if

$$\bar{R} = n^{-1} \sum_{j=1}^{n} \left(\frac{Y_j}{X_j} \right); \quad \bar{X} = n^{-1} \sum_{j=1}^{n} X_j; \quad \bar{Y} = n^{-1} \sum_{j=1}^{n} Y_j$$

then the expected value of

$$\bar{R}\bar{x} + \frac{N-1}{N} \cdot \frac{n}{n-1} (\bar{Y} - \bar{R}\bar{X})$$

(where $\bar{x} = N^{-1}\sum_{i=1}^{N} x_i$ is the population mean value of the character X) is equal to

$$\bar{y} = N^{-1} \sum_{i=1}^{N} y_i$$

HINT: Note that $E(\overline{RX}) = n^{-2}E\left[\left(\sum_{j=1}^{n} X_j\right)\left[\sum_{j'=1}^{n}\left(\frac{Y_{j'}}{X_{j'}}\right)\right]\right]$

$$= n^{-1}E\left[Y_j + X_j\left[\sum_{j=1}^{N}\left(\frac{y_i}{x_i}\right) - \left(\frac{Y_j}{X_j}\right)\right]\frac{n-1}{N-1}\right]$$

[The results of this exercise make it possible to construct an unbiased ratio estimator of $\bar{y}$, without any assumption about the form of the joint distribution of X and Y, provided the value of the population mean $(\bar{x})$ of the concomitant variable X is known. See Hartley and Ross, "Unbiased Ratio Estimators," *Nature*, **174** (1954).]

10. A series of randomized block experiments is planned to estimate differences between average yields of m different methods of production. These experiments will take place in a number (p) of different places. At each place the number (b) of blocks is to be the same. It may be assumed that differences in average yield between different methods of production can be represented by parameters, but there is a random interaction between method of production and place of experiment.

   Discuss the problem of optimal choice of sampling scheme structure—that is, the best choice of m, p, and b—subject to the following conditions:
   (a) total number of observations (mpb) is constant;
   (b) total cost $(= c_1 mpb + c_2 p)$ is constant.
   In particular show how the choice of sampling scheme depends on the ratio of interaction (Methods $\times$ Places) variance to the variance of the residual component.

11. A population is divided into 2 strata. The number of individuals in each of these strata is known to be large and approximately M_1 and M_2. Preliminary random samples of size n are taken from each stratum and the sample standard deviations of a measured character (X) calculated for each stratum.

   Denoting these sample standard deviations by S_1, S_2, respectively, the sizes of final samples to be chosen randomly from the 2 strata are calculated from the formulas

$$m_0\left(\frac{M_1 S_1}{M_1 S_1 + M_2 S_2}\right), \quad m_0\left(\frac{M_2 S_2}{M_1 S_1 + M_2 S_2}\right)$$

respectively. As an estimator of population mean value of X, the statistic

$$X = \frac{\sum_{i=1}^{2} M_i(\text{sample arithmetic mean for } i\text{th stratum})}{M_1 + M_2}$$

is to be used.

Assuming that X is normally distributed in each stratum, with standard deviations σ_1, σ_2, respectively, show that the probability that the variance of X is less than the variance X would have if the sample sizes were proportional to stratum size [that is, equal to $m_0 M_1/(M_1 + M_2), m_0 M_2/(M_1 + M_2)$, respectively] is equal to $(1 - \alpha)$ if

$$F_{n-1, n-1, 1-\alpha/2} = \max\left(\frac{\sigma_1^2}{\sigma_2^2}, \frac{\sigma_2^2}{\sigma_1^2}\right)$$

(Note that this condition does not depend on the value of M_1/M_2.)

12. Suppose that the primary objective of a sampling inquiry is to estimate the *standard deviation* of a character X, rather than the mean. Discuss the possible advantages of stratification in such an inquiry.

13. In the county of Zilshire there are 5 urban areas. Demographic and geographical information about these urban areas is set out in the table below:

URBAN AREA	NUMBER OF PRIVATE HOUSES	RESIDENT POPULATION	MILES FROM ZILCHESTER
Zilchester	41,000	150,000	
Zilchford	20,000	50,000	40
Zilchton	6,500	20,000	40
Zilwich	25,000	90,000	60
Zilchwater	13,000	30,000	80

It is desired to estimate the overall average value of contents per occupant of the houses in these urban areas. This is to be done by selecting samples of houses from each area and sending a skilled investigator to estimate total value of contents and record the number of occupants. All the investigators live in the county capital town of Zilchester, and will travel to and from the other towns assigned to them each day. Each investigator can visit 6 houses a day in Zilchester, or 5 in any one of the other urban areas.

It has been agreed to base calculations on the following assumptions.

(i) The standard deviation of average value of contents per occupant is twice as large in Zilchester as in each of the other towns.

(ii) Each investigator is paid 25 units per day, plus a traveling allowance (if working away from Zilchester) of 0.08 units/mi.

In what ratios (approximately) should be the number of houses included in the sample from each of the 5 urban areas?

(you may assume that the sample will be of sufficient size to require at least 10 houses from each urban area.)

14. In Exercise 13, how would your answer be modified if you were required to estimate the proportion of houses in which there were more than 2 persons per 100 ft^2 floor space, as well as the average value of contents per occupant?

Appendix

TABLE A

Cumulative Normal Distribution*

$$\Phi(x) = \frac{1}{\sqrt{2\pi}} \int_{-\infty}^{x} e^{-u^2/2}\, du$$

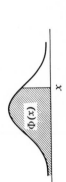

x	0.00	0.01	0.02	0.03	0.04	0.05	0.06	0.07	0.08	0.09
0.0	0.50000	0.50399	0.50798	0.51197	0.51595	0.51994	0.52392	0.52790	0.53188	0.53586
0.1	0.53983	0.54380	0.54776	0.55172	0.55567	0.55962	0.56356	0.56749	0.57142	0.57535
0.2	0.57926	0.58317	0.58706	0.59095	0.59483	0.59871	0.60257	0.60642	0.61026	0.61409
0.3	0.61791	0.62172	0.62552	0.62930	0.63307	0.63683	0.64058	0.64431	0.64803	0.65173
0.4	0.65542	0.65910	0.66276	0.66640	0.67003	0.67364	0.67724	0.68082	0.68439	0.68793
0.5	0.69146	0.69497	0.69847	0.70194	0.70540	0.70884	0.71226	0.71566	0.71904	0.72240
0.6	0.72575	0.72907	0.73237	0.73565	0.73891	0.74215	0.74537	0.74857	0.75175	0.75490
0.7	0.75804	0.76115	0.76424	0.76730	0.77035	0.77337	0.77637	0.77935	0.78230	0.78524
0.8	0.78814	0.79103	0.79389	0.79673	0.79955	0.80234	0.80511	0.80785	0.81057	0.81327
0.9	0.81594	0.81859	0.82121	0.82381	0.82639	0.82894	0.83147	0.83398	0.83646	0.83891
1.0	0.84134	0.84375	0.84614	0.84850	0.85083	0.85314	0.85543	0.85769	0.85993	0.86214
1.1	0.86433	0.86650	0.86864	0.87076	0.87286	0.87493	0.87698	0.87900	0.88100	0.88298
1.2	0.88493	0.88686	0.88877	0.89065	0.89251	0.89435	0.89617	0.89796	0.89973	0.90147
1.3	0.90320	0.90490	0.90658	0.90824	0.90988	0.91149	0.91309	0.91466	0.91621	0.91774
1.4	0.91924	0.92073	0.92220	0.92364	0.92507	0.92647	0.92786	0.92922	0.93056	0.93189
1.5	0.93319	0.93448	0.93574	0.93699	0.93822	0.93943	0.94062	0.94179	0.94295	0.94408
1.6	0.94520	0.94630	0.94738	0.94845	0.94950	0.95053	0.95154	0.95254	0.95352	0.95449
1.7	0.95543	0.95637	0.95728	0.95818	0.95907	0.95994	0.96080	0.96164	0.96246	0.96327
1.8	0.96407	0.96485	0.96562	0.96638	0.96712	0.96784	0.96856	0.96926	0.96995	0.97062
1.9	0.97128	0.97193	0.97257	0.97320	0.97381	0.97441	0.97500	0.97558	0.97615	0.97670

2.0	0.97725	0.97778	0.97831	0.97882	0.97932	0.97982	0.98030	0.98077	0.98124	0.98169
2.1	0.98214	0.98257	0.98300	0.98341	0.98382	0.98422	0.98461	0.98500	0.98537	0.98574
2.2	0.98610	0.98645	0.98679	0.98713	0.98745	0.98778	0.98809	0.98840	0.98870	0.98899
2.3	0.98928	0.98956	0.98983	0.99010	0.99036	0.99061	0.99086	0.99111	0.99134	0.99158
2.4	0.99180	0.99202	0.99224	0.99245	0.99266	0.99286	0.99305	0.99324	0.99343	0.99361
2.5	0.99379	0.99396	0.99413	0.99430	0.99446	0.99461	0.99477	0.99492	0.99506	0.99520
2.6	0.99534	0.99547	0.99560	0.99573	0.99585	0.99598	0.99609	0.99621	0.99632	0.99643
2.7	0.99653	0.99664	0.99674	0.99683	0.99693	0.99702	0.99711	0.99720	0.99728	0.99736
2.8	0.99744	0.99752	0.99760	0.99767	0.99774	0.99781	0.99788	0.99795	0.99801	0.99807
2.9	0.99813	0.99819	0.99825	0.99831	0.99836	0.99841	0.99846	0.99851	0.99856	0.99861
3.0	0.99865	0.99869	0.99874	0.99878	0.99882	0.99886	0.99889	0.99893	0.99897	0.99900
3.1	0.99903	0.99906	0.99910	0.99913	0.99916	0.99918	0.99921	0.99924	0.99926	0.99929
3.2	0.99931	0.99934	0.99936	0.99938	0.99940	0.99942	0.99944	0.99946	0.99948	0.99950
3.3	0.99952	0.99953	0.99957	0.99957	0.99958	0.99960	0.99961	0.99962	0.99964	0.99965
3.4	0.99966	0.99968	0.99969	0.99970	0.99971	0.99972	0.99973	0.99974	0.99975	0.99976
3.5	0.99977	0.99978	0.99978	0.99979	0.99980	0.99981	0.99981	0.99982	0.99983	0.99983
3.6	0.99984	0.99985	0.99985	0.99986	0.99986	0.99987	0.99987	0.99988	0.99988	0.99989
3.7	0.99989	0.99990	0.99990	0.99990	0.99991	0.99991	0.99992	0.99992	0.99992	0.99992
3.8	0.99993	0.99993	0.99993	0.99994	0.99994	0.99994	0.99994	0.99995	0.99995	0.99995
3.9	0.99995	0.99995	0.99996	0.99996	0.99996	0.99996	0.99996	0.99996	0.99997	0.99997

* Reproduced with permission from Pearson and Hartley, *Biometrika Tables for Statisticians*, Vol. 1 (1958), pp. 104–108.

TABLE B

Percentage Points of the χ^2
Distribution

Values of $\chi^2_{\nu,P}$ such that

$$P = \frac{1}{2^{\nu/2}\Gamma(\nu/2)} \int_0^{\chi^2_{\nu,P}} y^{\nu/2-1} e^{-y/2} \, dy$$

ν \ P	0.005	0.010	0.025	0.050	0.100	0.250	0.500
1	0.00004	0.00016	0.00098	0.00393	0.01579	0.1015	0.4549
2	0.0100	0.0201	0.0506	0.1026	0.2107	0.5754	1.386
3	0.0717	0.1148	0.2158	0.3518	0.5844	1.213	2.366
4	0.2070	0.2971	0.4844	0.7107	1.064	1.923	3.357
5	0.4117	0.5543	0.8312	1.145	1.610	2.675	4.351
6	0.6757	0.8721	1.2373	1.635	2.204	3.455	5.348
7	0.9893	1.239	1.690	2.167	2.833	4.255	6.346
8	1.344	1.646	2.180	2.733	3.490	5.071	7.344
9	1.735	2.088	2.700	3.325	4.168	5.899	8.343
10	2.156	2.558	3.247	3.940	4.865	6.737	9.342
11	2.603	3.053	3.816	4.575	5.578	7.584	10.34
12	3.074	3.571	4.404	5.226	6.304	8.438	11.34
13	3.565	4.107	5.009	5.892	7.041	9.299	12.34
14	4.075	4.660	5.629	6.571	7.790	10.17	13.34
15	4.601	5.229	6.262	7.261	8.547	11.04	14.34
16	5.142	5.812	6.908	7.962	9.312	11.91	15.34
17	5.697	6.408	7.564	8.672	10.09	12.79	16.34
18	6.265	7.015	8.231	9.390	10.86	13.68	17.34
19	6.844	7.633	8.907	10.12	11.65	14.56	18.34
20	7.434	8.260	9.591	10.85	12.44	15.45	19.34
21	8.034	8.897	10.28	11.59	13.24	16.34	20.34
22	8.643	9.542	10.98	12.34	14.04	17.24	21.34
23	9.260	10.20	11.69	13.09	14.85	18.14	22.34
24	9.886	10.86	12.40	13.85	15.66	19.04	23.34
25	10.52	11.52	13.12	14.61	16.47	19.94	24.34
26	11.16	12.20	13.84	15.38	17.29	20.84	25.34
27	11.81	12.88	14.57	16.15	18.11	21.75	26.34
28	12.46	13.56	15.31	16.93	18.94	22.66	27.34
29	13.12	14.26	16.05	17.71	19.77	23.57	28.34
30	13.79	14.95	16.79	18.49	20.60	24.48	29.34
40	20.71	22.16	24.43	26.51	29.05	33.66	39.34
50	27.99	29.71	32.36	34.76	37.69	42.94	49.33
60	35.53	37.48	40.48	43.19	46.46	52.29	59.33
70	43.28	45.44	48.76	51.74	55.33	61.70	69.33
80	51.17	53.54	57.15	60.39	64.28	71.14	79.33
90	59.20	61.75	65.65	69.13	73.29	80.62	89.33
100	67.33	70.06	74.22	77.93	82.36	90.13	99.33

* Reproduced with permission from Pearson and Hartley, *Biometrika Tables for Statisticians*, Vol. 1, pp. 130–131 (1958).

v \ P	0.750	0.900	0.950	0.975	0.990	0.995	0.999
1	1.323	2.706	3.841	5.024	6.635	7.879	10.83
2	2.773	4.605	5.991	7.378	9.210	10.60	13.82
3	4.108	6.251	7.815	9.348	11.34	12.84	16.27
4	5.385	7.779	9.488	11.14	13.28	14.86	18.47
5	6.626	9.236	11.07	12.83	15.09	16.75	20.52
6	7.841	10.64	12.59	14.45	16.81	18.55	22.46
7	9.037	12.02	14.07	16.01	18.48	20.28	24.32
8	10.22	13.36	15.51	17.53	20.09	21.96	26.12
9	11.39	14.68	16.92	19.02	21.67	23.59	27.88
10	12.55	15.99	18.31	20.48	23.21	25.19	29.59
11	13.70	17.28	19.68	21.92	24.72	26.76	31.26
12	14.85	18.55	21.03	23.34	26.22	28.30	32.91
13	15.98	19.81	22.36	24.74	27.69	29.82	34.53
14	17.12	21.06	23.68	26.12	29.14	31.32	36.12
15	18.25	22.31	25.00	27.49	30.58	32.80	37.70
16	19.37	23.54	26.30	28.85	32.00	34.27	39.25
17	20.49	24.77	27.59	30.19	33.41	35.72	40.79
18	21.60	25.99	28.87	31.53	34.81	37.16	42.31
19	22.72	27.20	30.14	32.85	36.19	38.58	43.82
20	23.83	28.41	31.41	34.17	37.57	40.00	45.32
21	24.93	29.62	32.67	35.48	38.93	41.40	46.80
22	26.04	30.81	33.92	36.78	40.29	42.80	48.27
23	27.14	32.01	35.17	38.08	41.64	44.18	49.73
24	28.24	33.20	36.42	39.36	42.98	45.56	51.18
25	29.34	34.38	37.65	40.65	44.31	46.93	52.62
26	30.43	35.56	38.89	41.92	45.64	48.29	54.05
27	31.53	36.74	40.11	43.19	46.96	49.64	55.48
28	32.62	37.92	41.34	44.46	48.28	50.99	56.89
29	33.71	39.09	42.56	45.72	49.59	52.34	58.30
30	34.80	40.26	43.77	46.98	50.89	53.67	59.70
40	45.62	51.80	55.76	59.34	63.69	66.77	73.40
50	56.33	63.17	67.50	71.42	76.15	79.49	86.66
60	66.98	74.40	79.08	83.30	88.38	91.95	99.61
70	77.58	85.53	90.53	95.02	100.4	104.2	112.3
80	88.13	96.58	101.9	106.6	112.3	116.3	124.8
90	98.65	107.6	113.1	118.1	124.1	128.3	137.2
100	109.1	118.5	124.3	129.6	135.8	140.2	149.4

TABLE C

PERCENTAGE POINTS OF STUDENT'S t-DISTRIBUTION

Values of $t_{\nu,P}$ such that

$$P = \int_{-\infty}^{t_{\nu,P}} \frac{1}{\sqrt{\nu\pi}} \frac{\Gamma\left(\dfrac{\nu+1}{2}\right)}{\Gamma\left(\dfrac{\nu}{2}\right)} \left(1+\frac{t^2}{\nu}\right)^{-\frac{\nu+1}{2}} dt$$

ν \ P	0.750	0.900	0.950	0.975	0.990	0.995	0.999	0.9995
1	1.000	3.078	6.314	12.706	31.821	63.657	318.31	636.62
2	0.816	1.886	2.920	4.303	6.965	9.925	22.326	31.598
3	0.765	1.638	2.353	3.182	4.541	5.841	10.213	12.924
4	0.741	1.533	2.132	2.776	3.747	4.604	7.173	8.610
5	0.727	1.476	2.015	2.571	3.365	4.032	5.893	6.869
6	0.718	1.440	1.943	2.447	3.143	3.707	5.208	5.959
7	0.711	1.415	1.895	2.365	2.998	3.499	4.785	5.408
8	0.706	1.397	1.860	2.306	2.896	3.355	4.501	5.041
9	0.703	1.383	1.833	2.262	2.821	3.250	4.297	4.781
10	0.700	1.372	1.812	2.228	2.764	3.169	4.144	4.587
11	0.697	1.363	1.796	2.201	2.718	3.106	4.025	4.437
12	0.695	1.356	1.782	2.179	2.681	3.055	3.930	4.318
13	0.694	1.350	1.771	2.160	2.650	3.012	3.852	4.221
14	0.692	1.345	1.761	2.145	2.624	2.977	3.787	4.140
15	0.691	1.341	1.753	2.131	2.602	2.947	3.733	4.073
16	0.690	1.337	1.746	2.120	2.583	2.921	3.686	4.015
17	0.689	1.333	1.740	2.110	2.567	2.898	3.646	3.965
18	0.688	1.330	1.734	2.101	2.552	2.878	3.610	3.922
19	0.688	1.328	1.729	2.093	2.539	2.861	3.579	3.883
20	0.687	1.325	1.725	2.086	2.528	2.845	3.552	3.850
21	0.686	1.323	1.721	2.080	2.518	2.831	3.527	3.819
22	0.686	1.321	1.717	2.074	2.508	2.819	3.505	3.792
23	0.685	1.319	1.714	2.069	2.500	2.807	3.485	3.767
24	0.685	1.318	1.711	2.064	2.492	2.797	3.467	3.745
25	0.684	1.316	1.708	2.060	2.485	2.787	3.450	3.725
26	0.684	1.315	1.706	2.056	2.479	2.779	3.435	3.707
27	0.684	1.314	1.703	2.052	2.473	2.771	3.421	3.690
28	0.683	1.313	1.701	2.048	2.467	2.763	3.408	3.674
29	0.683	1.311	1.699	2.045	2.462	2.756	3.396	3.659
30	0.683	1.310	1.697	2.042	2.457	2.750	3.385	3.646
40	0.681	1.303	1.684	2.021	2.423	2.704	3.307	3.551
60	0.679	1.296	1.671	2.000	2.390	2.660	3.232	3.460
120	0.677	1.289	1.658	1.980	2.358	2.617	3.160	3.373
∞	0.674	1.282	1.645	1.960	2.326	2.576	3.090	3.291

Reproduced from Pearson and Hartley, *Biometrika Tables for Statisticians*, Vol. 1 (1958), p. 138, and from Table III, Fisher and Yates, *Statistical Tables for Biological, Agricultural and Medical Research*, Oliver and Boyd, Edinburgh, 1953, and by permission of authors and publishers.

PERCENTAGE POINTS OF THE F-DISTRIBUTION*

Values of $F_{\nu_1,\nu_2,P}$ such that $P = \dfrac{1}{B(\nu_1/2,\,\nu_2/2)} \displaystyle\int_0^{\nu_1 F_{\nu_1,\nu_2,P}/\nu_2} g^{\nu_1/2-1}(1+g)^{-(\nu_1+\nu_2)/2}\,dg$

P = 0.95

ν_2 \ ν_1	1	2	3	4	5	6	7	8	9	10	12	15	20	24	30	40	60	120	∞
1	161.4	199.5	215.7	224.6	230.2	234.0	236.8	238.9	240.5	241.9	243.9	245.9	248.0	249.1	250.1	251.1	252.2	253.3	254.3
2	18.51	19.00	19.16	19.25	19.30	19.33	19.35	19.37	19.38	19.40	19.41	19.43	19.45	19.45	19.46	19.47	19.48	19.49	19.50
3	10.13	9.55	9.28	9.12	9.01	8.94	8.89	8.85	8.81	8.79	8.74	8.70	8.66	8.64	8.62	8.59	8.57	8.55	8.53
4	7.71	6.94	6.59	6.39	6.26	6.16	6.09	6.04	6.00	5.96	5.91	5.86	5.80	5.77	5.75	5.72	5.69	5.66	5.63
5	6.61	5.79	5.41	5.19	5.05	4.95	4.88	4.82	4.77	4.74	4.68	4.62	4.56	4.53	4.50	4.46	4.43	4.40	4.36
6	5.99	5.14	4.76	4.53	4.39	4.28	4.21	4.15	4.10	4.06	4.00	3.94	3.87	3.84	3.81	3.77	3.74	3.70	3.67
7	5.59	4.74	4.35	4.12	3.97	3.87	3.79	3.73	3.68	3.64	3.57	3.51	3.44	3.41	3.38	3.34	3.30	3.27	3.23
8	5.32	4.46	4.07	3.84	3.69	3.58	3.50	3.44	3.39	3.35	3.28	3.22	3.15	3.12	3.08	3.04	3.01	2.97	2.93
9	5.12	4.26	3.86	3.63	3.48	3.37	3.29	3.23	3.18	3.14	3.07	3.01	2.94	2.90	2.86	2.83	2.79	2.75	2.71
10	4.96	4.10	3.71	3.48	3.33	3.22	3.14	3.07	3.02	2.98	2.91	2.85	2.77	2.74	2.70	2.66	2.62	2.58	2.54
11	4.84	3.98	3.59	3.36	3.20	3.09	3.01	2.95	2.90	2.85	2.79	2.72	2.65	2.61	2.57	2.53	2.49	2.45	2.40
12	4.75	3.89	3.49	3.26	3.11	3.00	2.91	2.85	2.80	2.75	2.69	2.62	2.54	2.51	2.47	2.43	2.38	2.34	2.30
13	4.67	3.81	3.41	3.18	3.03	2.92	2.83	2.77	2.71	2.67	2.60	2.53	2.46	2.42	2.38	2.34	2.30	2.25	2.21
14	4.60	3.74	3.34	3.11	2.96	2.85	2.76	2.70	2.65	2.60	2.53	2.46	2.39	2.35	2.31	2.27	2.22	2.18	2.13
15	4.54	3.68	3.29	3.06	2.90	2.79	2.71	2.64	2.59	2.54	2.48	2.40	2.33	2.29	2.25	2.20	2.16	2.11	2.07
16	4.49	3.63	3.24	3.01	2.85	2.74	2.66	2.59	2.54	2.49	2.42	2.35	2.28	2.24	2.19	2.15	2.11	2.06	2.01
17	4.45	3.59	3.20	2.96	2.81	2.70	2.61	2.55	2.49	2.45	2.38	2.31	2.23	2.19	2.15	2.10	2.06	2.01	1.96
18	4.41	3.55	3.16	2.93	2.77	2.66	2.58	2.51	2.46	2.41	2.34	2.27	2.19	2.15	2.11	2.06	2.02	1.97	1.92
19	4.38	3.52	3.13	2.90	2.74	2.63	2.54	2.48	2.42	2.38	2.31	2.23	2.16	2.11	2.07	2.03	1.98	1.93	1.88
20	4.35	3.49	3.10	2.87	2.71	2.60	2.51	2.45	2.39	2.35	2.28	2.20	2.12	2.08	2.04	1.99	1.95	1.90	1.84
21	4.32	3.47	3.07	2.84	2.68	2.57	2.49	2.42	2.37	2.32	2.25	2.18	2.10	2.05	2.01	1.96	1.92	1.87	1.81
22	4.30	3.44	3.05	2.82	2.66	2.55	2.46	2.40	2.34	2.30	2.23	2.15	2.07	2.03	1.98	1.94	1.89	1.84	1.78
23	4.28	3.42	3.03	2.80	2.64	2.53	2.44	2.37	2.32	2.27	2.20	2.13	2.05	2.01	1.96	1.91	1.86	1.81	1.76
24	4.26	3.40	3.01	2.78	2.62	2.51	2.42	2.36	2.30	2.25	2.18	2.11	2.03	1.98	1.94	1.89	1.84	1.79	1.73
25	4.24	3.39	2.99	2.76	2.60	2.49	2.40	2.34	2.28	2.24	2.16	2.09	2.01	1.96	1.92	1.87	1.82	1.77	1.71
26	4.23	3.37	2.98	2.74	2.59	2.47	2.39	2.32	2.27	2.22	2.15	2.07	1.99	1.95	1.90	1.85	1.80	1.75	1.69
27	4.21	3.35	2.96	2.73	2.57	2.46	2.37	2.31	2.25	2.20	2.13	2.06	1.97	1.93	1.88	1.84	1.79	1.73	1.67
28	4.20	3.34	2.95	2.71	2.56	2.45	2.36	2.29	2.24	2.19	2.12	2.04	1.96	1.91	1.87	1.82	1.77	1.71	1.65
29	4.18	3.33	2.93	2.70	2.55	2.43	2.35	2.28	2.22	2.18	2.10	2.03	1.94	1.90	1.85	1.81	1.75	1.70	1.64
30	4.17	3.32	2.92	2.69	2.53	2.42	2.33	2.27	2.21	2.16	2.09	2.01	1.93	1.89	1.84	1.79	1.74	1.68	1.62
40	4.08	3.23	2.84	2.61	2.45	2.34	2.25	2.18	2.12	2.08	2.00	1.92	1.84	1.79	1.74	1.69	1.64	1.58	1.51
60	4.00	3.15	2.76	2.53	2.37	2.25	2.17	2.10	2.04	1.99	1.92	1.84	1.75	1.70	1.65	1.59	1.53	1.47	1.39
120	3.92	3.07	2.68	2.45	2.29	2.17	2.09	2.02	1.96	1.91	1.83	1.75	1.66	1.61	1.55	1.50	1.43	1.35	1.25
∞	3.84	3.00	2.60	2.37	2.21	2.10	2.01	1.94	1.88	1.83	1.75	1.67	1.57	1.52	1.46	1.39	1.32	1.22	1.00

0.95

$F_{\nu_1,\nu_2,0.95}$

* Reproduced with permission from Pearson and Hartley, *Biometrika Tables for Statisticians*, Vol. 1 (1958), pp. 159–163.

TABLE D (*Continued*)

Values of $F_{\nu_1, \nu_2, P}$ such that $P = \dfrac{1}{B(\nu_1/2,\ \nu_2/2)} \displaystyle\int_0^{\nu_1 F_{\nu_1, \nu_2, P}/\nu_2} g^{\nu_1/2 - 1}(1 + g)^{-(\nu_1 + \nu_2)/2}\, dg$

$P = 0.975$

$F_{\nu_1, \nu_2,\ 0.975}$

ν_2 \ ν_1	1	2	3	4	5	6	7	8	9	10	12	15	20	24	30	40	60	120	∞
1	647.8	799.5	864.2	899.6	921.8	937.1	948.2	956.7	963.3	968.6	976.7	984.9	993.1	997.2	1001	1006	1010	1014	1018
2	38.51	39.00	39.17	39.25	39.30	39.33	39.36	39.37	39.39	39.40	39.41	39.43	39.45	39.46	39.46	39.47	39.48	39.49	39.50
3	17.44	16.04	15.44	15.10	14.88	14.73	14.62	14.54	14.47	14.42	14.34	14.25	14.17	14.12	14.08	14.04	13.99	13.95	13.90
4	12.22	10.65	9.98	9.60	9.36	9.20	9.07	8.98	8.90	8.84	8.75	8.66	8.56	8.51	8.46	8.41	8.36	8.31	8.26
5	10.01	8.43	7.76	7.39	7.15	6.98	6.85	6.76	6.68	6.62	6.52	6.43	6.33	6.28	6.23	6.18	6.12	6.07	6.02
6	8.81	7.26	6.60	6.23	5.99	5.82	5.70	5.60	5.52	5.46	5.37	5.27	5.17	5.12	5.07	5.01	4.96	4.90	4.85
7	8.07	6.54	5.89	5.52	5.29	5.12	4.99	4.90	4.82	4.76	4.67	4.57	4.47	4.42	4.36	4.31	4.25	4.20	4.14
8	7.57	6.06	5.42	5.05	4.82	4.65	4.53	4.43	4.36	4.30	4.20	4.10	4.00	3.95	3.89	3.84	3.78	3.73	3.67
9	7.21	5.71	5.08	4.72	4.48	4.32	4.20	4.10	4.03	3.96	3.87	3.77	3.67	3.61	3.56	3.51	3.45	3.39	3.33
10	6.94	5.46	4.83	4.47	4.24	4.07	3.95	3.85	3.78	3.72	3.62	3.52	3.42	3.37	3.31	3.26	3.20	3.14	3.08
11	6.72	5.26	4.63	4.28	4.04	3.88	3.76	3.66	3.59	3.53	3.43	3.33	3.23	3.17	3.12	3.06	3.00	2.94	2.88
12	6.55	5.10	4.47	4.12	3.89	3.73	3.61	3.51	3.44	3.37	3.28	3.18	3.07	3.02	2.96	2.91	2.85	2.79	2.72
13	6.41	4.97	4.35	4.00	3.77	3.60	3.48	3.39	3.31	3.25	3.15	3.05	2.95	2.89	2.84	2.78	2.72	2.66	2.60
14	6.30	4.86	4.24	3.89	3.66	3.50	3.38	3.29	3.21	3.15	3.05	2.95	2.84	2.79	2.73	2.67	2.61	2.55	2.49
15	6.20	4.77	4.15	3.80	3.58	3.41	3.29	3.20	3.12	3.06	2.96	2.86	2.76	2.70	2.64	2.59	2.52	2.46	2.40
16	6.12	4.69	4.08	3.73	3.50	3.34	3.22	3.12	3.05	2.99	2.89	2.79	2.68	2.63	2.57	2.51	2.45	2.38	2.32
17	6.04	4.62	4.01	3.66	3.44	3.28	3.16	3.06	2.98	2.92	2.82	2.72	2.62	2.56	2.50	2.44	2.38	2.32	2.25
18	5.98	4.56	3.95	3.61	3.38	3.22	3.10	3.01	2.93	2.87	2.77	2.67	2.56	2.50	2.44	2.38	2.32	2.26	2.19
19	5.92	4.51	3.90	3.56	3.33	3.17	3.05	2.96	2.88	2.82	2.72	2.62	2.51	2.45	2.39	2.33	2.27	2.20	2.13
20	5.87	4.46	3.86	3.51	3.29	3.13	3.01	2.91	2.84	2.77	2.68	2.57	2.46	2.41	2.35	2.29	2.22	2.16	2.09
21	5.83	4.42	3.82	3.48	3.25	3.09	2.97	2.87	2.80	2.73	2.64	2.53	2.42	2.37	2.31	2.25	2.18	2.11	2.04
22	5.79	4.38	3.78	3.44	3.22	3.05	2.93	2.84	2.76	2.70	2.60	2.50	2.39	2.33	2.27	2.21	2.14	2.08	2.00
23	5.75	4.35	3.75	3.41	3.18	3.02	2.90	2.81	2.73	2.67	2.57	2.47	2.36	2.30	2.24	2.18	2.11	2.04	1.97
24	5.72	4.32	3.72	3.38	3.15	2.99	2.87	2.78	2.70	2.64	2.54	2.44	2.33	2.27	2.21	2.15	2.08	2.01	1.94
25	5.69	4.29	3.69	3.35	3.13	2.97	2.85	2.75	2.68	2.61	2.51	2.41	2.30	2.24	2.18	2.12	2.05	1.98	1.91
26	5.66	4.27	3.67	3.33	3.10	2.94	2.82	2.73	2.65	2.59	2.49	2.39	2.28	2.22	2.16	2.09	2.03	1.95	1.88
27	5.63	4.24	3.65	3.31	3.08	2.92	2.80	2.71	2.63	2.57	2.47	2.36	2.25	2.19	2.13	2.07	2.00	1.93	1.85
28	5.61	4.22	3.63	3.29	3.06	2.90	2.78	2.69	2.61	2.55	2.45	2.34	2.23	2.17	2.11	2.05	1.98	1.91	1.83
29	5.59	4.20	3.61	3.27	3.04	2.88	2.76	2.67	2.59	2.53	2.43	2.32	2.21	2.15	2.09	2.03	1.96	1.89	1.81
30	5.57	4.18	3.59	3.25	3.03	2.87	2.75	2.65	2.57	2.51	2.41	2.31	2.20	2.14	2.07	2.01	1.94	1.87	1.79
40	5.42	4.05	3.46	3.13	2.90	2.74	2.62	2.53	2.45	2.39	2.29	2.18	2.07	2.01	1.94	1.88	1.80	1.72	1.64
60	5.29	3.93	3.34	3.01	2.79	2.63	2.51	2.41	2.33	2.27	2.17	2.06	1.94	1.88	1.82	1.74	1.67	1.58	1.48
120	5.15	3.80	3.23	2.89	2.67	2.52	2.39	2.30	2.22	2.16	2.05	1.94	1.82	1.76	1.69	1.61	1.53	1.43	1.31
∞	5.02	3.69	3.12	2.79	2.57	2.41	2.29	2.19	2.11	2.05	1.94	1.83	1.71	1.64	1.57	1.48	1.39	1.27	1.00

$P = 0.99$ ν_1 / ν_2	1	2	3	4	5	6	7	8	9	10	12	15	20	24	30	40	60	120	∞
1	4052	4999.5	5403	5625	5764	5859	5928	5982	6022	6056	6106	6157	6209	6235	6261	6287	6313	6339	6366
2	98.50	99.00	99.17	99.25	99.30	99.33	99.36	99.37	99.39	99.40	99.42	99.43	99.45	99.46	99.47	99.47	99.48	99.49	99.50
3	34.12	30.82	29.46	28.71	28.24	27.91	27.67	27.49	27.35	27.23	27.05	26.87	26.69	26.60	26.50	26.41	26.32	26.22	26.13
4	21.20	18.00	16.69	15.98	15.52	15.21	14.98	14.80	14.66	14.55	14.37	14.20	14.02	13.93	13.84	13.75	13.65	13.56	13.46
5	16.26	13.27	12.06	11.39	10.97	10.67	10.46	10.29	10.16	10.05	9.89	9.72	9.55	9.47	9.38	9.29	9.20	9.11	9.02
6	13.75	10.92	9.78	9.15	8.75	8.47	8.26	8.10	7.98	7.87	7.72	7.56	7.40	7.31	7.23	7.14	7.06	6.97	6.88
7	12.25	9.55	8.45	7.85	7.46	7.19	6.99	6.84	6.72	6.62	6.47	6.31	6.16	6.07	5.99	5.91	5.82	5.74	5.65
8	11.26	8.65	7.59	7.01	6.63	6.37	6.18	6.03	5.91	5.81	5.67	5.52	5.36	5.28	5.20	5.12	5.03	4.95	4.86
9	10.56	8.02	6.99	6.42	6.06	5.80	5.61	5.47	5.35	5.26	5.11	4.96	4.81	4.73	4.65	4.57	4.48	4.40	4.31
10	10.04	7.56	6.55	5.99	5.64	5.39	5.20	5.06	4.94	4.85	4.71	4.56	4.41	4.33	4.25	4.17	4.08	4.00	3.91
11	9.65	7.21	6.22	5.67	5.32	5.07	4.89	4.74	4.63	4.54	4.40	4.25	4.10	4.02	3.94	3.86	3.78	3.69	3.60
12	9.33	6.93	5.95	5.41	5.06	4.82	4.64	4.50	4.39	4.30	4.16	4.01	3.86	3.78	3.70	3.62	3.54	3.45	3.36
13	9.07	6.70	5.74	5.21	4.86	4.62	4.44	4.30	4.19	4.10	3.96	3.82	3.66	3.59	3.51	3.43	3.34	3.25	3.17
14	8.86	6.51	5.56	5.04	4.69	4.46	4.28	4.14	4.03	3.94	3.80	3.66	3.51	3.43	3.35	3.27	3.18	3.09	3.00
15	8.68	6.36	5.42	4.89	4.56	4.32	4.14	4.00	3.89	3.80	3.67	3.52	3.37	3.29	3.21	3.13	3.05	2.96	2.87
16	8.53	6.23	5.29	4.77	4.44	4.20	4.03	3.89	3.78	3.69	3.55	3.41	3.26	3.18	3.10	3.02	2.93	2.84	2.75
17	8.40	6.11	5.18	4.67	4.34	4.10	3.93	3.79	3.68	3.59	3.46	3.31	3.16	3.08	3.00	2.92	2.83	2.75	2.65
18	8.29	6.01	5.09	4.58	4.25	4.01	3.84	3.71	3.60	3.51	3.37	3.23	3.08	3.00	2.92	2.84	2.75	2.66	2.57
19	8.18	5.93	5.01	4.50	4.17	3.94	3.77	3.63	3.52	3.43	3.30	3.15	3.00	2.92	2.84	2.76	2.67	2.58	2.49
20	8.10	5.85	4.94	4.43	4.10	3.87	3.70	3.56	3.46	3.37	3.23	3.09	2.94	2.86	2.78	2.69	2.61	2.52	2.42
21	8.02	5.78	4.87	4.37	4.04	3.81	3.64	3.51	3.40	3.31	3.17	3.03	2.88	2.80	2.72	2.64	2.55	2.46	2.36
22	7.95	5.72	4.82	4.31	3.99	3.76	3.59	3.45	3.35	3.26	3.12	2.98	2.83	2.75	2.67	2.58	2.50	2.40	2.31
23	7.88	5.66	4.76	4.26	3.94	3.71	3.54	3.41	3.30	3.21	3.07	2.93	2.78	2.70	2.62	2.54	2.45	2.35	2.26
24	7.82	5.61	4.72	4.22	3.90	3.67	3.50	3.36	3.26	3.17	3.03	2.89	2.74	2.66	2.58	2.49	2.40	2.31	2.21
25	7.77	5.57	4.68	4.18	3.85	3.63	3.46	3.32	3.22	3.13	2.99	2.85	2.70	2.62	2.54	2.45	2.36	2.27	2.17
26	7.72	5.53	4.64	4.14	3.82	3.59	3.42	3.29	3.18	3.09	2.96	2.81	2.66	2.58	2.50	2.42	2.33	2.23	2.13
27	7.68	5.49	4.60	4.11	3.78	3.56	3.39	3.26	3.15	3.06	2.93	2.78	2.63	2.55	2.47	2.38	2.29	2.20	2.10
28	7.64	5.45	4.57	4.07	3.75	3.53	3.36	3.23	3.12	3.03	2.90	2.75	2.60	2.52	2.44	2.35	2.26	2.17	2.06
29	7.60	5.42	4.54	4.04	3.73	3.50	3.33	3.20	3.09	3.00	2.87	2.73	2.57	2.49	2.41	2.33	2.23	2.14	2.03
30	7.56	5.39	4.51	4.02	3.70	3.47	3.30	3.17	3.07	2.98	2.84	2.70	2.55	2.47	2.39	2.30	2.21	2.11	2.01
40	7.31	5.18	4.31	3.83	3.51	3.29	3.12	2.99	2.89	2.80	2.66	2.52	2.37	2.29	2.20	2.11	2.02	1.92	1.80
60	7.08	4.98	4.13	3.65	3.34	3.12	2.95	2.82	2.72	2.63	2.50	2.35	2.20	2.12	2.03	1.94	1.84	1.73	1.60
120	6.85	4.79	3.95	3.48	3.17	2.96	2.79	2.66	2.56	2.47	2.34	2.19	2.03	1.95	1.86	1.76	1.66	1.53	1.38
∞	6.63	4.61	3.78	3.32	3.02	2.80	2.64	2.51	2.41	2.32	2.18	2.04	1.88	1.79	1.70	1.59	1.47	1.32	1.00

TABLE D (Continued)

Values of $F_{\nu_1, \nu_2, P}$ such that $P = \dfrac{1}{B(\nu_1/2,\, \nu_2/2)} \displaystyle\int_0^{\nu_1 F_{\nu_1,\nu_2,P}/\nu_2} g^{\nu_1/2-1}(1+g)^{-(\nu_1+\nu_2)/2}\, dg$

P = 0.999

(Figure: density curve with upper-tail area 0.999 shaded; horizontal axis labelled $F_{\nu_1,\nu_2,0.999}$)

ν_2 \ ν_1	1	2	3	4	5	6	7	8	9	10	12	15	20	24	30	40	60	120	∞
1	4053*	5000*	5404*	5625*	5764*	5859*	5929*	5981*	6023*	6056*	6107*	6158*	6209*	6235*	6261*	6287*	6313*	6340*	6366*
2	998.5	999.0	999.2	999.2	999.3	999.3	999.4	999.4	999.4	999.4	999.4	999.4	999.4	999.5	999.5	999.5	999.5	999.5	999.5
3	167.0	148.5	141.1	137.1	134.6	132.8	131.6	130.6	129.9	129.2	128.3	127.4	126.4	125.9	125.4	125.0	124.5	124.0	123.5
4	74.14	61.25	56.18	53.44	51.71	50.53	49.66	49.00	48.47	48.05	47.41	46.76	46.10	45.77	45.43	45.09	44.75	44.40	44.05
5	47.18	37.12	33.20	31.09	29.75	28.84	28.16	27.64	27.24	26.92	26.42	25.91	25.39	25.14	24.87	24.60	24.33	24.06	23.79
6	35.51	27.00	23.70	21.92	20.81	20.03	19.46	19.03	18.69	18.41	17.99	17.56	17.12	16.89	16.67	16.44	16.21	15.99	15.75
7	29.25	21.69	18.77	17.19	16.21	15.52	15.02	14.63	14.33	14.08	13.71	13.32	12.93	12.73	12.53	12.33	12.12	11.91	11.70
8	25.42	18.49	15.83	14.39	13.49	12.86	12.40	12.04	11.77	11.54	11.19	10.84	10.48	10.30	10.11	9.92	9.73	9.53	9.33
9	22.86	16.39	13.90	12.56	11.71	11.13	10.70	10.37	10.11	9.89	9.57	9.24	8.90	8.72	8.55	8.37	8.19	8.00	7.81
10	21.04	14.91	12.55	11.28	10.48	9.92	9.52	9.20	8.96	8.75	8.45	8.13	7.80	7.64	7.47	7.30	7.12	6.94	6.76
11	19.69	13.81	11.56	10.35	9.58	9.05	8.66	8.35	8.12	7.92	7.63	7.32	7.01	6.85	6.68	6.52	6.35	6.17	6.00
12	18.64	12.97	10.80	9.63	8.89	8.38	8.00	7.71	7.48	7.29	7.00	6.71	6.40	6.25	6.09	5.93	5.76	5.59	5.42
13	17.81	12.31	10.21	9.07	8.35	7.86	7.49	7.21	6.98	6.80	6.52	6.23	5.93	5.78	5.63	5.47	5.30	5.14	4.97
14	17.14	11.78	9.73	8.62	7.92	7.43	7.08	6.80	6.58	6.40	6.13	5.85	5.56	5.41	5.25	5.10	4.94	4.77	4.60
15	16.59	11.34	9.34	8.25	7.57	7.09	6.74	6.47	6.26	6.08	5.81	5.54	5.25	5.10	4.95	4.80	4.64	4.47	4.31
16	16.12	10.97	9.00	7.94	7.27	6.81	6.46	6.19	5.98	5.81	5.55	5.27	4.99	4.85	4.70	4.54	4.39	4.23	4.06
17	15.72	10.66	8.73	7.68	7.02	6.56	6.22	5.96	5.75	5.58	5.32	5.05	4.78	4.63	4.48	4.33	4.18	4.02	3.85
18	15.38	10.39	8.49	7.46	6.81	6.35	6.02	5.76	5.56	5.39	5.13	4.87	4.59	4.45	4.30	4.15	4.00	3.84	3.67
19	15.08	10.16	8.28	7.26	6.62	6.18	5.85	5.59	5.39	5.22	4.97	4.70	4.43	4.29	4.14	3.99	3.84	3.68	3.51
20	14.82	9.95	8.10	7.10	6.46	6.02	5.69	5.44	5.24	5.08	4.82	4.56	4.29	4.15	4.00	3.86	3.70	3.54	3.38
21	14.59	9.77	7.94	6.95	6.32	5.88	5.56	5.31	5.11	4.95	4.70	4.44	4.17	4.03	3.88	3.74	3.58	3.42	3.26
22	14.38	9.61	7.80	6.81	6.19	5.76	5.44	5.19	4.99	4.83	4.58	4.33	4.06	3.92	3.78	3.63	3.48	3.32	3.15
23	14.19	9.47	7.67	6.69	6.08	5.65	5.33	5.09	4.89	4.73	4.48	4.23	3.96	3.82	3.68	3.53	3.38	3.22	3.05
24	14.03	9.34	7.55	6.59	5.98	5.55	5.23	4.99	4.80	4.64	4.39	4.14	3.87	3.74	3.59	3.45	3.29	3.14	2.97
25	13.88	9.22	7.45	6.49	5.88	5.46	5.15	4.91	4.71	4.56	4.31	4.06	3.79	3.66	3.52	3.37	3.22	3.06	2.89
26	13.74	9.12	7.36	6.41	5.80	5.38	5.07	4.83	4.64	4.48	4.24	3.99	3.72	3.59	3.44	3.30	3.15	2.99	2.82
27	13.61	9.02	7.27	6.33	5.73	5.31	5.00	4.76	4.57	4.41	4.17	3.92	3.66	3.52	3.38	3.23	3.08	2.92	2.75
28	13.50	8.93	7.19	6.25	5.66	5.25	4.93	4.69	4.50	4.35	4.11	3.86	3.60	3.46	3.32	3.18	3.02	2.86	2.69
29	13.39	8.85	7.12	6.19	5.59	5.18	4.87	4.64	4.45	4.29	4.05	3.80	3.54	3.41	3.27	3.12	2.97	2.81	2.64
30	13.29	8.77	7.05	6.12	5.53	5.12	4.82	4.58	4.39	4.24	4.00	3.75	3.49	3.36	3.22	3.07	2.92	2.76	2.59
40	12.61	8.25	6.60	5.70	5.13	4.73	4.44	4.21	4.02	3.87	3.64	3.40	3.15	3.01	2.87	2.73	2.57	2.41	2.23
60	11.97	7.76	6.17	5.31	4.76	4.37	4.09	3.87	3.69	3.54	3.31	3.08	2.83	2.69	2.55	2.41	2.25	2.08	1.89
120	11.38	7.32	5.79	4.95	4.42	4.04	3.77	3.55	3.38	3.24	3.02	2.78	2.53	2.40	2.26	2.11	1.95	1.76	1.54
∞	10.83	6.91	5.42	4.62	4.10	3.74	3.47	3.27	3.10	2.96	2.74	2.51	2.27	2.13	1.99	1.84	1.66	1.45	1.00

* Multiply these entries by 100.

TABLE E

PERCENTAGE POINTS OF STUDENTIZED RANGE*

Values of $q_{k,\nu,1-\alpha}$ such that $1 - \alpha = \int_0^{q_{k,\nu,1-\alpha}} p_Q(q)dq$ (Q is defined as W/S)

The range of a sample of size k, from a normal population, divided by an independent estimate of the standard deviation of the population based on ν degrees of freedom, will exceed the values tabulated with probability α.

$$\alpha = 0.05$$

ν \ k	2	3	4	5	6	7	8	9	10	11	12
1	17.97	26.98	32.82	37.08	40.41	43.12	45.40	47.36	49.07	50.59	51.96
2	6.08	8.33	9.80	10.88	11.74	12.44	13.03	13.54	13.99	14.39	14.75
3	4.50	5.91	6.82	7.50	8.04	8.48	8.85	9.18	9.46	9.72	9.95
4	3.93	5.04	5.76	6.29	6.71	7.05	7.35	7.60	7.83	8.03	8.21
5	3.64	4.60	5.22	5.67	6.03	6.33	6.58	6.80	6.99	7.17	7.32
6	3.46	4.34	4.90	5.30	5.63	5.90	6.12	6.32	6.49	6.65	6.79
8	3.26	4.04	4.53	4.89	5.17	5.40	5.60	5.77	5.92	6.05	6.18
10	3.15	3.88	4.33	4.65	4.91	5.12	5.30	5.46	5.60	5.72	5.83
12	3.08	3.77	4.20	4.51	4.75	4.95	5.12	5.27	5.39	5.51	5.61
14	3.03	3.70	4.11	4.41	4.64	4.83	4.99	5.13	5.25	5.36	5.46
16	3.00	3.65	4.05	4.33	4.56	4.74	4.90	5.03	5.15	5.26	5.35
18	2.97	3.61	4.00	4.28	4.49	4.67	4.82	4.96	5.07	5.17	5.27
20	2.95	3.58	3.96	4.23	4.45	4.62	4.77	4.90	5.01	5.11	5.20
30	2.89	3.49	3.85	4.10	4.30	4.46	4.60	4.72	4.82	4.92	5.00
60	2.83	3.40	3.74	3.98	4.16	4.31	4.44	4.55	4.65	4.73	4.81

TABLE E (Continued)

$\alpha = 0.01$

k v	2	3	4	5	6	7	8	9	10	11	12
1	90.03	134.0	164.3	185.6	202.2	215.8	227.2	237.0	245.6	253.2	260.0
2	14.04	19.02	22.29	24.72	26.63	28.20	29.53	30.68	31.69	32.59	33.40
3	8.26	10.62	12.17	13.33	14.24	15.00	15.64	16.20	16.69	17.13	17.53
4	6.51	8.12	9.17	9.96	10.58	11.10	11.55	11.93	12.27	12.57	12.84
5	5.70	6.98	7.80	8.42	8.91	9.32	9.67	9.97	10.24	10.48	10.70
6	5.24	6.33	7.03	7.56	7.97	8.31	8.61	8.87	9.10	9.30	9.48
8	4.75	5.64	6.20	6.62	6.96	7.24	7.47	7.68	7.86	8.03	8.18
10	4.48	5.27	5.77	6.14	6.43	6.67	6.87	7.05	7.21	7.36	7.49
12	4.32	5.05	5.50	5.84	6.10	6.32	6.51	6.67	6.81	6.94	7.06
14	4.21	4.89	5.32	5.64	5.88	6.08	6.26	6.41	6.54	6.66	6.77
16	4.13	4.79	5.19	5.49	5.72	5.92	6.08	6.22	6.35	6.46	6.56
18	4.07	4.70	5.09	5.38	5.60	5.79	5.94	6.08	6.20	6.31	6.41
20	4.02	4.64	5.02	5.29	5.51	5.69	5.84	5.97	6.09	6.19	6.28
30	3.89	4.45	4.80	5.05	5.24	5.40	5.54	5.65	5.76	5.85	5.93
60	3.76	4.28	4.59	4.82	4.99	5.13	5.25	5.36	5.45	5.53	5.60

* Reproduced with permission of Professor E. S. Pearson, from "Extended and Corrected Tables of the Upper Percentage Points of the Studentized Range," computed by Joyce M. May, *Biometrika*, **39** (1952) and from J. Pachares, "Table of the Upper 10% Points of the Studentized Range," *Biometrika*, **46** (1959).

TABLE F
CRITICAL VALUES FOR TESTING OUTLIERS*
(X_1 is the least value)

STATISTIC	NUMBER OF MEANS, n	CRITICAL VALUES† $\alpha = 0.05$	$\alpha = 0.01$
$V_{10} = \dfrac{X_2 - X_1}{X_n - X_1}$	3	0.941	0.988
	4	0.765	0.889
	5	0.642	0.780
	6	0.560	0.698
	7	0.507	0.637
$V_{11} = \dfrac{X_2 - X_1}{X_{n-1} - X_1}$	8	0.554	0.683
	9	0.512	0.635
	10	0.477	0.597
$V_{21} = \dfrac{X_3 - X_1}{X_{n-1} - X_1}$	11	0.576	0.679
	12	0.546	0.642
	13	0.521	0.615
$V_{22} = \dfrac{X_3 - X_1}{X_{n-2} - X_1}$	14	0.546	0.641
	15	0.525	0.616
	16	0.507	0.595
	17	0.490	0.577
	18	0.475	0.561
	19	0.462	0.547
	20	0.450	0.535
	21	0.440	0.524
	22	0.430	0.514
	23	0.421	0.505
	24	0.413	0.497
	25	0.406	0.489
	26	0.399	0.486
	27	0.393	0.475
	28	0.387	0.469
	29	0.381	0.463
	30	0.376	0.457

*Reproduced with permission from Dixon and Massey, *Introduction to Statistical Analysis*, McGraw-Hill Book Co., 1951, and W. J. Dixon, "Ratios involving extreme values," *Annals of Mathematical Statistics*, 1951, pp. 68–78.

†These are for a "one-sided" test. For a "two-sided" test, $\alpha = 0.10$ and 0.02.

TABLE G

Normal Scores (Expected Values of Order Statistics from Normal Population $N(0, 1)$*)

i \ n	2	3	4	5	6	7	8	9	10	11	12
1	0.564	0.846	1.029	1.163	1.267	1.352	1.424	1.485	1.539	1.586	1.629
2		0.000	0.297	0.495	0.642	0.757	0.852	0.932	1.001	1.062	1.116
3				0.000	0.202	0.353	0.473	0.572	0.656	0.729	0.793
4						0.000	0.153	0.275	0.376	0.462	0.537
5								0.000	0.123	0.225	0.312
6										0.000	0.103

i \ n	13	14	15	16	17	18	19	20	21	22	23	24	25
1	1.668	1.703	1.736	1.766	1.794	1.820	1.844	1.867	1.89	1.91	1.93	1.95	1.97
2	1.164	1.208	1.248	1.285	1.319	1.350	1.380	1.408	1.43	1.46	1.48	1.50	1.52
3	0.850	0.901	0.948	0.990	1.029	1.066	1.099	1.131	1.16	1.19	1.21	1.24	1.26
4	0.603	0.662	0.715	0.763	0.807	0.848	0.886	0.921	0.95	0.98	1.01	1.04	1.07
5	0.388	0.456	0.516	0.570	0.619	0.665	0.707	0.745	0.78	0.82	0.85	0.88	0.91
6	0.190	0.267	0.335	0.396	0.451	0.502	0.548	0.590	0.63	0.67	0.70	0.73	0.76
7	0.000	0.088	0.165	0.234	0.295	0.351	0.402	0.448	0.49	0.53	0.57	0.60	0.64
8			0.000	0.077	0.146	0.208	0.264	0.315	0.36	0.41	0.45	0.48	0.52
9					0.000	0.069	0.131	0.187	0.24	0.29	0.33	0.37	0.41
10							0.000	0.062	0.12	0.17	0.22	0.26	0.30
11									0.00	0.06	0.11	0.16	0.20
12											0.00	0.05	0.10
13													0.00

Note that the expected value of the ith order statistics is equal to *minus* the expected value of the $(n - i + 1)$th order statistic.

* Reproduced with permission from Pearson and Hartley, *Biometrika Tables for Statisticians*, Vol. 1 (1958), p. 175, and Fisher and Yates, *Statistical Tables for Biological, Agricultural, and Medical Research*, Oliver and Boyd, 1953.

TABLE H
Power Functions for ANOVA Tests*

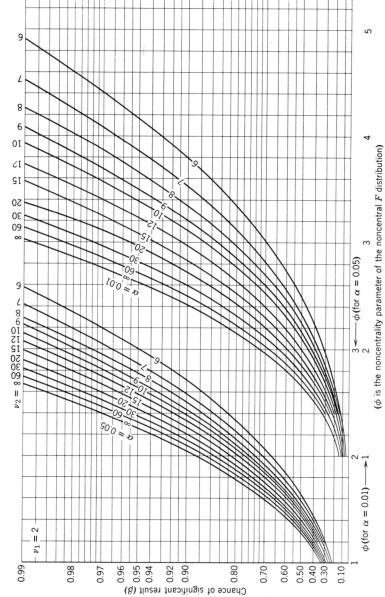

(φ is the noncentrality parameter of the noncentral F distribution)

* Reproduced with permission from Pearson and Hartley, *Biometrika Tables for Statisticians*, Vol. 1 (1958).

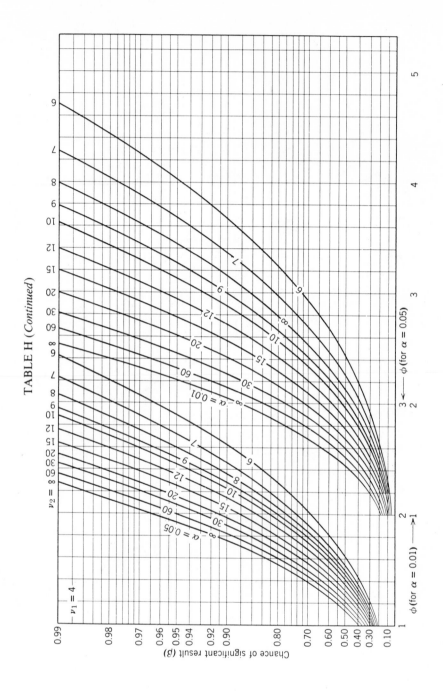

TABLE H (*Continued*)

TABLE H (*Continued*)

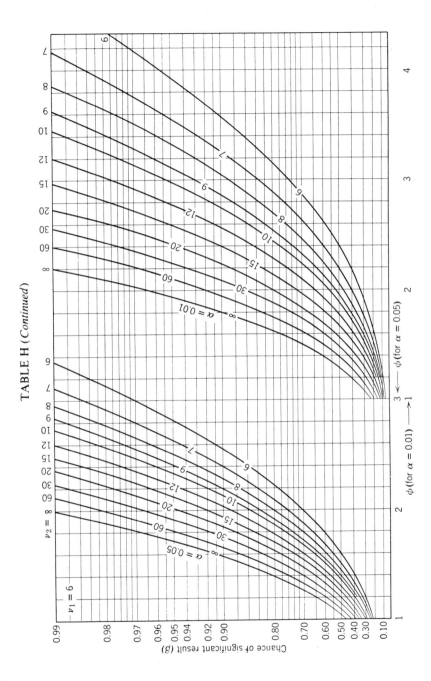

TABLE I

Chart for Determining the Power Function of the t-Test*

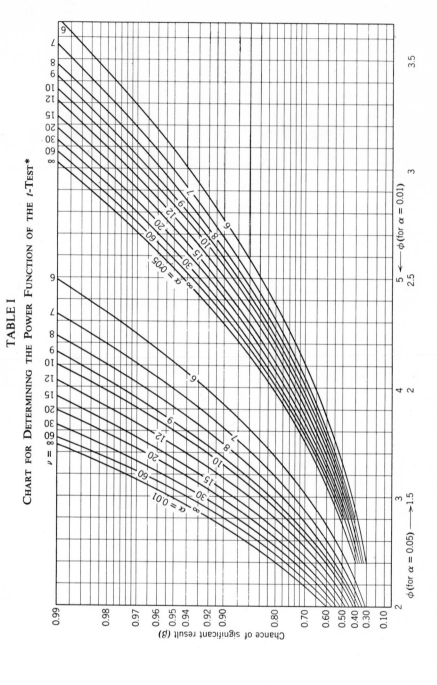

TABLE OF $K_{q,\nu_2,\nu_R,\alpha}$ SUCH THAT

$$-\left\{\nu_R - \tfrac{1}{2}(q-\nu_2+1)\right\} \ln U_{q,\nu_2,\nu_R,\alpha} = K_{q,\nu_2,\nu_R,\alpha}\chi^2_{q,\nu_2,1-\alpha}$$

	$q=3$							
	$\nu_2=4$		$\nu_2=6$		$\nu_2=8$		$\nu_2=10$	
$M=\nu_R-q+1\backslash^{\alpha}$	0.05	0.01	0.05	0.01	0.05	0.01	0.05	0.01
1	1.422	1.514	1.535	1.649	1.632	1.763	1.716	1.862
2	1.174	1.188	1.241	1.282	1.302	1.350	1.359	1.413
3	1.099	1.107	1.145	1.167	1.190	1.216	1.232	1.262
4	1.065	1.070	1.099	1.113	1.133	1.150	1.167	1.187
5	1.046	1.050	1.072	1.082	1.100	1.112	1.127	1.141
6	1.035	1.037	1.056	1.063	1.078	1.087	1.101	1.112
8	1.022	1.023	1.036	1.041	1.052	1.058	1.068	1.075
12	1.011	1.012	1.019	1.021	1.028	1.031	1.038	1.042
20	1.004	1.005	1.008	1.009	1.012	1.013	1.017	1.019
30	1.002	1.002	1.004	1.004	1.006	1.007	1.009	1.009
60	1.001	1.001	1.001	1.001	1.002	1.002	1.002	1.003
∞	1.000	1.000	1.000	1.000	1.000	1.000	1.000	1.000

	$q=4$							
	$\nu_2=4$		$\nu_2=6$		$\nu_2=8$		$\nu_2=10$	
1	1.451	1.550	1.517	1.628	1.583	1.704	1.644	1.774
2	1.194	1.229	1.240	1.279	1.286	1.330	1.331	1.379
3	1.114	1.132	1.148	1.168	1.183	1.207	1.218	1.244
4	1.076	1.088	1.102	1.115	1.130	1.146	1.159	1.176
5	1.055	1.063	1.076	1.085	1.099	1.109	1.122	1.134
6	1.042	1.048	1.059	1.066	1.078	1.086	1.097	1.107
8	1.027	1.030	1.038	1.043	1.052	1.058	1.067	1.073
12	1.014	1.015	1.020	1.023	1.029	1.031	1.038	1.041
20	1.006	1.006	1.009	1.010	1.013	1.014	1.017	1.019
30	1.003	1.003	1.004	1.005	1.006	1.007	1.009	1.009
60	1.001	1.001	1.001	1.001	1.002	1.002	1.003	1.003
∞	1.000	1.000	1.000	1.000	1.000	1.000	1.000	1.000

$$q = 5$$

$M = \nu_R - q + 1 \backslash^{\alpha}$	$\nu_2 = 4$		$\nu_2 = 6$		$\nu_2 = 8$		$\nu_2 = 10$	
	0.05	0.01	0.05	0.01	0.05	0.01	0.05	0.01
1	1.483	1.589	1.514	1.625	1.556	1.672	1.600	1.721
2	1.216	1.253	1.245	1.284	1.280	1.321	1.315	1.359
3	1.130	1.150	1.154	1.175	1.182	1.204	1.211	1.235
4	1.089	1.101	1.108	1.121	1.131	1.145	1.155	1.171
5	1.065	1.074	1.081	1.090	1.100	1.110	1.120	1.131
6	1.050	1.056	1.063	1.070	1.079	1.087	1.097	1.105
8	1.032	1.036	1.042	1.046	1.054	1.059	1.067	1.073
12	1.017	1.019	1.023	1.025	1.030	1.033	1.038	1.041
20	1.007	1.008	1.010	1.011	1.013	1.015	1.018	1.019
30	1.003	1.004	1.005	1.005	1.007	1.007	1.009	1.010
60	1.001	1.001	1.001	1.002	1.002	1.002	1.003	1.003
∞	1.000	1.000	1.000	1.000	1.000	1.000	1.000	1.000

Solutions of Selected Exercises

Chapter 13

1. (a) $X_{ti} = \alpha + \mu_t + Z_{ti}$.

   (b)

SOURCE	S.S.	D.F.	M.S.	M.S. RATIO	
Method	7.60	3	2.53	2.46 (N.S.)	
Residual	10.26	10	1.03		$(F_{3, 10, 0.95} = 3.71)$
Total	17.86	13			

2. (a) M.S.: Residual, 2839.97 (16 D.F.); Paints 12808.05 (3 D.F.).

   (b) $S_1 = 4770$; $S_2 = 26720$; $S_3 = 9970$; $S_4 = 3980$. N.S.

   (c) 115.5, 216.5 (based on Residual); 123.1, 208.9 (Paint 1 data only).

3.

SOURCE	D.F.	S.S.	MEAN SQUARE
Between Locations	2	28.29	14.14
Residual (within locations)	21	48.29	2.30
Total	23	76.57	

   (Use 80 as arbitrary origin.)

Location	I	II	III
Estimated mean (Brix degrees)	80.70	83.29	81.46

   Estimated standard deviation of each mean is $\sqrt{2.30/8} = 0.54$ Brix degrees. Hence Location II appears to be associated with higher Brix degrees of molasses.

6. (a) By linear comparison, S.S. is $(53 + 38)^2/54$ and ratio is $153.35/26.17 = 5.86 > F_{1, 24, 0.95}$ ($= 4.26$).

   (b) 95% confidence limits on stocks A, B, C are 98.6, 105.6; 100.2, 107.2; 94.4, 101.4.

7.

SOURCE	S.S.	D.F.	M.S.
Operators	346.2	4	86.55*
Residual	395.0	15	26.33
Total	741.2	19	

$\hat{\sigma}^2 = 26.33$ (articles)^2. 95% confidence limits for each operator are [observed mean for operator $\pm t_{15,0.975}(\sqrt{26.33/4})\times 10$]. (NOTE: limits are for long-run average; not for individual 10-unit periods.)

Operator	1	2	3	4	5
Limits	660, 770	665, 775	705, 815	760, 870	645, 755

For an operator *chosen at random* the limits are

$$\left[(\text{observed, overall mean}) \pm t_{4,0.975}(\sqrt{86.55/5})\times 10\right],$$

i.e., 627,857. (Note that these limits are considerably wider than those for individual operators.)

8. (a) $X_{ti} = \mu + \alpha_t + \beta_i + Z_{ti}$.
   (b)

SOURCE	D.F.	S.S.		M.S.	
Alloys					
$\quad$ A vs. B and C	1	406.70		406.70	
$\quad$ B vs. C	1	114.08		114.08	
	2		520.78		260.39
Temperatures	5		38041.11		7608.22***
Residual	10		7447.22		744.72
$\quad$ (Interaction)					
Total	17		46009.11		

(c) No significant differences among alloys. Residual is large, indicating possible inaccuracies in measurement. Mold temperature is strongly significant, as can be seen by inspection.

9.

SOURCE	S.S.	D.F.	M.S.
(a) B vs. A and C	373.8	1	373.8
(b) A vs. C	147.0	1	147.0
Error	7447.2	10	744.7

The yield does not differ significantly.

10. (a) $X_{ti} = \mu + U_t + Z_{ti}$.

(b)

SOURCE	S.S.	D.F.	M.S.	M.S. RATIO
Days	103.6	3	34.53	21.56***
Within Days	25.6	16	1.60	
Total	129.2	19		(Unit: 0.1%)

(c) $\widehat{\sigma^2} = 0.0160$; $\widehat{\sigma_t^2} = (34.53 - 1.60)/500 = 0.066$.

11. (a) $Y_{ij} = \alpha + \mu_i + \delta_j + Z_{ij}$.

(b)

SOURCE	S.S.	D.F.	M.S.	M.S. RATIO	
Machines	13444.8	3	4481.6	20.5**	
Days	2146.2	4	536.6	2.45	
Residual	2626.2	12	218.9		$(F_{3,12,0.99} = 5.95)$
Total	18217.2	19			

13. (a) $X_{ij} = \mu + \beta_i + \tau_j + Z_{ij}$.

(b)

SOURCE	S.S.	D.F.	M.S.
Cars	35.49	4	8.87***
Days	10.82	9	1.20*
Residual	16.69	36	$0.46 = \hat{\sigma}^2$
(Cars × Days)			
Total	63.00	49	

15.

ANOVA TABLE

SOURCE	D.F.		S.S.		M.S.	
Weeks	1		16.3		16.3	
Waxes	2		190.2		95.1	
Flake vs. Cake		1		28.2		28.2
Between B and C		1		162.0		162.0
Weeks × Waxes	2		142.2		71.1	
Duplicate measurements	6		227.0		37.8	
Total	11		575.7			

If the interaction Weeks × Waxes is represented by a random term, the Interaction mean square (71.1) should be used as a Residual; if the interaction is represented by a parameter the "Duplicates" mean square (37.8) should be used as a Residual.

There is no evidence of variation from week to week or for Flake vs. Cake.

Although the mean square between B and C is large, the ratio $162.0/37.8 = 4.3$ is not markedly significant ($F_{1,6,0.95} = 5.99$; $F_{1,6,0.90} = 3.78$).

17.

SOURCE	S.S.	D.F.	M.S.	E.M.S.
Batches	54900	3	18300	$\sigma^2 + 3\sigma_V^2 + 6\sigma_U^2$
Samples (within batches)	5300	4	1325	$\sigma^2 + 3\sigma_V^2$
Residual	14333.3	16	896	σ^2
Total	74533.3	23		

$\widehat{\sigma^2} = 896,\ \widehat{\sigma_V^2} = 143,\ \widehat{\sigma_U^2} = 2829$

Batches mean square is significant at 2.5% level [compared with Samples (Within Batches)]. Samples not significant compared with Residual.

Estimate of Residual variance is 896.

Estimate of variance of Samples component is
$(1325 - 896)/3 = 143$.

Estimate of variance of Batches component is
$(18300 - 1325)/6 = 2829$.

[Calculation facilitated by coding, using $(X - 650)/10$.]

18.

SOURCE	D.F.	S.S.	M.S.	M.S. RATIO
Types	1	0.001250	0.001250	1.646
Locations	2	0.348922	0.174461	229.8***
Interaction	2	0.000343	0.000171	0.225
Residual	12	0.009111	0.000759	
Total	17	0.359626		

The smallness of the interaction is noteworthy. Other features were evident from data without need for formal analysis.

19. (a)–(c)

SOURCE	S.S.	D.F.	M.S.	M.S. RATIO
Adhesives (H)	127.51	2	58.75	3.36*
Assemblies (A)	4.97	2	2.48	0.14
Interaction	196.09	4	49.02	2.80*
Residual	630.00	36	17.50	
Total	958.58	44		

$F_{2,36,0.95} = 3.28$

$F_{4,36,0.95} = 2.50$

Adhesives and Interaction effects both significant at 5% level.

(d) At 5% level, a significant difference is noted.

20. (a) $Y_{tij} = \alpha + \gamma_t + \phi_i + (\gamma\phi)_{ti} + Z_{tij}$.

SOURCE	D.F.	S.S.	M.S.
Furnaces	2	86.08	43.04
Compounds	3	264.46	88.15
Interaction	6	22.92	3.82
Residual	12	22.50	2.12
Total	23	398.96	

(b) To look for interaction effect and the possible effect of compound over all three furnaces.

(c) Yes, since mean square ratio $= 41.48 > F_{2, 12, 0.999}$.

21. (a) $X_{tij} = \mu + D_t + \phi_i + (D\phi)_{ti} + Z_{tij}$.

(b)

SOURCE	S.S.	D.F.	M.S.	M.S. RATIO	E.M.S.
Densitometers	44.33	3	14.78	6.18**	$\sigma^2 + 2\sigma_{ti}^2 + 6\sigma_t^2$
Films	1.75	2	0.87	0.40	$\sigma^2 + 2\sigma_{ti}^2 + 6\Sigma F_i^2$
$D \times \phi$	3.92	6	0.65	0.30	$\sigma^2 + 2\sigma_{ti}^2$
Residual	26.00	12	2.17		σ^2
Total	76.00	23			

(c) $\widehat{\sigma^2} = 2.17$, $\widehat{\sigma_{ti}^2} = 0$, $\widehat{\sigma_t^2} = 2.36 [= (14.78 - 0.65)/6]$.

(d) Densitometer variance is very significantly different from 0. (Note that $D \times \phi$ is not significant.)

22.

SOURCE	S.S.	S.S.	D.F.	M.S.
Content		207.25		
Linear	204.80		1	204.80**
Quadratic	2.25		1	2.25
Other	0.20		1	0.20
Temperature		67.25		
Linear	18.05		1	18.05
Quadratic	49.0		1	49.0
Other	0.25		1	0.25
Residual		89.25	9	9.92
Total		363.75	15	

There is linear effect caused by copper content.

25. (a) 126.69×10^{-10}.

(b) Yes, M. S. ratio $= 3.68 > F_{7,8,0.95}$.

(c) 1.764044, 1.764336.

(Based on homoscedasticity; possibly limits should be wider.)

(d) $X_{tij} = \mu + \alpha_t + \beta_i + (\alpha\beta)_{ti} + Z_{tij}$, $S^2 = 163.33 \times 10^{-10}$.

ANOVA TABLE ($Y_{tij} = [X_{tij} - 1.764] \times 10^5$)

SOURCE	S.S.	D.F.	M.S.	M.S. RATIO
Silver	2000.67	2	1000.33	6.12*
Iodine	161.33	1	161.33	0.99
S×I	408.67	2	204.33	1.25
Residual	980.00	6	163.33	
Total	3550.67	11		

30.

SOURCE	D.F.	S.S.	M.S.
Batches	4	18.67	4.67
Sample (Within Batch)	10	27.53	2.75
Residual	15	91.30	6.09
Total	29	137.5	

(b) Estimates of Within and Among Batch variances are 0. $\hat{\sigma}^2 = 6.09$. From first two M.S. alone, estimated Among Batch variance is 0.32.

(c) 95% confidence limits on σ^2 are 3.32 and 14.58.

36. Putting $g_0 = s_0$ and using

$$\frac{\partial (s_0, s_k, \ldots, s_1)}{\partial (g_0, g_k, \ldots, g_1)} = g_1^k \prod_{t=2}^{k} (1 + g_t)^{t-1}$$

we find $p(g_0, g_1, \ldots, g_k)$ and hence $p(g_1, \ldots, g_k)$. In Table 13.29, we can regard S_t as the sum of squares for the tth order polynomial; $S_0 + \Sigma_{j=t+1}^{k} S_j$ as the corresponding Residual. The result of the exercise shows that the successive test criteria are independent when the null hypothesis is true.

Chapter 14

1. (a) Estimated weight loss of each paint, independent of runs, is as follows:

Paint	A	B	C	D
Weight loss	14.46	16.08	21.83	23.96

(b) At 0.5% level of significance we conclude that paints are different with respect to abrasion loss.

(c) At 0.5% level of significance, we conclude that runs are significantly different.

3.

SOURCE	S.S.	D.F.	M.S.
Specimens (unadjusted) (Blocks)	381.05	9	
Curing times (adjusted)	432.4	4	108.1
Residual	203.1	6	33.85
Total	1016.55	19	

Curing time is not a significant factor.

4.

SOURCE	D.F.	S.S.	M.S.
Manufacturer (adjusted)	5	0.43	0.08
Condition (unadjusted)	9	(0.65)	
Residual	15	1.62	0.11
Total	29	2.70	

6.

SOURCE	S.S.	D.F.	M.S.
Temperature			
Linear	18.05	1	18.05
Quadratic	49.00	1	49.00
Other	0.20	1	0.20
Laboratory	21.25	3	7.08
Composition			
Linear	168.20	1	168.20*
Quadratic	2.25	1	2.25
Other	9.85	1	9.85
Residual	95.00	6	15.83
Total	363.75	15	

The quadratic effect of temperature is not strong enough to be considered significant, but might very well be an important factor in future experimentation.

7.

SOURCE	S.S.	D.F.	M.S.
Rows	2.5	3	0.83
Columns	12.5	3	4.17
Letters	15.5	3	5.17
Subscripts	4.5	3	1.50
Residual	11.0	3	3.67
Total	46.0	15	

Each factor not significant.

8.

SOURCE	D.F.	S.S.	M.S.	M.S. RATIO
Temperature	4	15100.56	3775.14	26.26***
Duration	4	3195.36	798.84	5.56*
Reactor	4	199.76	49.99	
Operator	4	254.16	63.54	
Residual	8	1149.92	143.74	
Total	24	19899.76		$(F_{4,8,0.95}=3.84)$

Temperature effect is highly significant. Thus all temperatures do not have the same effect on the extent of conversion. The temperature means are 28.2, 42.8, 62.6, 84.6, 93.6% conversion. The estimated standard deviation of each mean is 16.9%. No other effect is significant at the 5% level. Hence we cannot say that there is any difference among durations or operators or reactors.

9.

SOURCE	D.F.	S.S.	M.S.	M.S. RATIO
Runs (adjusted)	3	147.25	49.08	45.44**
Runs (unadjusted)		(288.92)		
Paints (adjusted)	3	164.58	54.86	50.80***
Positions	2	8.17	4.09	3.79
Residual	3	3.25	1.08	
Total	11	464.92		

$$F_{3,3,0.95}=9.28 \qquad F_{3,3,0.99}=29.5 \qquad F_{3,3,0.995}=47.5$$

10.

SOURCE	S.S.	D.F.	M.S.
Machines (crude)	72.96		
Compositions (adjusted)	75.90	6	12.65**
Compositions (crude)	118.96		
Machines (adjusted)	29.90	6	4.98*
Shifts	1.15	2	0.575
Residual	6.28	6	1.047
Total	156.29	20	

11.

SOURCE	D.F.	S.S.	M.S.	M.S. RATIO
Blocks (unadjusted)		(196.95)		
Blocks (adjusted)	6	82.29	13.71	10.63***
Treatments (unadjusted)		(608.28)		
Treatments (adjusted)	6	493.62	82.27	63.78***
Rows	2	8.66	4.33	3.36
Residual	6	7.72	1.29	
Total	20	706.95		

$F_{6,6,0.995} = 11.1$

$F_{2,6,0.95} = 5.14$

Blocks (engines) as well as the 7 different gasolines are significantly different. (It would be advisable to tabulate the adjusted means of gasolines and engines.)

13. (a) $X_{(tij)} = \mu + \alpha_t + \beta_i + \gamma_j + Z_{(tij)}$.

 (b) $\hat{\sigma} = 1.45$.

 (c)

SOURCE	D.F.	S.S.	M.S.	M.S. RATIO
Films (A)	4	164.24	41.06	19.65**
Cameras (B)	4	21.84	5.46	2.61
Flashbulbs (C)	4	136.64	34.16	16.34**
Residual	12	25.12	2.09	
Total	24	347.84		

Films effect and Flashbulbs effect are highly significant.

 (d) $\bar{X}_{(4..)} \pm t_{12,0.975} 1.45/\sqrt{5}$. Limits are 0.682, 0.710 in original units.

14. $Y_{ijk} = \mu + \alpha_i + \beta_j + \gamma_k + (\alpha\beta)_{ij} + (\alpha\gamma)_{ik} + (\beta\gamma)_{jk} + Z_{ijk}$.

SOURCE	S.S.	D.F.	M.S.	M.S. RATIO
Catalysts (A)	144.5	1	—	8.03
Concentration (B)	112.5	1	—	6.25
Temperature (C)	578.0	1	—	32.11
$A \times B$	4.5	1	—	0.25
$B \times C$	2.0	1	—	0.11
$A \times C$	128.0	1	—	7.1
Error ($A \times B \times C$)	18.0	1	—	
Total	987.5	7		

Though none of the effects is significant, a larger experiment is desirable to see whether there are real main effects, and possibly $A \times C$ interaction.

15. (a) $Y_{ijkl} = \mu + \alpha_i + \beta_{j(i)} + \gamma_{k(ij)} + \delta_{l(ijk)} + Z_{l\alpha(ijk)}$ where $i = 1,2$; $j = 1,2$; $k = 1,2$; $l = 1,2$; $\alpha = 1$ and α, β, γ, δ, and Z are random normal variables.

(c)

SOURCE	D.F.	S.S.	M.S.	E.M.S.
Runs	1	0.014	0.01400	$\sigma^2 + 2\sigma_\delta^2 + 4\sigma_\beta^2 + 8\sigma_\alpha^2$
Bags (within runs)	2	0.005	0.0025	$\sigma^2 + 2\sigma_\delta^2 + 4\sigma_\beta^2$
Samples (within bags)	4	0.001	0.00025	$\sigma^2 + 2\sigma_\delta^2$
Determinations (within samples)	8	0.002	0.00025	σ^2
Total	15	0.022		

$\widehat{\sigma^2} = 2.5 \times 10^{-4}$;

$\widehat{\sigma_\delta^2} = 0$;

$\widehat{\sigma_\beta^2} = (0.00250 - 0.00025)/4 = 0.00225/4 = 0.00056 = 5.6 \times 10^{-4}$;

$\widehat{\sigma_\alpha^2} = (0.0140 - 0.0025)/8 = 0.0115/8 = 0.00144 = 14.4 \times 10^{-4}$(all in coded units^2).

16. (a) $X_{tijk} = \mu + \alpha_t + U_{i(t)} + V_{j(ti)} + Z_{k(tij)}$ where t, i, j denote material, batch, sample, respectively.

(b) $\widehat{\sigma^2} = 15{,}933.37$, $\widehat{\sigma_V^2} = 980.71$, $\widehat{\sigma_U^2} = 0$.

(c) Yes, M.S. ratio $= 122{,}403.7/18{,}875.5 = 6.48 > F_{2,9,0.95}$. If U and V are pooled, M.S. ratio is

$$\frac{122{,}403.7}{13{,}333.3} = 9.18 > F_{2,15,0.99}$$

17.

ANOVA

SOURCE	S.S.	D.F.	M.S.
Laboratories	0.0192	1	0.0192
Catalysts	0.1533	2	0.0767
Pressure	0.2517	1	0.2517
Laboratories × Catalysts	0.2069	2	0.1035
Catalysts × Pressure	0.1646	2	0.0823
Laboratories × Pressure	0.0007	1	0.0007
Laboratories × Pressure × Catalysts	0.0545	2	0.0273
Residual	1.0992	24	0.0458
Total	1.9501	35	

(a) The catalysts do not have any significant effect. Pressure is significant at the 5% level but not at the 1% level.

(b) Yes, laboratories are consistent.

18. (a) $X_{ijk} = \alpha + \beta_i + \gamma_j + \delta_k + (\beta\gamma)_{ij} + (\beta\delta)_{ik} + (\gamma\delta)_{jk} + Z_{ijk}$

$\qquad i = 1, 2, \ldots, 6; \quad j = 1, 2, 3, 4; \quad k = 1, 2, 3.$

SOURCE	S.S.	D.F.	M.S.
Time (B)	0.007970	5	0.001594
pH (C)	0.063283	3	0.021094
Pressure (D)	0.077796	2	0.038898
$B \times C$	0.001921	15	0.000128
$B \times D$	0.034164	10	0.003416
$C \times D$	0.028804	6	0.004601
Residual	0.009825	30	0.0003275
	0.223763	71	

(b) Estimate of the residual variance is 0.0003275.

(c) All the main effects are significant. So also are the interactions. This brings up the question of whether the proper yield is being measured. A transformation may very well be in order. Further, it may be of value to consider quadratic and higher-order effects of the quantitative factors.

19. (a) $S_A^2 = 1.17$, $S_B^2 = 1.04$; (c) 17.25.

(b)

ANOVA

SOURCE	D.F.	S.S.	M.S.	M.S. RATIO
Resin	1	444.08	444.08	9.61*
Batch	6	277.17	46.20	2.70
Cure	1	3,570.75	3,507.75	209.
Resin × Cure	1	70.09	70.09	4.10
Batch × Cure	6	102.50	17.08	15.53**
Residual	32	35.33	1.10	
Total	47	4499.98		

21. (a) $X_{ijk} = \alpha + \beta_i + \gamma_j + (\beta\gamma)_{ij} + U_{k(j)} + Z_{ijk}$

$i = 1, 2, 3, 4$ test stations; $j = 1, 2, 3, 4$ setting stations; $k = 1, 2, 3$ regulators.

ANOVA

(b), (c)

SOURCE	S.S.	D.F.	M.S.
Among test stations B	2.77	3	0.923**
Among setting stations C	0.47	3	0.157*
$B \times C$	0.53	9	0.59
Among regulators within setting stations	2.13	8	0.266**
Residual error	1.06	24	0.044
	6.95	47	

23. (a) $X_{tijk} = \alpha + U_t + V_{i(t)} + W_{j(ti)} + \beta_k + (U\beta)_{tk} + (V\beta)_{ik(t)} + Z_{tijk}$

ANOVA

SOURCE	S.S.	D.F.	M.S.
Batches (U_t)	2.19	2	1.10
Bags ($V_{i(t)}$)	4.02	6	0.67
Samples ($W_{j(ti)}$)	1.74	9	0.19
Laboratories (β_k)	0.10	2	0.05
Batches × Laboratories	0.27	4	0.07
Bags × Laboratories	0.59	12	0.05
Residual	1.37	18	0.076
Total	10.28	53	

(b) Estimates
Residual variance = 0.037
Among bags = 0.068
Among samples = 0.061
Among batches = 0.024

(c) Laboratories are not significant.

24.

SOURCE	S.S.	D.F.	M.S.
Years	2.69	2	1.35
Quarters	0.12	1	0.12
Size	114.90	2	57.45
Companies within size categories	14.39	6	2.40
Years × Quarters	0.22	2	0.11
Years × Size	0.47	4	0.12
Years × Companies	0.34	12	0.03
Quarters × Size	0.07	2	0.04
Quarters × Companies	0.10	6	0.02
Years × Quarters × Size	0.05	4	0.01
Residual	0.39	12	0.03
	133.74	53	

Apart from Size, which is clearly of major importance, the only substantial effects appear to be the Years and Companies main effects.

26.

SOURCE	D.F.	10^4(S.S.)	M.S.
Films	4	192.52	0.00481
Cameras	4	269.32	0.00673
Bulb brands	4	5.12	0.00013
Filter types	4	484.32	0.01211
Interactions	8	53.04	0.00066
Residual (duplicates)	25	50.00	0.00020
Total	49	1054.32	

Brand of bulb appears to have little effect. The Interactions mean square is not significant (although rather large) compared with the Residual mean square, but the apparent significance of Films, Cameras, and Filter types would have been less if duplicates had not been available and the Interactions mean square used as Residual. (A possible fallacy here is that the duplicate may not be a true replicate.)

Chapter 15

2. Denote agitation speed by A, concentration by B, and temperature by C, and let the lower level of each factor correspond to the smaller numerical value. Coding the data by subtracting 85, the Yates and ANOVA tables are as follows:

YATES' TABLE

				ANOVA SOURCE	D.F.	S.S.($=$M.S.)
(1)	-1.1	$\cdots$	13.4	A	1	25.2
a	-6.6	$\cdots$	-14.2	B	1	15.7
b	2.2	$\cdots$	11.2	C	1	59.4
ab	1.3	$\cdots$	-0.4	$A \times B$	1	0.02
c	5.1	$\cdots$	21.8	$A \times C$	1	0.25
ac	3.7	$\cdots$	-1.4	$B \times C$	1	15.7
bc	7.6	$\cdots$	-11.2	$A \times B \times C$	1	11.5
abc	1.2	$\cdots$	-9.6			

The data appear to indicate that there is a real difference associated with change in temperature (Mean yield at 30°C, 83.9%; at 50°C, 89.4%). The small size of the experiment makes it difficult to choose a Residual sum of squares with confidence. If $(A \times B)$ and $(A \times C)$ are used (mean square 0.13 with 2 degrees of freedom) the other 5 mean squares are all significant. Choice of the two *smallest* mean squares, however, should be allowed for in the analysis.

3.

SOURCE	S.S.	D.F.	M.S.	M.S. RATIO
Days (D)	197.81	2	98.90	31.20***
Operators (O)	4.92	2	2.46	0.78
Machines (M)	22.70	2	11.35	3.58*
DO	11.75	4	2.94	0.93
DM	3.97	4	0.99	0.31
OM	17.86	4	4.46	1.41
DOM	30.47	8	3.81	1.20
Residual	85.50	27	3.17	
Total	374.98	53		

4.

SOURCE	D.F.	S.S.	M.S.
Catalyst	1	1080.0	1080.0
Laboratory	1	12.0	12.0
Pressure	i	360.4	360.4
Catalyst × Laboratory	1	610.0	610.0
Catalyst × Pressure	1	234.4	234.4
Laboratory × Pressure	1	3.4	3.4
Catalyst × Laboratory × Pressure	1	92.0	92.0
Residual	16	8812.6	550.8
Total	23	11205.0	

In contradistinction to Exercise 2 there is now available a clearly defined Residual mean square. No effect is significantly large compared with the Residual (though two are remarkably *small*). It would be interesting to compare the variances between replications for various combinations of factor levels.

5.

	1	2	3	4	EFFECT		1	2	3	4	EFFECT
(1)	159	280	598	1246	I	c	170	−38	38	50	C
a	121	318	648	−118	A	ac	122	−36	64	30	AC
b	177	292	−74	102	B	bc	176	−48	2	26	BC
ab	141	356	−44	54	AB	abc	180	4	52	50	ABC

6.

SOURCE	S.S.	D.F.	M.S.	M.S. RATIO
RPM (A)	3752.10	2	1876.05	297.31***
SAE Grades (B)	101.07	2	50.54	8.01**
Temperature (C)	133.31	2	66.66	10.56**
Oils (D)	17.87	2	8.94	1.42
AB	8.78	4	2.20	
AC	5.79	4	1.45	
AD	2.79	4	0.70	
BC	3.68	4	0.92	
BD	10.76	4	2.69	
CD	0.35	4	0.09	
ABC	2.16	8	0.27	
ABD	1.68	8	0.21	
ACD	1.56	8	0.20	
BCD	1.31	8	0.16	
Residual	100.91	16	6.31	
Total	4144.12	80		

8. The sums of squares of the main effects A, B, C, D, and E are 20.25, 42.25, 2.25, 12.25, and 0.25, respectively. In comparing these with the pooled third-order interactions (which we use as the Residual mean square) we find that none of the main effects is significant.

10.

SOURCE	D.F.	S.S.	SOURCE	D.F.	S.S.
P	1	700336.1	$P \times B \times A$	1	144.5
B	1	1444150.1	$P \times B \times C$	1	5618.0
A	1	297992.0	$P \times A \times C$	1	66.1
C	1	48050.0	$B \times A \times C$	1	38226.1
$P \times B$	1	252050.0	Blocks		
$P \times A$	1	22366.1	Repns.	1	720.0
$B \times A$	1	570846.1	Within		
$P \times C$	1	24531.1	Repns.	2	21062.5
$B \times C$	1	15.1	Residual	14	110125.2
$A \times C$	1	13944.5			
			Total	31	3550243.5

Experiment in 2 complete replications: Blocks 1, 2 and 3, 4. *PBAC* confounded with replication.

P, B, A, $P \times B$, and $P \times A$ are clearly significant. C, $P \times A$, $P \times C$, and $B \times A \times C$ may indicate real effects. C has a relatively small effect.

11.

YATES' TABLES

	REPLICATE I			REPLICATE II		
TREATMENT	YIELD	TOTAL		YIELD	TOTAL	EFFECT
(1)	6 ···	151		3 ···	154	
a	5	7		6	12	A
b	8	37		9	36	B
ab	7	9		12	−14	AB
c	7	25		9	32	C
ac	11	5		12	10	AC
bc	7	11		13	2	BC
abc	17	11		17	0	ABC
d	6	15		6	−8	D
ad	8	−17		7	−14	AD
bd	11	17		12	−6	BD
abd	12	−3		6	−16	ABD
cd	10	−7		5	−10	CD
acd	4	−27		11	8	ACD
bcd	17	7		14	8	BCD
abcd	15 ···	−1		12 ···	−2	ABCD

Sum of squares for A (unconfounded) is $(7+12)^2/32 = 11.3$
Sum of squares for ABC (confounded in I) is $0^2/16 = 0.0$
Sum of squares for BCD (confounded in II) is $7^2/16 = 3.1$

ANOVA TABLE

SOURCE	D.F.	S.S.	M.S.	SOURCE	D.F.	S.S.	M.S.
A	1	11.3		BC	1	5.3	
B	1	166.5	166.5***	BD	1	2.2	
C	1	101.5	101.5***	CD	1	9.0	
D	1	1.5		ABC	1	0.0	
AB	1	0.8		ABD	1	0.6	
AC	1	1.5		ACD	1	4.0	
AD	1	30.0	30.0**	BCD	1	3.1	
				ABCD	1	0.3	
				Replications	1	0.3	
				Blocks within			
				replications	6	97.4	16.2
				Residual	9	38.8	4.3
				Total	31	474.0	

The significance of the AD interaction is interesting, in view of the nonsignificance of the main A and D effects.

12. Using L for laboratories, T for temperature, and C for compositions, the treatments are (in order LTC)

002	100	201	303
013	111	210	312
020	122	223	321
031	133	232	330

13. Consider this as a 3-factor cross-classification with Thermometers (A), Analysts (B), and Replicates (C). Consider also A, B, and C as random variables. Although the direct analysis would be simpler, the Yates' algorithm with pseudolinear and quadratic effects is used to further illustrate this method. Following the Yates table, the appropriate summations of S.S. and D.F. are obtained. The data have been coded by subtracting 171.0 and dividing by 0.5. The divisor for the M.S. Ratio of A and B is 2.11.

(1) THERMOMETERS ANALYSTS REPLICATES	(2)	(3)	(4)	(5)	(6) EFFECT	(7) DIVISOR	(8) S.S.
0 0 0	6	11	30	102			
1 0 0	4	6	36	−22	A_L	$2 \times 9 \times 1$	26.89
2 0 0	1	13	36	−6	A_Q	$2 \times 27 \times 1$	0.67
0 1 0	4	13	−10	−3	B_L	$2 \times 9 \times 1$	0.50
1 1 0	2	10	−4	8	$A_L B_L$	$4 \times 3 \times 1$	5.33
2 1 0	0	13	−8	0	$A_Q B_L$	$4 \times 9 \times 1$	0
0 2 0	5	15	0	21	B_Q	$2 \times 27 \times 1$	8.17
1 2 0	4	11	−6	8	$A_L B_Q$	$4 \times 9 \times 1$	1.78
2 2 0	4	10	0	12	$A_Q B_Q$	$4 \times 27 \times 1$	1.33
0 0 1	5	−5	2	6	C_L	$2 \times 9 \times 1$	2.00
1 0 1	5	−4	0	2	$A_L C_L$	$4 \times 3 \times 1$	0.33
2 0 1	3	−1	−5	0	$A_Q C_L$	$4 \times 9 \times 1$	0
0 1 1	4	−2	4	−7	$B_L C_L$	$4 \times 3 \times 1$	4.08
1 1 1	4	−2	2	−2	$A_L B_L C_L$	$8 \times 1 \times 1$	0.50
2 1 1	2	0	2	−4	$A_Q B_L C_L$	$8 \times 3 \times 1$	0.67
0 2 1	4	−3	2	−9	$B_Q C_L$	$4 \times 9 \times 1$	2.25
1 2 1	5	−4	0	2	$A_L B_Q C_L$	$8 \times 3 \times 1$	0.17
2 2 1	4	−1	−2	12	$A_Q B_Q C_L$	$8 \times 9 \times 1$	2.00

(1)	(2)	(3)	(4)	(5)	(6)	(7)	(8)
THERMOMETERS							
ANALYSTS							
REPLICATES					EFFECT	DIVISOR	S.S.
0 0 2	7	−1	12	−6	C_Q	$2 \times 27 \times 1$	0.67
1 0 2	4	0	6	−10	$A_L C_Q$	$4 \times 9 \times 1$	2.78
2 0 2	4	1	3	12	$A_Q C_Q$	$4 \times 27 \times 1$	1.33
0 1 2	5	−2	2	−3	$B_L C_Q$	$4 \times 9 \times 1$	0.25
1 1 2	5	−2	2	2	$A_L B_L C_Q$	$8 \times 3 \times 1$	0.17
2 1 2	1	−2	4	0	$A_Q B_L C_Q$	$8 \times 9 \times 1$	0
0 2 2	4	3	0	3	$B_Q C_Q$	$4 \times 27 \times 1$	0.03
1 2 2	3	−4	0	2	$A_L B_Q C_Q$	$8 \times 9 \times 1$	0.03
2 2 2	3	1	12	12	$A_Q B_Q C_Q$	$8 \times 27 \times 1$	0.67

Total　62.60

SOURCE	S.S.	D.F.	M.S.	M.S. RATIO
Thermometers (A)	27.56	2	13.78	6.53
Analysts (B)	8.67	2	4.34	2.06
Replicates (C)	2.67	2	1.34	2.53
AB	8.44	4	2.11	3.98*
AC	4.44	4	1.11	2.09
BC	6.61	4	1.65	3.11
ABC (Residual)	4.21	8	0.53	
Total	62.60	26	(Unit: 0.5)	

$$\widehat{\sigma^2} = 0.53 \times (0.5)^2 = 0.13$$

$$\widehat{\sigma_A^2} = 1.23 \times (0.5)^2 = 0.31$$

14. See *Fractional Factorial Experiment Designs for Factors at Three Levels*, NBS Applied Math. Series 54, Plans 27.7.3 through 27.7.27.

15.

BLOCK I	BLOCK II	BLOCK III	BLOCK IV
a	(1)	(1)	ab
b	ab	a	ac
c	ac	bc	b
abc	bc	abc	c
First Replicate		Second Replicate	

16. *ABCDE, CDEFG, ABFG*
Using as confounding interactions *ACF*, *BDG*, and *ABCDFG*:

BLOCK 1			BLOCK 2	BLOCK 3		BLOCK 4	
(1)	*aef*	*ab*	*bef*	*abce*	*bcf*	*ce*	*acf*
abcd	*bcdef*	*cd*	*acdef*	*de*	*adf*	*abde*	*bdf*
abfg	*beg*	*fg*	*aeg*	*cefg*	*acg*	*abcefg*	*bcg*
cdfg	*acdeg*	*abcdfg*	*bcdeg*	*abdefg*	*bdg*	*defg*	*adg*

17. Using the defining contrasts *ABCDE*, *ABFG*, *CDEFG*, *ACF*, *BDEF*, *BCG*, and *ADEG*, the elements are

(1)	*aef*	*abfg*	*abce*
acg	*cefg*	*bcf*	*beg*
bdg	*abdefg*	*adf*	*acdeg*
abcd	*bcdef*	*cdfg*	*de*

18. Using the defining equation is $x_1 + x_2 + x_3 + x_4 = 0 \bmod 5$ and the confounding equation $x_1 + x_2 + x_3 = 0$ the principal block is

0000	1400	2300	3200	4100
0140	1040	2440	3340	4240
0230	1130	2030	3430	4330
0320	1220	2120	3020	4420
0410	1310	2210	3110	4010

25. (*a*) Model is

$$X_{tijl} = \alpha + \pi_t + \omega_i + (\pi\omega)_{ti} + G_{tij} + Z_{tijl} \qquad (t,i,j,l = 1,2,3)$$

where π = Propellant Temperature parameter
ω = Slant Range parameter
G = Group random variable
$(\pi\omega)$ = Interaction ($P \times S$) term

(*b*)

SOURCE	D.F.	S.S.	M.S.
Propellant Temperature (*T*)	2	1568.3	784.2
Slant Range (*R*)	2	1540.2	770.1
$T \times R$	4	23.9	6.0
Groups Within $T \times R$ Combinations	18	1875.8	104.2
Residual (Within Groups)	54	7710.0	142.8
Total	80	12718.2	

(c) If $(T \times R)$ is regarded as a parameter, each of the first three mean squares should be compared with the Groups mean square. We have $F_{2,18,0.99} = 6.01$ so the Propellant Temperature and Slant Range effects are significant at the 1% level. The smallness of the Interaction mean square indicates that these effects are additive, so we simply give estimates of means for each level of these two factors.

	R:	1	2	3		T:	1	2	3
Mean azimuth of miss distance		-7.7	-2.9	3.0			2.9	-2.6	-7.9

Estimated standard deviation of each of these estimates: $\sqrt{\dfrac{142.8}{27}} = 2.3$

Consideration might be given to pooling the Groups and Residual sums of squares. There is certainly no evidence of variation from group to group for a given combination of propellant temperature and slant range. Possible differences between variances might also be of interest in this investigation.

27. The whole set of observations can be ranked and then the rank orders can be analyzed formally as if they were the observed results in the experiment. However, the natural null hypothesis distribution is not appropriate, because of correlations between rank orders in the same block. Nevertheless this effect is not likely to be very important.

An alternative method is based on assigning a separate set of rank orders *within each block*. A simple null hypothesis distribution can be used, but rather more information may be lost with this method.

28. If the experiment is unconfounded the remarks on Exercise 27 are applicable. The same situation applies when dealing with single fractionally replicated experiments.

For confounded experiments, the second method described for Exercise 27 is usually preferable.

Chapter 16

1. Let X_m denote the number of occurrences of the attribute among the first m items observed. The procedure is:

$$\text{Accept } p_0 = 0.01 \text{ if } \left(\frac{0.02}{0.01} \right)^{X_m} \left(\frac{0.98}{0.99} \right)^{m-X_m} \leqslant \frac{0.02}{0.99}$$

(That is, if $X_m \leqslant 0.0144m - 5.55$)

$$\text{Accept } p_1 = 0.02 \text{ if } \left(\frac{0.02}{0.01} \right)^{X_m} \left(\frac{0.98}{0.99} \right)^{m-X_m} \geqslant \frac{0.98}{0.01}$$

(That is, if $X_m \geqslant 0.0144m + 6.52$)

Otherwise, examine a further item.

2. When the actual population proportion possessing the attribute is p, the a.s.n. is approximately

$$\frac{P\log 98 + (1-P)\log(2/99)}{\log(98/99) + p\log(99/49)}$$

where

$$P \doteqdot \frac{1-(2/99)^h}{98^h - (2/99)^h}$$

and h is the nonzero root of the equation

$$2^h p + \left(\frac{98}{99}\right)^h (1-p) = 1$$

(if such a root exists).

The simplest way to compute points on the a.s.n. curve is to choose a value of h, then find the corresponding values of p and P and so calculate the approximate a.s.n. value. Useful reference points are

$$p = p_0 = 0.01; \quad h = 1; \quad P \doteqdot 0.01; \quad \text{a.s.n.} \doteqdot 1231$$
$$p = p_1 = 0.02; \quad h = -1; \quad P \doteqdot 0.98; \quad \text{a.s.n.} \doteqdot 1128$$

$$p = \frac{\log(99/98)}{\log(99/49)} = 0.0144, \quad \text{there is no nonzero } h,$$

$$P \doteqdot \frac{\log(99/2)}{\log(99 \times 49)} = 0.46 \quad \text{and the a.s.n. is approximately}$$

$$\frac{P[\log 98]^2 + (1-P)[\log(2/99)]^2}{p[\log 2]^2 + (1-p)[\log(98/99)]^2} = 2550$$

The a.s.n.'s are rather large because the test has to discriminate between 2 rather small values of p, with high sensitivity.

3. Let the first m observations be $X_1, X_2, \ldots, X_m$. The procedure is

$$\text{Accept } H_0 \text{ if } \exp\left[\sum_{i=1}^{m} \left\{(X_i - 1.36)^2 - (X_i - 1.43)^2\right\}\right] \leq \frac{0.01}{0.95}$$

$$\left(\text{that is, } \sum_{i=1}^{m} X_i \leq 1.395m - 32.53\right)$$

$$\text{Accept } H_1 \text{ if } \exp\left[\sum_{i=1}^{m} \left\{(X_i - 1.36)^2 - (X_i - 1.43)^2\right\}\right] \geq \frac{0.99}{0.05}$$

$$\left(\text{that is, } \sum_{i=1}^{m} X_i \geq 1.395m + 21.33\right)$$

Otherwise take a further observation.

4. Denoting the population mean be θ, the OC is approximately

$$\beta = \frac{(99/5)^h - 1}{(99/5)^h - (1/95)^h}$$

where $h = (2.79 - 2\theta)/(0.07)$. (If $h = 0$, $\beta = \log(99/5)/\log(99 \times 19) = 0.396$.)
The a.s.n. is approximately

$$\frac{\beta \ln(1/95) + (1 - \beta) \ln(99/5)}{0.07(2\theta - 2.79)}$$

5. Procedure is:

$$\text{Accept } H_0 \text{ if } \left(\frac{3.6}{4.1}\right)^m \exp\left[\sum_{i=1}^m \left\{\left(\frac{X_i - 3.6}{3.6}\right)^2 - \left(\frac{X_i - 4.1}{4.1}\right)^2\right\}\right] \leqslant \frac{0.01}{0.99}$$

$$\left(\text{that is, if } 0.0176 \sum_{i=1}^m X_i^2 - 0.06775 \sum_{i=1}^m X_i \leqslant 0.130m - 4.595\right)$$

$$\text{Accept } H_1 \text{ if } 0.0176 \sum_{i=1}^m X_i^2 - 0.06775 \sum_{i=1}^m X_i \geqslant 0.130m + 4.595$$

Otherwise, make a further observation.

6. With the same notation as in Exercise 1, the procedure is

$$\text{Accept } D_0 \text{ if } \frac{\binom{D_1}{X_m}\binom{N - D_1}{m - X_m}}{\binom{D_0}{X_m}\binom{N - D_0}{m - X_m}} \leqslant \frac{\alpha_1}{1 - \alpha_0}$$

$$\text{Accept } D_1 \text{ if } \frac{\binom{D_1}{X_m}\binom{N - D_1}{m - X_m}}{\binom{D_0}{X_m}\binom{N - D_0}{m - X_m}} \geqslant \frac{1 - \alpha_1}{\alpha_0}$$

Otherwise examine a further individual.
(Note that if $X_m > D_0$, $(D_0 - X_m)!$ should be interpreted here as 0 so that D_1 is accepted; and similarly if $X_m < D_1 + m - N$, D_0 is accepted.)

7. The "continuation region" in this case is

$$\frac{\alpha_1}{2(1 - \alpha_0)} < \frac{(X_m + 3 - m)(X_m + 2 - m)}{(4 - X_m)(3 - X_m)} < \frac{1 - \alpha_1}{2\alpha_0}$$

Denoting the central term by $f_m(X_m)$ we have

$$f_1(0) = \tfrac{1}{6}; \qquad f_1(1) = 1$$
$$f_2(0) = 0; \qquad f_2(1) = \tfrac{1}{3}; \qquad f_2(2) = 3$$
$$f_3(0) = f_3(1) = 0; \qquad f_3(2) = 1; \qquad f_3(3) = \infty$$

So, unless $\alpha_1/2(1-\alpha_0) > \tfrac{1}{6}$ or $(1-\alpha_1)/2\alpha_0 < 3$, the only decisions that will be reached will, in fact, be *deductive* (as opposed to *inductive*), in that they will follow directly from the observed results on the assumption that the only possible values for D are $D_0 = 2$ and $D_1 = 4$. (This situation rapidly becomes unlikely to arise as N increases.)

8. Limit for retiming is $\bar{X} \leqslant 275 - 36.84 m^{-1}$.
9. Continuation region is

$$-0.0785 < \sum_{i=1}^{m} (X_i - 8.28) < 0.0785$$

m	1	2	3	4	5
$\sum_{i=1}^{m} (X_i - 8.28)$	0.00	-0.03	-0.07	-0.06	-0.11

The procedure leads to the choice $\theta \leqslant 8.25$, at the fifth observation.

10. Let D_m denote the number of defectives among the first m articles examined. The required inequalities are:

$$\text{Acceptance: } D_m \leqslant 0.0247m - 7.18$$

$$\text{Rejection: } D_m \geqslant 0.0247m + 10.95$$

11. A sample size of at least 30 is needed.

$$OC \doteqdot \frac{1 - \left(\dfrac{5}{99}\right)^h}{95^h - \left(\dfrac{5}{99}\right)^h}$$

where h is the nonzero root (if there is one) of the equation

$$(1.5)^h p + \left(\frac{97}{98}\right)^h (1-p) = 1$$

($100p$ is the percent defective in the lot.)

$$\text{a.s.n.} \doteqdot \frac{-(OC) \times 1.9777 + [1 - (OC)] \times 1.2967}{0.17609p - 0.00445(1-p)}$$

12. (a) For a given value of N

$$\Pr\left[|\bar{X}-\xi|<D|N\right]>1-\frac{\sigma^2}{ND^2}$$

Hence

$$\Pr\left[|\bar{X}-\xi|<D\right]>1-\frac{\sigma^2}{D^2}E\left(\frac{1}{N}\right)\geqslant1-\frac{\sigma^2}{D^2M}E\left([X_1-X_2]^{-2}\right)$$

Since

$$(X_1-X_2)^{-2}=\sigma^{-2}(Y_1-Y_2)^{-2}$$

$$E\left([X_1-X_2]^{-2}\right)=\sigma^{-2}\lambda$$

and

$$\Pr\left[|\bar{X}-\xi|<D\right]>1-\frac{\lambda}{D^2M}$$

(b) The result does not depend on normality of the X's. Field of application might be extended by relaxing the condition that $E[(Y_i-Y_j)^{-2}]$ is known, or by using a different method of calculating N.

13. Assuming normality of the error distribution the sequential procedure can be based on the sequence $V_1, V_2, V_3, \ldots$ of "sums of squares" (about arithmetic mean) of the sets of 5 observations. If Joe Zilch takes all the observations (hypothesis H_0) each V will be distributed as $0.0004\chi_4^2$ (if measurements are in ohms). If not, every third V will be distributed as $\sigma^2\chi_4^2$ (where $\sigma>0.02$), while each of the other V's will still have a $0.0004\chi_4^2$ distribution. It is not known which of these three possible alternatives (H', H'', H''' corresponding to J. Zilch, Jr., taking the 9:00 A.M., NOON, or 3:00 P.M. observations, respectively) is to be considered.

A procedure based on combining

$$S\left(H_0,\alpha_0;\ H',\tfrac{1}{3}\alpha_1\right)$$
$$S\left(H_0,\alpha_0;\ H'',\tfrac{1}{3}\alpha_1\right)$$
$$S\left(H_0,\alpha_0;\ H''',\tfrac{1}{3}\alpha_1\right)$$

will be suitable for practical purposes.

14. Let $X_i=1$ if event occurs at ith trial; $X_i=0$ otherwise. Continuation region is

$$\frac{\alpha_1}{1-\alpha_0}<\prod_{i=1}^{m}\left[\frac{p_1^{X_i}(1-p_1)^{1-X_i}}{p_0^{X_i}(1-p_0)^{1-X_i}}\right]<\frac{1-\alpha_1}{\alpha_0}$$

If $\alpha_0 = p_0$ and $\alpha_1 = 1 - p_1$, then

$$\frac{\alpha_1}{1-\alpha_0} = \frac{1-p_1}{1-p_0} \quad \text{and} \quad \frac{1-\alpha_1}{\alpha_0} = \frac{p_1}{p_0}$$

But these are just the two possible values for the likelihood ratio when $m = 1$. If $X_1 = 1$, the value is p_1/p_0 and the hypothesis $p = p_1$ is accepted; if $X_1 = 0$, the value is $(1-p_1)/(1-p_0)$ and the hypothesis $p = p_0$ is accepted. In either case the procedure terminates at the first trial. Since the limiting values must be attained *exactly*, the standard formulas are exact in this case, as may be verified by direct calculation.

16. (a) This can be shown using the methods described in Chapter 5. The expected value and standard deviation can, of course, be calculated directly.

(b) The continuation region is

$$\frac{\alpha_1}{1-\alpha_0} < \prod_{j=1}^{m}\left[\frac{\sigma_0}{\sigma_1}\exp\left\{-\frac{u_j^2}{2}\left(\frac{1}{\sigma_1^2} - \frac{1}{\sigma_0^2}\right)\right\}\right] < \frac{1-\alpha_1}{\alpha_0}$$

or

$$\ln\left(\frac{\alpha_1}{1-\alpha_0}\right) + m\ln\left(\frac{\sigma_1}{\sigma_0}\right) < \frac{1}{2}(\sigma_0^{-2} - \sigma_1^{-2})\sum_{j=1}^{m} u_j^2 < \ln\left(\frac{1-\alpha_1}{\alpha_0}\right) + m\ln\left(\frac{\sigma_1}{\sigma_0}\right)$$

(c) h should satisfy the equation

$$\frac{1}{\sqrt{2\pi}\,\sigma}\int_{-\infty}^{\infty}\left[\frac{\sigma_0}{\sigma_1}\exp\left\{-\frac{1}{2}u^2\left(\frac{1}{\sigma_1^2} - \frac{1}{\sigma_0^2}\right)\right\}\right]^{h}\exp\left(-\frac{u^2}{2\sigma^2}\right)du = 1$$

This can be rewritten

$$\frac{1}{\sqrt{2\pi}\,\sigma}\left(\frac{\sigma_0}{\sigma_1}\right)^{h}\int_{-\infty}^{\infty}\exp\left\{-\frac{u^2}{2}\left[h\left(\frac{1}{\sigma_1^2} - \frac{1}{\sigma_0^2}\right) + \frac{1}{\sigma^2}\right]\right\}du = 1$$

or

$$\frac{1}{\sigma}\left(\frac{\sigma_0}{\sigma_1}\right)^{h}\left[h\left(\frac{1}{\sigma_1^2} - \frac{1}{\sigma_0^2}\right) + \frac{1}{\sigma^2}\right]^{-\frac{1}{2}} = 1$$

This is equivalent to the result shown in the exercise.

(d) a.s.n. is approximately

$$\frac{\left[1-\left(\dfrac{\alpha_1}{1-\alpha_0}\right)^h\right]\ln\left(\dfrac{\alpha_1}{1-\alpha_0}\right)+\left[\left(\dfrac{1-\alpha_1}{\alpha_0}\right)^h-1\right]\ln\left(\dfrac{1-\alpha_1}{\alpha_0}\right)}{\left[\left(\dfrac{1-\alpha_1}{\alpha_0}\right)^h-\left(\dfrac{\alpha_1}{1-\alpha_0}\right)^h\right]\left[\ln\left(\dfrac{\sigma_0}{\sigma_1}\right)-\dfrac{1}{2}\left(\dfrac{\sigma}{\sigma_1}\right)^2+\dfrac{1}{2}\left(\dfrac{\sigma}{\sigma_0}\right)^2\right]}$$

where h satisfies the equation in (c).

17. Assume observations taken one at a time $X_1, X_2, \ldots$

(a) The likelihood ratio is

$$\frac{P(X_1,\ldots,X_m|\theta_1)}{P(X_1,\ldots,X_m|\theta_0)}=\left(\frac{\theta_1}{\theta_0}\right)^{\sum_1^m X_i}e^{-m(\theta_1-\theta_0)}$$

The continuation region is (assuming $\theta_1 > \theta_0$)

$$\frac{\ln\dfrac{\beta}{1-\alpha}+m(\theta_1-\theta_0)}{\ln(\theta_1/\theta_0)}<\sum_{i=1}^m X_i<\frac{\ln\dfrac{1-\beta}{\alpha}+m(\theta_1-\theta_0)}{\ln(\theta_1/\theta_0)}$$

(b) The likelihood ratio is

$$\frac{p(X_1,\ldots,X_m|\theta_1)}{p(X_1,\ldots,X_m|\theta_0)}=\left(\frac{\theta_0}{\theta_1}\right)^m e^{-(\theta_1^{-1}-\theta_0^{-1})\sum_{i=1}^m X_i}$$

The continuation region is (assuming $\theta_1 > \theta_0$)

$$\frac{\ln\dfrac{\beta}{1-\alpha}+m\ln(\theta_1/\theta_0)}{\theta_0^{-1}-\theta_1^{-1}}<\sum_{i=1}^m X_i<\frac{\ln\dfrac{1-\beta}{\alpha}+m\ln(\theta_1/\theta_0)}{\theta_0^{-1}-\theta_1^{-1}}$$

18. (a) h satisfies

$$\sum_{j=0}^{\infty}\frac{\theta^j}{j!}e^{-\theta}\cdot\left(\frac{\theta_1}{\theta_0}\right)^{hj}\cdot e^{h(\theta_0-\theta_1)}=1$$

that is,

$$e^{\theta(\theta_1/\theta_0)^h-\theta+h(\theta_0-\theta_1)}=1$$

Hence

$$\theta\left(\frac{\theta_1}{\theta_0}\right)^h-\theta+h(\theta_0-\theta_1)=0$$

or

$$\left(\frac{\theta_1}{\theta_0}\right)^h = 1 + h\left(\frac{\theta_0}{\theta}\right)\left(\frac{\theta_1}{\theta_0} - 1\right)$$

(b) h satisfies

$$\theta^{-1}\int_0^\infty e^{-x/\theta} \cdot \left(\frac{\theta_0}{\theta_1}\right)^h e^{-hx(\theta_1^{-1}-\theta_0^{-1})}dx = 1$$

or

$$\theta^{-1}\left(\frac{\theta_0}{\theta_1}\right)^h \left(\theta^{-1} + h\left[\theta_1^{-1} - \theta_0^{-1}\right]\right)^{-1} = 1$$

that is,

$$\left(\frac{\theta_0}{\theta_1}\right)^h = 1 + h\left(\frac{\theta}{\theta_0}\right)\left[\frac{\theta_0}{\theta_1} - 1\right]$$

In (a),

$$y = \frac{\theta_1}{\theta_0}; \qquad x = \frac{\theta_0}{\theta}$$

In (b),

$$y = \frac{\theta_0}{\theta_1}; \qquad x = \frac{\theta}{\theta_0}$$

21. Let

$$p_1 = \Pr[X_i > \theta_0'' | \theta_0''] = \tfrac{1}{2}$$

$$p_2 = \Pr[\theta_0 < X_i < \theta_0'' | \theta_0''] = \frac{1}{\sqrt{2\pi}}\int_{-1.96}^0 e^{-\frac{1}{2}u^2}du = 0.475$$

$$p_3 = \Pr[\theta_0' < X_i < \theta_0 | \theta_0''] = \frac{1}{\sqrt{2\pi}}\int_{-3.92}^{-1.96} e^{-\frac{1}{2}u^2} \doteq 0.025$$

$$p_4 = \Pr[X_i < \theta_0' | \theta_0''] \doteq 0$$

(actually $p_4 = 0.000044$ and $p_3 = 0.024956$).
We then form the table below:

n	PROBABILITY OF NO DECISION AFTER n SAMPLES	PROBABILITY, AT nTH SAMPLE, OF NOT ACCEPTING $\theta = \theta_0''$
1	$p_2 + p_3$	p_4
2	$2p_2 p_3$	$p_4(p_2 + p_3) + p_3^2$
3	$p_2 p_3(p_2 + p_3)$	$p_4(2p_2 p_3) + p_3^2 p_2$
4	$2p_2^2 p_3^2$	$p_4[p_2 p_3(p_2 + p_3)] + p_3^2(p_2 p_3)$

and so on. [Note that for no decision to be reached observations must be alternately in the intervals $(\theta_0', \theta_0)(\theta_0, \theta_0'')$.] Hence probability of *not* accepting

$\theta = \theta_0''$ is

$$p_4\Big[1+(p_2+p_3)\{1+p_2p_3+(p_2p_3)^2+\cdots\}+2\{p_2p_3+(p_2p_3)^2+\cdots\}\Big]$$

$$+p_3^2\{1+p_2+p_2p_3+p_2^2p_3+p_2^2p_3^2+\cdots\}$$

$$=p_4\left[1+\frac{p_2+p_3+2p_2p_3}{1-p_2p_3}\right]+p_3^2\cdot\frac{1+p_2}{1-p_2p_3}$$

Putting in the (approximate) numerical values of the p's, we obtain

$$(0.025)^2\cdot\frac{1.475}{0.988}\doteqdot 0.001$$

Also

$$\text{a.s.n.}\doteqdot 1+(p_2+p_3)+2p_2p_3+p_2p_3(p_2+p_3)+2p_2^2p_3^2+\cdots$$

$$=1+\frac{p_2+p_3+2p_2p_3}{1-p_2p_3}$$

Putting in the numerical values of the p's gives

$$\text{a.s.n.}=1+\frac{0.5+0.02375}{1-0.011875}$$

$$=1+\frac{0.52375}{0.988125}=1.53$$

22. The argument in the solution to Exercise 21 depends only on the values of the p's, not on the form of distribution of the random variables representing the observations.

In particular the a.s.n. is a maximum, for a given value of p_2, if $p_3=1-p_2$. So

$$\text{maximum a.s.n.}=1+\frac{1+2p_2(1-p_2)}{1-p_2(1-p_2)}$$

And since $p_2(1-p_2)\leqslant\frac{1}{4}$

$$\text{maximum a.s.n.}\leqslant 1+\frac{1+\frac{1}{2}}{1-\frac{1}{4}}=3$$

A drawback of this procedure is that θ_0' and θ_0'' *must* differ from θ_0 by 1.96σ whether this is a convenient choice or not.

23. Assume that H_0 and H_1 each specify the observations to be represented by independent, identically distributed random variables.

Using standard formulas, the average sample number is

$$\omega_0 \cdot \frac{\alpha_0 \ln\left(\dfrac{1-\alpha_1}{\alpha_0}\right) + (1-\alpha_0)\ln\left(\dfrac{\alpha_1}{1-\alpha_0}\right)}{E(Z|H_0)}$$

$$+ \omega_1 \cdot \frac{(1-\alpha_1)\ln\left(\dfrac{1-\alpha_1}{\alpha_0}\right) + \alpha_1 \ln\left(\dfrac{\alpha_1}{1-\alpha_0}\right)}{E(Z|H_1)}$$

[where $Z = \ln$ (likelihood ratio, for a single observation, of H_1 to H_0)].
α_0, α_1 have to be chosen to minimize this quantity, subject to the conditions
$\omega_0\alpha_0 + \omega_1\alpha_1 = 1$, $0 \leq \alpha_0 < 1$, and $0 < \alpha_1 < 1$. Putting $\alpha_0 = (\alpha - \omega_1\alpha_1)/\omega_0$, we obtain
a function of α_1, to be minimized with respect to α_1.

24. If $E(Z|H_1) = -E(Z|H_0)$ (>0) then we have to minimize

$$(\omega_1 - \alpha)\ln\left[\frac{\omega_0(1-\alpha_1)}{\alpha - \omega_1\alpha_1}\right] + (\alpha - \omega_0)\ln\left[\frac{\omega_0\alpha_1}{\omega_0 - \alpha + \omega_1\alpha_1}\right]$$

with respect to α_1.
Differentiating with respect to α_1, and equating the result to 0, we obtain

$$\alpha_1(1-\alpha_0)(\omega_1 - \alpha)^2 = \alpha_0(1-\alpha_1)(\omega_0 - \alpha)^2$$

Putting $\alpha = \omega_0\alpha_0 + \omega_1\alpha_1$, this equation becomes

$$\omega_1^2\alpha_1(1-\alpha_0)(1-\alpha_1)^2 - 2\omega_0\omega_1\alpha_0\alpha_1(1-\alpha_0)(1-\alpha_1) + \omega_0^2\alpha_0^2\alpha_1(1-\alpha_0)$$

$$= \omega_1^2\alpha_0\alpha_1^2(1-\alpha_1) - 2\omega_0\omega_1\alpha_0\alpha_1(1-\alpha_0)(1-\alpha_1) + \omega_0^2\alpha_0(1-\alpha_0)^2(1-\alpha_1)$$

or

$$\omega_1^2\alpha_1(1-\alpha_1)(1-\alpha_0-\alpha_1) = \omega_0^2\alpha_0(1-\alpha_0)(1-\alpha_0-\alpha_1)$$

It is necessary to check that $\alpha_0 + \alpha_1 \neq 1$ (and also that the solution does indeed
provide a minimum) to obtain the required result.
The condition $E(Z|H_1) = -E(Z|H_0)$ is satisfied in the case stated in the
exercise. The result also holds in other cases.

25. The methods used are similar to those in Exercises 23 and 24. The function to
be minimized is now

$$\omega_0 c_0 \alpha_0 + \omega_1 c_1 \alpha_1 + c \left[\omega_0 \cdot \frac{\alpha_0 \ln\left(\dfrac{1-\alpha_1}{\alpha_0}\right) + (1-\alpha_0)\ln\left(\dfrac{\alpha_1}{1-\alpha_0}\right)}{E(Z|H_1)} \right.$$

$$\left. + \omega_1 \cdot \frac{(1-\alpha_1)\ln\left(\dfrac{1-\alpha_1}{\alpha_0}\right) + \alpha_1 \ln\left(\dfrac{\alpha_1}{1-\alpha_0}\right)}{E(Z|H_1)} \right]$$

Noting that

$$\omega_0 c_0 \alpha_0 + \omega_1 c_1 \alpha_1 = c_0(\alpha - \omega_1 \alpha_1) + \omega_1 c_1 \alpha_1$$

$$= c_0 \alpha + \omega_1 (c_1 - c_0) \alpha_1$$

we find that under the conditions of Exercise 18, the derivative of the loss function with respect to α_1 is equal to 0 when

$$\omega_1(c_1 - c_0) + \frac{1}{\omega_0 E(Z|H_1)} \left[\frac{(\omega_1 - \alpha)^2}{\alpha_0(1-\alpha_1)} - \frac{(\omega_0 - \alpha)^2}{\alpha_1(1-\alpha_0)} \right] = 0$$

or

$$\omega_1(c_1 - c_0) + \frac{(1-\alpha_0-\alpha_1)\left[\omega_1^2 \alpha_1(1-\alpha_1) - \omega_0^2 \alpha_0(1-\alpha_0)\right]}{\alpha_0 \alpha_1 (1-\alpha_0)(1-\alpha_1)\omega_0 E(Z|H_1)} = 0$$

27. Consider the situation after the first Z is observed. If $Z \leqslant a$ observations cease with (i) true; the probability of this is $\Pr[Z \leqslant a]$. If $a < Z < b$, then we are starting a new sequence with a, b replaced by $a - Z, b - Z$. The conditional probability that (i) is true when observations cease is $P(a - Z, b - Z)$. The first result follows. The second result is obtained by direct algebra.
The formula $P(a, b) = (e^{bh} - 1)/(e^{bh} - e^{ah})$ is the approximate formula used in Wald s.p.r.t.'s. The result shows that a different value of h would be needed for different values of a and b, so the formula cannot be exact for all a and b.

Chapter 17

1. The two factories have the same variance–covariance matrices

$$10^{-4} \times \begin{pmatrix} 9 & 15.75 & 11.7 \\ 15.75 & 49 & 37.8 \\ 11.7 & 37.8 & 36 \end{pmatrix}$$

for measurements of X_1, X_2, and X_3. The inverse matrix is

$$10^4 \times \begin{pmatrix} 0.2559 & -0.0952 & 0.0168 \\ -0.0952 & 0.1429 & -0.1190 \\ 0.0168 & -0.1190 & 0.1473 \end{pmatrix}$$

In the linear discriminant $\lambda_1 \overline{X}_1 + \lambda_2 \overline{X}_2 + \lambda_3 \overline{X}_3$

$$\lambda_1 : \lambda_2 : \lambda_3 = -0.0123 : 0.0079 : -0.0045$$

Taking $F = 79\overline{X}_2 - 123\overline{X}_1 - 45\overline{X}_3$, we have

$$E(F|\text{Factory 1}) = -61.88; \quad E(F|\text{Factory 2}) = -68.83; \quad \text{var}(F) = 6.955n^{-1},$$

where $n =$ number of samples from which means $\overline{X}_1, \overline{X}_2, \overline{X}_3$ were calculated. Discriminant assigns to Factory 1 if $F > -65.355$. Probability of correct assignment is

$$\Phi\left(3.475\sqrt{\frac{n}{6.955}}\right)$$

Probability is greater than 0.99 if

$$\frac{3.475}{2.637}\sqrt{n} > 2.3263$$

Least value of n is 4.

2. (a) Optimum discrimination is obtained by assigning to Factory 1 if $F > F_0$, where F_0 maximizes

$$\tfrac{2}{3}\left\{1 - \Phi\left(0.379(F_0 + 61.88)\sqrt{n}\right)\right\} + \tfrac{1}{3}\Phi\left(0.379(F_0 + 68.83)\sqrt{n}\right)$$

F_0 then satisfies the equation

$$\tfrac{1}{2}n(0.379)^2\left[(F_0 + 61.88)^2 - (F_0 + 68.83)^2\right] = \ln 2$$

so

$$F_0 = -65.355 - \frac{0.693}{\sqrt{n}}$$

Probability of correct decision is

$$P_n = \tfrac{2}{3}\left\{1 - \Phi(-1.318\sqrt{n} - 0.263)\right\} + \tfrac{1}{3}\Phi(1.318\sqrt{n} - 0.263)$$

In this case it is found that a sample of size 3 gives $P_3 \doteqdot 0.99$.

(b)

MEASUREMENTS	DISCRIMINATOR	VARIANCE $(= n^{-1}[\text{DIFFERENCE IN MEANS}])$
X_1, X_2	$44\overline{X}_2 - 119\overline{X}_1$	$5.64n^{-1}$
X_1, X_3	$20\overline{X}_3 - 71\overline{X}_1$	$2.64n^{-1}$
X_2, X_3	$33\overline{X}_2 - 37\overline{X}_3$	$1.03n^{-1}$

X_1 and X_2 are the best pair of measurements to take. To obtain at least 99% chance of correct decision

$$\tfrac{1}{2}\sqrt{5.64n} > 2.3263$$

The least possible value for n is again 4 (as in Exercise 1). (If either of the other pairs of measurements are used considerably larger values of n are needed.)

3. Let $F_i = 79X_{2i} - 123X_{1i} - 45X_{3i}$. Then

$$F_i|\text{Factory 1} \quad \frown N(-61.88, 6.955)$$

$$F_i|\text{Factory 2} \quad \frown N(-68.83, 6.955)$$

From (16.36) we get the continuation region

$$\frac{6.955}{7.95} \ln\left(\frac{\alpha_1}{1-\alpha_0}\right) - 65.355m < \sum_{i=1}^{m} F_i$$

$$< \frac{6.955}{7.95} \ln\left(\frac{1-\alpha_1}{\alpha_0}\right) - 65.355m$$

or, more conveniently,

$$0.875 \ln\left(\frac{\alpha_1}{1-\alpha_0}\right) < \sum_{i=1}^{m} (F_i + 65.355) < 0.875 \ln\left(\frac{1-\alpha_1}{\alpha_0}\right)$$

[Note the average sample size would be so small (for common values of α_0, α_1) that the usual s.p.r.t. approximations would not be very exact.]

4. The expected cost per set of n units is

$$n[(\text{cost of measurements on each unit})$$

$$+ C \times (\text{probability of misclassification})]$$

The expected cost *per unit* is shown in the following table.

VARIABLE(S) USED	EXPECTED COST PER UNIT	
X_1	$2 + Cg(0.667\sqrt{n})$	
X_2	$1 + Cg(0.143\sqrt{n})$	
X_3	$4 + Cg(0.083\sqrt{n})$	
X_1, X_2	$3 + Cg(1.172\sqrt{n})$	(using solution
X_1, X_3	$6 + Cg(0.812\sqrt{n})$	to Exercise 1)
X_2, X_3	$5 + Cg(0.507\sqrt{n})$	
X_1, X_2, X_3	$7 + Cg(1.319\sqrt{n})$	(using solution to Exercise 2)

In this table $g(y) = 1 - \Phi(y)$.

Because $a + Cg(b) > a' + Cg(b')$ if $a > a'$ and $b < b'$, there is no need to consider X_3 alone, or X_1 and X_3, or X_2 and X_3.

(*i*) With $n = 1$, expected costs per unit are

$$
\begin{array}{ll}
X_1 & : \quad 2 + 0.2525\,C \\
X_2 & : \quad 1 + 0.4432\,C \\
X_1, X_2 & : \quad 3 + 0.1206\,C \\
X_1, X_2, X_3 & : \quad 7 + 0.0936\,C
\end{array}
$$

The least expected cost per unit is obtained by using

$$
\begin{array}{ll}
X_2 & \text{if } C < 5.2 \\
X_1 & \text{if } 5.2 < C < 7.6 \\
X_1, X_2 & \text{if } 7.6 < C < 148.1 \\
X_1, X_2, X_3 & \text{if } C > 148.1
\end{array}
$$

The answer required is therefore 148.1.

(*ii*) The calculations follow similar lines for other values of n. As n increases the values of g decreases and the "cross-over" values of C increase.

(It is interesting to note the considerable increase in discriminatory power attained by using X_2 in addition to X_1. This arises from the fact that the differences in expected values for the 2 variables are of opposite signs, whereas the correlation between the variables is positive.)

5. The theory needed to solve this problem is as follows (assuming correlations of Exercise 1): Let Factory 2 standard deviations be $\sqrt{k}$ times those of Factory 1. Then (denoting observed characters by $X_1, X_2, \ldots, X_q$) the likelihood ratio (with Factory 1 likelihood as numerator) is

$$
\left\{ \frac{|\mathbf{C}|^{1/2}}{(2\pi)^{(1/2)q}} \exp\left[-\frac{1}{2} \sum_{i=1}^{q} \sum_{j=1}^{q} c_{ij}(x_i - \xi_{i1})(x_j - \xi_{j1}) \right] \right\}
$$

$$
\times \left\{ \frac{|\mathbf{C}|^{1/2}}{(2\pi k)^{(1/2)q}} \exp\left[-\frac{1}{2k} \sum_{i=1}^{q} \sum_{j=1}^{q} c_{ij}(x_i - \xi_{i2})(x_j - \xi_{j2}) \right] \right\}^{-1}
$$

where $\mathbf{C} = (c_{ij})$ is the inverse of the variance–covariance matrix for Factory 1, and $(\xi_{1l}, \xi_{2l}, \ldots, \xi_{ql})$ are the expected values of $X_1, X_2, \ldots, X_q$, respectively, for Factory l ($l = 1, 2$). The discriminant function should be a monotonic function of

$$
\sum_{i=1}^{q} \sum_{j=1}^{q} c_{ij}\left\{ (X_i - \xi_{i1})(X_j - \xi_{j1}) - k^{-1}(X_i - \xi_{i2})(X_j - \xi_{j2}) \right\}
$$

$$
= k^{-1}(k-1) \sum_{i=1}^{q} \sum_{j=1}^{q} c_{ij}(X_i - \xi_i')(X_j - \xi_j') - \sum_{i=1}^{q} \sum_{j=1}^{q} c_{ij}\delta_{ij}
$$

where

$$\xi_i' = k(k-1)^{-1}(\xi_{i1} - k^{-1}\xi_{i2});$$

$$\delta_{ij} = (k-1)^{-1}(\xi_{i1} - \xi_{i2})(\xi_{j1} - \xi_{ji})$$

A convenient form of discriminant function to use is

$$F = \sum_{i=1}^{q} \sum_{j=1}^{q} c_{ij}(X_i - \xi_i')(X_j - \xi_j')$$

If no knowledge of prior probabilities is available, a guaranteed probability of error can be obtained by choosing F_0 so that $\Pr[F < F_0 | 2] = \Pr[F > F_0 | 1]$ and (if $k > 1$) assigning to Factory 1 if $F < F_0$, to Factory 2 otherwise (if $k < 1$, assigning to Factory 1 if $F > F_0$, otherwise to Factory 2). If Factory 1 is the producer, F is distributed as noncentral χ^2 with q degrees of freedom and noncentrality parameter

$$\Delta = \sum_{i=1}^{q} \sum_{j=1}^{q} c_{ij}(\xi_{i1} - \xi_i')(\xi_{j1} - \xi_j')$$

$$= (k-1)^{-1} \sum_{i=1}^{q} \sum_{j=1}^{q} c_{ij}\delta_{ij}$$

If Factory 2 is the producer, F is distributed as $k \times$(noncentral χ^2 with q degrees of freedom and noncentrality parameter $k\Delta$).
So the critical value, F_0, is chosen so that

$$\Pr[\chi_q'^2(\Delta) > F_0] = \Pr[\chi_q'^2(k\Delta) < k^{-1}F_0]$$

(Note that the same equation is used whether $k < 1$ or $k > 1$). Tables of the noncentral χ^2 distribution (G. E. Haynam, Z. Govindarajulu and F. C. Leone, *Selected Tables in Mathematical Statistics*, (H. L. Harter and D. B. Owen, eds), Vol 1, American Mathematical Society, Providence, R.I., 1970, pp. 1–78) may be used to obtain the value of F_0. Approximate solutions may be obtained by using approximations to the noncentral χ^2 distribution. If prior probabilities ω_1 and ω_2 ($=1-\omega_1$) for Factories 1 and 2, respectively, are known, then the optimum discriminant can be determined. This assigns to Factory 1 if

$$k^{(1/2)q}\exp\left[-\frac{1}{2}\left\{k^{-1}(k-1)\sum_{i=1}^{q}\sum_{j=1}^{q}C_{ij}(x_i-\xi_i')(x_j-\xi_j')-(k-1)\Delta\right\}\right] > \frac{\omega_2}{\omega_1}$$

This is equivalent to

$$F \gtrless \frac{2k}{k-1}\ln\left(\frac{\omega_1 k^{(1/2)q}}{\omega_2}\right) + k\Delta$$

according as $k \gtrless 1$. It is interesting to note that if $\omega_1 = \omega_2 = \frac{1}{2}$ the limiting value of F is $(k-1)^{-1}kq\ln k + k\Delta$. If $k < 1$ the probability of error is

$$\omega_1 \Pr[\chi_q'^2(\Delta) < F_0'] + (1-\omega_1)\Pr[\chi_q'^2(k\Delta) > k^{-1}F_0']$$

with

$$F_0' = \frac{2k}{k-1} \ln\left(\frac{\omega_1 k^{(1/2)q}}{\omega_2} \right) + k\Delta$$

6. (a) The "best discriminator" will be based on $F(=79X_2 - 123X_1 - 45X_3)$ of Exercise 1. In default of any additional information the procedure to be used would be the following:

Assign to Factory 1 if $F > -65.355$;
otherwise assign to Factory 2.

(b) Assuming this procedure used, and denoting the actual (but unknown) proportion of samples from Factory 1 by p, the expected value of the proportion *assigned* to Factory 1 is

$$p \cdot \Pr[F > -65.355 | \text{Factory 1}]$$

$$+ (1-p)\Pr[F > -65.355 | \text{Factory 2}] = 0.0938 + 0.8124p$$

This is not, in general, equal to p, and so the proportion $80/100$ is not, in general, an unbiased estimate of p.

(c) $\hat{p} = (0.80 - 0.0938)/0.8124 = 0.869$ is an unbiased estimate of p. Using this estimate as if it were the true value of p, optimum discrimination is obtained by assigning to Factory 1 if $F > F_1$ where

$$0.869 \exp\left[-\frac{1}{2} \frac{(F_1 + 61.88)^2}{6.95} \right] = 0.131 \exp\left[-\frac{1}{2} \frac{(F_1 + 68.83)^2}{6.95} \right].$$

This gives

$$F_1 = -65.355 - \ln\left(\frac{0.869}{0.131} \right)$$

$$= -67.247$$

7. (a) The discriminator F, obtained in the solution of Exercise 5, should be used (with $k = \frac{1}{4}$, of course).

(b) If the criterion ($F > F_0$ indicating assignment to Factory 2), giving a guaranteed probability of error (P), is used, the expected proportion of assignments to Factory 1 is

$$\omega(1-P) + (1-\omega)P = \omega + P - 2\omega P$$

where $\omega =$ proportion of product from Factory 1. Equating 0.80 to this value, we obtain as an estimator of ω

$$\hat{\omega} = \frac{0.80 - P}{1 - 2P}$$

(c) The method of the last part of the answer to Exercise 5, putting $\omega_1 = \hat{\omega}$ (and $\omega_2 = 1 - \hat{\omega}$), may be used.

8. Comparing

$$10^4\big[0.2559(1.39-1.37)^2 - 0.1904(1.39-1.37)(1.75-1.72)$$

$$+\cdots -0.2380(1.75-1.72)(0.61-0.65)\big]=6.21$$

with the χ^2 distribution with 3 degrees of freedom, there is no strong evidence of departure from the expected values for Factory 1 ($\chi^2_{3,0.90}=6.25$). (Note that this is just a test of expected values, not specifying any particular alternative, such as Factory 2.)

9. n should be chosen such that $6.21n>\chi^2_{3,0.99}(=11.34)$. Least value of n is 2. (Factory 2 is not specified as alternative in this test.)

10. The inequality $T^2 \leqslant T_0^2$ can be regarded as defining a region in the "space" of expected values. The confidence region is ellipsoidal.

11. (a) $X_{ti}=\alpha+U_t+Z_{ti}=\alpha+Y_{ti}$ ($i=1,2,\ldots,n_t$). The joint distribution of the Y_{ti}'s is multinormal with expected values $(0,0,\ldots,0)$ and variance–covariance matrix

$$\begin{bmatrix} \sigma^2+\sigma'^2 & \sigma'^2 & \cdots & \sigma'^2 \\ \sigma'^2 & \sigma^2+\sigma'^2 & \cdots & \sigma'^2 \\ \cdot & \cdot & & \cdot \\ \cdot & \cdot & & \cdot \\ \sigma'^2 & \sigma'^2 & \cdots & \sigma^2+\sigma'^2 \end{bmatrix}$$

where $\sigma^2=\operatorname{var}(Z_{ti})$; $\sigma'^2=\operatorname{var}(U_t)$ for all t and i.

(b) The hypothesis specifies that the variance–covariance matrix is $\sigma^2\times\mathbf{I}$, where

$$\mathbf{I}=\begin{bmatrix} 1 & 0 & \cdots & 0 \\ 0 & 1 & \cdots & 0 \\ \cdot & \cdot & & \cdot \\ \cdot & \cdot & & \cdot \\ 0 & 0 & \cdots & 1 \end{bmatrix}$$

(c) The joint probability density function of $X_{11},\ldots,X_{kn_k}$ is

$$(2\pi)^{-(1/2)/N}(\sigma^2)^{-(1/2)(N-k)}\prod_{t=1}^{k}(\sigma^2+n_t\sigma'^2)^{-1/2}$$

$$\cdot\exp\left[-\frac{1}{2\sigma^2}\sum_{t=1}^{k}\sum_{i=1}^{n_t}(x_{ti}-\bar{x}_{t.})^2 - \sum_{t=1}^{k}\frac{n_t(\bar{x}_{t.}-\alpha)^2}{2(\sigma^2+n_t\sigma'^2)}\right]$$

If $\sigma'^2=0$, the maximum likelihood estimators of α and σ^2 are $\bar{X}_{..}$ and

$$N^{-1}\sum_{t=1}^{k}\sum_{i=1}^{n_t}(X_{ti}-\bar{X}_{..})^2=\hat{\sigma}^2$$

and the maximized likelihood is $(2\pi\hat{\sigma}^2)^{-(1/2)N}e^{-(1/2)N}$. If σ'^2 is not assumed equal to 0, the maximum likelihood estimators $\hat{\alpha}$, $\hat{\sigma}$, $\hat{\sigma}'$ satisfy the equations

$$\sum_{t=1}^{k} \frac{n_t(\bar{X}_{t.}-\hat{\alpha})}{\widehat{\sigma^2+n_t\sigma'^2}} = 0$$

$$\frac{N-k}{\hat{\sigma}^2} + \sum_{t=1}^{k} \frac{1}{\widehat{\hat{\sigma}^2+n_t\sigma'^2}} = \frac{1}{\hat{\sigma}} \sum_{t=1}^{k} \sum_{i=1}^{n_t} \left(X_{ti}-\bar{X}_{t.}\right)^2 + \sum_{t=1}^{k} \frac{n_t(\bar{X}_{t.}-\hat{\alpha})^2}{\left(\widehat{\hat{\sigma}^2+n_t\sigma'^2}\right)^2}$$

$$\sum_{t=1}^{k} \frac{n_t}{\widehat{\hat{\sigma}^2+n_t\sigma'^2}} = \sum_{t=1}^{k} \frac{n_t^2(\bar{X}_{t.}-\hat{\alpha})^2}{\left(\widehat{\hat{\sigma}^2+n_t\sigma'^2}\right)^2}$$

These equations are complicated to solve, in general, but if $n_1=n_2=\cdots=n_k=n$ the simple solutions

$$\hat{\alpha}=\bar{X}_{..}; \qquad \hat{\sigma}^2=(N-k)^{-1}\sum_{t=1}^{k}\sum_{i=1}^{m}\left(X_{ti}-\bar{X}_{t.}\right)^2$$

$$\widehat{\hat{\sigma}^2+n\sigma'^2}=(nk^{-1})\sum_{t=1}^{k}\left(\bar{X}_{t.}-\bar{X}_{..}\right)^2$$

are obtained, provided the value of $\hat{\sigma}'^2$ is positive. (Otherwise $\hat{\sigma}'$ is taken as 0.) The maximized likelihood is

$$(2\pi)^{-(1/2)N}(\hat{\sigma}^2)^{-(1/2)(N-k)}(\widehat{\hat{\sigma}^2+n\sigma'^2})^{-(1/2)k}e^{-(1/2)N}$$

The likelihood ratio is

$$\left(\frac{\hat{\sigma}^2}{\hat{\hat{\sigma}}^2}\right)^{(1/2)(N-k)} \cdot \left(\frac{\widehat{\hat{\sigma}^2+n\sigma'^2}}{\hat{\hat{\sigma}}^2}\right)^{(1/2)k}$$

Noting that

$$N\hat{\hat{\sigma}}^2=(N-k)\hat{\sigma}^2+k\widehat{\sigma'^2}$$

we see that the likelihood ratio is a function of $\widehat{\sigma'^2}/\hat{\sigma}^2$, or equivalently, of mean square ratio in the analysis of variance. Remembering that $\hat{\sigma}'=0$ if

$$(nk^{-1})\sum_{t=1}^{k}\left(\bar{X}_{t.}-\bar{X}_{..}\right)^2<(N-k)^{-1}\sum_{t=1}^{k}\sum_{i=1}^{n}\left(X_{ti}-\bar{X}_{t.}\right)^2$$

we can see that the likelihood ratio approach gives effectively the standard analysis of variance test in this case.

12. A linear discriminant function is a linear function of characters measured on an individual, the value of which is used to assign the individual to 1 of 2 (or more) populations.

The probability of correct classification (for individuals chosen from either population) is

$$\Phi\left(\frac{11.99-10.71}{1.10}\right)=\Phi(1.164)=0.87772$$

Hence the expected number misclassified (among 500 measured) is $500\times 0.12228=61.14$.

If the proportion of individuals in Π_1 is ω_1, then the expected number classified as belonging to Π_1, out of 500 measured, is

$$500\left[0.87772\omega_1+0.12228(1-\omega_1)\right]=377.72\omega_1+61.14$$

Equating this to the observed number (380) assigned to Π_1, we get the value

$$\hat{\omega}_1=\frac{380-61.14}{377.72}=0.84417$$

If this were the true value of ω_1, the expected proportion of misclassifications would be minimized by choosing F_0 as critical value for F, where

$$\hat{\omega}_1\exp\left[-\frac{1}{2}\left(\frac{F_0-10.71}{1.10}\right)^2\right]=(1-\hat{\omega}_1)\exp\left[-\frac{1}{2}\left(\frac{F_0-13.27}{1.10}\right)^2\right]$$

whence

$$F_0=\frac{2(1.10)^2\ln\left(\frac{0.87772}{0.12228}\right)+(13.27)^2-(10.71)^2}{2(13.27-10.71)}$$

$$=\frac{2\ln\left(\frac{0.87772}{0.12228}\right)}{2.56}+11.99=12.92$$

13. Note that the answers to (a) and (b) are the same. They are unaffected by the method of subdivision among the non-Zilch factories.

Let ξ_0 denote the vector of means for the Zilch factory; and let ξ_1, ξ_2 denote the vectors of means for factories 1 and 2, respectively. Also let C denote the inverse of the common variance–covariance matrix.

Optimal assignment is obtained by choosing the source for which $(X-\xi_i)\times C(X-\xi_i)'$ is smallest. The individual will be assigned to Zilch if

$$(\xi_0-\xi_j)CX'>\tfrac{1}{2}(\xi_0 C\xi_0'-\xi_j C\xi_j')\qquad\text{for}\qquad j=1,2.$$

If the individual is, indeed, produced by Zilch the probability of correct assignment is

$$P = \Pr\left[(G_1 > \tfrac{1}{2}(\xi_0 - \xi_1)C(\xi_0 - \xi_1)') \cap (G_2 > \tfrac{1}{2}(\xi_0 - \xi_2)C(\xi_0 - \xi_2)') \right]$$

where

$$G_j = (\xi_0 - \xi_j)CY' \qquad (j = 1, 2)$$

and Y has a multinormal distribution with expected value 0 and variance–co-variance matrix C^{-1}. The joint distribution of G_1 and G_2 is bivariate normal with expected values $0, 0$; variances $(\xi_0 - \xi_1)C(\xi_0 - \xi_1)'$, $(\xi_0 - \xi_2)'$; and covariance $(\xi_3 - \xi_1)C(\xi_0 - \xi_2)'$.

For a given value of $E(X_1)$, the probability P can be evaluated from tables of the bivariate normal integral. Limits between which $E(X_1)$ should lie for $P < 0.40$ can then be found by trial and error. [It is clear that P is a minimum when $E(X_1) = 1.37$ or 1.41.]

14. Let X_j denote the number of occurrences of the jth word in 2000θ words of manuscript, and let $m_j(A)$, $m_j(B)$ represent the expected number of oc-currences in 2000 words for A, B, respectively. Then X_j has a Poisson distribution with expected value $\theta m_j^{(A)}$ or $\theta m_j^{(B)}$ and the optimal discriminant is of the form:

$$\text{if } \sum_{j=1}^{k} X_j \ln\left(\frac{m_j^{(B)}}{m_j^{(A)}} \right) > \phi, \qquad \text{assign to } B$$

15. If $k = 8$ it is reasonable to assume

$$\sum_{j=1}^{8} X_j \ln\left(\frac{m_j^{(B)}}{m_j^{(A)}} \right)$$

approximately normally distributed with expected value

$$\theta \sum_{j=1}^{8} m_j \ln\left(\frac{m_j^{(B)}}{m_j^{(A)}} \right)$$

and standard deviation

$$\left\{ \theta \sum_{j=1}^{8} m_j \left[\ln\left(\frac{m_j^{(B)}}{m_j^{(A)}} \right) \right]^2 \right\}^{1/2}$$

where $m_j = m_j^{(A)}$ or $m_j^{(B)}$, according to the population from which the measured individual originates.

The present calculations can be carried out in terms of common (base 10) rather than natural (base e) logarithms, as will now become apparent. Using the notation $c = \ln 10$ we have

	POPULATION A	POPULATION B
EXPECTED VALUE	$-6.5465c\theta$	$-2.6442c\theta$
STANDARD DEVIATION	$1.4233c\sqrt{\theta}$	$1.1971c\sqrt{\theta}$

The rule will be

assign to B if
$$0.2050X_1 + 0.0354X_2 - 0.2241X_3 - 0.1651X_4$$
$$+ 0.0359X_5 - 0.1918X_6 - 0.3323X_7$$
$$+ 0.0345X_8 > c\phi$$

(in notation of the solution to Exercise 14). The probability of correct assignment is approximately

$$\Phi\left(\frac{\phi + 6.5465\theta}{1.4233\sqrt{\theta}}\right) \qquad \text{for individuals from population } A$$

$$1 - \Phi\left(\frac{\phi + 2.6442\theta}{1.1971\sqrt{\theta}}\right) \qquad \text{for individuals from population } B$$

(Note that these values do not depend on c.)
  These probabilities are equal if

$$\frac{\phi + 6.5465\theta}{1.4233\sqrt{\theta}} = -\frac{\phi + 2.6442\theta}{1.1971\sqrt{\theta}}$$

whence

$$\phi = -\frac{(2.6442 \times 1.4233) + (6.5465 \times 1.1971)}{1.4233 + 1.1971}\theta = -4.4269\theta$$

The guaranteed probability of correct assignment is $\Phi(1.4892\sqrt{\theta}\,)$. To make this 95% we need

$$\theta = \left(\frac{1.6449}{1.4892}\right)^2 = \underline{1.22}$$

(so that a sample of about 2440 words is needed).

Chapter 18

1. $y = 72.5 - 0.187x_1 + 1.187x_2$

ANOVA TABLE

SOURCE	S.S.	D.F.	M.S.
Due to regression	73.2499	2	36.6249
Remainder	368.6249	5	73.7249
Total	441.8749	7	

2. $y = 74.706 + 10.765x_1 + 5.471x_2$

ANOVA TABLE

SOURCE	S.S.	D.F.	M.S.	M.S. RATIO
Due to regression	800.6806	2	400.8403	20.775**
Remainder	77.1764	4	19.2941	
Total	878.8571	6		

3. $y = 25 + 1.4x_1 + 3.6x_2 + 6.1x_3 + 3.4x_4$

SOURCE	S.S.	D.F.	M.S.
$\Sigma(y - \bar{y})^2$	8026	12	
Due to B_0	7500	1	
Due to B_1	15.125	1	
Due to B_2	105.125	1	
Due to B_3	300.125	1	
Due to B_4	91.125	1	
Residual sum of squares	14.500	7	
Lack of Fit sum of squares	5.750	4	1.4
Error sum of squares	8.750	3	2.9

5. $y = 24.0401 - 4.7420x_1 - 1.1885x_2 + 0.2285x_3 - 5.4262x_1^2 - 0.8837x_2^2 - 5.1593x_3^2 - 1.6738x_1x_2 - 1.1462x_1x_3 - 0.9722x_2x_3$

SOURCE	S.S.	D.F.	M.S.
First-order terms	327.1	3	109.0
Second-order terms	760.3	6	126.7
Lack of fit	40.1	5	8.0
Error	52.6	5	10.5
Total	1180.1	19	

6. $(\pm 1, \pm 1)$, $(\pm 1.414, 0)$, $(0, \pm 1.414)$, 8 points at $(0, 0)$.

7.

x_1	x_2	x_3	x_4	x_5	x_1	x_2	x_3	x_4	x_5
-1	1	1	1	1	1	1	1	1	-1
1	-1	1	1	1	-1	-1	1	1	-1
1	1	-1	1	1	-1	1	-1	1	-1
-1	-1	-1	1	1	1	-1	-1	1	-1
1	1	1	-1	1	-1	1	1	-1	-1
-1	-1	1	-1	1	1	-1	1	-1	-1
-1	1	-1	-1	1	1	1	-1	-1	-1
1	-1	-1	-1	1	-1	-1	-1	-1	-1

x_1	x_2	x_3	x_4	x_5	
± 2.000	0	0	0	0	
0	± 2.000	0	0	0	
0	0	± 2.000	0	0	
0	0	0	± 2.000	0	
0	0	0	0	± 2.000	
0	0	0	0	0	
0	0	0	0	0	10 points

INNER CIRCLE

x_1	0.250	0.177	0	-0.177	-0.250	-0.177	0	0.177
x_2	0	0.177	0.250	0.177	0	-0.177	-0.250	-0.177

OUTER CIRCLE

x_1	1.000	0.500	-0.500	-1.000	-0.500	0.500
x_2	0	0.866	0.866	0	-0.866	-0.866

9. $y = 28.19 + 1.53x_1 + 8.78x_2 + 2.31x_3 + 11.23x_1^2 + 10.85x_2^2 + 3.11x_3^2 - 7.28x_1x_2 - 0.81x_1x_3 - 11.06x_2x_3.$

10. $y = 82.18 - 1.05x_1 - 6.11x_2 + 0.92x_1^2 - 4.63x_2^2 - 1.19x_1x_2.$

SOURCE	S.S.	D.F.	M.S.
Due to B_0	76,225.08	1	
First-order terms	307.58	2	
Second-order terms	184.52	3	
Lack of fit	157.09	3	52.36
Error	5.95	3	1.98
Total	76,880.22	12	

13. If the experimental points are plotted they spell ZILCH [see diagram (a)].
   By replacing the points $(0,4)$, $(0,2)$, and $(-2,0)$ by $(-1,3)$, $(1,3)$, and $(0,1)$ we obtain a symmetrical design [see diagram (b)]. For this design $[a_1a_2]=0$ if either a_1 or a_2 (or both) is odd. Also

$$[20]=34; \quad [40]=82; \quad [02]=144; \quad [04]=1584; \quad [22]=136$$

All other quantities $[a_1a_2]$ with $1 \leqslant a_1+a_2 \leqslant 4$ are 0.

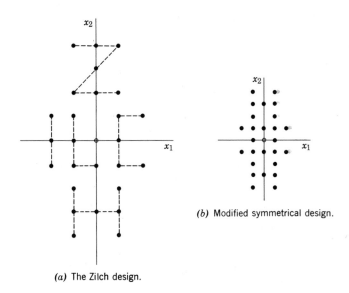

(b) Modified symmetrical design.

(a) The Zilch design.

If the regression is fitted in the form

$$\hat{y}_x = B_0^* + B_1x_1 + B_2x_2 + B_{11}\left(x_1^2 - \frac{34}{26}\right) + B_{22}\left(x_2^2 - \frac{144}{26}\right) + B_{12}x_1x_2$$

the appropriate **C** matrix is

$$
\begin{array}{c}
B_0^* \\
B_1 \\
B_2 \\
B_{11} \\
B_{22} \\
B_{12}
\end{array}
\left[
\begin{array}{cccccc}
26 & 0 & 0 & 0 & 0 & 0 \\
0 & 34 & 0 & 0 & 0 & 0 \\
0 & 0 & 144 & 0 & 0 & 0 \\
0 & 0 & 0 & 37.54 & -52.31 & 0 \\
0 & 0 & 0 & -52.31 & 789.46 & 0 \\
0 & 0 & 0 & 0 & 0 & 136
\end{array}
\right]
$$

This is "almost orthogonal." We have

$$\mathbf{C}^{-1} = \begin{bmatrix} 0.0385 & 0 & 0 & 0 & 0 & 0 \\ 0 & 0.0294 & 0 & 0 & 0 & 0 \\ 0 & 0 & 0.0069 & 0 & 0 & 0 \\ 0 & 0 & 0 & 0.0294 & 0.0020 & 0 \\ 0 & 0 & 0 & 0.0020 & 0.0014 & 0 \\ 0 & 0 & 0 & 0 & 0 & 0.0074 \end{bmatrix}$$

and so

$$\text{var}(\hat{y}_x) = \sigma^2 \left[0.0385 + 0.0294 x_1^2 + 0.0069 x_2^2 + 0.0294 \left(x_1^2 - \frac{34}{26} \right)^2 \right.$$

$$\left. + 0.0014 \left(x_2^2 - \frac{144}{26} \right)^2 + 0.0074 x_1^2 x_2^2 + 0.0040 \left(x_1^2 - \frac{34}{26} \right) \left(x_2^2 - \frac{144}{26} \right) \right]$$

Chapter 19

1. (a) Stratum sample sizes are $8, 8, 14, 17, 2, 51$ (variance of estimator of total X: $6,156,919$).
   (b) Stratum sample sizes are $4, 7, 5, 42, 2, 40$ (variance of estimator of total Y: $49,636,179$).
   (Numbers may vary by ± 1 from those shown.)
2. A proportional sample would have the constitution $4, 9, 14, 41, 5, 27$. The variance of the estimator of total X would be $8,872,560$, and the variance of the estimator of total Y would be $58,376,545$. By taking a sort of "mean" of the optimum samples for (a) and (b) of Exercise 1, say, with constitution $6, 7, 10, 30, 2, 45$ variance of estimator of total X is $6,657,096$ and variance of estimator of total Y is $54,216,596$.
3. If the sample sizes in Exercise 1 are multiplied by 6 then the "sample sizes" for strata I and VI of Exercise 1 (a) exceed the sizes of the corresponding strata. So we would take all 340 individuals from these 2 strata and the remaining 260 from strata II–V in the optimum proportions for estimating total X.
4. Dividing values in the first table of the answer to Exercise 1 by the corresponding value of $\sqrt{\text{cost}}$ gives the following figures

| | (SIZE × STANDARD DEVIATION) × (COST)$^{-1/2}$ | |
STRATUM	X	Y
I	2000	1700
II	1000	2500
III	1875	1875
IV	4500	21000
V	400	1000
VI	5400	10000
	15175	38075

Optimal structures should be proportional to the figures shown. The corresponding costs are proportional to the figures shown below:

STRATUM	I	II	III	IV	V	VI	TOTAL
X	2000	4000	7500	4500	900	33750	52650
Y	6800	10000	7500	47250	2250	90000	163800

For variable X, therefore, an amount $350 \times 2000/52650 = 13.3$ should be spent on Stratum I corresponding to about 13 individuals; in Stratum II $350 \times 4000/52650 = 26.6$ should be spent corresponding to about 7 individuals. From calculations of this kind the following sample structures are obtained

$$\begin{array}{lcccccc} \text{For } X: & 13 & 7 & 12 & 30 & 3 & 36 \\ \text{For } Y: & 4 & 5 & 4 & 44 & 2 & 21 \end{array}$$

5. Since in each stratum the standard deviation of Y is at least as big as that of X, the estimator of total Y must have a variance at least as big as the variance of the estimator of total X. Hence the required sample will be the sample optimal for Y with costs equal to $3.75, 6, 6, 2.4375, 3.375, 11.4375$.

6. This is a more complicated case than that of Exercise 5. In strata where it is cheaper to measure X it might be better to measure X only, and to measure Y only where costs are equal.

    Generally, since the estimator of Y has the greater variance (when strata sample sizes are the same), it will be an improvement if more Y's (and fewer X's) are measured when the costs of measuring each are about equal.

7. The chosen stratum should be (a) large and (b) variable. On these grounds, Stratum VI seems to be the most promising.

    To assess the improvement, it would be necessary to make some assumption about the standard deviations (of X and Y) within the two new strata.

8. The mean and variance in stratum (t) are

$$\xi_t = \frac{\int_{x_{t-1}}^{x_t} xf(x)\,dx}{\int_{x_{t-1}}^{x_t} f(x)\,dx} \quad \text{and} \quad V_t = \frac{\int_{x_{t-1}}^{x_t} x^2 f(x)\,dx}{\int_{x_{t-1}}^{x_t} f(x)\,dx} - \xi_t^2$$

respectively. It is required to minimize

$$\sum_{t=1}^{k} V_t \int_{x_{t-1}}^{x_t} f(x)\,dx = V$$

$$\frac{\partial V}{\partial x_t} = (V_t - V_{t+1})f(x_t) + \frac{\partial V_t}{\partial x_t} \int_{x_{t-1}}^{x_t} f(x)\,dx + \frac{\partial V_{t+1}}{\partial x_t} \int_{x_t}^{x_{t+1}} f(x)\,dx$$

$$= \left[\xi_t^2 - 2x_t\xi_t - \xi_{t+1}^2 + 2x_t\xi_{t+1} \right] f(x_t)$$

So

$$\frac{\partial V}{\partial x_t} = 0 \quad \text{when } x_t = \tfrac{1}{2}(\xi_t + \xi_{t+1})$$

It can be verified that these values of $x_t (t = 1, \ldots, k-1)$ do, in fact, minimize V.

9. From the "hint,"

$$E(\bar{R}\bar{X}) = n^{-1}\left[\bar{y} + \left\{\bar{x}\sum_{i=1}^{N}\left(\frac{y_i}{x_i}\right) - \bar{y}\right\}\frac{n-1}{N-1}\right]$$

$$= \frac{1}{n(N-1)}\left[(N-n)\bar{y} + (n-1)\bar{X}\sum_{i=1}^{N}\left(\frac{y_i}{x_i}\right)\right]$$

So

$$E\left[\bar{R}\bar{x} + \frac{N-1}{N}\cdot\frac{n}{n-1}(\bar{Y} - \bar{R}\bar{X})\right]$$

$$= \bar{x}E(\bar{R}) + \frac{N-1}{N}\cdot\frac{n}{n-1}\left[E(\bar{Y}) - E(\bar{R}\bar{X})\right]$$

$$= \bar{x}N^{-1}\sum_{i=1}^{N}\left(\frac{y_i}{x_i}\right) + \frac{N-1}{N}\cdot\frac{n}{n-1}\left[\bar{y}\left\{1 - \frac{N-n}{n(N-1)}\right\} - (n-1)\bar{x}\sum_{i=1}^{N}\left(\frac{y_i}{x_i}\right)\right]$$

$$= \bar{y}$$

10. The variance of the difference between mean yields for two methods of production is

$$2\left[\frac{\sigma'^2}{p} + \frac{\sigma^2}{bp}\right]$$

where

$$\sigma'^2 = \text{variance of Methods} \times \text{Places interaction}$$

and

$$\sigma^2 = \text{residual variance (Methods} \times \text{Blocks Within Places)}$$

(1) To minimize $\sigma'^2/p + \sigma^2/bp$ subject to the condition mpb constant, we evidently make p as large as possible.

(2) To minimize $\sigma'^2/p + \sigma^2/bp$ subject to condition $c_1'pb + c_2p = K$, write $pb = (K - c_2p)/c_1'$; then $(\sigma'^2/p) + \sigma^2c_1'/(K - c_2p)$ has to be minimized with respect to p. The appropriate value of p satisfies

$$-\left(\frac{\sigma'}{p}\right)^2 + c_1'c_2\left(\frac{\sigma}{K - c_2p}\right)^2 = 0;$$

that is,

$$p = K\left[c_2 + \sqrt{c_1'c_2}\,\frac{\sigma}{\sigma'}\right]^{-1}$$

An integer value of p, near the value given by this formula will be suitable; b is chosen so that $c_1' pb + c_2 p = K$.

It is not usually possible to satisfy the last equation exactly with integer values of p and b. (Note that $c_1' = c_1 m$.)

11. If $V(m_1, m_2) = (M_1 + M_2)^{-2} [M_1^2 \sigma_1^2 / m_1 + M_2^2 \sigma_2^2 / m_2]$, then putting

$$m_1 = m_0 \left(\frac{M_1 S_1}{M_1 S_1 + M_2 S_2} \right), \qquad m_2 = m_0 \left(\frac{M_2 S_2}{M_1 S_1 + M_2 S_2} \right)$$

we find $V = (M_1 + M_2)^{-2} m_0^{-1} (M_1 S_1 + M_2 S_2)[M_1 \sigma_1^2 / S_1 + M_2 \sigma_2^2 / S_2]$. For proportional sampling, with

$$M_1 = \frac{m_0 M_1}{M_1 + M_2}; \qquad m_2 = \frac{m_0 M_2}{M_1 + M_2}$$

we have

$$V = (M_1 + M_2)^{-1} m_0^{-1} \left[M_1 \sigma_1^2 + M_2 \sigma_2^2 \right]$$

The first value of V is less than the second value if

$$(M_1 S_1 + M_2 S_2)(M_1 \sigma_1^2 S_1^{-1} + M_2 \sigma_2^2 S_2^{-1}) < (M_1 + M_2)(M_1 \sigma_1^2 + M_2 \sigma_2^2)$$

or $\sigma_1^2 (S_2 / S_1) + \sigma_2^2 (S_1 / S_2) < \sigma_1^2 + \sigma_2^2$. So S_1 / S_2 must lie between 1 and σ_1^2 / σ_2^2. Now S_1^2 / S_2^2 is distributed as $(\sigma_1^2 / \sigma_2) F_{n-1, n-1}$. So the required probability is

$$\Pr\left[\frac{\sigma_1^2}{\sigma_2^2} F_{n-1, n-1} \text{ between } 1 \text{ and } \frac{\sigma_1^4}{\sigma_2^4} \right]$$

that is,

$$\Pr\left[F_{n-1, n-1} \text{ between } \frac{\sigma_2^2}{\sigma_1^2} \text{ and } \frac{\sigma_1^2}{\sigma_2^2} \right]$$

This probability equals $1 - \alpha$ if

$$F_{n-1, n-1, 1 - \frac{1}{2}\alpha} = \max\left(\frac{\sigma_1^2}{\sigma_2^2}, \frac{\sigma_2^2}{\sigma_1^2} \right)$$

12. The population variance is (in the notation of Section 18.1 and 18.2)

$$M_0^{-1} \sum_{i=1}^{k} M_i \left[\sigma_i^2 + \left(\xi_i - \bar{\xi} \right)^2 \right]$$

An unbiased estimator of this quantity is

$$M_0^{-1} \sum_{i=1}^{k} M_i \left[(1 - m_i^{-1}) - M_i (m_i M_0)^{-1} V_i + \left(\bar{X}_i - \bar{X} \right)^2 \right]$$

where

$$V_i = (m_i - 1)^{-1} \sum_{j=1}^{m_i} \left(X_{ij} - \bar{X}_i \right)^2 \quad \text{and} \quad \bar{X} = M_0^{-1} \sum_{i=1}^{k} M_i \bar{X}_i$$

The variance of this estimator is approximately

$$M_0^{-2} \sum_{i=1}^{k} \frac{M_i^2 \sigma_i^4}{m_i} (\beta_{2i} - 1)$$

where β_{2i} is the second moment ratio in the ith stratum. To minimize this approximate variance, subject to the condition $\sum_{i=1}^{k} m_i$ fixed, m_i should be taken proportional to $M_i \sigma_i^2 \sqrt{\beta_{2i} - 1}$. If $\sum_{i=1}^{k} m_i c_i$ is fixed, m_i should be proportional to

$$M_i \sigma_i^2 \sqrt{\frac{\beta_{2i} - 1}{c_i}}$$

Apart from the fact that an approximate formula for the variance of the estimator is used, it should be noted that the quantity estimated is the population variance, and not the population standard deviation. Also, even when the σ_i's are known (at least up to a proportionality factor), values of β_{2i} may well not be available. However, even in such cases, choosing m_i proportional to $M_i \sigma_i^2$ is very often a reasonable procedure, with good chances of advantage.

13. The sampling units are houses. The population is not relevant in the present context, though "number of occupants" might be used as a concomitant variable if fuller information were available.

Costs of sampling (units per house) are as follows:

Zilchester	4.17
Zilchford	6.28
Zilchton	6.28
Zilwich	6.92
Zilchwater	7.56

Optimum proportions are in the ratios:

$$\frac{82,000}{\sqrt{4.17}} : \frac{20,000}{\sqrt{6.28}} : \frac{6,500}{\sqrt{6.28}} : \frac{25,000}{\sqrt{6.92}} : \frac{13,000}{\sqrt{7.56}}$$

These ratios are equivalent to

$$0.618 : 0.123 : 0.040 : 0.146 : 0.073$$

14. Considerations would be similar to those in Exercise 13, except that the relative size of standard deviations of the new variable (equal to 1 if more than 2 persons per 100 ft^2, and to 0 otherwise) in the 5 urban areas needs to be taken into account.

Index